Erosion and Sediment Control Handbook

Erosion and Sediment Control Handbook

Steven J. Goldman
California Tahoe Conservancy

Katharine Jackson
Consultant

Taras A. Bursztynsky, P.E.
Association of Bay Area Governments

McGraw-Hill, Inc.
New York St. Louis San Francisco Auckland Bogotá
Caracas Lisbon London Madrid Mexico Milan
Montreal New Delhi Paris San Juan Singapore
Sydney Tokyo Toronto

Library of Congress Cataloging-in-Publication Data

Goldman, Steven J.
Erosion and sediment control handbook.

Includes bibliographies and index.
1.Sediment control—United States—Handbooks, manuals, etc. 2.Soil erosion—United States—Handbooks, manuals, etc. 3.Soil conservation—United States—Handbooks, manuals, etc. I.Jackson, Katharine. II.Bursztynsky, Taras A. III.Title.
TC423.G645 1986 627′.52 85-15371
ISBN 0-07-023655-0

34567890 HDHD 998765432

ISBN 0-07-023655-0

The editors for this book were Joan Zseleczky and Galen H. Fleck; the designer was Mark E. Safran; and the production supervisor was Thomas G. Kowalczyk. It was set in Century Schoolbook by University Graphics, Inc.

Printed and bound by Halliday Lithograph

To my mother and father
Ruth Lachow and Leslie A. Goldman

Contents

Preface

When land is disturbed for construction, road building, mining, logging, or other activities, the soil erosion rate increases from 2 to 40,000 times. Millions of tons of this soil end up in our rivers, lakes, and reservoirs. Each year, taxpayers, land developers, and property owners spend billions of dollars cleaning up sediment and repairing eroded streambanks, gullied hillsides, washed-out roads, mud-choked drains, and other erosion damage.

Most erosion and sediment problems can be avoided. This handbook describes how to prevent or greatly reduce erosion by proper planning and by using simple, low-cost control measures. It is designed for a variety of readers. Engineers, architects, landscape architects, and others responsible for preparing site plans will find it especially useful. It will also be helpful to public agency officials who review and approve land development plans and to construction site inspectors and foremen who examine control measures in the field. Landscape contractors should find the material on vegetative soil stabilization methods of particular interest. The handbook can also be used as a college textbook for courses in environmental planning, engineering, soil science, grading and drainage design, and related fields. Review questions and sample problems have been included for classroom use.

This handbook is an outgrowth of a series of professional seminars that the authors have presented numerous times over the past several years. During the late 1970s, many of the cities and counties of the San Francisco Bay Area asked us for help in solving their erosion problems. They said both they and land developers lacked expertise on this subject. They saw a need for technical standards and specifications for erosion and sediment control measures. In response to this need, we developed a manual of standards for control measures. Subsequently, many cities and counties adopted the manual as an official document. Later, we developed a professional seminar on erosion control. The *Erosion and Sediment Control Handbook* was originally prepared as a reference book for this seminar.

During the preparation of the seminar and handbook, we studied control measure performance on scores of construction sites. We were struck by the similarity of erosion problems from place to place and by the high failure rate of the various control measures. The problems we saw in California were similar to the problems others were experiencing throughout the United States. The failure of control measures was due in part to improper design and installation and also to a failure to use common sense. A row of unstaked straw bales with gaps between them will obviously not stop sediment. Sandbags placed in a gully will not keep the gully from enlarging. We took hundreds of photographs of erosion and sediment control measures in the field, particularly during and after storms. Some measures were working, but many were not. A large number of the annotated photographs have been included in this handbook to illustrate both good and faulty designs. Often we can learn more from our failures than we can from our successes.

The handbook is organized into chapters on each of the major subject areas of erosion control. Chapter 1 describes erosion problems, including both environmental impacts and economic costs. It also describes the erosion process. Chapter 2 outlines ten basic strategies for preventing erosion and trapping sediment. Each of the ten techniques is illustrated with examples. Chapter 3 describes how to set up and enforce an effective erosion and sediment control program in a city or county, and it outlines the key features of an effective erosion and sediment control ordinance. (A complete model ordinance is provided as an appendix.) Chapter 3 also includes detailed data on private sector costs for constructing control measures and public agency costs for implementing a control program. This chapter will be of particular interest to local-government officials and to planning and public works department staff responsible for controlling erosion problems in their jurisdictions.

Chapter 4 shows how to use the rational formula to estimate runoff volumes for sizing temporary erosion and sediment control measures. Emphasis is on the use of short-duration rainfall intensity data. A list of state climatologists in 48 states, together with descriptions of the specific types of data available in most of those states, is provided. Chapter 4 also covers the use of Manning's equation and the continuity equation for sizing drainageways and provides several example calculations with step-by-step solutions. Chapter 5 describes how to use the universal soil loss equation to estimate the volume of sediment likely to be eroded from a construction site. Again, sample problems and step-by-step solutions are provided. The soil loss calculations can be used to compute the storage volume requirements of sediment basins and traps and to evaluate the erosion potential of a site and the effectiveness of a proposed control plan.

Chapters 6, 7, and 8 are the primary reference chapters in the handbook; each gives detailed specifications for designing and installing control measures. These chapters will be particularly useful to engineers and landscape architects. Chapter 6 describes how to use plants and mulches to protect graded areas from erosion. It rates the effectiveness and cost-effectiveness of seven common revegetation techniques, describes the characteristics of plants that are effective in controlling erosion, and lists sources of information on suitable plant types throughout the world. Chapter 7 provides guidelines for designing water conveyance and energy dissipation structures such as dikes, swales, lined waterways, pipe slope drains, and riprap aprons. It includes sizing charts, sample designs, and practical tips for designers and builders. Chapter 8 shows how to design and construct sediment basins and traps, straw bale dikes, and silt fences. These devices are intended to prevent eroded soil from leaving a site (sediment control) as contrasted with the measures in Chapters 6 and 7, which are intended to prevent erosion. Chapter 8 also discusses soil particle settling processes and their relations to the design of effective sediment basins. After studying the performance of numerous sediment basins, we found that the key design factor was the surface area of the basin in relation to the size of soil particles on a site. The positioning of the basin inlet and outlet was also very important. Most of the basins we observed had flaws in their designs which resulted in very low trap efficiencies; some basins even produced more sediment than they captured. With proper design, these costly errors could have been avoided.

The first part of Chapter 9 shows how to develop an erosion and sediment control plan for a site. The step-by-step planning process is illustrated with examples from an actual project. The last part of the chapter gives a method, designed for public officials, for evaluating the adequacy of an erosion and sediment control plan, and it includes a plan review checklist. Chapter 10 covers control measure maintenance. Common failure points of control measures are described and illustrated with annotated photographs from dozens of construction sites. Designed for site inspectors and project foremen, this final chapter also provides tips on what to look for on a site and how to repair damaged control measures.

Appendix A is a model grading and erosion and sediment control ordinance. This model ordinance combines the provisions of a commonly used grading ordinance (Chapter 70 of the *Uniform Building Code*) with effective erosion and sediment control provisions. Ordinances based on this model have been adopted by many cities and counties. Appendix B provides tables and charts for designing grass-lined waterways. Appendix C provides answers to the calculation problems in the review questions at the ends of Chapters 4, 5, 7, and 8. These problems and solutions are intended to serve as teaching aids.

In summary, the *Erosion and Sediment Control Handbook* is a comprehensive, how-to-do-it guide with everything you need in one volume. We have tried to present the material straightforwardly by using plain, simple language. Photographs, illustrations, checklists, charts, and tables are used wherever possible. Much hard-to-find but necessary information, such as cost data, erosion control product specifications, and sources of local rainfall data, has been provided.

Sample problems with step-by-step solutions have been used extensively throughout because we feel strongly that this is one of the best ways to learn a subject.

ACKNOWLEDGMENTS

First, I would like to thank my coauthors, Katharine Jackson and Taras Bursztynsky. This handbook was built on the foundation of many years of hard work by many individuals from coast to coast. The publication developed by Mark Boysen and the staff of the Soil Conservation Service in College Park, Maryland, *Standards and Specifications for Soil Erosion and Sediment Control in Developing Areas,* became the cornerstone of our erosion control program in the San Francisco Bay Area and, later, of this handbook. Another excellent publication, the *Virginia Erosion and Sediment Control Handbook,* prepared by Gerard Seeley, Jr., and the staff of the Virginia Soil and Water Conservation Commission, provided many of the design drawings of control measures that appear in this handbook. Burgess Kay, of the Department of Agronomy and Range Science at the University of California at Davis, played a major role in the preparation of the chapter on vegetative soil stabilization methods and supplied most of the photographs used in that chapter.

Special thanks go to Patrick Baker of the Alameda County Public Works Agency for reviewing the manuscript and providing many helpful suggestions. Mr. Baker also helped write portions of Chapters 3 and 9. Piero Ruggeri of Bissell & Karn, Inc. Civil Engineers and Robert Crowell of Cagwin & Dorward Landscape Contractors and Engineers have contributed many practical ideas. Attorneys Kenneth Moy and Marc Lampe were instrumental in developing the model erosion control ordinance. Ross Turner prepared many of the illustrations. Stuart Chaitkin converted hundreds of English units to metric equivalents. Galen Fleck edited the manuscript. Present and former staff members at the Association of Bay Area Governments—Emy Chan, Emily Pimentel, Michael McMillan, Marci Loss, and Ann Berry—all made significant contributions. Thanks are also due to Revan Tranter and Eugene Leong, Executive Director and Deputy Executive Director, respectively, of the Association of Bay Area Governments, for providing the opportunity to work on this project. Final thanks go to my wife Harriet for her support and encouragement and to my sons Joshua and Jonathan for sustaining my spirits.

STEVEN J. GOLDMAN

chapter 1

The Erosion Problem and Its Causes

1.1 PURPOSE OF THIS HANDBOOK

The purpose of this handbook is to describe the means of controlling erosion and sediment from land-disturbing activities. The principal land-disturbing activity addressed is construction. However, much of the material presented in this handbook is also applicable to controlling erosion and sediment from mining, logging, agriculture, and other activities. In fact, many of the basic control measures described here (such as dikes, swales, vegetation, and mulching) were originally developed for agricultural applications. This handbook provides information for:

1. Evaluating erosion potential
2. Developing a regulatory program for erosion and sediment control
3. Planning for erosion and sediment control
4. Designing and constructing erosion and sediment control measures

The emphasis is on the fourth item. Design and construction details are vitally important to successful control of erosion and sediment. The technology for erosion and sediment control is well developed. Much of it is common sense. The problem is to apply the technology correctly and thereby control the problem at its source. Unless the control measures are designed and installed properly, they will not work. Improper design often results in problems more severe than those that were to be avoided.

1.2 THE PROBLEM

Each year an estimated 80 million tons of sediment is washed from construction sites into the lakes, rivers, and waterways of the United States. (9)* Although this sediment is only a fraction of the total sediment load, it is the major source of pollution of many lakes and streams that drain small watersheds in which development is occurring. The needless destruction of nature and the consequent burden on the taxpayer for cleanup could be avoided if cities, counties, and land developers implemented simple erosion control practices.

1.2a Sources of Erosion

The primary sources of erosion in the United States are agriculture, silviculture (logging), mining, and construction. Although agriculture produces the largest percentage of the total sediment load, construction causes the most concentrated form of erosion. When land is disturbed by construction activities, soil erosion increases from 2 to 40,000 times the preconstruction erosion rate. (11, 13) Erosion rates from construction sites are typically 10 to 20 times those from agricultural lands and they can be 100 times as high. (10) Wolman and Schick (13) found sediment yields on an open construction site in Maryland of 220 tons/acre (492 t/ha) compared to 1.7 tons/acre (3.8 t/ha) in a stable urban area nearby. In the San Francisco Bay Area the authors found the average rate of erosion in all land uses (grazing, agriculture, forests, etc.) was about 3.5 tons/(acre)(year) [7.8 t/(ha)(yr)], whereas the erosion rate from construction sites was 52 to 70 tons/(acre)(year) [116 to 157 t/(ha)(yr)] and sometimes higher. Erosion rates from construction sites were typically 20 times the average rates. Although a wide variation in erosion rates is reported in the literature, it is clear that construction causes a large increase in erosion. One need only observe a bare graded slope before and after a single storm to verify that fact.

1.2b Impacts of Erosion and Sedimentation

Erosion and sedimentation cause both environmental and economic impacts. Both are important, but it is often only an economic impact that spurs a jurisdiction to take action. Environmental impacts are harder to see. They tend to build slowly and not produce dramatic results for many years, when it may be too late to correct the problem. In the following section some of the more common impacts are described.

*Numbers in parentheses are those of references listed at the end of the chapter.

Environmental Impacts

- Eroded soil contains nitrogen, phosphorus, and other nutrients. When carried into water bodies, these nutrients trigger algal blooms that reduce water clarity, deplete oxygen, lead to fish kills, and create odors. For example, construction in the Lake Tahoe basin has seriously diminished the quality of the lake. Sixteen years ago you could see 21 feet (6.5m) farther down into the lake than you can today. There has also been a dramatic increase in algae attached to shoreline rocks. (2, 14)
- Erosion of streambanks and adjacent areas destroys streamside vegetation that provides aquatic and wildlife habitats.
- Excessive deposition of sediments in streams blankets the bottom fauna, "paves" stream bottoms, and destroys fish spawning areas.
- Turbidity from sediment reduces in-stream photosynthesis, which leads to reduced food supply and habitat.
- Suspended sediment abrades and coats aquatic organisms.
- Erosion removes the smaller and less dense constituents of topsoil. These constituents, clay and fine silt particles and organic material, hold nutrients that plants require. The remaining subsoil is often hard, rocky, infertile, and droughty. Thus, reestablishment of vegetation is difficult and the eroded soil produces less growth.

Economic Impacts

Many economic impacts are hard to quantify. How does one set a value on loss of aquatic habitat or diminished water clarity? Other impacts may be readily quantified. For example, the cost of a silted-up reservoir may be the costs of dredging and disposing of the accumulated sediment.

In the following section a variety of economic impacts are described. During the recent development of an erosion control program for the San Francisco Bay Area, research by the authors revealed numerous specific examples of actual costs directly attributable to erosion. Some of these examples are presented below to illustrate the magnitude of the problem. Similar economic impacts can be found throughout the country.

1. Excessive sediment accumulation reduces reservoir storage capacity and more frequent sediment removal is required.
   - Cull Canyon Reservoir is located at the bottom of a 4000-acre (1600-ha) watershed in suburban Alameda County, California. Less than 10 percent of the watershed has been urbanized. Eleven years after construction of the reservoir, 400,000 yd^3 (306,000 m^3) of sediment was removed at a cost of about $1 million. Regular maintenance dredging is still required.
   - Lake Temescal is a small recreational lake in Oakland, California. Highway and home construction in the hills above the lake caused 40 ft (12.2 m) of sediment to accumulate in the once 65-ft- (19.8-m-) deep water body. The

local park district was recently forced to spend $750,000 to dredge 47,000 yd^3 (36,000 m^3) of sediment from the lake.

- The cost of dredging San Pablo Reservoir, in Contra Costa County, California, was estimated at $12 million.
- Reported dredging costs range from $2.50 to $16 yd^3 ($2 to $12/$m^3$). Based on a typical dredging and disposal cost of $10/$yd^3$ ($13/$m^3$), the cost of removing sediment from San Francisco Bay Area lakes and reservoirs alone (not counting the bay) would be $30 million per year.
- The cost of building new reservoirs to replace lost reservoir capacity also is high. Increasing land values and lack of available sites are making this alternative much less feasible.
- Sediment deposited into streams reduces flow capacity, interferes with navigation, and increases the risks of flooding. Regular maintenance dredging is thus required.
- $2 million is spent every 15 years to remove sediment from the Napa River in northern California. In 1980, $500,000 was budgeted for removal of 125,000 yd^3 (95,000 m^3) of sediment from the navigation channel.
- The Alameda County Flood Control and Water Conservation District projected a cost of $25 million in the 10-year period from 1980 to 1990 to clean up 1.8 million yd^3 (1.4 million m^3) of accumulated sediment.

2. Erosion severely diminishes the ability of the soil to support plant growth. To restore this ability costs money, although restoration is not always undertaken.

- The value of plant nutrients lost is estimated to be $6/$yd^3$ ($7.85/$m^3$) of eroded topsoil. (7)
- Another way to estimate the cost of eroded topsoil is replacement cost. At a typical cost of $20/$yd^3$ ($26/$m^3$), for commercial topsoil, the cost to replace the 80 million tons/year lost from construction sites in the United States would be about $1.6 billion per year. The cost to replace the lost nutrients alone would be about $480 million per year.

Fig. 1.1 Erosive splash of raindrop striking bare soil. (8)

1.3 THE EROSION PROCESS

Erosion is essentially a two-part process. One part is the loosening of soil particles caused largely by raindrop impact (Fig. 1.1). Other causes are freezing-and-thawing and wetting-and-drying cycles. The other part of the process is the transportation of soil particles, largely by flowing water. The major types of erosion are briefly described in the following section.

1.3a Types of Erosion

Splash Erosion

When vegetative cover is stripped away, the soil surface is directly exposed to raindrop impact (Fig. 1.1). On some soils, a very heavy rain may splash as much as 100 tons/acre (224 t/ha) of soil. (2) Some of the splashed particles may rise as high as 2 ft (0.6 m) above the ground and move up to 5 ft (1.5 m) horizontally. (2) If the soil is on a slope, gravity will cause the splashed particles to move downhill (Fig. 1.2).

When raindrops strike bare soil, the soil aggregates are broken up. Fine particles and organic matter are separated from heavier soil particles. This pounding action destroys the soil structure. A hard crust often forms when the soil dries. This crust inhibits water infiltration and plant establishment, and runoff and future erosion are thereby increased.

Splash erosion is closely related to raindrop size. Large raindrops have a much greater impact than small raindrops. Raindrop size is discussed further in Sec. 1.3b.

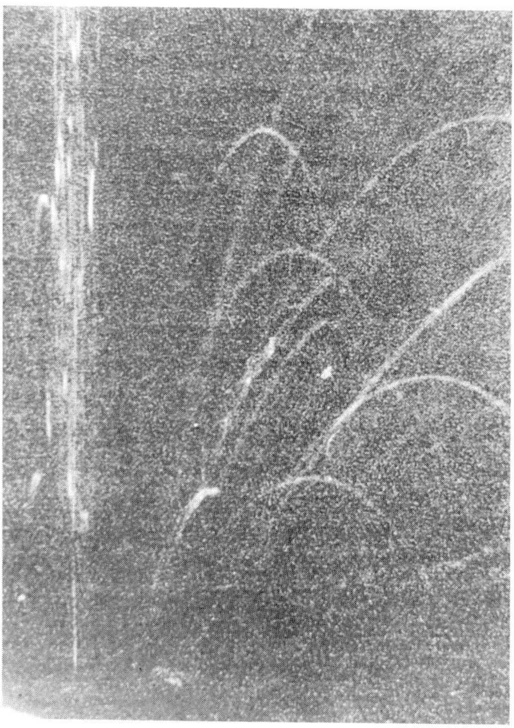

Fig. 1.2 Splashed soil particles on slopes move downhill. (Courtesy of Burgess Kay)

Sheet Erosion

Sheet erosion is caused by shallow "sheets" of water flowing over the soil surface (Fig. 1.3). These very shallow moving sheets of water are seldom the detaching agent, but the flow transports soil particles that have been detached by raindrop impact. The shallow surface flow rarely moves as a uniform sheet for more than a few feet before concentrating in the surface irregularities. (11)

Rill Erosion

Rill erosion begins when shallow surface flow starts to concentrate in low spots in the soil surface (Fig. 1.3). As the flow changes from sheet flow to deeper flow in these low areas, the velocity and turbulence of flow increase. The energy of this concentrated flow is able to both detach and transport soil particles. This action begins to cut tiny channels called *rills*. Rills are small but well-defined channels that are at most only a few inches deep. (11)

Gully Erosion

Gully formation is a complex process that is not fully understood. Some gullies are formed when runoff cuts rills deeper and wider or when the flows from several rills come together and form a large channel. Gullies can enlarge in both uphill and downhill directions. Water flowing over the headwall of a gully causes undercutting. In addition, large chunks of soil can fall from a gully headwall in a process called mass-wasting. This soil is later carried away by stormwater runoff.

A heavy rain can transform a small rill into a major gully almost overnight. The gully shown in Fig. 1.4 was formed during only two moderate storms. Once

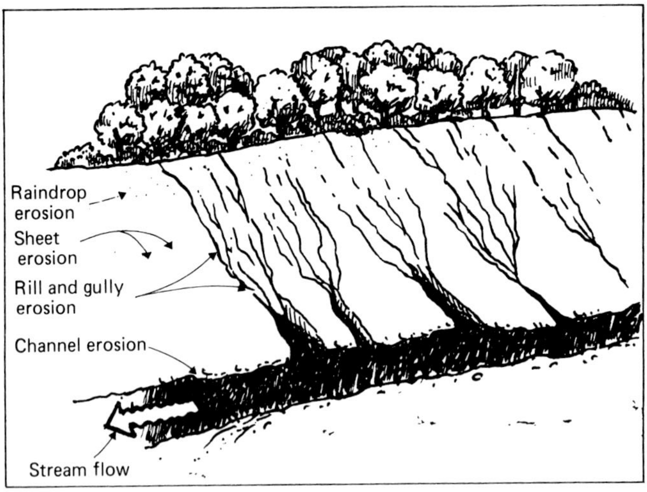

Fig. 1.3 Types of erosion. (Adapted from 1)

Fig. 1.4 Gully formed during two storms.

a gully is created, it is very difficult to stop it from growing, and it is costly to repair.

Channel Erosion

Channel erosion occurs when bank vegetation is disturbed or when the volume or velocity of flow in a stream is increased. Natural streams have adjusted over time to the quantity and velocity of runoff that normally occur in the watershed. The vegetation and rocks lining the banks are sufficient to prevent erosion under these steady-state conditions. However, when a watershed is altered by removing vegetation, by increasing the amount of impervious surfaces, or by paving tributaries, stream flows also are changed. Typical changes are an increase in the peak flow during storms and an increase in stream velocity. Either of these changes can destroy the equilibrium of the stream and cause channel erosion to begin. Common points where erosion occurs are at stream bends and at constrictions, such as those where bridges cross a stream. Erosion may also begin at the point where a storm drain or culvert discharges into a stream. Repair of eroded streambanks is difficult and costly.

1.3b Erosion Factors

The four principal factors in soil erosion are climate, soil characteristics, topography, and ground cover. They form the basis of the universal soil loss equation (USLE). The USLE is a method for quantifying the interaction of the factors to estimate the tonnage of soil loss per year. In the following section, each of the four factors is described qualitatively. How to quantify each factor for specific sites and calculate the potential erosion rate is described in Chap. 5.*

Climate

Climate affects erosion potential both directly and indirectly. In the direct relation, rain is the driving force of erosion. Raindrops dislodge soil particles, and runoff carries the particles away. The erosive power of rain is determined by rainfall intensity (inches or millimeters of rain per hour) and droplet size. Table 1.1 shows the kinetic energy of rainfalls of various intensities. The table shows that 6-mm (0.24-in) raindrops falling from a cloudburst have over 2000 times as much kinetic energy per unit time as a drizzle with 1-mm (0.04-in) raindrops.

A highly intense rainfall of relatively short duration can produce far more erosion that a long-duration storm of low intensity. Also, storms with large raindrops are much more erosive than misty rains with small droplets. Rainfall intensity and duration and droplet size are determined by geographic location. Figure 1.5 shows the pattern of storm intensity across the United States. The most intense rainfalls occur in the southeastern states.

Note the close correlation between the maps of rainfall intensity (Fig. 1.5) and the rainfall erosion factors R in the USLE (Fig. 5.2). With the exception of localized mountainous areas along the Pacific coast, the southeastern states also receive the greatest annual precipitation. Though yearly rainfalls over 100 in

*The USLE actually has five factors. However the fifth factor, called the erosion control practice factor, or P factor, relates only to the roughness of the ground surface as determined by grading or tillage techniques. This factor has a relatively small effect on the soil loss estimate produced by using the USLE. In one perspective, the P factor could be considered a component of both ground cover and topography. (See Chap. 5 for a further discussion of the P factor.)

TABLE 1.1 Kinetic Energy of Rainfalls of Various Intensities and Droplet Sizes (3, after 4)

Rainfall	Intensity, in/hr	Median diameter, mm	Velocity of fall, ft/sec	Drops per ft^2/sec	Kinetic energy, ft·lb/$(ft^2)(hr)$
Fog	0.005	0.01	0.01	6,264,000	4.04×10^{-8}
Mist	0.002	0.1	0.7	2,510	7.94×10^{-5}
Drizzle	0.01	0.96	13.5	14	0.148
Light rain	0.04	1.24	15.7	26	0.797
Moderate rain	0.15	1.60	18.7	46	4.241
Heavy rain	0.60	2.05	22.0	46	23.47
Excessive rain	1.60	2.40	24.0	76	74.48
Cloudburst	4.00	2.85	25.9	113	216.9
Cloudburst	4.00	4.00	29.2	41	275.8
Cloudburst	4.00	6.00	30.5	12	300.7

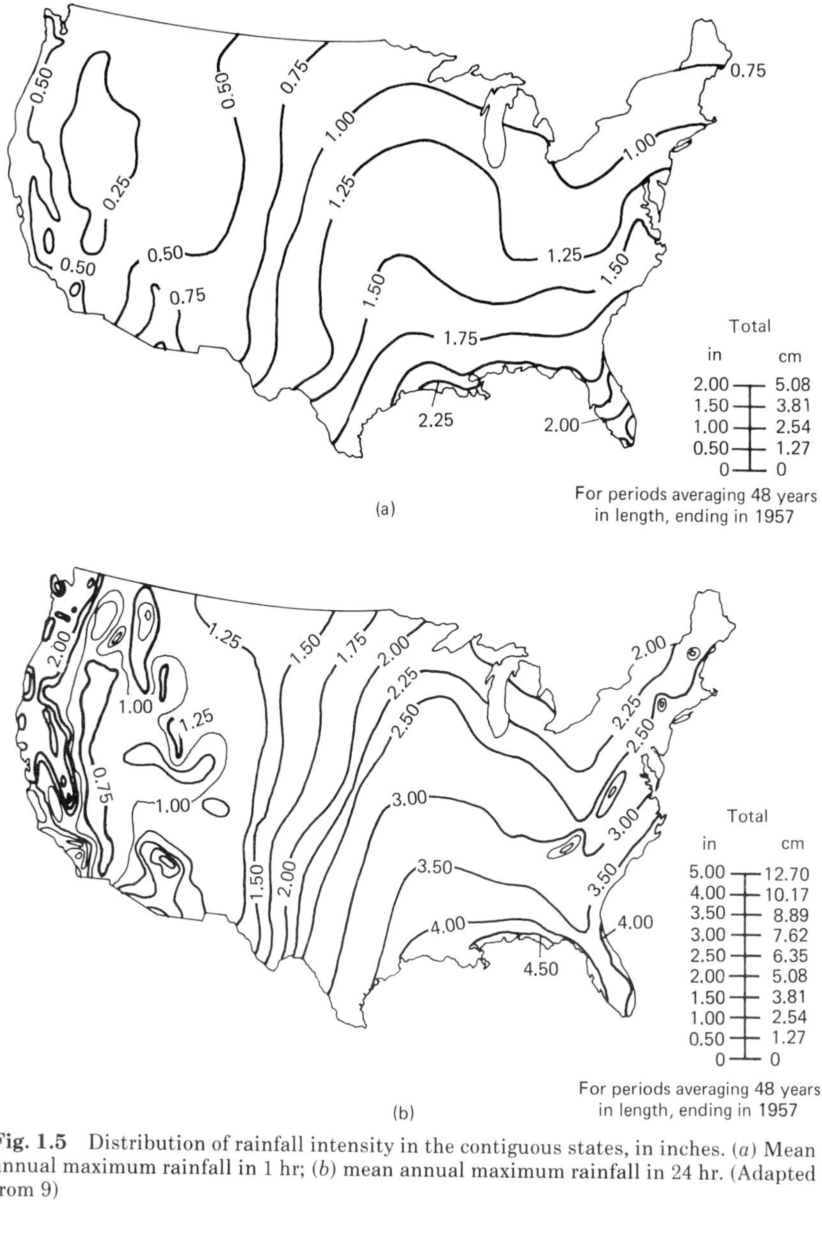

Fig. 1.5 Distribution of rainfall intensity in the contiguous states, in inches. (*a*) Mean annual maximum rainfall in 1 hr; (*b*) mean annual maximum rainfall in 24 hr. (Adapted from 9)

(2540 mm) commonly occur in the Pacific northwest, storms in that area tend to be of low intensity with a very fine droplet size.

The indirect relation between climate and erosion is more subtle. The yearly pattern of rainfall and temperature by and large determines both the extent and the growth rate of vegetation. As will be seen later, vegetation is the most important form of erosion control. Climates with relatively mild year-round temperatures and frequent, regular rainfall (as in the southeastern United States and the British Isles) are highly favorable to plant growth. Vegetation grows rapidly and provides a complete ground cover, which protects the soil from erosion. If land is cleared for construction, it can be revegetated easily if the revegetation is properly done.

Cold climates, such as the higher elevations of the Sierra Nevada, Cascade Range, and Rocky Mountains, and dry climates, such as the vast desert areas of the southwestern United States, are far less favorable to plant growth and thus are much more susceptible to erosion. In each of those climatic extremes, the natural vegetation required a very long period of time to become established. It lives in a fragile balance with its environment. Because the climate is so harsh, it is very difficult to reestablish any plant cover that is disturbed. Rainfall is very infrequent in deserts; but when it does occur, it is typically very intense. Erosion rates are often high because there is little ground cover to protect the soil. In cold climates the growing season is very short. Plant reestablishment is a difficult and costly process. Even with the best of efforts, success cannot be assured. (12)

The relation of climate to the timing of land disturbance and reseeding for erosion control is discussed in detail in Chaps. 2 and 6.

Soil Characteristics

The following four soil characteristics are important in determining soil erodibility:

- Texture
- Organic matter content
- Structure
- Permeability

Although each characteristic is discussed individually below, the four characteristics are integrally related. Chapter 5 contains a more thorough discussion of this topic.

Soil *texture* refers to the sizes and proportions of the particles making up a particular soil. Sand, silt, and clay are the three major classes of soil particles. Soils high in sand content are said to be coarse-textured. Because water readily infiltrates into sandy soils, the runoff, and consequently the erosion potential, is relatively low. Soils with a high content of silts and clays are said to be fine-textured or heavy. Clay, because of its stickiness, binds soil particles together and makes a soil resistant to erosion. However, once the fine particles are eroded by heavy rain or fast-flowing water, they will travel great distances before settling. Even with the sediment control measures described in Chap. 8, it is extremely difficult to remove clay particles from flowing water. Typically, particles of clay and fine silt will settle in a large, calm water body, such as a bay, lake, or reservoir, at the bottom of a watershed. Thus, silty and clayey soils are frequently the worst water polluters. Soils that are high in silt and fine sand and low in clay and organic matter are generally the most erodible. (5, 11) Well-drained sandy and rocky soils are the least erodible.

Organic matter consists of plant and animal litter in various stages of decomposition. Organic matter improves soil structure and increases permeability, water-holding capacity, and soil fertility. Organic matter in an undisturbed soil or in a mulch covering a disturbed site reduces runoff and, consequently, erosion potential. Mulch on the surface also reduces the erosive impact of raindrops.

Soil *structure* is the arrangement of soil particles into aggregates. A granular structure is the most desirable one. Soil structure affects the soil's ability to absorb water. When the soil surface is compacted or crusted, water tends to run off rather than infiltrate. Erosion hazard increases with increased runoff. Loose, granular soils absorb and retain water, which reduces runoff and encourages plant growth.

Soil *permeability* refers to the ability of the soil to allow air and water to move through the soil. Soil texture and structure and organic matter all contribute to permeability. Soils with high permeability produce less runoff at a lower rate than soils with low permeability, which minimizes erosion potential. The higher water content of a permeable soil is favorable for plant growth, although it may reduce slope stability in some situations.

By identifying erodible soil types early in the planning process, the site planner can know what portions of the site require the most diligent erosion control efforts. Mulching and vegetating exposed soils and minimizing the area of exposure of highly erodible soils are effective techniques for preventing soil movement. Sediment control structures, such as sediment basins, that prevent sediment from being washed off sites also are necessary, since some soil movement is inevitable.

Topography

Slope length and slope steepness are critical factors in erosion potential, since they determine in large part the velocity of runoff. The energy (and thus the erosive potential) of flowing water increases as the square of the velocity (Sec. 2.1g). Long, continuous slopes allow runoff to build up momentum. The high-velocity runoff tends to concentrate in narrow channels and produce rills and gullies.

The shape of a slope also has a major bearing on erosion potential. The base of a slope is more susceptible to erosion than the top, because runoff has more momentum and is more concentrated as it approaches the base. Constructing a convex slope magnifies this problem, whereas a concave slope reduces it (Fig. 1.6). Leaving a relatively flat area at the base of a slope not only reduces erosion but also allows sediment from the upper portions of the slope to settle out.

Figure 1.6*b* illustrates a problem that may occur at the top of a concave slope.

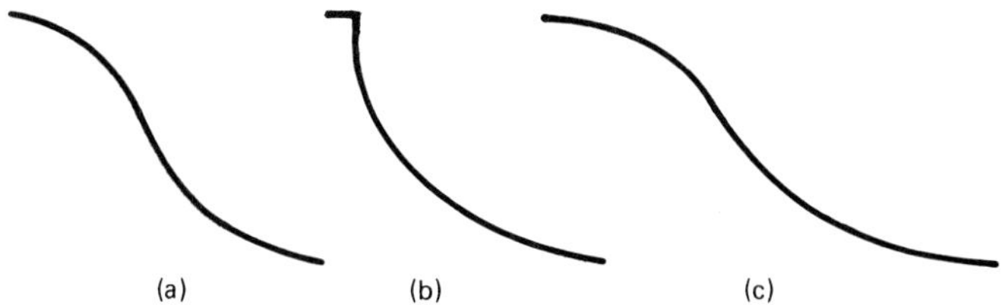

Fig. 1.6 Effect of slope shape on erosion potential. (*a*) Convex slope; (*b*) concave slope; (*c*) stable slope. (Adapted from 5)

If there is an abrupt change from flat upland to concave downslope, water flowing over the top of the slope may undercut the slope. Rounding the upper slope minimizes this problem and creates a more stable slope, which has the appearance of a natural hillside (Fig. 1.6*c*).

Slope orientation can also be a factor in determining erosion potential. In northern latitudes, south-facing slopes are hotter and drier than other slope orientations. In drier climates, vegetation is sparser on such slopes and reestablishment of vegetation there may be relatively difficult. Conversely, northern exposures tend to be cooler and more moist; but they also receive less sun, which results in slower plant growth.

Ground Cover

The term "ground cover" refers principally to vegetation, but it also includes surface treatments placed by man such as mulches, jute netting, wood chips, and crushed rock. Vegetation is without question the most effective form of erosion control. No man-made products can approach it in long-term durability and effectiveness. Vegetation

- Shields the soil surface from the impact of falling rain
- Slows the velocity of runoff
- Holds soil particles in place
- Maintains the soil's capacity to absorb water

Nonvegetative ground covers can perform only the first two of the above functions. The role of vegetation is discussed in more detail in Secs. 2.1c and 2.1d and in Chap. 6.

REFERENCES

1. Beckett Jackson Raeder, Inc., *Michigan Soil Erosion and Sedimentation Control Guidebook,* Michigan Department of Natural Resources, Bureau of Water Management, Lansing, Michigan, 1975.
2. H. O. Buckman, and N. C. Brady, *The Nature and Properties of Soils,* The Macmillan Company, New York, 1969.

3. D. H. Gray, and A. T. Leiser, *Biotechnical Slope Protection and Erosion Control,* Van Nostrand Reinhold, New York, 1982.
4. H. W. Lull, *Soil Compaction on Forest and Range Lands,* U.S. Department of Agriculture Miscellaneous Publication 768, Washington, D.C., 1959.
5. T. R. Mills and M. L. Clar, *Erosion and Sediment Control in Surface Mining in the Eastern U.S.,* U.S. Environmental Protection Agency, Washington, D.C., 1976.
6. Tahoe Regional Planning Agency and the Association of Bay Area Governments, *How to Protect Your Property from Erosion—A Guide for Homebuilders in the Lake Tahoe Basin,* TRPA, South Lake Tahoe, Calif., 1982.
7. U.S. Department of Agriculture, *Soil and Water Resources Conservation Act: Program Report and Environmental Impact Statement, 1980 Review Draft,* U.S.D.A., Washington, D.C., 1980.
8. U.S. Department of Agriculture, Soil Conservation Service, *Standards and Specifications for Soil Erosion and Sediment Control in Developing Areas,* U.S.D.A. SCS, College Park, Md., 1975.
9. U.S. Department of the Interior, Geological Survey, *The National Atlas of The United States of America,* GPO, Washington, D.C., 1970.
10. U.S. Environmental Protection Agency, Office of Water Program Operations, *Report to Congress: Nonpoint Source Pollution in the U.S.,* U.S. EPA, Washington, D.C., 1984.
11. Virginia Soil and Water Conservation Commission, *Virginia Erosion and Sediment Control Handbook,* 2d ed., Richmond, VA., 1980.
12. C. A. White and A. L. Franks, *Demonstration of Erosion and Sediment Control Technology, Lake Tahoe Region of California,* Final Report, California State Water Resources Control Board, U.S. EPA, Cincinnati, 1978.
13. M. G. Wolman, and A. P. Schick, "Effects of Construction on Fluvial Sediment, Urban and Suburban Areas of Maryland," *Water Resources Research,* vol. 3, pp. 451–464, 1967.
14. C. R. Goldman, *The Greening ofLake Tahoe: A Serious Increase in Algal Growth Accompanied by an Alarming Decline in Transparency Continues in Lake Tahoe,* Environmental Science and Engineering Lecture, UCLA, Tahoe Research Group, University of California, Davis, Calif., May 31, 1985.

chapter 2

Principles of Erosion and Sediment Control

Erosion and sediment control is basically a commonsense process that can be summarized as follows:

1. Fit development to the terrain.
2. Time grading and construction to minimize soil exposure.
3. Retain existing vegetation whenever feasible.
4. Vegetate and mulch denuded areas.
5. Divert runoff away from denuded areas.
6. Minimize length and steepness of slopes.
7. Keep runoff velocities low.
8. Prepare drainageways and outlets to handle concentrated or increased runoff.
9. Trap sediment on site.
10. Inspect and maintain control measures.

These 10 principles and the relation between them are discussed in this chapter. The application of the principles is illustrated by the sample plan in Chap. 9.

Before beginning this chapter it is helpful to understand the difference between erosion control and sediment control. Erosion is the wearing away of soil; sediment is soil that has already been eroded. Erosion control is, therefore, the prevention or minimizing of erosion. Sediment control is the trapping of suspended soil particles. Erosion control is clearly the preferred approach. Sediment control is necessary because some erosion is unavoidable.

2.1 TEN BASIC PRINCIPLES

2.1a Principle 1: Fit Development to the Terrain

The best way to minimize the risk of creating erosion and sedimentation problems by construction is to disturb as little of the land surface as possible. Minimizing land disturbance means minimizing grading. When a gridded street pattern with broad terraces for house pads is rigidly imposed on a hillside, a great amount of earth must be moved. Typically, hilltops are scraped away and the material removed is used to fill valleys and low points. Because this type of development radically alters the stability of the natural hillside, erosion potential is greatly increased. When development is tailored to the natural contours of the land, little grading is necessary and erosion potential is consequently lower.

Figure 2.1 illustrates development fit to the terrain. The streets generally follow the natural contours (and are not visible from below). Trees, together with other vegetation between houses, have been preserved. In Fig. 2.2 development does not fit the terrain. Streets run directly downhill. None of the existing topography, vegetation, or drainage system has been preserved. Unfortunately, the latter example is much more typical of development today than the first example is.

Fig. 2.1 Development that is fit to the terrain.

Fig. 2.2 Development that is not fit to the terrain.

To fit development to the terrain, apply the following general guidelines. They are discussed in more detail in Chap. 9.

1. Analyze the natural characteristics of a site carefully before designing a project.
2. Align roads along contours rather than straight up and down hills.
3. Locate building pads on the flattest portions of a site.
4. Keep disturbed areas small (Fig. 2.3).

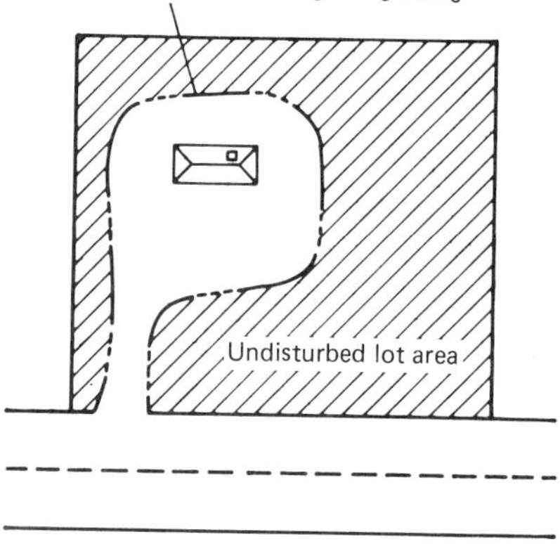

Fig. 2.3 Disturbed areas should be small.

Not all of these guidelines will apply in all cases. On some sites, hillside homes may be desirable for views or other reasons and flat areas may be desirable park sites. Figure 9.4 illustrates how buildings can be adapted to slopes to minimize slope disturbance.

2.1b Principle 2: Time Grading and Construction to Minimize Soil Exposure

There are two aspects to this principle. One relates to the staging of construction to minimize the size of exposed areas and the length of time the areas are exposed. The other relates to timing the grading to coincide with a dry season or a period of lower erosion potential.

Stage the grading so that only small areas are exposed to erosion at any one time; only the areas that are actively being developed should be exposed. As soon as grading is complete in one area, seed and mulch the exposed soils.

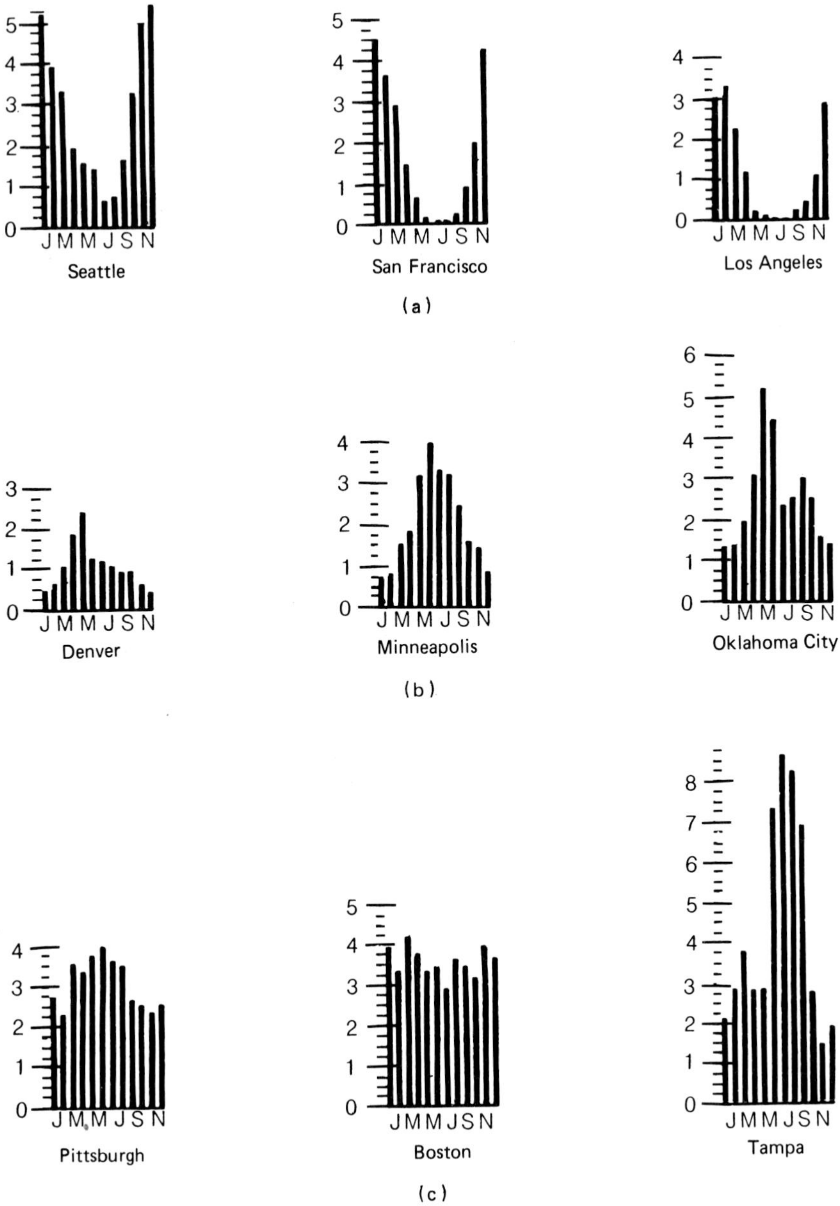

Fig. 2.4 Monthly precipitation at selected cities (*a*) on the west coast, (*b*) in the central states, and (*c*) in the Atlantic and gulf coast states. (2)

In many parts of the country there is a dry season or a time of year when erosion potential is relatively low. Figure 2.4 shows monthly precipitation at selected U.S. cities. The Pacific coast states have a very pronounced dry season, particularly from the California-Mexico border to about the center of Oregon. In most of California only about 10 percent of the annual precipitation falls between May 1 and October 31. (1) Thus May is an ideal month in which to begin grading in that area. If construction can be completed during the 6-month dry season, temporary erosion control measures may not be necessary. If grass is seeded in September and early October, it will germinate with the light autumn rains and grow large enough by December to prevent erosion from the heavy winter rains. Thus the cost of constructing sediment basins and traps, dikes, and ditches and the cost of sediment removal and gully repair can be avoided.

In the central, Atlantic, and gulf coast states significant rainfall occurs in every month of the year. More important, however, the heaviest rainfalls tend to occur during certain months. Rainfall intensity is a key factor in erosion potential; both the i value in the rational formula and the R factor in the universal soil loss equation (see Chaps. 4 and 5) are determined by rainfall intensity, *not* rainfall totals. In Virginia, for example, 70 to 80 percent of the annual rainfall energy occurs between May and September. (3)

In the remaining months the rainfall energy potential is much lower. Therefore, to minimize erosion potential, says Gerard Seeley, Jr., Chief Engineer of the Virginia Soil and Water Conservaton Commission, "a construction manager can try to schedule his land disturbing operations during the least erosive times of the year and attempt to restabilize disturbed areas before the highly erosive rainfalls are likely to occur." (3) He should also pay special attention to installing and maintaining erosion and sediment control measures during the most erosive rainfall months.*

Many parts of the country are blanketed by snow during winter months. When the snow melts in late winter and early spring, erosion potential is greatly increased. Snowfall produces a constant flow of runoff which can carry away bare soil. In addition, the soft, wet ground is easily turned into mud by construction vehicles. In these areas, grading should be avoided during the spring thaw period.

2.1c Principle 3: Retain Existing Vegetation Wherever Feasible

Vegetation is the most effective form of erosion control; very little erosion occurs on a soil covered with undisturbed natural vegetation. Reestablishing vegetation can be a difficult and costly process. Even after restabilization, disturbed soils usually erode at faster than the natural rate for many years. The authors have found that it typically takes about 5 years for the rate of erosion from a construction site to approximate the preconstruction rate.

* That we have used masculine pronouns in such impersonal passages is due to the limitations of English and not to our ignorance of the important contributions to the field made by female engineers, planners, and other land development professionals.

If possible, strip only the area where construction will actually occur—the foundation site and an access area immediately around it (Fig. 2.3), street and driveway lines, and cut and fill slopes. Try to integrate existing trees and other natural vegetation into the site improvement plan. It is not necessary to clear an entire site before beginning construction, although that is the common practice in many areas.

2.1d Principle 4: Vegetate and Mulch Denuded Areas

Seed and mulch disturbed soils as soon as possible after grading is completed. Mulch helps seedlings to become established and protects the soil until vegetation takes over the job. A properly revegetated soil will be protected from erosion indefinitely without any need for human attention.

Soils may be planted with temporary or permanent vegetation. Temporary vegetation provides quick, continuous ground cover until permanent landscaping is installed. Temporary vegetation may also be used as permanent cover.

In most areas of the country, seeds can be planted from late winter until mid-autumn (roughly March 1 to November 1). If soil will be exposed during the winter months, protective measures other than vegetation must be used. (See Chap. 6 for details on seeding and mulching.)

Figure 2.5 shows a stabilized, well-vegetated fill slope. This slope was seeded and mulched in a timely manner before the first rain. Figure 2.6 shows a slope below a house pad that was not seeded. The gullies were created by only a few relatively light rains.

Fig. 2.5 Well-vegetated fill slope

Fig. 2.6 Gully erosion on unseeded slope.

2.1e Principle 5: Divert Runoff Away from Denuded Areas

When vegetative cover is removed from land, the soil becomes highly susceptible to erosion. Runoff from areas above those that have been denuded should not be allowed to cross the exposed soils, particularly when the denuded areas are on slopes. Use dikes or ditches to divert upland runoff away from a disturbed area to a stable outlet (Figs. 2.7 to 2.9). The two most common applications of these diversion devices are to intercept runoff on cut or fill slopes and to prevent runoff from entering a disturbed area, such as a group of building pads. In the latter

Fig. 2.7 Diversion ditch above cut slope.

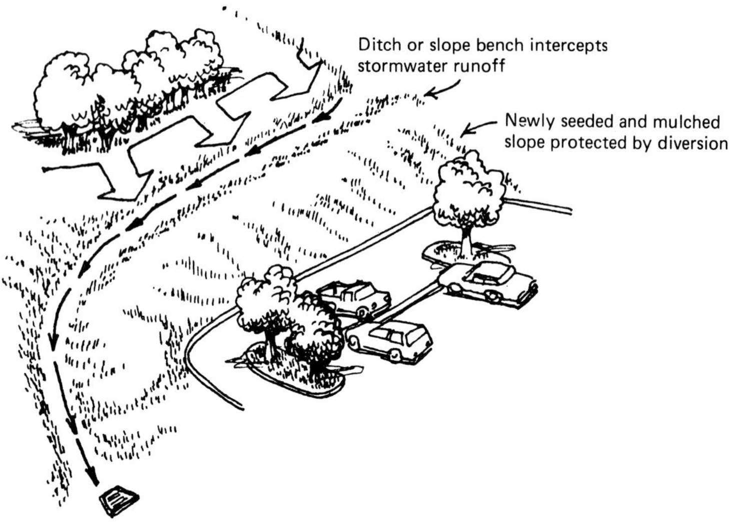

Fig. 2.8 Diverted runoff drains to a stable outlet.

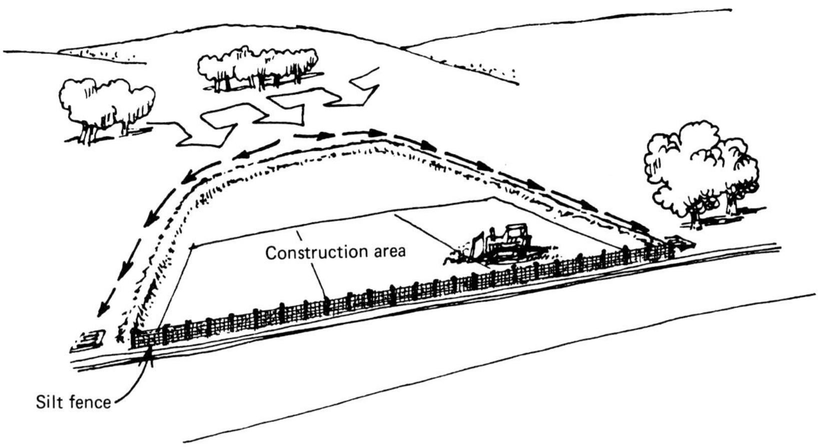

Fig. 2.9 Perimeter dike prevents runoff from crossing a construction area.

example, a perimeter dike may entirely encircle the disturbed area and route the runoff to a protected outlet (Fig. 2.9).

2.1f Principle 6: Minimize Length and Steepness of Slopes

Slope length and steepness are among the most critical factors in determining erosion potential. Increasing slope length and steepness increases the velocity of runoff, which greatly increases its erosive energy (see Principle 7). If slope steepness is doubled while other factors are held constant, soil loss potential is increased 2½ times (see Table 5.5). If both slope steepness and length are doubled, soil loss potential is nearly 4 times greater.

To prevent erosive velocities from occurring on long or steep slopes, break up the slopes with terraces at regular intervals. Terraces will slow down the runoff and provide a place for small amounts of sediment to settle out. Slope benches are usually constructed with ditches along them or are back-sloped at a gentle angle toward the hill (Figs. 2.5 and 2.10). These benches and ditches intercept runoff before it can reach an erosive velocity and divert it to a stable outlet. The slopes of these cross-slope channels should be gentle, and the channels should be protected with erosion-resistant linings if the velocities in the channels will exceed the tolerance of the bare soil surfaces (see Chap. 7).

2.1g Principle 7: Keep Runoff Velocities Low

The energy of flowing water increases as the square of the velocity. If the velocity doubles, the erosive energy quadruples and the water can theoretically move particles 64 times larger by volume. Channel velocities can be kept low by lining drainageways with rough surfaces such as vegetation or riprap, by designing broad, shallow flow areas, and by constructing check dams at frequent intervals. Concrete channels, although efficient and easy to maintain, remove runoff so quickly that channel erosion and flooding often result downstream (Fig. 2.11). Grass- or riprap-lined channels are less likely to cause those problems, because they more nearly duplicate natural streams (Fig. 2.12).

Fig. 2.10 A ditch intercepts runoff at midslope.

Fig. 2.11 A smooth, straight concrete channel carries runoff at high velocity.

Fig. 2.12 A grass lining reduces runoff velocity in a channel.

Overland flow velocities can be kept low by minimizing slope steepness and length (Principle 6) and by providing a rough surface for runoff to cross. Driving a bulldozer up and down a slope (called trackwalking) creates tread marks parallel to slope contours. These miniature terraces both slow runoff velocity and provide flat places for vegetation to take hold (Fig. 6.11). Raking or disking the soil surface before seeding also keeps runoff velocities down and increases plant establishment rates. Vegetation, once established, will further reduce runoff rates.

2.1h Principle 8: Prepare Drainageways and Outlets to Handle Concentrated or Increased Runoff

Development changes the characteristics of runoff. Creating impervious surfaces, such as streets, sidewalks, and rooftops, increases the volume of runoff. Removal of plant cover and compaction of soil by construction traffic also increase the volume of runoff, because much less water is able to percolate into the soil after disturbance. Crusting of the bare soil surface caused by raindrop impact further augments the runoff.

The above changes in the ground surface increase the velocity of the runoff also. Pavement and rooftops have much smoother surfaces than grass or shrubs. Thus water runs off these materials faster and reaches drainageways sooner than before development, which further increases peak flows in streams draining developed areas.

These changes in the amount and rate of runoff must be dealt with in erosion control planning. To prevent channel erosion from occurring:

- Calculate peak flows and velocities for all drainageways that drain a project site (see Chap. 4).

Fig. 2.13 Riprap used to protect a storm drain outlet.

- Design drainageways to withstand the peak flows without erosion.
- Select and install lining materials that are appropriate for the peak flows (see Chap. 7).
- Deenergize concentrated flows at outlets by constructing energy dissipators such as riprap aprons (Fig. 2.13).

If development substantially changes the natural drainage conditions in a watershed, merely protecting the drainage channels on a project site may not be sufficient to prevent erosion. The peak runoff rate may have been increased to the point at which the natural stream well below the developed area can no longer handle the flow. The site planner should assess the ability of off-site waterways that receive drainage from the site to carry peak flows after development. These off-site receiving channels may require enlargement or protective linings. Alternatively, if a project can be so designed that runoff from developed areas is percolated into the soil on-site (as is required in the Lake Tahoe basin, see Sec. 9.1c, Minimize Impervious Areas), no off-site channel enlargement or protection should be necessary.

2.1i Principle 9: Trap Sediment on Site

Some erosion during construction is unavoidable. The function of a sediment barrier is to prevent sediment from leaving a site after the soil has been eroded from its place of origin. Sediment-laden runoff should be detained on-site so that the soil particles can settle out before the runoff enters receiving waters.

The most common sediment barriers are sediment basins and traps (Fig. 2.14), straw bale dikes, and silt fences. Locate sediment basins and traps at low

Fig. 2.14 A sediment trap.

points below disturbed areas. Use earth dikes or swales to route drainage from disturbed areas into the basins. Straw bale dikes and silt fences can be placed below small disturbed areas on gentle to moderate slopes. Storm runoff temporarily ponds up behind these barriers, which allows sediment to settle out. Gradually the water seeps out, leaving the silt behind.

Though sediment barriers can be used to trap eroded soil particles, erosion prevention with vegetation and mulch is a much better form of sediment control. Once soil is washed into suspension, only a fraction of it can be trapped on-site. When soils are high in clay and fine silt, very little of the suspended soil can be recaptured (Secs. 5.1a, 8.2c). Even though sediment basins can retain larger soil particles on-site, this soil has been eroded from its original location. The remaining soil is less able to support plant growth. In addition, the soil impounded in the basins often must be disposed of, sometimes at high cost. Retaining existing vegetation and seeding disturbed areas as soon as possible after grading are the best methods of both erosion and sediment control.

2.1j Principle 10: Inspect and Maintain Control Measures

Inspection and maintenance of control measures are vital to the success of an erosion and sediment control program. Most control measures require regular maintenance. Problems often develop during a single storm. Some problems, if left untreated, can result in more erosion damage than might have occurred without any control measures. For example, a breach in a sediment basin dam may lead to a large deposition of sediment below. A dike or swale with a low spot in it may cause a serious gully problem (see Figs. 10.1 and 10.7). Straw bale dikes are frequently dislodged or separated by stormwater flows. When a wire holding a bale together breaks, the bale will quickly disintegrate.

Figure 2.15 shows a sediment trap just after it was constructed, and Fig. 2.16 shows the same trap after only one storm. Without cleaning, this trap will no longer catch sediment. Sediment-laden runoff will flow directly into the storm drain inlet.

Inspect control measures frequently, particularly during storms, to ensure that they are working properly (see Chap. 11). Correct problems as soon as they develop. A routine end-of-day check is strongly advised.

REVIEW QUESTIONS

1. How does fitting development to the terrain lessen the erosion potential?
2. When is the best time to begin grading?
3. What are the two main purposes of diversion devices?
4. What advantages do grass- or riprap-lined channels have over paved channels?
5. Why is it important to retain as much existing vegetation as possible?

Fig. 2.15 Newly constructed sediment trap.

Fig. 2.16 Sediment trap after one storm; cleaning is required.

6. What is the purpose of a sediment basin?
7. Why is vegetation a better control measure than a sediment basin?
8. Why is it important to inspect control measures?
9. How often should control measures be checked?
10. Under what circumstances might a well-designed energy dissipator at a storm drain outlet fail to prevent erosion downstream?

REFERENCES

1. U.S. Department of Commerce, National Oceanic and Atmospheric Administration, Environmental Data Service, *Climate of California,* Climatography of the United States No. 60, NOAA, Asheville, N.C., 1977.
2. U.S. Department of Commerce, National Oceanic and Atmospheric Administration, Environmental Data Service, *Climates of the United States,* GPO, Washington, D.C., 1973.
3. Virginia Soil and Water Conservation Commission, *E&S Bulletin,* no. 18, winter, 1982–1983, p. 2.

chapter 3

Developing and Enforcing an Erosion and Sediment Control Program

3.1 ELEMENTS OF A SUCCESSFUL EROSION AND SEDIMENT CONTROL PROGRAM

Three key elements must be present for an erosion and sediment control program to be successful: recognition of need, adequate ordinance and administrative procedures, and technical expertise. These elements are briefly described below.

3.1a Recognition of Need

For an erosion and sediment control program to be successful, elected officials, public agency management and staff, developers, design consultants, soils engineers, and contractors must all recognize the need for erosion control. Each of these groups will see the need from a different perspective. For example, an elected official may balance resistance to overregulation against savings of taxpayers' dollars by avoiding unnecessary costs for maintenance of public facilities. The developer may balance the cost of erosion and sediment control measures against the cost of delays which may be incurred in resisting the public agency's requirements for the control measures. The agency management may balance the increased staffing requirements against permit fees and public facilities maintenance costs. Indifference or hostility by any of these people or groups will undermine implementation of a successful erosion and sediment control program.

3.1b Adequate Ordinance and Administrative Standards and Procedures

Recognition of need on the part of elected officials and public agency management should lead to an adequate erosion and sediment control (or grading) ordinance and adequate administrative standards and procedures. The ordinance provides the legal basis for requiring controls. Without it implementation would be voluntary and hence ineffective in the absence of 100 percent developer cooperation. Administrative procedures and standards provide for even-handed application of the ordinance, tell the developer what to expect, and give enforcement staff a means of assuring implementation. All these are essential to effective implementation.

3.1c Technical Expertise

Technical expertise is required for the site planner to develop an effective erosion and sediment control plan and for municipal agency staff to review and approve such a plan. Technical expertise is also necessary for the developer and the contractor to implement the measures called for by the plan and to understand the plan well enough to react effectively to unanticipated or changing field conditions. Technical expertise is necessary for the agency inspector to know whether facilities are being installed correctly, whether the contractor is reacting adequately to site conditions, and when to permit or require field modifications of plans. Technical expertise comes with training and experience. Both are necessary. Maryland recognized the importance of technical expertise when it amended its sediment control law to require all persons responsible for land-disturbing activities on a construction project to have a certificate of attendance at a state-approved erosion and sediment control training program. (5)

All of the above three elements (recognition of need, adequate ordinance and administrative standards, and technical expertise) are *essential* if erosion and sediment control efforts are to be routinely successful.

3.2 EROSION AND SEDIMENT CONTROL LAWS AND ORDINANCES

In most areas of the country, grading and land development are regulated by the local jurisdiction, typically a county or city. It therefore makes sense for erosion and sediment control also to be regulated at the local level. A local grading ordinance or erosion and sediment control ordinance is probably the most effective means of implementing and enforcing erosion and sediment controls.

In recognition of the importance of local ordinances and regulations, many states have passed laws mandating local programs. State erosion and sediment control laws are discussed in the next section, which is followed by a discussion of local programs. A model grading, erosion, and sediment control ordinance is provided in Appendix A.

In addition to state and local erosion and sediment control laws and ordinances, some federal and state laws peripherally relate to this subject; examples are the Clean Water Act and the National Environmental Policy Act. These and other laws are briefly discussed at the end of this section.

3.2a State Laws

Eighteen states and the District of Columbia have enacted erosion and sediment control laws (Table 3.1), and several other states are considering similar bills. Some of the laws are weakened by long lists of exemptions. Pennsylvania's law, for example, exempts developments under 25 acres (10 ha). Maine's and Vermont's laws are weakened by being partially voluntary. (9)

TABLE 3.1 States with Construction Sediment Control Laws (9)

Connecticut	Maine	Ohio
Delaware	Maryland	Pennsylvania
Georgia	Michigan	Rhode Island
Hawaii	Montana	South Dakota
Illinois	New Jersey	Vermont
Iowa	North Carolina	Virginia

Maryland's Sediment Control Law—A First

Maryland's sediment control law is notable for several reasons. It was the first statewide erosion control legislation; it was signed into law on Earth Day, April 22, 1970. Its key provisions are summarized below:

1. Counties and municipalities are required to adopt grading and sediment control ordinances and to implement and enforce sediment control programs.
2. No county or municipality may issue a grading or building permit until the developer "(1) submits a grading and sediment control plan approved by the appropriate soil conservation district," and "(2) certifies that all land clearing, construction, and development will be done under the plan." (5)
3. If a county or municipality has not adopted or is not implementing an acceptable sediment control program, the state can order the local jurisdiction not to issue a building or grading permit or it can take over administration of the local sediment control program.
4. Any person who violates any provision of the law is guilty of a misdemeanor and is subject to a maximum fine of $5000 or one year in prison or both for each day on which a violation occurs. Any agency or interested person may seek an injunction against any person who violates or threatens to violate any provision of the law.

Points 3 and 4 give real teeth to the law. The Maryland Department of Natural Resources is responsible for establishing erosion and sediment control standards and for reviewing and approving local ordinances and programs. In the years following enactment of the state's Sediment Control Law, 23 Maryland counties and 151 municipalities adopted approved sediment control programs.

Amendments that were later added to the Maryland Sediment Control Law made significant changes. A 1979 amendment stated:

> After July 1, 1980 any applicant for sediment and erosion control plan approval shall certify to the appropriate jurisdiction that any responsible personnel (foreman, superintendent, or project engineer who is in charge of onsite clearing and grading operations or sediment control associated with a construction project) . . . will have a certificate of attendance at a Department of Natural Resources approved training program for the control of sediment and erosion before beginning the project.

Experience had shown that technical training was vital to the success of an erosion control program. During the first half of 1980 the state scheduled one-day training sessions in each county of the state.

Because of recent cutbacks in local enforcement staffs and other enforcement problems at the local level, the Maryland legislature, in July 1984, amended the Sediment Control Law again. Effective April 1, 1985, the State's Department of Natural Resources has the sole responsibility to enforce the law except in counties or municipalities in which enforcement authority has been officially delegated. Counties or municipalities that wish to enforce the law themselves must first petition the secretary of Natural Resources for approval. Before approving a request, the secretary must first find that the enforcement capability within that jurisdiction is comparable with the state's "in terms of laws, procedures, manpower, equipment, and overall effectiveness." (5) The delegation of enforcement authority is effective for only 2 years, but it may be renewed by the state.

Maryland adopted the *Standards and Specifications for Soil Erosion and Sediment Control in Developing Areas* as the official standards for use in Maryland. This document, first published in 1969 and improved, expanded, and reprinted in 1975 and again in 1983 (6), was a joint effort of the State Water Resources Administration, the U.S. Soil Conservation Service office in College Park, Maryland, and the State Soil Conservation Committee. This excellent publication is now widely used throughout the country. Many of the technical specifications and reference tables used in this handbook and in the erosion and sediment control manuals published by the Association of Bay Area Governments (in California), the Virginia Soil and Water Conservation Commission, and other agencies were derived from the Maryland document.

In 1973, Virginia passed a law very similar to the Maryland Sediment Control Law. The Virginia law requires all soil and water conservation districts or, where there is no district, the county, city, or town to adopt an erosion and sediment control program. No land-disturbing activities are allowed without an approved erosion and sediment control plan. The Virginia Soil and Water Conservation Commission was directed to establish minimum standards for erosion and sediment control and to approve local programs.

However, unlike the Maryland law, the Virginia law does not give a state agency the authority to take over an inadequate local program. When a local ordinance is unacceptable, the state sends it back for revisions. As a result of the law, 171 local erosion and sediment control programs have been adopted in Virginia.

3.2b Local Ordinances

As stated earlier, local ordinances are the principal means of regulating land-disturbing activities. Erosion and sediment control requirements are commonly found in:

- Grading ordinances
- Erosion and sediment control ordinances
- Combined grading and erosion and sediment control ordinances
- Land development ordinances (sometimes called subdivision ordinances)

Montgomery County, Maryland, Leads the Way

In 1971, Montgomery County, Maryland, which borders Washington D.C., became the first county in the nation to enact a mandatory erosion and sediment control ordinance. (9) The ordinance requires a permit for land-disturbing activities, and a detailed erosion and sediment control plan is required as part of the permit application. If a violation of the plan occurs, the county may revoke the permit and issue a stop-work order that makes the person responsible subject to arrest and fines if the violation continues.

Key Features of an Effective Erosion and Sediment Control Ordinance

During the preparation of a water quality management plan for the San Francisco Bay Area, the authors conducted a nationwide survey of erosion and sediment control ordinances. The purpose was to develop a model ordinance that could be adopted by the counties and cities of the region. Review of existing ordinances and interviews with local officials from coast to coast revealed that certain key provisions were necessary to make an ordinance effective. Those key features are briefly discussed below. Each of the features has been incorporated in the Model Grading and Erosion and Sediment Control Ordinance which is reproduced in Appendix A. The ordinance section number in which a provision is found is given in parentheses.

1. *Water quality is an explicit goal of the ordinance.* Traditional grading ordinances were designed primarily for protection of life, limb, and property. This provision specifically informs persons proposing land-disturbing activities that water quality must be protected and provides a legal basis for prosecution when water pollution occurs. (Appendix A, Sec. 2)

2. *A permit is required for land-disturbing activities.* Exemptions are limited. All grading that may result in significant erosion or sedimentation is regulated. Anyone who wishes to do such grading must apply for a permit to be approved by a local authority (typically a municipal public works director). Exemptions should be kept to a minimum; typically, they are limited to agricultural activities and very small disturbances such as home gardening. Grading to construct a single home often causes erosion problems and should be controlled by the ordinance. (Appendix A, Secs. 10 to 12)

3. *Temporary and permanent erosion and sediment control plans that meet minimum standards must be submitted.* This provision, which provides a strong enforcement tool, may well be the most important requirement. Applicants must specify in writing how they will control erosion and sediment. The local authority can thus assess, in a timely manner, whether a plan makes sense, whether there are sufficient details, and whether the specifications meet accepted standards. Such erosion and sediment control standards should be specified in a state or locally adopted handbook, such as Maryland's. (6, 8) If the measures shown on the plan are not installed or are implemented contrary to the plan's specifications, there is an easily documentable violation. It is much easier to show that a plan was not implemented than it is to show that water pollution has occurred. (Appendix A, Secs. 17 and 18)

4. *Control measures must be installed before the rainy season begins.* This provision primarily applies to west coast communities because of the very pronounced wet and dry seasons that occur in the region (see Sec. 2.1b). Seeding bare areas and installing control measures by early fall greatly reduce the likelihood of erosion problems. (Appendix A, Sec. 21) Communities in other parts of the country may want to consider special controls during critical times of the year such as during the spring thaw or during months in which very intense rain storms are common.

5. *Regular permittee reports and site inspections are scheduled.* Permittees are required to notify the local authority at key times, such as just before grading begins and at the completion of rough grading. The local authority should inspect the site at these and other important times, such as during or immediately following storms. Because many California communities were facing severe financial and manpower shortages and were not able to inspect every project, the model ordinance made the inspections discretionary. Instead, the ordinance requires the permittee to file reports if there are any delays in the work schedule or if any work is not being done in conformance with the approved plans. If a report is not filed on time or if a report suggests a potential problem, the local authority has the option to inspect. The combination of permittee reports and selective inspections places some of the burden for monitoring projects on the developer and allows the inspector to concentrate on the most critical projects. (Appendix A, Secs. 31 and 32*b*)

6. *The local authority may require modification of an erosion and sediment control plan.* When an approved erosion and sediment control plan is found to be inadequate or the plan as implemented appears to be ineffective, modifications may be required. This provision is the legal basis for requiring plan changes after initial plan approval. (Appendix A, Sec. 32*a*)

7. *Strong enforcement mechanisms are available to the local authority.* A sequence of progressive provisions, such as suspension or revocation of permit, fines, and imprisonment, is included. The penalties must be severe enough to induce permittees to comply faithfully with the grading regulations. (Appendix A, Secs. 34 and 35)
8. *Permittee is required to provide security.* When a permittee's failure to properly install appropriate erosion control devices causes a threat of erosion and sedimentation, security in the form of a deposit or bond is available to the local jurisdiction to finance remedial work. Security must be sufficiently large and available on short notice if excessive erosion is to be avoided. (Appendix A, Secs. 22, 36, and 37)

The model ordinance in Appendix A is based on Chapter 70 of the *Uniform Building Code.* (4) Many cities have adopted Chapter 70 as their grading ordinance. The authors added erosion and sediment control provisions to Chapter 70 to make it a combined grading and erosion and sediment control ordinance. Ordinances similar to this model have been adopted by many cities and counties in the San Francisco Bay Area.

Unfortunately, traditional grading laws, including Chapter 70 of the *Uniform Building Code,* have failed to control erosion and sediment problems adequately. Many communities have recently amended or are currently revising their grading regulations and administrative procedures to strengthen their erosion control provisions. In Table 3.2 the key erosion and sediment control provisions in some recently adopted grading ordinances are compared with those in some typical older grading ordinances.

3.2c Other Laws

Numerous federal, state, and local laws indirectly pertain to erosion and sediment control. The federal Clean Water Act (formerly the National Water Pollution Control Act Amendments of 1972) required, in Section 208, the preparation of comprehensive water quality management plans for the major water basins in the United States. Such plans were prepared for most major metropolitan areas throughout the United States. Some of these plans contain provisions related to erosion control and other nonpoint pollution sources.

For example, the Water Quality Management Plan for the San Francisco Bay Area (1) calls for Bay Area cities and counties to adopt and enforce effective ordinances for erosion control and to implement the best management practices described in the Association of Bay Area Governments' *Manual of Standards for Surface Runoff Control Measures.* (2) The Lake Tahoe Basin Water Quality Plan, prepared pursuant to Section 208, sets forth policies and procedures for controlling erosion around the lake. Volume II of the plan, the *Handbook of Best Management Practices,* contains specifications for these erosion control measures. (7) Implementation of the Tahoe plan is achieved through various ordinances adopted by the Tahoe Regional Planning Agency. These ordinances apply to both the California and Nevada sides of the lake.

The Clean Water Act also regulates the discharge of "point sources" of pollutants to the waters of the United States. Under the Act, states are authorized

TABLE 3.2 Comparison of Erosion and Sediment Control Features of Traditional Grading Ordinances and Recent Erosion and Sediment Control Ordinances

Feature	Traditional grading ordinances	Recent ordinances with specific erosion and sediment control provisions
Purpose of ordinance	Public health, welfare, and property rights are protected.	In addition to protection of health and safety, preservation of the natural environment and water quality are explicit goals.
Exemptions from permit requirement	Exemptions for specific types of activities are available; general exemptions based on amount of material moved, area, and slope vary widely.	Exemptions for specific types of activities are available; general exemptions are limited to sites with minimal erosion and sedimentation potential.
Erosion control plans	Designated devices for stormwater management must be detailed on the grading plan. In some ordinances, erosion control devices must also be displayed on the grading plan. Technical standards for erosion control measures are limited in scope.	Separate erosion and sediment control plans are required for both interim (during construction) and final (postconstruction) phases. Design and construction details for the erosion control devices must be provided. Specifications for the control measures must meet accepted standards defined in the ordinance or in a manual referred to in the ordinance. Critical dates by which control measures must be installed are specified. These dates are tied to the rainy season.
Reports and inspections	Inspections are normally at the discretion of the local jurisdiction. Notice of specified stages of grading activity and/or progress reports are required from the permittee by some jurisdictions.	A scheme of reports and inspections to inform the local agency whether the permittee has installed erosion and sediment control devices in a proper and timely manner is specified.
Modification of erosion and sediment control plan	Not covered.	When an approved erosion and sediment control plan is later found to be inadequate or the plan as implemented appears to be ineffective, the permit administrator may require modification of the plan.
Enforcement	Suspension and revocation of permit and/or fines and penalties are available in some ordinances.	Comprehensive means of enforcement, including both suspension and revocation of permit, fines, and imprisonment penalties, are specified.

TABLE 3.2 *(Continued)*

Feature	Traditional grading ordinances	Recent ordinances with specific erosion and sediment control provisions
Security	A bond covering the grading plan is required. The amount is specified or is at the discretion of the permit administrator.	Security in the form of a cash deposit, letter of credit, and/or bond is required for both the grading plan and the erosion control plan. Security must be sufficiently large to cover the cost of remedial work and must be available on short notice.

to require discharge permits as part of the National Pollution Discharge Elimination System. Some states have water quality protection laws that supplement the federal law. California's Water Quality Control Act, for example, makes it unlawful to discharge polluted water into a waterway without first reporting it to or getting a permit from the regional water quality control board. When one northern California land developer allowed sediment from a construction site to pollute a local creek, the regional board in the area considered the problem a "pollutant discharge" and asked the state attorney general to prosecute under the Water Quality Control Act. The developer was thus liable for fines of up to $6000 for each day that a violation occurred. Though the case was eventually settled out of court, the developer could have been fined $498,000.

Maryland has a Water Pollution Control and Abatement Law which sets penalties of up to $10,000/day for discharging pollutants into the waters of the state. A 1961 attorney general's opinion in that state declared that sediment is a pollutant, which means that the water pollution control law can be invoked when erosion from a construction site occurs. The law has been used primarily to control pollution from agricultural activities, but it is sometimes co-cited in sediment pollution cases.

The National Environmental Policy Act (NEPA) requires an environmental impact statement for a project in which the federal government is either directly or indirectly involved. Some state laws patterned after NEPA (such as California's Environmental Quality Act) have similar requirements for state projects. These laws require environmental impact reports for many construction projects. The reports must contain detailed information on environmental effects of a project and recommendations on how to minimize negative impacts. In practice, however, these reports have provided only very general guidelines for controlling erosion and sediment on construction sites.

Other state laws may pertain to construction and other land-disturbing activities. For example, in California, the Fish and Game Code makes it unlawful to divert, obstruct, or substantially change the bed of a stream or lake without first getting permission from the Department of Fish and Game. In Maryland, any person proposing to change the course, cross section, or velocity of a stream must first obtain a permit from that state's Water Resources Administration.

Some state laws place restrictions on dams or water impoundments exceeding a certain size. Very large siltation basins could be regulated by such laws.

Many state laws and local ordinances place requirements on land subdivisions. Some localities have watercourse protection ordinances, hillside protection ordinances, coastal protection ordinances, or other specialized regulations that cover erosion control. In addition, general tort law or property law could be the basis of legal action when damages occur as a result of improper grading activities. Any persons contemplating land-disturbing activities should contact the local authorities and familiarize themselves with both state and local requirements before proceeding.

3.3 ENFORCEMENT

3.3a People and Communication Are the Keys

Regardless of whether a city or county has a strong erosion control ordinance and regulations, the key to preventing erosion and sedimentation is people. Two of the most important people in this process are the developer's construction site representative (superintendent) and the local authority's inspector. These two people are the ones directly responsible for preventing erosion on the job site. The inspector must see that the superintendent knows what control measures are required on a site. The superintendent must see that the contractors understand and do what is required. The environment will not be protected without the combined efforts of all parties concerned.

Often the developers will take only the actions which they feel the local authority will insist on. The inspector must make sure that the superintendent knows the minimum acceptable levels of performance and that doing less will not be tolerated. A superintendent can detect whether an inspector and his or her supervisors are serious about enforcement.

Developers and contractors will approach their projects with different attitudes and styles, but all of them have one thing in common: they're in it for the money. That is what makes enforcement necessary. Since the objectives of erosion and sediment control go beyond preventing damage to a site which would have to be repaired at developer or contractor expense, it is too much to expect developers or contractors to voluntarily spend money on measures which do not serve their limited purposes. Because delay of a project is extremely costly, a developer may implement a control measure for the sole reason that failure to do so would mean an inspector will hold up the job pending compliance.

3.3b Enforcement Tools

The tools of the inspector's trade, in addition to knowledge and intelligence, are the means by which the developer can be persuaded or forced to do what is required by approved plans and relevant ordinances. Punitive measures are not intended to be used, because their use means that implementation of the project plans has been unsuccessful. Rather, the developer must believe the inspector *will* use the available punitive enforcement tools if necessary. The exact nature of those tools will vary from jurisdiction to jurisdiction. Some of the tools, in their typical order of use, are the following:

1. *Inspection.* Just being there to see that the job is being done in accordance with approved plans and specifications is the inspector's first-line enforcement tool.
2. *Communication.* This was covered in some detail in Sec. 3.3a. Briefly, the simple fact that the developer's representative and the inspector are routinely talking to each other will help to ensure that there are no misunderstandings about what is expected.
3. *Negotiation.* The inspector and the developer's representative should work together to resolve conflicts before further, stronger action becomes necessary.
4. *Warning.* When conflicts cannot be resolved amicably, the inspector must warn the developer's representative of the consequences of noncompliance. The warning should be in the form of a written notice citing the pertinent ordinance, describing the violation and what must be done to correct it, and setting a deadline for correction.
5. *Stop work.* If warnings are disregarded, the inspector (by this time, with the authorization of a supervisor) can shut down the job, in most jurisdictions, until compliance is obtained.
6. *Performance bonds or deposits.* The grading or erosion control ordinance should contain provisions that enable the jurisdiction to unilaterally bring the project site into compliance, or at least stabilize it, by using its own resources or independent contractors and funds for such purposes deposited in advance by the developer. These funds should be used as soon as it becomes apparent that the developer will not otherwise act within a reasonable time. (A reasonable time may be very short if the situation occurs during the rainy season and the noncompliance affects erosion and sediment controls.)
7. *Court actions.* Finally, the matter may be referred to the appropriate authorities for prosecution in court to invoke whatever criminal penalties are authorized by the jurisdiction's grading or erosion and sediment control ordinance.

3.3c Routine Enforcement Procedures

If site development is proceeding smoothly and everyone is trying to achieve the same thing, little enforcement is required other than routine inspections and communication to head off potential problems. The following routine procedures are recommended:

Preconstruction or Pre-rainy-season Meeting

It is strongly recommended that at least one meeting be held on the job site with the developer's team to make sure everyone concerned understands and will carry out the erosion and sediment control plan. A meeting is particularly important on large projects on steep terrain. In most parts of the country the meeting should be held before construction begins.

On the west coast, particularly in California, there is a long dry season from May until September. Projects which begin during this dry season usually do not need erosion controls until mid-September. In this area, a pre-rainy-season meeting should be held in August or September, about a month before the date erosion and sediment control measures are scheduled to be in place. More lead time is required for large, difficult sites than for small, easily controlled sites. This prewinter meeting should be held *every year* during which the grading and building permits are active unless the local authority has determined that the site is stable and will remain so. If construction will be starting during the rainy season, the meeting should be held before work begins.

Attendance at the Meeting

The meeting should be attended by:

- The developer or a representative authorized to commit the developer to action
- The developer's on-site superintendent
- The grading contractor's on-site superintendent
- The consultant civil engineer or landscape architect responsible for development of the erosion and sediment control plan
- The consultant soils engineer if that person's input affected the erosion sediment control plan in any important way
- The chief inspector from the city or county
- The city or county inspector for the job
- The plan checker who reviewed the erosion and sediment control plan for the city or county

Many of these people do not need to attend a particular meeting if it can be foreseen that their input will not be needed, but at a minimum, the city or county inspector for the job and the developer or the developer's on-site superintendent must be there.

Preparation for the Meeting

Before the meeting, the city or county job inspector should go over the plan in detail with the plan checker to be sure the plan is understood. Also, the developer's and contractor's on-site superintendents should meet with the consultant responsible for development of the erosion and sediment control plan to be sure they understand the plan. If the meeting is a pre-rainy-season meeting (west coast), the latter group should also prepare a *realistic* updated schedule of activities affecting the plan through the winter (e.g., completion of grading, storm drainage, paving, etc.) and the schedule of implementation of erosion and sediment control measures. Several copies of these schedules, the grading plans, and the erosion and sediment control plans should be brought to the meeting.

At the Meeting

The meeting should result in a mutual understanding of the erosion and sediment control plan, the identification of critical measures and sensitive areas, how runoff will be managed (what drainage goes where), and the dates when key activities will be started or finished. The erosion and sediment control plan should be updated and modified at this time as necessary to reflect current site conditions and scheduling. Modifications should appear on all copies of the plan.

Inspections

The following inspections are recommended as a minimum:

1. *Initial inspection.* Inspect when the site is staked for grading but grading has not begun.
2. *Rough grading inspections.* Inspect at key points during rough grading, including stripping, keying, compaction, and installation of subsurface drains.
3. *Erosion and sediment control compliance inspection.* Inspect the site immediately after sediment basins, dikes, and other control measures are scheduled to be installed. Inspections at fall seeding deadlines are particularly important. It is strongly recommended that an inspector be present at the construction site when seeding is done to review the seed tags and guarantee the application of seed, fertilizer, and mulch according to the composition and rate given in the specifications.
4. *Final grading inspection.* Inspect when all grading, drainage, paving, planting, and permanent erosion and sediment control structures are complete.
5. *Final stabilization inspection.* Inspect when the site is supposed to be stable (i.e., after vegetation has had time to become established). Additional inspections should be made on large or difficult (erosion-prone) sites as follows:
   - Before all forecast major storms to be sure erosion and sediment control measures are in working order. On the west coast an inspection should be made in the fall before the first forecast rain of the season.
   - During or immediately after all major storms to check performance of control measures and correct problems.

Inspectors should always carry erosion and sediment control plans on site visits and compare what they see on the ground to what is shown on the plans. Some things to watch for during any inspection are:

- Rills or gullies on finished or unfinished slopes
- Sediment-laden runoff flowing into storm drains, watercourses, or lakes
- Sediment deposits in streets, storm drains, watercourses, lakes, or on adjacent property
- Debris or soil deposits near a storm drain, watercourse, or lake
- Vegetation removed from an area not designated for grading
- Bare, unprotected soil, particularly on slopes, at the start of or during the rainy season
- Runoff flowing down unstabilized slopes or to facilities not designed to handle it (often because of the absence of a top-of-slope diversion)
- Erosion at storm drain, swale, or pipe outlets
- Sediment traps or silt fences filled with sediment and requiring cleaning
- Runoff flowing under, around, or through a diversion structure (e.g., water undercutting a straw bale dike or silt fence, spilling through a breach in a dike, or overflowing from a low spot in a swale)
- Control measures shown on the erosion control plan but not present on the site
- Control measures not installed according to the plans

Chapter 10 provides numerous illustrations of the above conditions and describes corrective actions.

Reports

Under some circumstances, an inspection may not be necessary, as on a small, flat site that presents no known problems. Some communities may not be able to afford the staff to inspect all sites regularly. In such cases a written report from the developer may suffice. This reporting procedure will not only conserve staff time for the more critical projects but will save developers money as well. By using a standardized form, such as the sample shown in Table 3.3, the permittee informs the responsible agency of the status of grading and erosion control activities on the project. The permittee must alert the agency to any delays in or problems with implementation of the approved plan. If a report indicates the likelihood of a problem, the local authority may choose to make an inspection.

The report form must be tailored to the specific ordinance and procedural requirements of the local authority. Reports should certify the conditions on the site at the time and be signed by a representative of the developer with the authority and expertise to make the certification.

TABLE 3.3 Sample Permittee Report Form

Status of Erosion and Sediment Control Activities

1. Date:

2. Project name: Permit No.

3. Location:

4. Name and title of person preparing this report:

5. Is the project ahead of or behind schedule as specified in the approved Grading Plan?

   Yes ______ No ______

   If yes, explain:

6. Are there any other departures from the approved Site Map and Grading Plan which may affect implementation of the Erosion and Sediment Control Plan as scheduled?

   Yes ______ No ______

   If yes, explain:

7. Indicate possible delays in installing utilities or in obtaining materials, machinery, services, or manpower necessary to implement the Erosion and Sediment Control Plan as scheduled.

8. Describe the progress of or delays in the installation of individual control measures (including each planned sediment basin or trap and application of seed and mulch to specified slopes).

9. Describe any other departures from the implementation of the Erosion and Sediment Control Plan.

____

Signature of permittee or authorized agent

____

Title Phone

3.3d Enforcement in Problem Situations

A construction project which stays on schedule and is routine from start to finish is a rarity indeed. It would be impossible to deal with all the possible ways plans can get fouled up, but the following ways are some of the typical ones.

Scheduling Problems

Many west coast construction projects are scheduled during the dry season to avoid weather-related delays and the need for temporary drainage and erosion controls. Typically, a project schedule calls for curbs, gutters, and storm drains to be installed and streets to be paved before the rainy season begins. Unfortunately, projects never seem to proceed according to plan and delays are inevitable. If, for example, a delay forces postponement of the improvements from August until November, temporary drainage and erosion and sediment control measures must be planned and installed before the rainy season begins.

A change in the sequence of construction also should be carefully evaluated. For example, if a storm drain system is to convey silt-laden runoff to a sediment basin but a delay causes the storm drain to be completed well after the sediment basin, the result is a sediment basin that will collect no runoff.

Seeding can present critical scheduling problems. West coast developers frequently try to continue grading operations well into the fall if the rains have not stopped them. When the first rain is forecast, they phone their local hydroseeding contractors. Those contractors, now deluged with simultaneous requests for their services, can handle only a limited number of jobs in the short time period. Thus, many sites remain unprotected when the fall rains come. Developers should plan ahead and order seeding and mulching services well before fall so that contractors can provide their services in a timely manner.

Although some kinds of delays are not serious, others can have a critical effect on erosion and sediment control. If completion of grading, especially cut and fill slopes, is delayed into the fall, revegetation also will be delayed. If seeding is done too late in the fall, short days and cold weather will inhibit plant growth and the grass will be too young to provide protection from winter rains.

Some delays are beyond the control of either the developer or the local authority, but often a timely reminder (e.g., to order seed and mulch) can prevent an inadvertent delay caused by a lapse of memory or ignorance of the necessary lead time. Delays can also be intentional. A contractor or developer may feel that, given enough foot-dragging, the inspector will forget about some control measure.

Unanticipated Conditions

No erosion control plan, no matter how detailed, can tell you how to deal with every possible situation which may arise. A good plan will provide the flexibility to deal with unanticipated conditions, especially in an emergency.

Unanticipated conditions can result from a host of situations such as inaccurate topographic mapping, minor differences between the grading plan and the actual grading, incorrectly constructed drainage or sediment control facilities, structural failure of graded or natural slopes, and scheduling mix-ups or delays. Utility trenching can breach berms and ditches; vehicle traffic can destroy berms and ditches and remove vegetative cover from supposedly undisturbed areas; silt or discarded construction materials can block drainageways. Stockpiled materials and borrow areas also can be sources of problems.

When nonemergency situations seem to require temporary or permanent changes in the erosion and sediment control facilities, consult the plan checker and the person who prepared the plan. Any changes should be approved in advance by the local authority, and all copies of the plan should be changed accordingly. This process will avoid reactions to unanticipated conditions that, because plans are poorly understood, cause more problems than they solve.

In an emergency, as during a heavy rainstorm, field changes may have to be made to cope with unanticipated conditions or failure of drainage or sediment control structures. In such cases a thorough understanding of the plan and of the overall drainage on the site is very important.

For example, diverting runoff from an overflowing sediment trap may overload some other drainage facility and cause it to fail. Then the problems can snowball. Trying to figure out the plan in the pouring rain is not conducive to making good decisions, so both the contractor and the inspector must understand the plan thoroughly. If the inspector is not on the job site at the time, the contractor may have to act alone. On the other hand, the inspector, if there, may have to authorize the contractor's action on the spot. The inspector should also be familiar with all the standards for erosion and sediment control structures established by the responsible agency and with the agency's erosion and sediment control manual, if there is one. Such knowledge will enable the inspector to authorize emergency changes with confidence and even suggest courses of action to the superintendent or contractor. Any changes which have been made in the field without prior approval of the local authority should be reviewed and approved by the local authority and the plans changed accordingly. Alternatively, the field changes should be removed and the conditions which made them necessary corrected as soon as possible.

3.4 PUBLIC COSTS TO IMPLEMENT A CITY OR COUNTY EROSION AND SEDIMENT CONTROL PROGRAM

Implementing an effective erosion and sediment control program requires both manpower and money. This section provides information on the public costs of carrying out such a program and how to recover some of the costs through fees. Private sector costs for constructing erosion and sediment control measures are discussed in Sec. 3.5. The erosion and sediment control program cost estimates in this section are based on a regulatory program similar to the ones described in Sec. 3.2 or one that would be created by adopting an ordinance similar to the one in Appendix A. The basic features of such a program are:

- Any project that could have significant erosion potential is regulated.
- A detailed erosion and sediment control plan is required before a grading permit can be issued.
- Regular inspections are conducted to ensure compliance.

Case Study—Montgomery County, Maryland

Montgomery County, Maryland, has been administering a comprehensive erosion and sediment control program for more than a decade and has kept detailed records of costs. The county covers about 316 mi^2 (818 km^2) of flat and rolling terrain. The population is approximately 600,000. The county contains the communities of Bethesda, Silver Springs, Rockville, and Potomac—all suburbs of Baltimore and Washington. The area is relatively urbanized, and single-family homes predominate. The construction season is spring through summer, a period during which intermittent, high-intensity storms can cause significant erosion damage. Because of the frequency of spring and summer rainfall, the implementation of erosion control measures is important during construction to protect sites and their surrounding environment.

From 1975 to 1980 the average operating budget of Montgomery County's Sediment Control Section was approximately $180,000/year. This budget covered two plan reviewers, one chief inspector, and five other inspectors. In fiscal year 1979, a typical year, the section reviewed 1289 sediment control permit applications, issued 911 permits, and collected approximately $168,000 in fees.

Each permit required, on the average, 14 inspections per year to ensure good control. Large projects with sediment basins required seven inspections per basin, weekly or biweekly inspections during rough grading operations, and monthly inspections after completion of rough grading to ensure that measures were maintained in working order. Small projects, on the other hand, required as few as two inspections. Table 3.4 summarizes cost data from Montgomery County's sediment control program.

Setting Permit Fees for Cost Recovery

Many communities, faced with tight budgets, do not have readily available funds to implement erosion and sediment control programs. One solution is to make a program self-supporting. To do so, a city or county must collect enough fees to cover costs for program administration, erosion control plan reviews, project inspections, and other enforcement work.

Since the cost of reviewing a project is a function of project size, an equitable way to set the permit fee is to base the fee on the area disturbed. A survey conducted by the authors in 1980 found that erosion plan review and inspection costs ranged from a low of about $0.0006/$ft^2$ ($0.0065/$m^2$) of disturbed area for a very large project to a high of $0.0043/$ft^2$ ($0.046/$m^2$) of disturbed area for a sin-

TABLE 3.4 Data from Montgomery County, Maryland, Sediment Control Program (1979)*

	Housing units per project				
	1 unit	2–10 units	10–50 units	50–100 units	100+ units
Percent of projects	55	8	21	12	4
Sediment control plan review time per project	0.25 hr	2 hr	4 hr	10 hr (8–12 hr)	16 hr (8–16 hr)
Average number of inspections per project	4	6	12	20	30
Average time per inspection	0.25 hr	0.5 hr	0.75 hr	1 hr	2 hr
Total staff time per project	1.25 hr	5 hr	13 hr	30 hr	76 hr
Processing fee per square foot of disturbed area†	$40 min.	$0.002 ($87/acre)	$0.002 ($87/acre)	$0.002 ($87/acre)	$0.002 ($87/acre)
Fee range per project	$40	$40–200	$200–1000	$1000–2000	$2000+
Average fee per housing unit	$40	$20	$20	$20	$20

*Based on information furnished by R. Seely and D. Boswell, Montgomery County Sediment Control Section, 1980.

†Estimated area disturbed per unit = 10,000 ft^2 (930 m^2). For developments larger than 50 units, housing density per acre increased greatly because of the high proportion of condominium and apartment units.

gle home. On the basis of those costs, with adjustment for inflation, the authors developed Table 3.5 as a guide for setting permit fees.

The fee schedule in Table 3.5 was designed to cover *all* the costs of administering an erosion and sediment control program (plan review, inspections, enforcement, overhead, etc.) in a fairly large community with many new projects each year. The fees suitable for a particular community will vary with local salary rates, the number of projects reviewed per year, the number of inspections per project, and the efficiency of staff in evaluating projects.

TABLE 3.5 Suggested Permit Fees for Recovering Costs of an Erosion and Sediment Control Program

Area disturbed	Suggested permit fee
Less than 25,000 ft^2 (2300 m^2)	$50–100
25,000–100,000 ft^2 (2300–9300 m^2)	$0.002/ft^2 ($0.02/m^2)
Greater than 100,000 ft^2 (9300 m^2)	$0.001/ft^2 ($0.01/m^2)

Many smaller communities have neither the resources nor the growth necessary to support a full-time erosion control staff. Such communities may wish to consider other options:

- Contracting for plan review and inspection services with the county or a nearby large city that has a trained staff
- Using the services of the local soil conservation district
- Hiring a consultant with erosion control expertise to review plans and inspect sites as needed.

3.5 PRIVATE SECTOR COSTS FOR EROSION AND SEDIMENT CONTROL MEASURES

Erosion and sediment control measures add to the costs of land development. Control plans must be prepared. Certain structures, such as sediment basins, must be engineered. Then there are material and labor costs.

Costs for erosion and sediment control are often hard to separate from other land development costs. With good planning, permanent drainage facilities such as swales, storm drains, and inlets can double as erosion and sediment control measures, and the cost of a separate temporary system (see Chaps. 7 and 9) can thereby be avoided. Permanent landscaping, if installed in a timely manner, can obviate the need for temporary plantings. Thus, erosion control costs can be minimal. Finally, by proper design and construction of erosion and sediment control measures, these costs can be avoided: regrading and reseeding slopes, rebuilding roads, cleaning mud from reservoirs, streets, and private property, and repairing other erosion damage.

Erosion and sediment control measure costs are highly variable. The following list outlines some of the key variables:

- Site steepness and erodibility
- Proximity of materials (such as stone for riprap and straw for mulch)
- Local labor costs
- Timing of construction (e.g., to coincide with the dry season)
- Extent of local requirements

The Montgomery County, Maryland, Sediment Control Section estimated that, in 1980, developer costs for erosion and sediment controls were $300 to $400 per housing unit. The authors found developer costs in the San Francisco Bay Area ranged from $118 to $676 per housing unit. Table 3.6 presents the results of a survey conducted in 1980 in northern California. The survey focused on hillside construction because erosion control is most critical on slopes. All the projects listed in the table were moderate to very large subdivisions. (Several developers of smaller subdivisions also were interviewed. Unfortunately, the small-scale developers generally did not keep itemized accounts of construction expenses, particularly for erosion control.) The measures employed by the developers varied greatly, depending on the steepness of the terrain and the requirements of the city or county.

In the spring of 1983 the Maryland Water Resources Administration conducted a study of the private sector costs for erosion and sediment control. (3) Contractors were interviewed to determine their criteria for preparing bids in the Baltimore-Washington metropolitan area. The cost figures in Table 3.7 were based primarily on those interviews and on a review of bid documents for state construction projects. Because of the high variability in costs, the following assumptions were made in developing the cost figures in the table:

1. Project locations were reasonably accessible and within standard transportation distances.
2. The sites were relatively open; that is, they were not broken up with trees, rock outcroppings, or buildings.
3. Slopes were not severe.

The costs represent the total costs to the developer, including labor, equipment, materials, maintenance, removal at end of project (if necessary), overhead, and profit. (3)

REVIEW QUESTIONS

1. What is the primary legal basis for erosion and sediment control regulations in communities in the United States?
2. What are the key features of an effective erosion and sediment control ordinance?
3. Why have traditional grading ordinances not been adequate to control erosion and sediment from construction activities?
4. Why is it important to require an erosion and sediment control plan as part of a project application?
5. What is the purpose of a preconstruction meeting? What should be covered at this meeting? Who should attend it?
6. What five inspections are recommended as a minimum? When should they be made?
7. What is the purpose of a permittee report? How can it reduce enforcement costs?
8. What tools can an inspector use to force a developer to do what is required to prevent erosion?
9. What is the range of erosion and sediment control costs per housing unit?
10. How can an erosion and sediment control program be financed without cost to the local government?

TABLE 3.6 Costs of Erosion Control Measures on Hillside Residential Projects in Northern California*

Development/developer	Project location	Project size and slope	Erosion control measures implemented	Estimated costs, $	Estimated cost per unit
Briar Ridge	Cull Canyon—above	192 units	Hydromulching	26,000	
Centex Homes	Castro Valley,	approx. 80 acres,	4 sediment traps	6,300	
	Alameda County	5–30% slopes	40 storm drain inlet cages	12,000	
			Admin. + engineering	600	
				44,900	$234
Highland Glen	City of Hayward—hill	Phase I—350 units,	Hydromulching	54,900	
Centex Homes	above Ziley Creek,	approx. 140 acres,	Engineered sediment basin		
	Alameda County	5–30% slopes	(permanent w/spillway + access)	90,000	
			Admin. + engineering	9,000	
				153,900	$440
Jensen Ranch	Tract 4468	Phase I—280 units,	Hydroseeding	75,000	
Phase I–Castro Heights	Hayward hills,	over 500 acres,	Hydromulching	20,000	
Shappell Industries	Alameda County	5–30% slopes	18 sediment traps	72,000	
			Trap maintenance	15,000	
			Admin. + engineering	7,200	
				189,200	$676
Lakeridge Estates	Tracts 3921 and 4204	Phases I/II—80 units,	Hydromulching, sediment basins		
Orlando Homes	off Kelley St. in	5–30% slopes	and other measures	24,000	$300
	Hayward hills,				
	Alameda County				
Canyon Lake	Hayward hills off	72 units,	Hydromulching	9,600	
Pro Land	Kelley St.,	14 acres,	3 sediment traps	3,700	
	Alameda County	5–30% slopes	Dike traps	9,700	
			Straw bale dike	6,300	
			Other measures	7,400	
			Admin. + engineering	5,000	

Orindawoods Harold Smith Co.	Orinda, Contra Costa County	368 units, 190 acres, 5–40% slopes	Hydromulching (native grass mix + wildflowers)	86,000	
			12 sediment traps	72,000	
			Trap maintenance	8,000	
			Other measures	5,000	
			Admin. + engineering	7,700	
				178,000	$486
Brian Ranch Harold Smith Co.	Stone Valley–Alamo, Contra Costa County	Phase I—108 units, 626 acres, 5–30% slopes	Hydromulching	15,000	
			Other measures	5,000	
			Admin. + engineering	35,000	
				55,500	$169
Blackhawk Blackhawk Constr. Co.	Danville, Contra Costa County	Phase III—985 units, 1600 acres, 5–30% slopes	Hydromulching	27,000	
			2 sediment basins	25,000	
			Other measures	20,000	
			Maintenance + basin removal	40,000	
			Admin. + engineering	4,200	
				116,200	$118
McBail Construction Co.	Crockett, off Pomona St., Contra Costa County	127 units, 46 acres, 5–30% slopes	Hydroseeding	25,000	
			2 sediment basins	20,000	
			Other measures including basin repair and cleanup	15,000	
			Admin. + engineering	2,000	
				62,000	$488
Centex Homes	Hercules, Contra Costa County	1100 units, 315 acres, 1% avg. slope	Hydroseeding	123,200	
			Other measures	10,000	
			Admin. + engineering	5,000	
				138,200	$126

*The costs are based on 1979 dollars, the year when most of these measures were installed. Cost data was furnished by the developers. Administrative and engineering fees were assumed to be 10 percent of the cost of sediment basins, traps, and other engineered control measures.

TABLE 3.7 Unit Costs of Temporary Erosion and Sediment Control Measures in Maryland (3)*

Measures	Cost range	Average cost	Comments
Earth dike		\$3.50/ft (\$11.50/m)	Seeding, fertilizing and mulching adds \$0.50/ft (\$1.64/m).
Swale		\$4.20/ft (\$13.80/m) or \$7.00/yd^3 (\$9.20/m^3)	Seeding, fertilizing, and mulching adds \$0.50/ft (\$1.64 /m).
Straw bale dike	\$1.50–5.00/ft (\$4.90–16.40/m)	\$3.25/ft (\$10.70/m)	
Silt fence	\$2.00–6.00/ (\$6.60–19.70/m)	\$3.50/ft (\$11 .50/m)	
Pipe slope drain	\$7.00–15.00/ft (\$23.00–49.20/m)	\$10.50/ft (\$34.40/m)	Cost depends on pipe diameter.
Stabilized construction entrance	\$1000–4000	\$2000	50 ft (15 m) long with 2-in (5-cm) stone and filter cloth. Cost includes maintenance.
Riprap lining	\$25–30/yd^2 (\$30–36/m^2)		Costs based on nongrouted lining 1 ft (0.3 m) thick. Costs vary with rock size and distance from quarry.
Sediment trap with pipe outlet	\$3400–6700		Cost depends on trap size. Lowest cost in range based on 1-acre (0.4-ha) drainage area. Highest cost based on 5-acre (2-ha) drainage area. About 75% of cost is for construction and installation of riser and barrel.
Sediment trap with stone outlet	\$675–2550		Costs based on 1–5 acre (0.4–2 ha) drainage area range, as above. Excavation costs are the same as for pipe outlet sediment trap (\$425–2200).
Sediment basin		\$10,000–15,000	\$15,000 is estimated cost for basin with 30-acre (12-ha) drainage area, \$6500 of which is for riser, antivortex device, barrel, antiseep collars, and emergency spillway.

*Costs include labor, materials, equipment, maintenance, removal (if necessary), overhead and profit. Data is based on average costs in Maryland in 1983. West coast cost data on vegetation measures is provided in Chap. 6.

TABLE 3.7 *(Continued)*

Measures	Cost range	Average cost	Comments
Hydraulic seeding and mulching	$1000–1350/ acre ($2500–3400/ ha)		Costs vary with seed type and size of site. Estimates include fertilizer and anchoring of mulch. If wood fiber mulch is used, cost is at lower end of range.
Straw mulching		$275/acre ($690/ha) on flat area $400/acre $1000/ha) on a slope	Costs based on 1.5–2.0 tons/acre (3.4–4.5 t/ha), mechanically anchored.
Seeding and mulching with jute, excelsior, or synthetic netting	$1.25–2.75/yd² ($1.50–3.30/ m²)	$1.75/yd² ($2.10/m²)	Used primarily on critical areas. May be temporary or permanent measure.

REFERENCES

1. Association of Bay Area Governments, *San Francisco Bay Area, Environmental Management Plan, Volume I,* ABAG, Oakland, Calif., 1978.
2. Association of Bay Area Governments, "Manual of Standards for Surface Runoff Control Measures," *San Francisco Bay Area Environmental Management Plan, Appendix I,* ABAG, Oakland, Calif., 1980.
3. K. R. Kaumeyer and R. E. Benner, "Comparative Costs of Implementing Erosion and Sediment Control Practices in Maryland," unpublished paper, State of Maryland Water Resources Administration, Annapolis, Md., 1983.
4. International Conference of Building Officials, *Uniform Building Code,* ICBO, Whittier, Calif., 1982.
5. Maryland, *Annotated Code of Maryland,* Title 8, Subtitle 11, Michie Co., Charlottesville, Va., 1974.
6. Maryland Water Resources Administration, U.S. Soil Conservation Service and State Soil Conservation Committee, *1983 Maryland Standards and Specifications for Soil Erosion and Sediment Control,* MWRA, Annapolis, Md., 1983.
7. Tahoe Regional Planning Agency, *Lake Tahoe Basin Water Quality Management Plan, Volume II, Handbook of Best Management Practices,* TRPA, South Lake Tahoe, Calif., 1978.
8. U.S. Department of Agriculture, Soil Conservation Service, *Standards and Specifications for Soil Erosion and Sediment Control in Developing Areas,* U.S.D.A., SCS, College Park, Md., 1975.
9. U.S. Environmental Protection Agency, Office of Water Program Operations, *Report to Congress: Nonpoint Source Pollution in the U.S.,* U.S. EPA, Washington, D.C., 1984.

chapter 4

Estimating Runoff

Estimating runoff and soil loss is a necessary first step in the design of erosion control facilities. Proper sizing of runoff conveyance facilities and sediment retention structures depends on knowing both the amount and rate of runoff and the amount of sediment expected to be carried in the runoff. Erosion and sediment control measures can then be designed to handle the anticipated flows adequately and without unnecessary investment in oversized facilities.

Erosion control facilities form part of the drainage system in a project. It is expected that the civil engineer designing the roads and storm drains will also be responsible for sediment basin and temporary channel design. A soils engineer should participate if considerable earth movement is involved and, in any event, should be asked to evaluate the entire erosion control plan. If a large settling basin is proposed, the soils engineer's opinion is most important to ensure that the structure is properly designed for the geologic conditions at the site.

This chapter describes a simple method for calculating runoff by using the rational method. Soil loss prediction is described in Chap. 5. Hydraulics of open channel flow and design of permanent drainage facilities are not covered. These subjects are the responsibility of the civil engineer. Portions of the drainage system design process relevant to erosion control are discussed in Chaps. 7 and 8. For more thorough treatment, consult appropriate engineering handbooks such as those listed in the references at end of this chapter.

4.1 ESTIMATING RUNOFF BY THE RATIONAL METHOD

In erosion control planning, there are several reasons for calculating runoff. They include:

- Sizing conveyance structures
- Selection of channel linings
- Sizing outlet protection (e.g., riprap aprons)
- Sizing retention structures (e.g., sediment basins)

This section discusses the rational method as a simple formula for finding peak and average runoff. Any other proven method is also acceptable and can be applied to an erosion control project.

4.1a The Equation

In the rational method a simple equation is used to compute discharge from small areas. Although the method has been applied to areas as large as 5 mi^2 (13 km^2), it is strongly recommended that it not be applied to areas larger than 200 acres (81 ha). (14) The equation is

$$Q = C \times i \times A$$

where Q = runoff rate, ft^3/sec
C = runoff coefficient, a factor chosen to reflect such watershed characteristics as topography, soil type, vegetation, and land use
i = precipitation intensity, in/hr
A = watershed area, acres

In English units, the dimensions of the various factors do not match but the conversion factor needed to make the units on both sides of the equation the same is nearly 1. A runoff rate of 1 acre·in/hr equals 1.008 ft^3/sec. The use of a conversion factor of 1/1.008 is unnecessary in so approximate a relation as the rational formula.

In metric units

$$Q\ (\text{m}^3/\text{sec}) = \frac{C \times i\ (\text{mm/hr}) \times A\ (\text{ha})}{360}$$

The factor 1/360 is necessary to make the metric units match:

$$\frac{1\ \text{m}}{1000\ \text{mm}} \times \frac{1\ \text{hr}}{3600\ \text{sec}} \times \frac{10{,}000\ \text{m}^2}{\text{ha}} = \frac{1}{360}$$

At the beginning of a storm, runoff from distant parts of a watershed will not have reached the discharge point where the watershed's runoff Q is monitored.

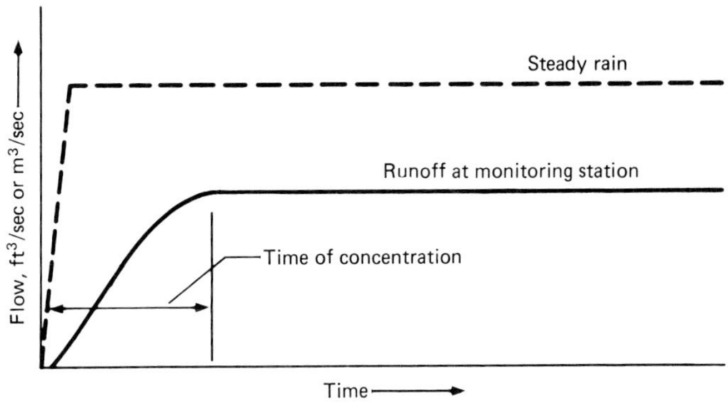

Fig. 4.1 Graphic representation of time of concentration.

After a period that is specific to each watershed, termed "time of concentration," a steady-state flow will occur as shown in Fig. 4.1. Time of concentration is defined as the time required for the runoff from the most remote part of the drainage area to reach the point under consideration. It is used to find the precipitation intensity i, and it predicts the peak runoff rate Q_{peak}.

The rational formula can also be used to find an average flow rate from a watershed. When an average flow rate is required, as in the design of sediment basins, the average rainfall intensity throughout the duration of a storm is used to find i. Using average intensity produces an estimate of average flow rate Q_{avg}, which is not related to the time of concentration.

4.1b Design Storm

We know from experience that rainfall amounts vary considerably from year to year. Every few years a large storm occurs. The *return period* is defined as the average number of years between storms of given duration and intensity. For example, at Fairfield, California, the rainfall records show that every 2 years you can expect a 6-hr rainfall that drops 1.16 in (29 mm) of rain. Every 100 years you can expect a 6-hr rainfall that drops 2.91 in (74 mm) of rain. Two years and 100 years are the return periods for the 6-hr rainfalls *of those magnitudes*. It is usually not desirable to design a structure to handle the greatest rainfall that has ever occurred.

It is often more economical to allow a periodic overflow than to design a very large structure that will never overflow. However, if human life would be threatened by an overflow, the structure should be sized to handle the largest storm expected. The rainfall duration and intensity (such as the 24-hr, 6-in storm) used to size a drainage facility is called the *design storm*. Many jurisdictions have specified storms for designing various structures. Those specifications usually apply to permanent structures such as storm drains.

Erosion control structures may be temporary or permanent. Concrete channels are often a permanent part of the drainage system of a fill slope. Sediment basins are usually temporary, although the risers may discharge directly into permanent storm drains. Unlined channels are frequently temporary; they are used when grading continues during the rainy season. Permanent structures must be designed according to the local standards. But when designing temporary structures, the planner can often choose a design storm. This choice should be based on a comparison of the risks and costs of hydraulic failure with the expense of building a structure sized to a larger design storm. If the cost of failure will not be high and the local jurisdiction has not specified otherwise, a storm with a 10-year return interval is recommended for sizing temporary erosion control measures.

The costs associated with failed systems vary with site factors such as the topography, climate, and geology, the amount of grading (cost of repairing washed-out fill slopes), the presence of natural drainageways on or near the site, and surrounding land uses, as well as local policy toward sedimentation of waterways and lakes. For example, a very thin layer of sediment deposited on grass may have no adverse affects, but a similar deposit on a roadway may cause hazardous driving conditions. Sediments deposited downslope on a construction site may present a minor problem, but if the sediment fills the newly built storm drain system and causes overflow onto a street, additional cleanup costs are imposed on both the developer and the local public works department. And if a sediment basin discharges into a natural creek, the local flood control district or public works department probably will be very concerned about siltation of the channel. In any location where the consequence of failure is high, temporary erosion and sediment control structures should be sized for a design storm larger than the 10-year return period storm.

One consideration favoring a more conservative choice of design storm is that, in many permanent structures, the effects of failure are mitigated by secondary systems. When a storm drain overflows, the excess can usually be contained by the curb and gutter system. If a terrace drain overflows, the runoff may flow over a vegetatively stabilized slope with only minor damage. But temporary measures often have no backup systems to limit damage if a failure does occur. If an interceptor swale at the top of a new fill slope fails, the recently planted ground cover will be washed away. Gullies can form to such an extent that the fill slope may have to be completely rebuilt. Collapse of an undersized temporary sediment basin could release a flood of water and a considerable amount of sediment to adjacent properties. Both public and private owners could sustain substantial damage under such circumstances, and the developer and the agency that approved the project could be held responsible.

Finally, it is especially important to prevent erosion in areas of high erosion potential. Such areas are likely to be associated with unstable slopes or with drainage problems which are already recognized as hazards by the soils engineer. Not only can erosion eventually threaten the stability of foundations and roadways, but loss of topsoil makes it much more difficult to establish permanent vegetation that will prevent erosion over the long term. A consistent local policy in favor of preventing erosion during development will reduce long-term maintenance and repair costs. Such a policy should be based on whether and where such high-risk areas exist. An appropriately longer storm return interval is advised in high-risk areas.

4.1c Use of Q_{peak}

The peak runoff, which is normally used to size drainage systems, is calculated when the capacity of a channel or other conveyance structure must be sufficient to carry all of the flow. In erosion control work, Q_{peak} is important not only to size conveyance facilities but also to:

- Check for potentially erosive velocities in unlined channels
- Select channel linings that will not erode
- Design outlet protection

In these cases, the rational method is applied by using a peak precipitation intensity:

$$Q_{peak} = C \times i_{peak} \times A$$

Peak precipitation intensity i_{peak} is determined by estimating the time of concentration for the drainage area and then finding the maximum rainfall intensity for that time duration and design storm return interval. For example, if the time of concentration for a watershed is 1 hr, you should use the peak 1-hr rainfall intensity in your calculations. The procedure for determining this time is explained in Sec. 4.1g.

4.1d Use of Q_{avg}

An average flow Q_{avg}, rather than peak flow, is used to find the required surface area of sediment basins and traps. The rational formula is still applied, except that an average precipitation intensity instead of the peak intensity is used:

$$Q_{avg} = C \times i_{avg} \times A$$

Average precipitation intensity i_{avg} is determined by taking the total rainfall for a specified storm return period and duration (e.g., 10-year, 6-hr storm) and dividing that total by the number of hours of duration:

$$i_{avg} = \frac{\text{total 6-hr rain}}{6}$$

A 6-hr storm duration is suggested. Sediment basins designed with a 6-hr storm strike a reasonable compromise between being somewhat undersized during storm peaks and being somewhat oversized during the rest of the storm.

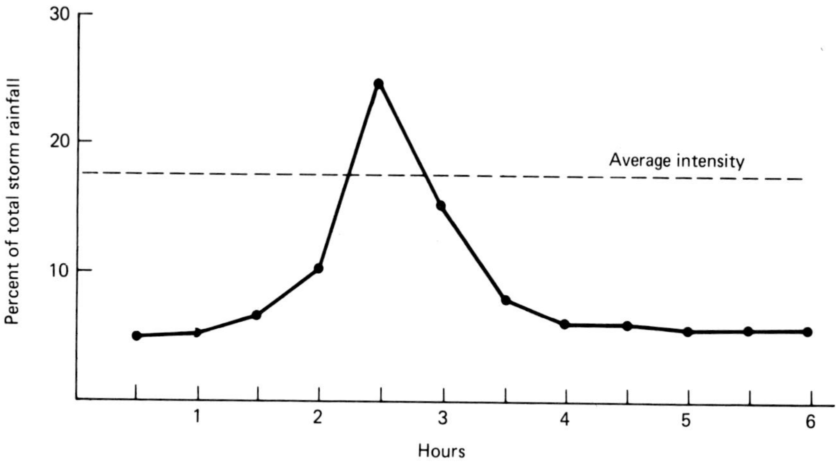

Fig. 4.2 Typical rainfall distribution and storm runoff during a 6-hr storm.

Sediment trapping efficiency is lower during the portion of a storm in which the immediate flow exceeds the average flow. The graph in Fig. 4.2 illustrates a typical rainfall distribution and storm runoff for a 6-hr storm. Peak intensity occurs at the top of the distribution curve; average intensity is indicated by a dashed line. Only during a small portion of the storm will the flow exceed average design flow. The graph was generated by placing the ½-hr incremental rainfall for a 6-hr storm in the sequence suggested by the U.S. Soil Conservation Service (SCS), 6-hr design storm distribution. (11, 12)

Use of average flow is a cost-saving measure. For the same design storm return interval, peak flow, which is based on a short time of concentration, is much larger than average flow of longer duration. A basin sized by using peak flow will thus be much larger than one sized by using the average flow. The sizing procedure is described in Chap. 8.

4.1e Runoff Coefficient C

The runoff coefficient C determines the portion of rainfall that will run off the watershed. It is based on the permeability and water-holding capacity of the various surfaces in the watershed. The C value can vary from close to zero to up to 1.0. A low C value indicates that most of the water is retained for a time on the site, as by soaking into the ground or forming puddles, whereas a high C value means that most of the rain runs off rapidly. Well-vegetated areas have low C values. Developed land, with its pavement, rooftops, and other impermeable surfaces, has a high C value. A high runoff coefficient produces higher runoff than does a low C value, and Q is directly proportional to C.

Table 4.1 lists C values for use in the rational formula. Select a C value within the range for land use. The designer must exercise judgment in selecting C from

TABLE 4.1 Rational Method *C* Values (13)

Land use	*C*	Land use	*C*
Business		Lawns	
Downtown areas	0.70–0.95	Sandy soil, flat, 2%	0.05–0.10
Neighborhood areas	0.50–0.70	Sandy soil, average, 2–7%	0.10–0.15
Residential		Sandy soil, steep, 7%	0.15–0.20
Single-family areas	0.30–0.50	Heavy soil, flat, 2%	0.13–0.17
Multi units, detached	0.40–0.60	Heavy soil, average, 2–7%	0.18–0.22
Multi units, attached	0.60–0.75	Heavy soil, steep, 7%	0.25–0.35
Suburban	0.25–0.40	Agricultural land, 0–30%	
Industrial		Bare packed soil	
Light areas	0.50–0.80	Smooth	0.30–0.60
Heavy areas	0.60–0.90	Rough	0.20–0.50
Parks, cemeteries	0.10–0.25	Cultivated rows	
		Heavy, soil, no crop	0.30–0.60
Playgrounds	0.20–0.35	Heavy soil with crop	0.20–0.50
Railroad yard areas	0.20–0.40	Sandy soil, no crop	0.20–0.40
Unimproved areas	0.10–0.30	Sandy soil with crop	0.10–0.25
Streets		Pasture	
Asphaltic	0.70–0.95	Heavy soil	0.15–0.45
Concrete	0.80–0.95	Sandy soil	0.05–0.25
Brick	0.70–0.85	Woodlands	0.05–0.25
Drives and walks	0.75–0.85	Barren slopes, >30%*	
		Smooth, impervious	0.70–0.90
Roofs	0.75–0.95	Rough	0.50–0.70

Note: The designer must use judgment to select the appropriate *C* value within the range. Generally, larger areas with permeable soils, flat slopes, and dense vegetation should have lowest *C* values. Smaller areas with dense soils, moderate to steep slopes, and sparse vegetation should be assigned highest *C* values.

*From Portland Cement Association, *Handbook of Concrete Culvert Pipe Hydraulics,* 1964, p. 45.

the range given by considering factors such as permeability, soil type, steepness, and vegetation.

For construction sites, when the soil is bare and the slope is less than 30 percent, use the agricultural values in the table and consider soil conditions and density of vegetation. For areas with temporary vegetative cover, select a value from the ranges for "cultivated rows"; for undisturbed areas under natural grass and shrub cover assign an appropriate "unimproved areas" C value between 0.10 and 0.30. If the slope gradient is greater than 30 percent, for example, 3:1 or 2:1, choose a value in the range 0.50–0.90 under "barren slopes." Soil depth or depth to impermeable rock influences the choice within the ranges given; the C value is higher for shallower soils. For sites with mixed land uses, compute a weighted average of the individual C values, as follows:

If area $A = x + y$, then

$$C \text{ (weighted)} = \frac{(x \times C_x) + (y \times C_y)}{A}$$

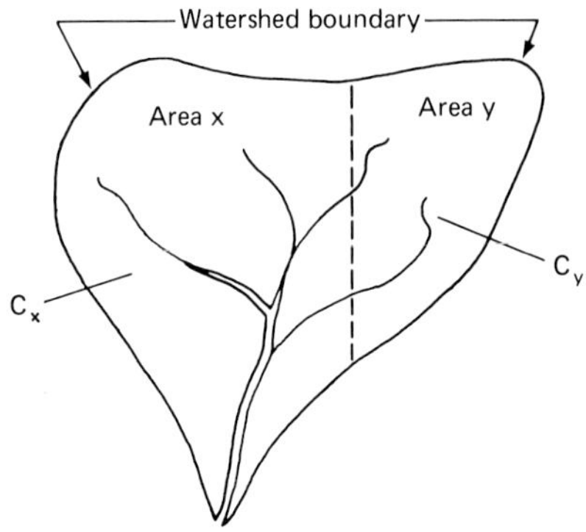

EXAMPLE 4.1

Given: A 15-acre (6-ha) site with clay soil and the slope and vegetative conditions shown below.

Find: The *C* value for each land use and a weighted *C* value for the entire site.

Solution:

Site condition	*C*
5 acres (2 ha): 3:1 gradient, fill (pervious), hydroseeded (rough surface)	0.50
10 acres (4 ha): 2% gradient, smooth, bare, packed soil no vegetation	0.30

The lower end of each range was selected because fill is usually more pervious than a cut slope in the case of the 5 acres (2 ha) and the slope is very flat on the 10-acre (5-ha) portion of the site.

$$C \text{ (weighted)} = \frac{(5 \text{ acres} \times 0.5) + (10 \text{ acres} \times 0.3)}{15 \text{ acres}} \quad \left[\frac{(2 \text{ ha} \times 0.5) + (4 \text{ ha} \times 0.3)}{6 \text{ ha}}\right]$$

$$= 0.37$$

4.1f Precipitation Intensity *i*

The *i* value, precipitation intensity, represents the rate of rainfall during a storm. It is calculated from depth and duration of rainfall, and it can vary tremendously with the average annual rainfall and the duration of the individual storm. Many states compile precipitation frequency, duration, and intensity data. In California, the Department of Water Resources (DWR) publishes precipitation data derived from records of rainfall data collected from more than 650 recording rain gauges throughout the state. (2)

TABLE 4.2 Addresses of State Climatologists in the United States
Address inquiries to "State Climatologist" at the addresses below.

Location	Type of data available*
ALABAMA K. E. Johnson Environmental & Energy Center The University of Alabama–Huntsville Huntsville, AL 35899	24-hr maximum rainfall for all cooperative observing sites, plus summaries of intensities for Mobile, Montgomery, Birmingham, and Huntsville.
ALASKA AEIDC/University of Alaska Alaska Climate Center 707 A Street Anchorage, AK 99501	U.S. Dept of Commerce Technical Papers Nos. 47 and 52. National Climatic Data Center tape with 15-min and hourly precipitation data for 1976–1978.
ARIZONA The Laboratory of Climatology Arizona State University Tempe, AZ 85287	NOAA publications, plus some additional data at county scales.
ARKANSAS Department of Geography Carnall Hall 104 University of Arkansas Fayetteville, AR 72701	Hourly precipitation and local climatological data.
CALIFORNIA California Dept. of Water Resources Division of Flood Management P.O. Box 388 Sacramento, CA 95802	5-, 10-, 15-, and 30-min, 1-, 2-, 3-, 5-, 12-, and 24-hr maximum precipitation for return periods of 2, 5, 10, 20, 25, 40, 50, 100, 200, 1000, and 10,000 years and PMP. Long-duration data is also available. The data, which is on microfiche, is from 689 recording rain gauges and 853 nonrecording gauges in California.
COLORADO Colorado Climate Center Dept. of Atmospheric Science Colorado State University Fort Collins, CO 80523	NOAA Atlas 2, Hydrometeorological Report Series—PMP, daily and hourly precipitation data.
CONNECTICUT Dept. of Renewable Natural Resources University of Connecticut, Box U-87 Storrs, CT 06268	Real-time data from automated weather stations around state. Statistical analyses of data from cooperating and first-order weather stations.
DELAWARE Department of Geography University of Deleware Newark, DE 19716	Office is unfunded; services are limited. Will answer questions and locate data sources if readily available.

Location	Type of data available*
FLORIDA Department of Meteorology Florida State University Tallahassee, FL 32306	Monthly and daily rainfall amounts only.
GEORGIA Institute of Natural Resources Ecology Building University of Georgia Athens, GA 30602	Monthly total rainfalls since earliest records for 23 places representative of statewide conditions. Moving averages and plots available on contract.
HAWAII Division of Water & Land Development Dept. of Land & Natural Resources P.O. Box 373 Honolulu, HI 96809	Hourly and daily rainfall data published and available from the National Climatic Data Center, Asheville, NC 28801.
IDAHO Agricultural Engineering Dept. University of Idaho Moscow, ID 83843	National Weather Service publications. Will prepare other data on a straight-time cost basis when available. Very short duration data is scarce.
ILLINOIS Illinois State Water Survey P.O. Box 5050, Station A Champaign, IL 61820	Annual data by section of state. Currently analyzing monthly data by region.
INDIANA Department of Agronomy Room 201-5 Poultry Science Bldg. Purdue University West Lafayette, IN 47907	All official climate data for Indiana from 1901 to date on computer disks and tapes, including hourly precipitation data.
IOWA Iowa Dept. of Agriculture Weather Service Municipal Airport, Room 10 Des Moines, IA 50321	*Climatology of Excessive Short Duration Rainfall in Iowa.* Climate of Iowa Series #6; *Iowa's Greatest 24-hr. Precipitation and Related Rainstorm Data,* Climate of Iowa Series #2 (an adaptation of U.S.W.B. Tech. Paper No. 40). These publications can be ordered for $2.50 per copy.
KANSAS Dept. of Physics—Caldwell Hall Kansas State University Manhattan, KS 66505	HPD publication and magnetic tape files for these data. Not funded as a service agency and will not provide off-campus service, but files are open to all who come to the office.
KENTUCKY Kentucky Climate Center Dept. of Geography & Geology Western Kentucky University Bowling Green, KY 42101	Hourly and daily rainfall data from which frequencies can be calculated.

Location	Type of data available*
LOUISIANA Dept. of Geography & Anthropology Lousiana State University Baton Rouge, LA 70803	
MAINE Pest Management Office Cooperative Extension Service 491 College Avenue Orono, ME 04473	
MARYLAND 1123A, Jull Hall University of Maryland College Park, MD 20742	
MASSACHUSETTS Dept. of Environmental Management Division of Water Resources 496 Park Street North Reading, MA 01864	
MICHIGAN MDA/Climatology 417 Natural Science Bldg. Michigan State University East Lansing, MI 48824	Hourly and excessive rainfall data for the Detroit metropolitan area (sponsored by South East Michigan Council of Governments) and for two agricultural watersheds near East Lansing. Also has federal publications TP-25, TP-40, and Hydro-35 for reference or to make a limited number of photocopies from.
MINNESOTA Minnesota Dept. of Natural Resources University of Minnesota 279 North Hall St. Paul, MN 55108	File copies of data from the National Climatic Data Center.
MISSISSIPPI Dept. of Geology & Geography Mississippi State Mississippi State, MS 39762	Expects to publish a weekly precipitation probability analysis in late 1985.
MISSOURI Dept. of Atmospheric Science University of Missouri–Columbia 701 Hitt Street Columbia, MO 65211	Daily rainfall values for 120 locations for the period 1948–1981.
MONTANA Plant & Soil Science Department Montana State University Bozeman, MT 59717	

Location	Type of data available*
NEBRASKA CAMAC 239 L. W. Chase Hall University of Nebraska Lincoln, NE 68583-0728	Hourly and daily rainfall data, probabilities of x amount of rain in a given time period, U.S. Weather Bureau maps of rainfall frequencies and intensities, National Climatic Data Center published summaries.
NEVADA Geography Dept. University of Nevada, Reno Reno, Nevada 89557	
NEW HAMPSHIRE Dept. of Geography University of New Hampshire Durham, NH 03824	
NEW JERSEY Dept. of Meteorology & Physical Oceanography Cook College, Rutgers University P.O. Box 231 New Brunswick, NJ 08903	
NEW MEXICO P.O. Box 5702 New Mexico Dept. of Agriculture Las Cruces, NM 88003	*Precipation-Frequency Atlas of the Western U.S., Volume IV—New Mexico,* precipitation summaries for various New Mexico locations, which include 24-hr rainfall by month.
NEW YORK Northeast Regional Climate Center Box 21, Bradfield Hall Cornell University Ithaca, NY 14853	Daily totals on tape, hourly totals on paper.
NORTH CAROLINA Dept. of Marine, Earth & Atmos. Sciences P.O. Box 8208 North Carolina State University Raleigh, NC 27695-8208	Monthly and hourly precipitation data for approximately 45 stations, published by NOAA. Data can be obtained by writing to state climatologist or calling (919) 737-3056.
NORTH DAKOTA Soils Department North Dakota State University Fargo, ND 58105	
OHIO Dept. of Geography Ohio State University 103 Bricker Hall Columbus, OH 43210-1361	U.S. Weather Bureau Technical Paper No. 25 (1955).

Location	Type of data available*
OKLAHOMA Oklahoma Climatological Survey University of Oklahoma 710 Asp, Suite 8 Norman, OK 73019	U.S. Weather Bureau Technical Paper No. 40.
OREGON Climatic Research Institute Oregon State University Corvallis, OR 97331	Hourly rainfall amounts 1948–1982 (approx. 180 stations), daily rainfall amounts 1928–1982 (approx. 450 stations), monthly rainfall amounts 1900–1982 (approx. 400 stations).
PENNSYLVANIA No state climatologist at this time.	
RHODE ISLAND Dept. of Plant Sciences Room 313, Woodward Hall University of Rhode Island Kingston, RI 02881	Daily precipitation—50 years. Distribution—2 years. Published in *Climatological Data—New England.*
SOUTH CAROLINA S.C. Water Resources Commission 3830 Forest Drive P.O. Box 4440 Columbia, SC 29240	5 min to 24 hr rainfall amounts for various locations in state, plus daily and monthly rainfalls.
SOUTH DAKOTA Agricultural Engineering Dept. South Dakota State University Brookings, SD 57007	U.S. Weather Bureau Technical Paper No. 40. The values in this publication do not apply in the Black Hills area, where intensities have proved to be higher.
TENNESSEE Tennessee Valley Authority 310 Evans Building Knoxville, TN 37902	Daily rainfall totals for approx. 150 gauges in Tennessee, western Virginia, western North Carolina and northern Alabama.
TEXAS Meteorology Department Texas A&M University College Station, TX 77843	
UTAH Utah State Climatologist Utah State University, UMC-48 Logan, UT 84322	Hourly rainfall data published by National Climatic Data Center; estimated return periods for short-duration precipitation.
VERMONT Hills Building University of Vermont Burlington, VT 05401	For data, write to Northeast Regional Climate Center, Box 21, Bradfield Hall, Cornell University, Ithaca, NY 14853.

Location	Type of data available*
VIRGINIA Dept. of Environmental Sciences Clark Hall University of Virginia Charlottesville, VA 22903	Rainfall frequency and intensity data for approximately 30 locations.
WASHINGTON Office of the State Climatologist Western Washington University Bellingham, WA 98225	
WEST VIRGINIA Dr. Stanley J. Tajchman Forestry P.O. Box 6125 Morgantown, WV 26505-6125	Data and graphs for specific project sites only. For more general information, contact the National Climatic Data Center, Asheville, NC 28801, (704) 259-0682
WISCONSIN University of Wisconsin Extension 1353 Meteorology & Space Science Bldg. 1225 West Dayton St. Madison, WI 53706	Hourly and daily precipitation data on magnetic tape for all Wisconsin stations. Reference library of federal precipitation atlases and publications.
WYOMING No state climatologist at this time.	

*Source: Responses to a survey conducted by the authors in 1985.

In other states, the office of the state climatologist compiles similar rainfall data. Table 4.2 lists the addresses of state climatologists in 48 states. (Pennsylvania and Wyoming did not have state climatologists at the time of writing.) Local flood control and public works agencies can often provide many more rain gauge stations, but the intensity information may be limited or nonexistent.

The California DWR presents the precipitation values in tabular form by rain gauge for different return intervals and rainfall durations. Table 4.3 shows the values for one station in California. Microfiche sets of the tables, the only practical way of presenting 3687 pages of data, are updated annually and are available from the DWR Publications Office in Sacramento. (2) Detailed rainfall intensity data for specific subregions of the state, such as the report prepared by Rantz covering the San Francisco Bay region (9) may also be available.

Tabular precipitation data such as that shown in Table 4.3, is the easiest form of data from which to determine i values for use in the rational formula. In areas east of the Rocky Mountains, for which detailed tabulations are not available, maximum rainfalls of ½-, 1-, 2-, 3-, 6-, 12-, and 24-hr duration can be estimated by using the *Rainfall Frequency Atlas of the United States.* (6) The atlas contains maps of total rainfall for storms of those durations and return periods of 1, 2, 5, 10, 24, 50, and 100 years. Figures 4.3 and 4.4 are simplified versions of the atlas maps of the 10-year, 30-min rainfall and the 10-year, 6-hr rainfall, respectively.

TABLE 4.3 Precipitation Values for Rain Gauge 2933, Fairfield, California (2)

Return period, years	Maximum precipitation (in) for indicated duration: m = minutes; h = hours										
	5m	10m	15m	30m	1h	2h	3h	6h	12h	24h	C–yr
2	0.17	0.23	0.27	0.36	0.48	0.68	0.83	1.16	1.54	2.07	20.47
5	0.23	0.33	0.38	0.50	0.67	0.95	1.17	1.64	2.17	2.92	26.70
10	0.28	0.39	0.45	0.60	0.80	1.14	1.60	1.95	2.59	3.49	30.29
20	0.32	0.46	0.52	0.69	0.93	1.31	1.62	2.25	2.98	4.02	33.45
25	0.34	0.47	0.55	0.72	0.97	1.37	1.68	2.35	3.11	4.19	34.40
40	0.36	0.51	0.59	0.78	1.05	1.48	1.82	2.54	3.36	4.53	36.32
50	0.38	0.53	0.61	0.81	1.08	1.53	1.89	2.63	3.48	4.70	37.20
100	0.42	0.59	0.68	0.90	1.20	1.69	2.09	2.91	3.85	5.19	39.82
200	0.46	0.64	0.74	0.98	1.31	1.85	2.28	3.18	4.21	5.68	42.30
1,000	0.54	0.77	0.88	1.17	1.57	2.21	2.73	3.80	5.03	6.78	47.68
10,000	0.67	0.94	1.08	1.44	1.92	2.71	3.35	4.66	6.17	8.32	54.74
PMP	1.28	1.81	2.08	2.75	3.68	5.20	6.41	8.93	11.83	15.95	125.87

PMP = probable maximum precipitation

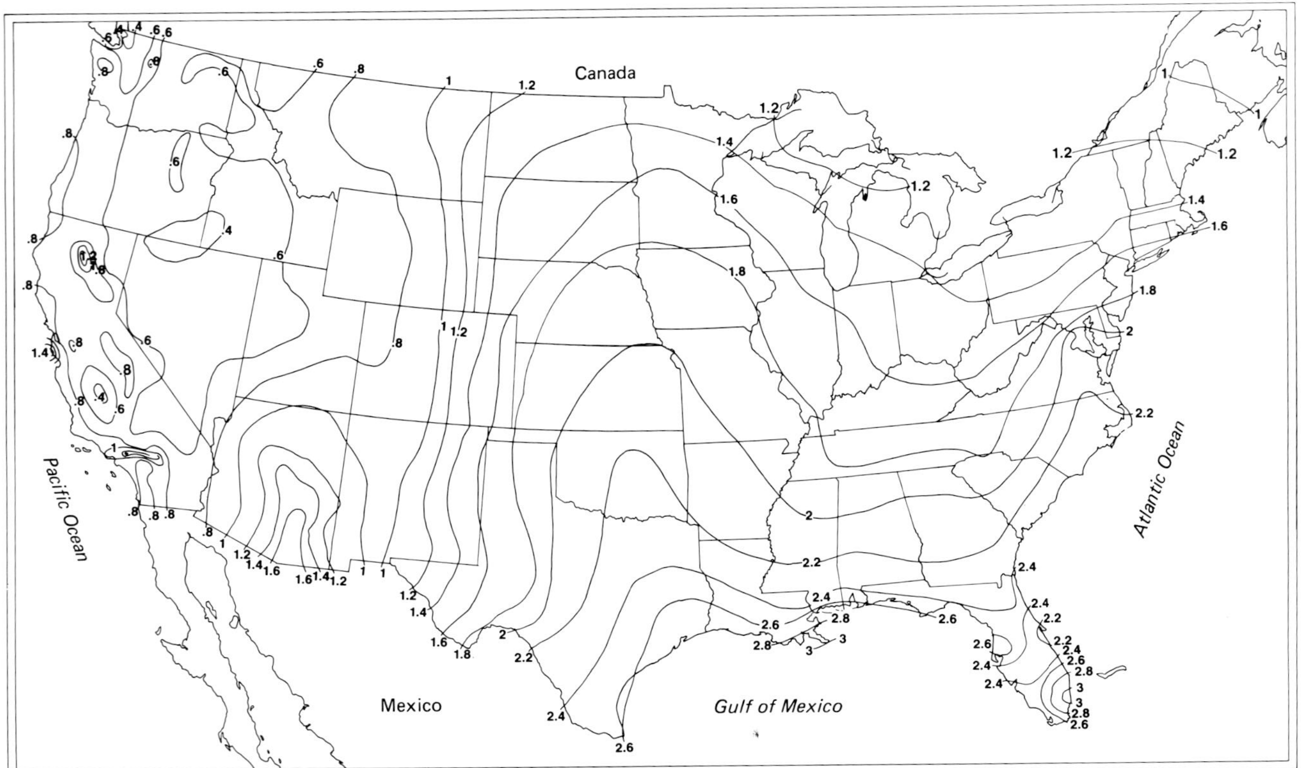

Fig. 4.3 Map of 10-year, 30-min rainfall, in inches. (Adapted from 6)

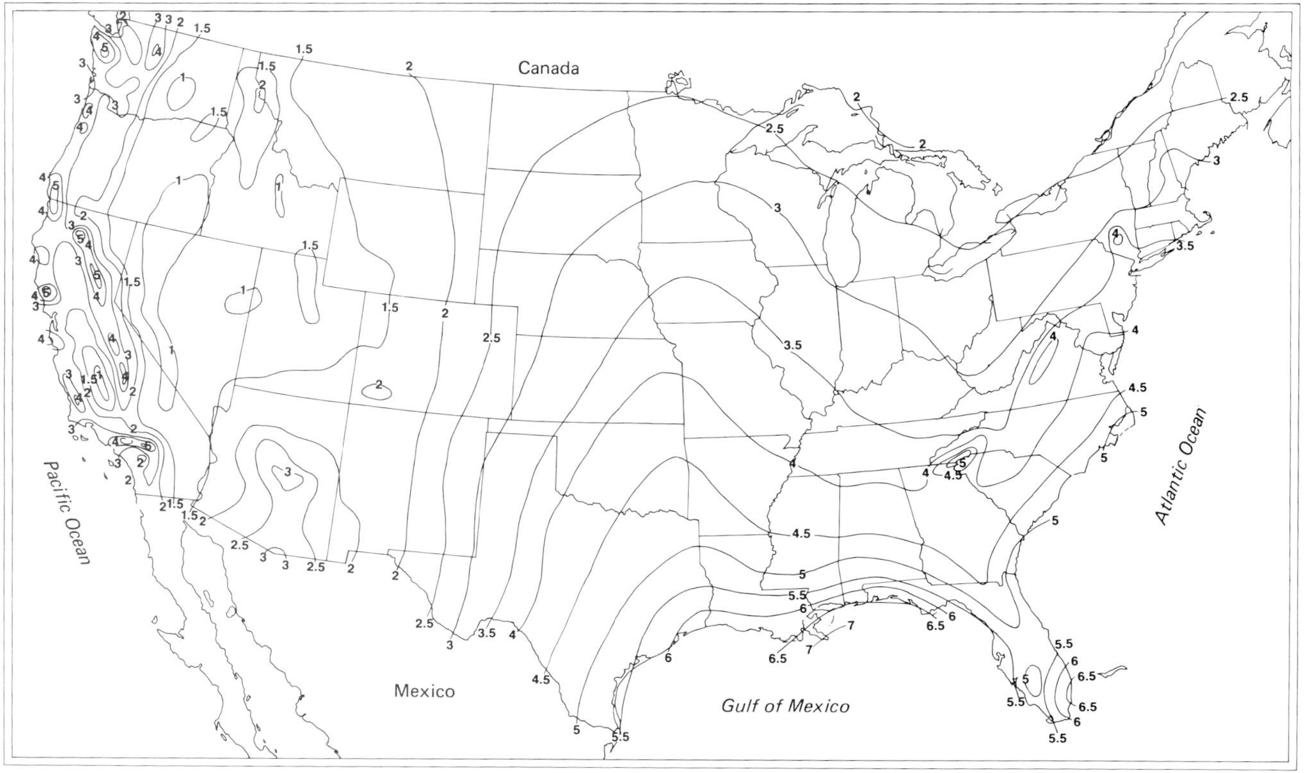

Fig. 4.4 Map of 10-year, 6-hr rainfall, in inches. (Adapted from 6)

Short-duration maximum rainfalls of 5 to 60 min in areas east of the Rocky Mountains can be estimated from NOAA Technical Memorandum NWS Hydro-35. (5) This document can be ordered from the National Technical Information Service, 5285 Port Royal Road, Springfield, VA 22161 (Publication No. PB 272-112). In the 11 western states, precipitation intensities can be estimated by using NOAA's *Precipitation Frequency Atlas of the Western States.* (8) This document, published in 11 separate volumes, can be ordered from the National Climatic Data Center, Federal Building, Asheville, NC 28801. It is available in both microfilm and paper versions, and the microfilm is substantially less expensive. Call the Data Center in Asheville, (704) 259-0682, for a price estimate before ordering.

The National Environmental Data Referral Service (NEDRES) publishes a *Data Base User's Guide.* The guide describes various types of atmospheric data sets that have been developed in specific areas of the United States, lists the responsible agency for each area, and tells you how to order the data. To order the guide, contact the National Environmental Data Referral Service Office, 3300 White Haven Street, N.W., Room 533, Washington, DC 20235 [(202) 634-7722]. On-line access to the data base also is available.

The following examples illustrate the use of the two types of precipitation data (tables and maps) for computing i values.

EXAMPLE 4.2

Given: A site in Fairfield, California.

Find: The average precipitation intensity i for a 10-year, 6-hr storm by using Table 4.3.

Solution: The precipitation data table for the Fairfield rain gauge (Table 4.3) is obtained from the California DWR microfiche set. The table shows that 1.95 in (50 mm) of rain can be expected from the desired storm. Dividing 9.5 by 6 hr gives an average intensity of 0.33 in/hr (8.4 mm/hr).

EXAMPLE 4.3

Given: A site in central Wisconsin.

Find: The maximum rainfall to be expected from a 10-year, 30-min rainfall and the average intensity i.

Solution: In Fig. 4.3, the 10-year, 30-min rainfall is between 1.4 and 1.6 in (36 and 41 mm) for central Wisconsin. Interpolating between the isohyets produces a maximum rainfall of 1.5 in (38 mm).

The average intensity i for this storm duration and frequency is found by dividing i by the duration of 0.5 hr for an average i of 3 in/hr (76 mm/hr).

EXAMPLE 4.4

Given: A site in Fairfield, California.

Find: The average intensity i of a rainfall with a duration of 45 min and return interval of 15 years.

Solution: Using Table 4.3, we find the precipitation for a 10-year, 30-min event to be 0.60 in (15 mm); a 20-year, 30-min event to be 0.69 in (18 mm); a 10-year, 1-hr event to be 0.80 in (20 mm); and a 20-year, 1-hr event to be 0.93 in (24 mm). By interpolating to find the 10-year, 45-min rainfall, we get 0.70 in (18 mm), and the 20-year, 45-min rainfall is 0.81 in (21 mm). By interpolating again between the 10- and 20-year values, we get a 15-year, 45-min rainfall of 0.75 in (19 mm). This is sufficient precision for erosion control design.

4.1g Time of Concentration T_c

As discussed earlier, Q_{avg} is used to size water retention structures and is based upon the total precipitation during a storm event of selected length. It is a straightforward task to select a 6-hr storm period for design, find the average i over the 6 hr, and calculate the average runoff.

For sizing flow conveyance structures, the peak flow Q_{peak} must be selected on the basis of site-specific factors that affect the time of concentration. If a channel is sized for a peak i averaged over a shorter duration than the time of concentration T_c, the channel will be oversized. This is because, for any return frequency event, the shorter-duration events (such as 5 or 10 min) will produce a greater i than will a longer-duration event (such as 30 min or 1 hr). Conversely, a channel sized for a Q based upon a rainfall duration exceeding the T_c will be undersized.

Remember that, for a rainfall duration less than T_c, not all of the rain falling upon the earth will be reflected at the measuring point in the channel. Time must be allowed for overland flow. But for a rainfall event with intensity averaged over T_c, at T_c the runoff measured in the channel will reflect the rate of rainfall over the site in question.

To calculate Q_{peak} we need to use an intensity i based upon the storm frequency of choice and a duration equal to the T_c. The time of concentration consists of two parts:

- Time for overland flow to reach a drainage channel
- Channel flow time from the point of entry into the channel to the point under construction

Because channel flow is much faster than overland flow, the total time of concentration does not necessarily represent the greatest distance traveled; it represents the longest travel time. The overland flow time is usually the major portion of the time of concentration.

Overland Flow Time

Overland flow time is a function of the length of travel path, average slope, and the rational method runoff coefficient C. A nomograph relating those three criteria (Fig. 4.5) can be used to estimate overland travel time. To use the nomograph, follow these steps:

- Determine the distance from the most remote part of the drainage area to the intercepting channel.

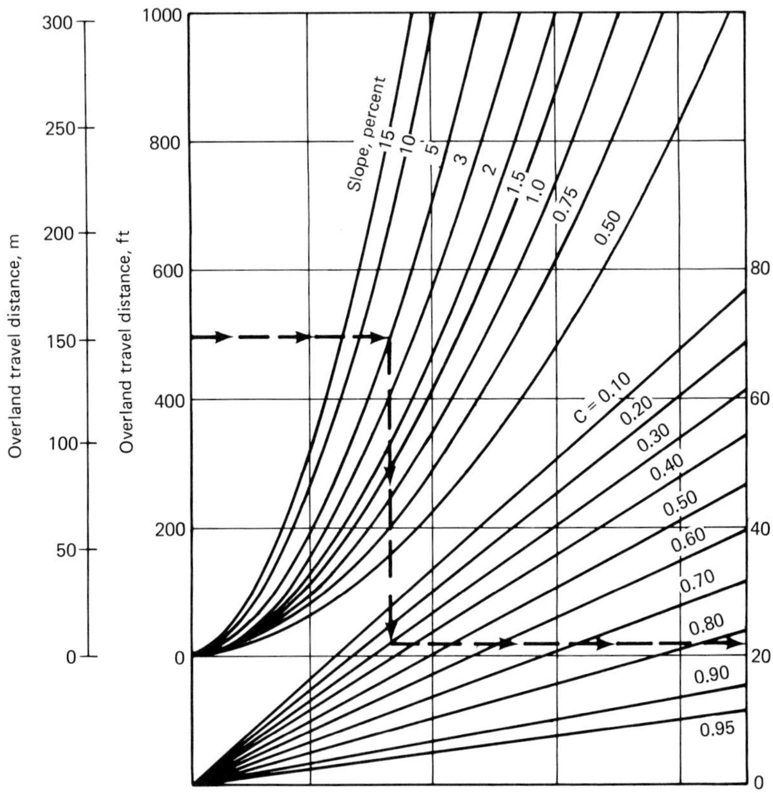

Fig. 4.5 A nomograph of overland flow time. (10) Enter left margin with slope length; move right to slope curve and down to *C* value; and find overland travel time on right margin.

- Calculate the average slope by computing the difference in altitude between the highest and lowest points of the flow path and dividing by the distance between those points.
- Find or compute the *C* value.
- Enter the graph on the left margin with the overland travel distance; move to the right to the correct slope curve; move down to the *C* value; and then move over to the right margin.
- Read the overland flow time from the right-hand scale.

EXAMPLE 4.5

Given: A site 500 ft (152 m) long with 5 percent average slope and a *C* value of 0.30.

Find: Overland flow time.

Solution: From Fig. 4.5, the estimated flow time is 22 min.

Channel Flow Time

Channel flow time can sometimes be ignored in calculating runoff. For example, when a short, paved channel collects water from a broad gentle slope and drains to a sediment basin, the channel travel time may be 2 min whereas the overland flow might be 20 min or more. There are two important reasons to know channel travel time: to assure adequate capacity and to be sure that the flow of water will not erode the channel. Frequently, temporary diversions and channels on construction sites either are not lined or are not sized for the expected flow. If the channel is unlined and the velocity exceeds the maximum permissible value for the soil type, the channel bed itself erodes, which further contributes to erosion damage on and off site. If the channel is not large enough to carry the flow, the runoff can overflow, form new channels that also erode, and severely damage unprotected slopes.

Two equations are used to calculate the flow in open channels: Manning's equation and the continuity equation. These equations should be familiar to all civil engineers. The *Manning equation* is:

$$V = \frac{1.49 \times r^{2/3} \times s^{1/2}}{n}$$

where V = velocity, ft/sec
n = Manning roughness coefficient, dimensionless
r = hydraulic radius, ft, at the depth of flow, i.e., channel cross-sectional area, ft^2, divided by wetted perimeter, ft, A/WP
s = average streambed slope, in decimals

In metric units

$$V = \frac{r^{2/3} \times s^{1/2}}{n}$$

where V is in m/sec, r is in m, and n and s are the same as above.

The *continuity equation* is:

$$Q = A \times V$$

where Q = flow in the channel, ft^3/sec (m^3/sec)
A = cross-sectional area of the flow, ft^2 (m^2)
V = velocity, ft/sec (m/sec)

The Manning roughness coefficient n reflects the condition of the channel. Tables of the coefficient can be found in handbooks used by drainage designers. Typical values useful in erosion control work are shown in Table 4.4. For riprap-lined channels, n can be determined from the following equation:

$$n = 0.0395 d_{50}{}^{1/6}$$

where n = Manning's roughness coefficient
d_{50} = median size stone in the mixture of riprap, ft (For d_{50} in meters, $n = 0.0481 d_{50}{}^{1/6}$)

Figure 7.29 solves this equation graphically.

TABLE 4.4 Manning Roughness Coefficients

n	Channel surface condition
0.013	Plastic sheet
0.015	Concrete or asphalt-lined channel
0.02	Ordinary earth, smoothly graded
0.025	Gravel-lined channel
0.030	Natural channel in good condition
0.040	Sod, depth of flow more than 6 in
0.10	Weed-choked natural channel

Note: The n value for sod, 0.040, is appropriate for a grassed waterway but changes slightly with the height of the grass (see Appendix B).

By far the easiest way to estimate channel travel time as a contributing part of time of concentration is to use the *Handbook of Hydraulics.* (1) Section 7 of the handbook covers steady uniform flow in open channels and contains numerous tables derived from the Manning equation.

The solution of a channel-sizing problem is an iterative process because a flow must be assumed before a depth and velocity can be calculated. However, the flow depends on the time of concentration, which, when it includes channel flow time, itself depends on flow. A two-part (or more, if channel size is altered) procedure is used to solve for T_c, V, and Q_{peak}:

- Find an initial flow Q_{peak} based on overland flow time.
- Use this initial Q_{peak}, the channel characteristics, and the tables in the *Handbook of Hydraulics* to solve for depth of flow.
- Use this value of depth of flow and Manning's equation to obtain an initial estimate of velocity.
- Find channel travel time by dividing the channel length by velocity:

$$\frac{L}{60V} = \text{travel time, min}$$

Now repeat this sequence; use the new total time of concentration equal to overland flow time plus channel travel time:

- With the new T_c, find new i value and Q_{peak}.
- Use the *Handbook of Hydraulics* to solve for depth of flow.
- Reapply Manning's equation to find the revised estimate of velocity in the channel.
- Check depth of flow and quantity against channel size and capacity. If channel is not adequate, redesign is necessary.
- Check velocity against maximum permissible velocity for the specific channel lining. Maximum velocities for unlined channels are listed in Table 7.1. Figures

7.36, 7.37, and 7.39 also allow a direct check for erosive velocities based on depth of flow and channel slope for several types of lining material.

If the *Handbook of Hydraulics* is not available, manual calculation is possible, although laborious. A calculator helps make the process somewhat less time-consuming. In this case, channel travel time is first calculated by assuming full flow depth, and actual depth and velocity are found by iterative applications of the equations. Section 4.1h contains examples of the use of the rational method to find Q_{peak} and examples of the ways discussed above to find the time of concentration T_c.

4.1h Examples of Use of the Rational Method

The following examples build on those already presented to illustrate how the individual components of the rational equation combine to yield estimates of Q_{peak}. Examples 4.6 and 4.7 develop a Q_{peak} based entirely on overland flow. Example 4.8 is a simple average flow calculation. Example 4.9 is a complete calculation of the adequacy of a channel. It includes computing a weighted C value and overland flow time, approximating channel flow time and checking for capacity and susceptibility of the channel to erosion, and using the *Handbook of Hydraulics* to help solve Manning's equation. Example 4.10 repeats the computation in Example 4.8 without using the *Handbook of Hydraulics.*

EXAMPLE 4.6

Given: A 3-acre (1.2-ha) site in southern Tennessee.
Slope length of 500 ft (152 m)
Average slope of 5 percent
C value of 0.30
T_c of 22 min
Detailed rain gauge data not available

Find: Peak intensity i_{peak} for a 10-year storm for sizing of conveyance channels.

Solution: Detailed rain gauge data is generally available at major metropolitan centers. If this data has not been compiled or is not available, the *Rainfall Frequency Atlas* (6)can be employed. Refer to the 10-year, 30-min map and find 1.8 in (46 mm) of rainfall over 30 min. For short-duration events, at each halving of the time from 30 min, we recommend a conservative assumption of 90 percent of the preceding rainfall. Thus for a 15-min event, the design rainfall would be 0.90 × 1.8 in = 1.62 in (41 mm) and for a 7.5-min event it would be 0.90 × 1.62 = 1.46 in (37 mm).

The T_c for this problem is 22 min. Interpolation between 15 min = 1.62 in (41 mm) and 30 min = 1.8 in (46 mm) produces a design rainfall of 22 min = 1.70 in (43 mm). The result is

$$i_{\text{peak}} = \frac{1.70\text{ in}}{22\text{ min}} \times 60\text{ min/hr} = 4.6\text{ in/hr} \qquad \frac{43\text{ mm}}{22\text{ min}} \times \frac{60\text{ min}}{\text{hr}} = 117\text{ mm/hr}$$

This i_{peak} will usually result in a flow estimation higher than is likely to be encountered. However, a channel sized to handle a flow with such a short T_c will generally serve a small area, and the total excess cost to the builder is likely to be very minor.

EXAMPLE 4.7 The example is in two parts. In the first part we find Q_{peak}.

Given: The site in southern Tennessee of Example 4.6.
C value of 0.30
i_{peak} = 4.6 in/hr (117 mm/hr) (from Example 4.6)
Area of 3 acres (1.2 ha)

Find: Peak flow Q_{peak} for a 10-year storm.

Solution:

$$Q_{peak} = C \times i_{peak} \times A$$
$$= 0.30(4.6)(3) \qquad [0.30(117 \text{ mm/hr})(1.2 \text{ ha}/360)]$$
$$= 4.14 \text{ ft}^3/\text{sec} \qquad (0.117 \text{ m}^3/\text{sec})$$

Peak flow is used to size channels and risers for sediment basins and to check for potential erosion of the channel bed.

In this part of the example we find Q_{avg}.

Given: The southern Tennessee site of Example 4.6.

Find: Average flow Q_{avg} for 10-year, 6-hr storm event.

Solution: First, find i_{avg}. From Fig. 4.4, for 10-year, 6-hr precipitation intensity:
10-year, 6-hr rainfall = 3.7 in (94 mm)
i_{avg} = 3.7 in/6 hr = 0.62 in/hr (16 mm/hr)
$Q_{avg} = C \times i_{avg} \times A$
Q_{avg} = 0.30(0.62)(3) [0.30(16 mm/hr)(1.2 ha)/360)]
= 0.56 ft^3/sec (0.016 m^3/sec)
This average flow rate could be used to size a sediment basin.

Basic, Step-by-Step Procedure for Applying the Rational Method to a Channel Flow Problem

1. Determine drainage area.
2. Determine the proper C value for the site.
3. Determine overland flow time. Enter Fig. 4.5 with the slope length, slope percent, and the C value to find overland flow time.
4. Find an initial i value based on overland flow time. Use Fig. 4.3 and calculate i or, preferably, use rain gauge data from a nearby weather station. Remember to convert inches (millimeters) of rain for the storm duration to inches per hour (millimeters per hour) to obtain intensity i.
5. Compute initial Q_{peak}:

$$Q_{peak} = C \times i \times A$$

6. Use the tables in the *Handbook of Hydraulics* (1) and Manning's equation to determine initial estimates of the depth and velocity of flow in the channel.

7. Compute channel travel time by dividing the channel length by velocity of flow. Convert to minutes.
8. Determine total time of concentration: add overland flow time to channel travel time.
9. Repeat steps 4 through 6 for a second approximation of Q, depth of flow, and velocity, starting with a new i value based on the total time of concentration from step 8.
10. Determine adequacy of channel size and lining: Compare depth of flow with channel size and compare velocity with the maximum permissible velocity in an unlined channel for the soil type in the channel. If depth of flow plus minimum freeboard is deeper than the channel, choose a larger channel and repeat the calculations to verify the new size. If the capacity, but not the lining, is adequate, select a more erosion resistant channel lining.

EXAMPLE 4.8 Channel Flow Calculations Using the *Handbook of Hydraulics* (1)

Given: The sample site and the channel cross section shown in Fig. 4.6.

Location, southeastern corner of Michigan
10-year, 30-min rainfall of 1.4 in (36 mm)
10-year, 1-hr rainfall of 1.6 in (41 mm)
Drainage area of 20 acres (8.1 ha)
Degree of development: rough graded bare earth on 70 percent of the watershed; the remaining 30 percent in natural condition
Overland travel path characteristics
Overland travel distance, 850 ft (259 m)
Average slope of overland travel path, 5 percent

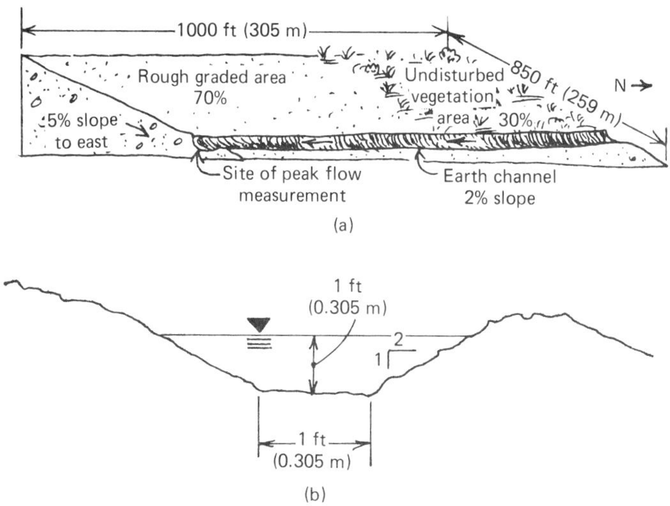

Fig. 4.6 Project area and channel geometry for Example 4.8.

Average channel characteristics
Length L, 1000 ft (305 m)
Slope s, 2 percent
Channel bed, bare earth
Roughness coefficient n, 0.02
Base width b, 1 ft (0.305 m)
Depth D, 1 ft (0.305 m)
Side slope z, 2:1

Channel section formulas are given in Appendix B. For trapezoidal channels with 2:1 side slopes,

Cross-sectional area: $A = bD + 2D^2$
Wetted perimeter: $\mathrm{WP} = b + 2D\sqrt{5}$
Hydraulic radius: $r = A/\mathrm{WP}$

Find: Peak discharge for a 10-year return interval storm, depth of flow in channel, and velocity of flow in channel.

Solution:

STEP 1. Drainage area is given; it is 20 acres (8.1 ha).

STEP 2. Weighted C value.
Using Table 4.1, a rough graded surface (70 percent of area) has a C of about 0.20. Undisturbed areas (30 percent of area) have a C value of 0.10.
The weighted C value is

$$C = \frac{(0.7 \times 0.2) + (0.3 \times 0.1)}{1.0} = 0.17$$

STEP 3. Overland flow time. Overland flow time should be calculated on a long slope. In this example of a rectangular property, the proper slope length is 850 ft (259 m). The slowest overland flow will occur in the vegetated area, where the C value is 0.10. If the south end of this slope had been significantly larger than the vegetated north end, we would have chosen the south end to calculate overland flow time.

Enter Fig. 4.5 with a length of 850 ft (259 m), a slope of 5 percent, and a C value of 0.10 and read an overland travel time of 37 min.

STEP 4. Initial i value based on overland travel time only. From the *Rainfall Frequency Atlas* (6) we find the 10-year, 30-min rainfall to be 1.4 in (36 mm) and the 10-year 1-hr rainfall to be 1.6 in (41 mm). Interpolating between the numbers produces a rainfall of 1.45 in (36.8 mm) in 37 min.

$$i = \frac{1.45\ \text{in}}{37\ \text{min}} \times \frac{60\ \text{min}}{\text{hr}} = 2.35\ \text{in/hr} \qquad \left(\frac{36.8\ \text{mm}}{37\ \text{min}} \times \frac{60\ \text{mm}}{\text{hr}} = 59.7\ \text{mm/hr}\right)$$

STEP 5. Initial Q_{peak}.
$C = 0.17$ (from step 2)
$i = 2.35$ in/hr (59.7 mm/hr) (from step 4)
$A = 20$ acres (8.1 ha)

$$Q_{\text{peak}} = C \times i \times A$$
$$= 0.17(2.35)(20) \qquad \left[\frac{0.17(59.7)(8.1)}{360}\right]$$
$$= 7.99\ \text{ft}^3/\text{sec} \qquad (0.228\ \text{m}^3/\text{sec})$$

STEP 6. Determine depth of flow. In the *Handbook of Hydraulics*, Table 7-11 (reproduced here as Table 4.5) provides the ratio of depth to base width after solving the following equation:

$$K' = \frac{Q \text{ (in ft}^3\text{/sec)} \times n}{b^{8/3} \times s^{1/2}} \qquad \left[\frac{1.486 \times Q \text{ (in m}^3\text{/sec)} \times n}{b^{8/3} \text{ (in m)} \times s^{1/2}}\right]$$

where b = base width of trapezoidal channel = 1 ft (0.305 m)

n = Manning's roughness coefficient = 0.02
Q = 7.99 ft^3/sec (0.228 m^3/sec)

Therefore, K' = 1.13.

Find 1.13 in Table 4.5 in the column "2:1 side slope" and read D/b, where D = depth of water in channel and D/b = 0.64. Since b = 1 ft (0.305 m), D = 0.64 ft (0.20 m).

Initial estimate of velocity. First, Table 4.6 (Table 7.1 in the *Handbook of Hydraulics*) provides the hydraulic radius by the formula

$$r = C_r \times D$$

where C_r = tabulated value

For D/b = 0.64, find C_r = 0.590 in Table 4.6 in column "2:1 side slope."

$$r = 0.59(0.64 \text{ ft}) = 0.38 \text{ ft} \qquad [0.59(0.20 \text{ m}) = 0.12 \text{ m}]$$

Second, find velocity using Manning's equation:

$$V = \frac{1.49 \times r^{2/3}\, s^{1/2}}{n}$$

$$V = \frac{1.49(0.52 \text{ ft})(0.1414)}{0.02} \qquad \left[\frac{(0.24 \text{ m})(0.1414)}{0.02}\right]$$

$$= 5.47 \text{ ft/sec} \qquad (1.7 \text{ m/sec})$$

STEP 7. Channel travel time, in minutes:

$$\frac{L}{60V} = \frac{1000 \text{ ft}}{(60 \text{ sec/min})(5.47 \text{ ft/sec})} = 3 \text{ min} \qquad \left[\frac{305 \text{ m}}{(60 \text{ sec/min})(1.7 \text{ m/sec})} = 3 \text{ min}\right]$$

STEP 8. New time of concentration: overland plus channel travel time.

$$T_c = 37 + 3 = 40 \text{ min}$$

This new time of concentration is about 8 percent greater than the initial estimate used to determine the i value.

STEP 9. Repeat steps 4, 5, and 6 to find second approximation of Q_{peak}, velocity, and depth of flow. Begin with a recalculation of i based upon the new T_c from step 8.

10-year, 30-min rainfall = 1.4 in (36 mm)
10-year, 1-hr rainfall = 1.6 in (41 mm)
By interpolation, 10-year, 40-min rainfall = 1.47 in (38 mm)

$$i = \frac{1.47 \text{ in}}{40 \text{ min}} \times \frac{60 \text{ min}}{\text{hr}} = 2.21 \text{ in/hr} \qquad \left(\frac{38 \text{ mm}}{40 \text{ min}} \times \frac{60 \text{ min}}{\text{hr}} = 57 \text{ mm}\right)$$

Revised Q_{peak}:

$$Q_{\text{peak}} = C \times i \times A$$

$$= 0.17(2.21 \text{ in})(20 \text{ acres}) \qquad \left[\frac{0.17(57 \text{ mm})(8.1 \text{ ha})}{360}\right]$$

$$= 7.5 \text{ ft}^3\text{/sec} \qquad (0.22 \text{ m}^3\text{/sec})$$

TABLE 4.5 Solution of Manning Equation for Trapezoidal Channels*

Values of K' in formula $Q = \frac{K'}{n} b^{8/3} S^{1/2}$ for Trapezoidal Channels† D = depth of water b = bottom width of channel

$\frac{D}{b}$	Side slopes of channel, ratio of horizontal to vertical									
	Vertical	¼–1	½–1	¾–1	1–1	1½–1	2–1	2½–1	3–1	4–1
.01	.00068	.00068	.00069	.00069	.00069	.00069	.00069	.00069	.00070	.00070
.02	.00213	.00215	.00216	.00217	.00218	.00220	.00221	.00222	.00223	.00225
.03	.00414	.00419	.00423	.00426	.00428	.00433	.00436	.00439	.00443	.00449
.04	.00660	.00670	.00679	.00685	.00691	.00700	.00708	.00716	.00723	.00736
.05	.00946	.00964	.00979	.00991	.01002	.01019	.01033	.01047	.01060	.01086
.06	.0127	.0130	.0132	.0134	.0136	.0138	.0141	.0143	.0145	.0150
.07	.0162	.0166	.0170	.0173	.0175	.0180	.0183	.0187	.0190	.0197
.08	.0200	.0206	.0211	.0215	.0219	.0225	.0231	.0236	.0240	.0250
.09	.0241	.0249	.0256	.0262	.0267	.0275	.0282	.0289	.0296	.0310
.10	.0284	.0294	.0304	.0311	.0318	.0329	.0339	.0348	.0358	.0376
.11	.0329	.0343	.0354	.0364	.0373	.0387	.0400	.0413	.0424	.0448
.12	.0376	.0393	.0408	.0420	.0431	.0450	.0466	0482	.0497	.0527
.13	.0425	.0446	.0464	.0480	.0493	.0516	.0537	.0556	.0575	.0613
.14	.0476	.0502	.0524	.0542	.0559	.0587	.0612	.0636	.0659	.0706
.15	.0528	.0559	.0585	.0608	.0627	.0662	.0692	.0721	.0749	.0805
.16	.0582	.0619	.0650	.0676	.0700	.0740	.0777	.0811	.0845	.0912
.17	.0638	.0680	.0716	.0748	.0775	.0823	.0866	.0907	.0947	.1026
.18	.0695	.0744	.0786	.0822	.0854	.0910	.0960	.1008	.1055	.1148
.19	.0753	.0809	.0857	.0899	.0936	.1001	.1059	.1115	.1169	.1277
.20	.0812	.0876	.0931	.0979	.1021	.1096	.1163	.1227	.1290	.1414
.21	.0873	.0945	.101	.106	.111	.120	.127	.135	.142	.156
.22	.0934	.1015	.109	.115	.120	.130	.139	.147	.155	.171
.23	.0997	.1087	.117	.124	.130	.141	.150	.160	.169	.187
.24	.1061	.1161	.125	.133	.140	.152	.163	.173	.184	.204
.25	.1125	.1236	.133	.142	.150	.163	.176	.188	.199	.222
.26	.119	.131	.142	.152	.160	.175	.189	.202	.215	.241
.27	.126	.139	.151	.162	.171	.188	.203	.218	.232	.260
.28	.132	.147	.160	.172	.182	.201	.217	.234	.249	.281
.29	.139	.155	.170	.182	.194	.214	.232	.250	.268	.302
.30	.146	.163	.179	.193	.205	.228	.248	.267	.287	.324
.31	.153	.172	.189	.204	.218	.242	.264	.285	.306	.347
.32	.160	.180	.199	.215	.230	.256	.281	.304	.327	.371
.33	.167	.189	.209	.227	.243	.271	298	.323	.348	.396
.34	.174	.198	.219	.238	.256	.287	.316	.343	.370	.423
.35	.181	.207	.230	.251	.269	.303	.334	.363	.392	.450
.36	.189	.216	.241	.263	.283	.319	.353	.385	.416	.478
.37	.196	.225	.252	.275	.297	.336	.372	.406	.440	.507
.38	.203	.234	.263	.288	.312	.353	.392	.429	.465	.537
.39	.211	.244	.274	.301	.326	.371	.413	.452	.491	.568
.40	.218	.253	.286	.315	.341	.389	.434	.476	.518	.600
.41	.226	.263	.297	.328	.357	.408	.456	.501	.546	.633
.42	.233	.273	.309	.342	.373	.427	.478	.526	.574	.668
.43	.241	.283	.321	.357	.389	.447	.501	.553	.603	.703
.44	.248	.293	.334	.371	.405	.467	.525	.580	.633	.740
.45	.256	.303	.346	.386	.422	.488	.549	.607	.664	.777

$\frac{D}{b}$	Side slopes of channel, ratio of horizontal to vertical									
	Vertical	¼–1	½–1	¾–1	1–1	1½–1	2–1	2½–1	3–1	4–1
.46	.264	.313	.359	.401	.439	.509	.574	.636	.696	.816
.47	.271	.323	.372	.416	.457	.531	.599	.665	.729	.856
.48	.279	.334	.385	.432	.474	.553	.625	.695	.763	.897
.49	.287	.344	.398	.447	.493	.575	.652	.725	.797	.939
.50	.295	.355	.412	.463	.511	.598	.679	.757	.833	.983
.51	.303	.366	.425	.480	.530	.622	.707	.789	.869	1.03
.52	.311	.377	.439	.496	.549	.646	.736	.822	.907	1.07
.53	.319	.388	.453	.513	.569	.671	.765	.856	.945	1.12
.54	.327	.399	.467	.531	.589	.696	.795	.891	.984	1.17
.55	.335	.410	.482	.548	.609	.722	.826	.926	1.025	1.22
.56	.343	.422	.497	.566	.630	.748	.857	.963	1.07	1.27
.57	.351	.433	.511	.584	.651	.775	.889	1.000	1.11	1.32
.58	.359	.445	.526	.602	.673	.802	.922	1.038	1.15	1.37
.59	.367	.456	.542	.621	.694	.830	.956	1.077	1.20	1.43
.60	.375	.468	.557	.640	.717	.858	.990	1.117	1.24	1.49
.61	.383	.480	.573	.659	.739	.887	1.02	1.16	1.29	1.54
.62	.391	.492	.588	.678	.762	.916	1.06	1.20	1.33	1.60
.63	.399	.504	.604	.698	.785	.946	1.10	1.24	1.38	1.66
.64	.408	.516	.620	.718	.809	.977	1.13	1.28	1.43	1.72
.65	.416	.529	.637	.738	.833	1.008	1.17	1.33	1.48	1.79
.66	.424	.541	.653	.759	.857	1.04	1.21	1.37	1.53	1.85
.67	.433	.553	.670	.780	.882	1.07	1.25	1.42	1.59	1.91
.68	.441	.566	.687	.801	.907	1.10	1.29	1.47	1.64	1.98
.69	.449	.579	.704	.822	.933	1.14	1.33	1.51	1.69	2.05
.70	.457	.592	.722	.844	.959	1.17	1.37	1.56	1.75	2.12
.71	.466	.604	.739	.866	.985	1.21	1.41	1.61	1.81	2.19
.72	.474	.617	.757	.889	1.012	1.24	1.46	1.66	1.86	2.26
.73	.483	.631	.775	.911	1.039	1.28	1.50	1.71	1.92	2.34
.74	.491	.644	.793	.934	1.067	1.31	1.54	1.77	1.98	2.41
.75	.499	.657	.811	.957	1.095	1.35	1.59	1.82	2.05	2.49
.76	.508	.670	.830	.981	1.12	1.39	1.63	1.87	2.11	2.57
.77	.516	.684	.849	1.005	1.15	1.43	1.68	1.93	2.17	2.65
.78	.525	.698	.868	1.029	1.18	1.46	1.73	1.99	2.24	2.73
.79	.533	.711	.887	1.053	1.21	1.50	1.78	2.05	2.30	2.81
.80	.542	.725	.906	1.078	1.24	1.54	1.83	2.10	2.37	2.90
.81	.550	.739	.925	1.10	1.27	1.58	1.88	2.16	2.44	2.98
.82	.559	.753	.945	1.13	1.30	1.62	1.93	2.22	2.51	3.07
.83	.567	.767	.965	1.15	1.33	1.67	1.98	2.28	2.58	3.16
.84	.576	.781	.985	1.18	1.36	1.71	2.03	2.34	2.65	3.25
.85	.585	.796	1.006	1.21	1.40	1.75	2.08	2.41	2.72	3.35
.86	.593	.810	1.03	1.23	1.43	1.79	2.14	2.47	2.80	3.44
.87	.602	.825	1.05	1.26	1.46	1.84	2.19	2.54	2.87	3.54
.88	.610	.839	1.07	1.29	1.49	1.88	2.25	2.60	2.95	3.63
.89	.619	.854	1.09	1.31	1.53	1.93	2.31	2.67	3.03	3.73
.90	.628	.869	1.11	1.34	1.56	1.98	2.36	2.74	3.11	3.83

*E. F. Brater and H. W. King, *Handbook of Hydraulics*, 6th ed., copyright © McGraw-Hill Book Company, 1976. Used with the permission of the McGraw-Hill Book Company.

†K' values in the table are based on Q, cubic feet per second, and b, in feet. In metric units,

$$K' = 1.486 \frac{Q \text{ (in m}^3\text{/sec)}}{n} b \text{ (in m)}^{8/3} \times s^{1/2}$$

TABLE 4.6 Hydraulic Radius of Trapezoidal Channels*

For Determining Hydraulic Radius r for Trapezoidal Channels of Various Side Slopes

Let $\frac{\text{depth of water}}{\text{bottom width of channel}} = \frac{D}{b}$ and C_r = tabulated value. Then $r = C_r D$.

$\frac{D}{b}$	Side slopes of channel, ratio of horizontal to vertical									
	Vertical	¼–1	½–1	¾–1	1–1	1½–1	2–2	2½–1	3–1	4–1
.00	1.000	1.000	1.000	1.000	1.000	1.000	1.000	1.000	1.000	1.000
.01	.980	.982	.983	.983	.982	.980	.976	.973	.969	.961
.02	.962	.965	.967	.967	.965	.961	.955	.948	.941	.927
.03	.943	.949	.951	.951	.949	.943	.935	.926	.916	.898
.04	.926	.933	.936	.936	.934	.926	.916	.905	.894	.872
.05	.909	.918	.922	.922	.920	.911	.899	.886	.874	.850
.06	.893	.903	.908	.909	.906	.896	.883	.869	.856	.830
.07	.877	.889	.895	.896	.893	.882	.868	.853	.839	.812
.08	.862	.876	.882	.883	.881	.869	.854	.839	.823	.795
.09	.847	.863	.870	.871	.869	.857	.841	.825	.809	.781
.10	.833	.850	.858	.860	.858	.845	.829	.812	.797	.767
.11	.820	.838	.847	.849	.847	.834	.818	.801	.784	.755
.12	.806	.826	.836	.838	.836	.824	.807	.790	.773	.744
.13	.794	.814	.825	.828	.826	.814	.797	.779	.763	.734
.14	.781	.803	.815	.819	.817	.804	.787	.770	.753	.724
.15	.769	.793	.805	.809	.807	.795	.778	.761	.744	.715
.16	.758	.782	.795	.800	.799	.786	.769	.752	.736	.707
.17	.746	.772	.786	.791	.790	.778	.761	.744	.728	.700
.18	.735	.762	.777	.782	.782	.770	.753	.736	.720	.693
.19	.725	.752	.768	.774	.774	.763	.746	.729	.713	.686
.20	.714	.743	.760	.767	.766	.755	.739	.722	.706	.679
.21	.704	.734	.752	.759	.759	.748	.732	.716	.700	.674
.22	.694	.726	.744	.751	.752	.741	.726	.709	.694	.668
.23	.685	.717	.736	.744	.745	.735	.720	.704	.688	.663
.24	.676	.709	.729	.737	.739	.729	.714	.698	.683	.658
.25	.667	.701	.722	.730	.732	.723	.708	.693	.678	.653
.26	.658	.693	.715	.724	.726	.717	.703	.688	.673	.649
.27	.649	.686	.708	.717	.720	.712	.698	.683	.668	.645
.28	.641	.678	.701	.711	.714	.707	.693	.678	.664	.641
.29	.633	.671	.695	.706	.709	.702	.688	.673	.660	.637
.30	.625	.664	.688	.700	.703	.697	.683	.669	.656	.633
.31	.617	.657	.682	.694	.698	.692	.679	.665	.652	.630
.32	.610	.651	.676	.689	.693	.687	.675	.661	.648	.627
.33	.602	.644	.670	.684	.688	.683	.671	.657	.645	.624
.34	.595	.638	.665	.678	.683	.678	.667	.654	.641	.621
.35	.588	.632	.659	.673	.678	.674	.663	.650	.638	.618
.36	.581	.626	.654	.668	.674	.670	.659	.647	.635	.615
.37	.575	.620	.648	.664	.669	.666	.655	.643	.632	.612
.38	.568	.614	.643	.659	.665	.662	.652	.640	.629	.610
.39	.562	.608	.638	.654	.661	.658	.649	.637	.626	.607
.40	.556	.603	.633	.650	.657	.655	.645	.634	.623	.605
.41	.549	.598	.629	.646	.653	.652	.642	.631	.621	.603
.42	.543	.592	.624	.641	.649	.648	.639	.629	.618	.600

$\frac{D}{b}$	Side slopes of channel, ratio of horizontal to vertical									
	Vertical	¼–1	½–1	¾–1	1–1	1½–1	2–2	2½–1	3–1	4–1
.43	.538	.587	.619	.637	.645	.645	.636	.626	.616	.598
.44	.532	.582	.615	.633	.641	.642	.633	.623	.613	.596
.45	.526	.577	.611	.629	.638	.639	.631	.621	.611	.594
.46	.521	.572	.606	.626	.635	.636	.628	.618	.609	.592
.47	.515	.568	.602	.622	.631	.633	.625	.616	.607	.591
.48	.510	.563	.598	.618	.628	.630	.623	.614	.605	.589
.49	.505	.558	.594	.615	.625	.627	.620	.611	.603	.587
.50	.500	.554	.590	.611	.621	.624	.618	.609	.601	.586
.51	.495	.550	.587	.608	.618	.622	.616	.607	.599	.584
.52	.490	.545	.583	.604	.615	.619	.613	.605	.597	.583
.53	.485	.541	.579	.601	.612	.617	.611	.603	.595	.581
.54	.481	.537	.576	.598	.610	.614	.609	.601	.594	.580
.55	.476	.533	.572	.595	.607	.612	.607	.600	.592	.578
.56	.472	.529	.568	.592	.604	.610	.605	.598	.590	.577
.57	.467	.525	.505	.589	.601	.607	.603	.596	.589	.576
.58	.463	.521	.562	.586	.598	.605	.601	.594	.587	.574
.59	.459	.518	.558	.583	.595	.603	.599	.593	.586	.573
.60	.455	.514	.555	.580	.593	.601	.597	.591	.584	.572
.61	.450	.510	.552	.577	.591	.599	.596	.589	.583	.571
.62	.446	.507	.549	.575	.588	.597	.594	.588	.581	.569
.63	.442	.504	.546	.572	.586	.595	.592	.586	.580	.568
.64	.439	.500	.543	.569	.584	.593	.590	.585	.579	.567
.65	.435	.497	.540	.567	.581	.591	.589	.583	.577	.566
.66	.431	.494	.537	.564	.579	.589	.587	.582	.576	.565
.67	.427	.490	.534	.562	.577	.587	.586	.580	.575	.564
.68	.424	.487	.432	.559	.575	.585	.584	.579	.574	.563
.69	.420	.484	.529	.557	573	.583	.583	.578	.573	.562
.70	.417	.481	.526	.555	.571	.582	.581	.577	.571	.561
.71	.413	.478	.524	.552	.569	.580	.580	.575	.570	.560
.72	.410	.475	.521	.550	.567	.578	.578	.574	.569	.559
.73	.407	.472	.518	.548	.565	.577	.577	.573	.568	.558
.74	.403	.469	.516	.546	.563	.575	.576	.572	.567	.558
.75	.400	.467	.514	.544	.561	.573	.574	.570	.566	.557
.76	.397	.464	.511	.542	.559	.572	.573	.569	.565	.556
.77	.394	.461	.509	.539	.557	.570	.572	.568	.564	.555
.78	.391	.458	.507	.537	.555	.569	.570	.567	.563	.554
.79	.388	.456	.504	.535	.554	.567	.569	.566	.562	.554
.80	.385	.453	.502	.533	.552	.566	.568	.565	.561	.553
.81	.382	.450	.500	.531	.550	.565	.567	.564	.560	.552
.82	.379	.448	.498	.530	.548	.564	.566	.563	.559	.551
.83	.376	.445	.495	.528	.547	.562	.565	.562	.558	.551

*E. F. Brater and H. W. King, *Handbook of Hydraulics*, 6th ed., copyright © McGraw-Hill Book Company, 1976. Used with the permission of the McGraw-Hill Book Company.

Revised K' and depth of flow D:

$$K' = \frac{Q \times n}{b^{8/3} \times s^{1/2}}$$

$$= \frac{(7.5\ \text{ft}^3/\text{sec})\ (0.02)}{(1\ \text{ft})(0.1414)} \qquad \left\{ \frac{1.486(0.22\ \text{m}^3/\text{sec})(0.02)}{[(0.305\ \text{m})^{8/3}](0.1414)} \right\}$$

$$= 1.06 \qquad (1.10)^*$$

From Table 4.5, $D/b = 0.62$ (0.63)*; thus, $D = 0.62$ ft (0.19 m).

It is apparent that the revised depth of flow is not significantly different from the first approximation. The resulting velocity also will not be significantly different. This example illustrates that channel flow times are much shorter than overland flow times for comparable distances. Thus, *unless* the channel length is very large, channel flow time may be ignored in calculating T_c. However, if channel length is significant, or channel slope is very small, then channel flow time must be considered.

STEP 10. Adequacy of channel. It will be very easy to construct this channel, since a 1-ft (0.305-m) depth will allow a safe freeboard of 0.38 ft (0.12 m), or 4.6 in (12 cm) above the peak flow.

However, a velocity of 5.47 ft/sec (1.7m/sec) will be erosive. Maximum permissible velocity on ordinary firm loam, carrying silt-laden runoff, is 3.5 ft^3/sec (1.1 m/sec), whereas on stiff clay the maximum allowable velocity is 5.0 ft/sec (1.5 m/sec) (see Table 7.1 for maximum permissible velocities in unlined channels). In this case, a larger channel or a channel lining of coarse gravel is needed to prevent channel bed erosion.† If a coarse gravel lining is used, the channel roughness factor n in Manning's equation changes from 0.02 to 0.03 and the calculated velocity drops to 3.6 ft/sec (1.1 m/sec). The channel size must again be checked for adequacy with the new velocity.

Modified Procedure for Solving Channel Flow Problem without the Handbook of Hydraulics

If the *Handbook of Hydraulics* is not available, the following procedure based on simultaneous equations can be used to calculate velocity. The step-by-step procedure used in Example 4.8 is modified: At step 4, instead of determining Q_{peak} on the basis of overland travel time only, assume full flow in the channel and estimate a channel travel time before computing Q_{peak}. Steps 1, 2, and 3 are the same as given earlier.

STEP 4. As a first approximation, assume channel flowing full. Find the velocity by using Manning's equation. Compute channel travel time by dividing length by velocity.

STEP 5. Sum overland travel time and channel travel time to obtain total time of concentration.

STEP 6. Determine the precipitation intensity.

STEP 7. Compute the peak discharge.

*Because of round-off error, the values of K' and D/b are slightly different when metric units are used.

†Figure 7.36, which shows maximum permissible flow depths for unlined channels of various slopes, indicates that a lining is needed for the channel in this example, regardless of the soil type. Therefore, a lining should be installed (Sec. 7.2e).

STEP 8. Determine the second approximation of velocity in the channel to evaluate erosivity.

EXAMPLE 4.9 Channel Flow Calculations without the *Handbook of Hydraulics*

Given: This example uses the same site as Example 4.8 uses. However, let's now change the drainage channel slope to 0.1 percent and see what happens.

Solution:

STEP 4. Channel travel time. Select a reasonable initial channel size, 1 ft (0.305 m) deep. Assuming the channel is flowing full, use Manning's equation to find the velocity of flow in the channel: $n = 0.02$.
Area:

$$A = bD + 2D^2$$
$$= 1(1) + 2(1) \qquad [0.305(0.305) + 2(0.305^2)]$$
$$= 3 \text{ ft}^2 \qquad (0.28 \text{ m}^2)$$

Wetted perimeter:

$$\text{WP} = b + 2D\sqrt{2^2 + 1}$$
$$= 1 + 2(1)(\sqrt{5}) \qquad [0.305 + 2(0.305)(\sqrt{5})]$$
$$= 5.5 \text{ ft} \qquad (1.67 \text{ m})$$

$$r = \frac{A}{\text{WP}} = \frac{3}{5.5} = 0.55 \text{ ft} \qquad \left(\frac{0.28}{1.67} = 0.17 \text{ m}\right)$$

$$r^{2/3} = 0.67 \qquad (0.307)$$

$$s = 0.001$$

$$s^{1/2} = 0.032$$

$$V = \frac{1.49 \times r^{2/3} \times s^{1/2}}{n}$$
$$= \frac{1.49\ (0.67)(0.032)}{0.02} \qquad \left[\frac{0.307(0.032)}{0.02}\right]$$
$$= 1.6 \text{ ft/sec} \qquad (0.49 \text{ m/sec})$$

Channel travel time:

$$\frac{L}{60V} = \frac{1000 \text{ ft}}{(60 \text{ sec/min})(1.6 \text{ ft/sec})}$$
$$= 10.4 \text{ min} \qquad \left[\frac{305 \text{ m}}{(60 \text{ sec/min}) \times (0.49 \text{ m/sec})} = 10.4 \text{ min}\right]$$

STEP 5. Time of concentration:

$$T_c = \text{overland flow time plus channel travel time}$$
$$= 37 + 10.4$$
$$= 47.4 \text{ min}$$

STEP 6. Precipitation intensity. From the 10-year, 30-min and 10-year, 1-hr rainfalls given in the *Rainfall Frequency Atlas,* we interpolate to find the 10-year, 47.4-min rainfall:

10-year, 30-min = 1.4 in (36 mm)

10-year, 1-hr = 1.6 in (41 mm)

10-year, 47.4 min = 1.52 in (39 mm)

$$i = \frac{1.52 \text{ in}}{47.4 \text{ min}} \times \frac{60 \text{ min}}{\text{hr}} = 1.92 \text{ in/hr} \qquad \left(\frac{39 \text{ mm}}{47.4 \text{ min}} \times \frac{60 \text{ min}}{\text{hr}} = 49 \text{ mm/hr}\right)$$

STEP 7. Peak discharge:

$$Q_{\text{peak}} = C \times i \times A$$

$$= 0.17(1.92)(20) \qquad \left[\frac{0.17(49 \text{ mm/hr})(8.1 \text{ ha})}{360}\right]$$

$$= 6.5 \text{ ft}^3\text{/sec} \qquad (0.19 \text{ m}^3\text{/sec})$$

The peak discharge from this drainage area is estimated to be 6.5 ft^3/sec (0.19 m^3/sec). By use of Manning's equation in step 4, the discharge through the full channel cross section was initially calculated to have a velocity of 1.6 ft/sec (0.49 m/sec). This maximum velocity provides a conservative estimate of peak discharge.

In actuality, a flow of 6.5 ft^3/sec (0.19 m^3/sec) through a channel of 3-ft^2 (0.028-m^2) cross-sectional area should have a velocity of 2.2 ft/sec (0.68 m/sec) [since $V = Q/A$ = 6.5 ft^3/sec ÷ 3 ft^2 = 2.2 ft/sec (0.19 m^3/sec ÷ 0.28 m^2 = 0.68 m/sec)]. This mismatch of velocities tells us that the channel would not flow full and the velocity would be somewhere between 1.6 and 2.2 ft/sec (0.49 and 0.68 m/sec).

STEP 8. In step 4 we assumed the channel was flowing full. The actual velocity in the channel can be estimated more precisely by reapplying Manning's equation. We need to determine the actual cross-sectional area of the channel when it is discharging 6.5 ft^3/sec (0.19 m^3/sec). Combining the equations gives us

$$V = \frac{1.49 \times r^{2/3} \times s^{1/2}}{n} \qquad Q = A \times V$$

results in

$$Q = A \times V = \frac{A \times 1.49 \times r^{2/3} \times s^{1/2}}{n} \qquad \left[Q \text{ (in m}^3\text{/sec)} = \frac{A \times r^{2/3} \times s^{1/2}}{n}\right]$$

From the preceding computations:

$$Q = 6.5 \qquad (0.19 \text{ m}^3\text{/sec})$$

$$\frac{1.49}{n} = 74.5 \qquad \left(\frac{1}{0.02} = 50\right)$$

$$s^{1/2} = 0.032$$

The equation now becomes

$$6.5 = A \times 74.5 \times r^{2/3} \times 0.032 \qquad (0.19 = A \times 50 \times r^{2/3} \times 0.032)$$

$$6.5 = A \times 2.38 \times r^{2/3} \qquad (0.09 = A \times 1.6 \times r^{2/3})$$

$$2.73 = A \times r^{2/3} \qquad (0.12 = A \times r^{2/3})$$

Since $r = \dfrac{A}{\text{WP}}$ (from step 4)

$$2.73 = A \times \left(\frac{A}{\text{WP}}\right)^{2/3} \qquad \left\{0.12 = A \times \left(\frac{A}{\text{WP}}\right)^{2/3}\right\}$$

$$2.73 = \frac{A^{5/3}}{\text{WP}^{2/3}} \qquad \left(0.12 = \frac{A^{5/3}}{\text{WP}^{2/3}}\right)$$

Since both the cross-sectional area and the wetted perimeter can be expressed in terms of the channel depth, the equation can be solved algebraically. Alternatively, the equation can be solved by iteration: substitute a reasonable number and adjust it according to the results.

For a trapezoidal channel with 2:1 side slopes, the area can be expressed in terms of the depth D as

$$A = bD + 2D^2 = 1 \times D + 2D^2 \qquad (0.305\text{ m} \times D + 2D^2)$$

The wetted perimeter can also be expressed in terms of depth as

$$\text{WP} = b + 2D\sqrt{z^2 + 1} = 1 + 2D\sqrt{5} \qquad (0.305\text{ m} + 2D\sqrt{5})$$

Choosing iteration as the solution method, we first try a depth of 0.8 ft (0.24 m).

$$A = 0.8 + 2(0.8^2) = 2.08 \qquad \{(0.305\text{ m})(0.24\text{ m}) + 2[(0.24\text{ m})^2] = 0.188\text{ m}^2\}$$

$$\text{WP} = 1 + 2(0.8)(\sqrt{5}) = 4.58 \qquad [0.305\text{ m} + 2(0.24\text{ m})(\sqrt{5}) = 1.38\text{ m}]$$

and substituting into $2.73 \times \text{WP}^{2/3} = A^{5/3}$ [$0.12 \times \text{WP}^{2/3}$ (in m) $= A^{5/3}$ (in m)] gives us

$$2.73(4.58^{2/3}) = 2.08^{5/3} \qquad \{0.12\,[(1.38\text{ m})^{2/3}] = (0.188\text{ m}^2)^{5/3}\}$$

$$7.53 > 3.39 \qquad (0.149 > 0.062)$$

Next, try a depth of 1.5 ft (0.46 m):

$$A = 1.5 + 2(1.5^2) = 6 \qquad \{0.46\text{ m} + 2(0.46\text{ m})^2 = 0.88\text{ m}^2\}$$

$$\text{WP} = 1 + 2(1.5)(\sqrt{5}) = 7.7 \qquad [0.305\text{ m} + 2(0.46\text{ m})(\sqrt{5}) = 2.36\text{ m}]$$

$$2.73(7.7^{2/3}) = 6^{5/3} \qquad \{0.12[(2.36\text{ m})^{2/3}] = (0.88\text{ m}^2)^{5/3}\}$$

$$10.6 < 19.8 \qquad (0.213 < 0.808)$$

Since the left side is now smaller rather than larger than the right side, we try several depths between 0.8 and 1.5 ft (0.24 and 0.46 m).

The actual channel depth when discharging 6.5 ft^3/sec (0.19 m^3/sec) would be 1.15 ft (0.35 m). The cross-sectional area with that depth is 3.79 ft^2 (0.35 m^2). The more precise channel velocity then is

$$Q = A \times V$$

$$6.5 = 3.79 \times V \qquad (0.19\text{ m}^3/\text{sec} = 0.35\text{ m}^2 \times V)$$

$$V = 1.7\text{ ft/sec} \qquad (0.54\text{ m/sec})$$

According to Table 7.1, unless the channel material were fine sand, this velocity would not be erosive to an unlined channel. However, Fig. 7.36, which was based on different experimental data and parameters, shows that a lining is probably needed in the channel. In deciding whether to install a lining, the prudent approach is to check both tables. If either table shows that a lining is needed, then a lining should be installed. Greater weight should be placed on the maximum depth of flow table (Fig. 7.36), since the maximum permissible velocity table (Table 7.1) assumes "aged" channels, and newly constructed channels on a construction site are unlikely to be in that relatively stable condition.

However, the 1-ft (0.305-m) channel depth assumed in Example 4.8 would not contain the peak flow with the flatter conditions of Example 4.9. Allowing for freeboard, the channel in this last example should be constructed with a depth of at least 1.5 ft (0.46 m).

4.2 OTHER METHODS FOR ESTIMATING RUNOFF

This handbook focuses on the use of the rational method to compute peak discharge from a small watershed. Brief descriptions of three other methods are presented below, and they are followed by a comparison of the methods. Any reliable method to compute runoff can be used for drainage system design.

4.2a Unit Hydrograph Method

The following description was taken from Rantz (10):

> The unit hydrograph is a widely used device for relating runoff to storm precipitation and is described in all standard hydrology texts. The unit hydrograph shows the time distribution of surface runoff resulting from a storm that produces 1 inch (25.4 mm) of rainfall excess over the watershed in some selected interval of time. Rainfall excess is defined as that part of the rainfall that is available to produce surface runoff, after the demands of infiltration and surface retention have been met. That part of the precipitation that infiltrates into the ground or is retained above ground is known as water loss.
>
> The time interval used for the 1 inch (25.4 mm) of rainfall excess varies with basin size and with the time response of runoff to rainfall. It may be as short as 1 minute for a small experimental plot or as long as 24 hours for a large slow-rising river, but in practice it generally ranges between 5 minutes and 6 hours. Given the unit hydrograph for a watershed and the precipitation distribution for a given storm, the hydrologist can produce the resulting hydrograph of surface runoff.
>
> The unit hydrograph for a gaged watershed can be derived from observed hydrographs and the record from a recording raingage in the basin. From the characteristics of the unit hydrographs for several gaged watersheds in a region, it is then possible to derive synthetic unit hydrographs for use with ungaged watersheds in the region.

Rantz continues with a discussion of design elements for synthetic unit hydrographs and develops some adjustments to the unit hydrograph for the effects of urbanization.

4.2b Hydrologic Basin Models

The use of hydrologic basin models for simulating runoff has been made possible by the widespread availability and sophistication of computers. This book does not address runoff models in detail. The reader may refer to references cited or to local engineering firms who use their own models. Models are more appropriate to large drainage areas and are associated more often with flood control projects than with individual construction sites.

Many hydrologic basin models for use in runoff simulation are under investigation, but the only ones in use at the time of the Rantz study (10) for computing storm runoff from small watersheds were the USGS Watershed Model (4) and various versions of the Stanford Watershed Model (3). Both the USGS and Stanford models:

- Use precipitation and pan evaporation as hydrometeorological inputs
- Maintain a water budget that is balanced at short intervals (usually every 15 min during storm periods)
- Require only a short period of runoff record for model calibration

Neither model requires the assumption of identical frequencies for peak discharge and peak precipitation rates, because both use historic sequences of storm precipitation over a period of years long enough to permit statistical frequency analysis of the derived discharge data.

4.2c SCS Method for Small Watersheds

The U.S. Soil Conservation Service (SCS) developed a set of charts for estimating the peak discharge from small areas. (7) The graphs were prepared by computer-processing national rainfall data and inputting the results into the runoff equation developed by Mockus in the SCS *National Engineering Handbook*. (11)

Peak discharges range from 5 to 2000 ft^3/sec (0.1 to 57 m^3/sec); drainage areas range from 5 to 2000 acres (2 to 809 ha); and 24-hr rainfalls range from 1 to 12 in (25 to 305 mm).

4.2d Comparison of Methods

The rational method is simple to apply. It provides a runoff rate for a small watershed no larger than 200 acres (81 ha) and it is applied separately to each drainage area on a site. Its principal disadvantage is that it provides only a peak or average discharge from a watershed, not a complete hydrograph, and therefore it cannot be used for routing multiple flows toward a single outlet. But erosion and sediment control structures are more effective when located throughout a drainage area rather than at a single collection point. So the rational method remains the simplest way to obtain estimates of peak discharge and average runoff rates for the design of erosion and sediment control structures.

The unit hydrograph method, although more complex in application than the rational method, has the advantage of providing a complete storm hydrograph rather than just the peak discharge. Thus, in dealing with a complex watershed, the storm hydrograph for each subwatershed can be computed independently by the unit hydrograph method for subsequent routing down the main channel. No such direct procedure for combining flood peaks from subwatersheds is possible with the rational method. The complete storm hydrograph is needed for flood control project design and for some water quality studies.

The chief weakness of the unit hydrograph method, according to Rantz, is that infiltration loss is difficult to determine. Rantz tentatively related infiltration to mean annual precipitation. He concluded that the unit hydrograph method would give better results than the rational method, especially on larger watersheds.

In erosion control, we are interested more in individual sites of limited area within a watershed than in the entire watershed. Although perhaps less accurate than the unit hydrograph method, the rational method does account for infiltration loss through the coefficient C. A complete storm hydrograph is not needed unless flood control or a water quality monitoring program is planned. Thus, in erosion control work, the rational method is usually sufficient.

Runoff simulation models are most useful for large-scale design projects. For erosion control planning, the site planner may apply an existing model to a construction site, but development of a new model just to obtain the peak discharge is impractical.

The SCS model for small watersheds is used by SCS offices nationwide. However, the factors used in this model (called curve numbers) do not provide the flexibility that the rational method C values do in evaluating site characteristics, since the SCS model was designed for rural areas. In addition, the SCS model cannot use precipitation depth-duration-frequency data, which, if available, improves the accuracy of the other methods.

In summary, the rational method provides sufficiently accurate, easy-to-obtain estimates of runoff from a construction site for use in erosion control planning. Other, more complex methods exist and may be used by those familiar with them.

REVIEW QUESTIONS

1. What is the rational method equation?
2. What are some of the advantages and disadvantages of the rational method equation?
3. What other methods for computing runoff are available?
4. Describe the three factors in the rational method equation.
5. How is design storm return interval chosen?
6. What is Q_{peak} used for? Q_{avg}?

7. What effect does slope steepness have on the C value? What effect does soil depth have?
8. What is the most reliable source of rainfall intensity data?
9. What local options for rainfall and intensity information may be available?
10. Describe how to determine the average and peak precipitation intensity. What information about the site do you need? What reference materials?
11. What is meant by the terms "time of concentration" and "overland flow"? For what computation are they necessary?
12. What are the two components of the time of concentration?
13. How is the overload flow time estimated? What site factors influence overland flow time?
14. How is channel flow time computed?
15. Why is it important to know the velocity in the channel? What are the alternatives if the calculated flow or velocity exceeds the capacity of a channel?
16. ***Given:*** A 4-acre (1.6-ha) site just north of New York City.
Longest overland flow path length is 200 ft (61 m).
Average slope is 10 percent.
Bare, smooth, packed earth.

Find: C value
Overland flow time T_c
Peak intensity i_{peak} for 10-year storm
Q_{peak}, 10-year storm

17. For the site in Question 16, find the average intensity and average runoff for a 10-year, 6-hr storm.

REFERENCES

1. E. F. Brater and H. W. King, *Handbook of Hydraulics,* 6th ed., McGraw-Hill Book Company, New York, 1976.
2. California Department of Water Resources, *Rainfall Depth-Duration-Frequency for California,* microfiche, DWR, Sacramento, Calif., 1981.
3. N. H. Crawford and R. K. Linsley, *Digital Simulation in Hydrology, Stanford Watershed Model IV,* Stanford University, Department of Civil Engineering Technical Report No. 39, Palo Alto, Calif., 1966.
4. D. R. Dawdy, R. W. Lichty, and J. M. Bergmann, *A Rainfall-Runoff Simulation Model for Estimation of Flood Peaks for Small Drainage Basins—A Progress Report,* U.S. Geological Survey Open-File Report, Menlo Park, Calif., 1970.
5. R. H. Frederick, V. A. Myers, and E. P. Auciello, *Five- to Sixty-Minute Precipitation Frequency for the Eastern and Central United States,* NOAA Technical Memorandum NWS Hydro-35 (prepared for Engineering Division, U.S. Department of Agriculture, Soil Conservation Service), U.S. Department of Commerce, National Oceanic and Atmospheric Administration, National Weather Service, Office of Hydrology, Silver Springs, Md., 1977.

6. D. M. Hershfield, *Rainfall Frequency Atlas of the United States,* Technical Paper No. 40, U.S. Department of Commerce, Weather Bureau, Washington, D.C., 1961.
7. K. M. Kent, *A Method for Estimating Volume and Rate of Runoff in Small Watersheds,* SCS Technical Paper No. 149, U.S. Department of Agriculture, Soil Conservation Service, GPO, Washington, D.C., 1973.
8. J. F. Miller, R. H. Frederick, and R. J. Tracey, *Precipitation Frequency Atlas of the Western States* (11 volumes), NOAA Atlas 2 (prepared for Engineering Division, U.S. Department of Agriculture, Soil Conservation Service), U.S. Department of Commerce, National Oceanic and Atmospheric Administration, National Weather Service, Office of Hydrology, GPO, 1973.
9. S. E. Rantz, *Precipitation Depth-Duration-Frequency Relations for the San Francisco Bay Region,* U.S. Geological Survey, Basic Data Contribution No. 25, U.S.G.S., Menlo Park, Calif., 1971.
10. S. E. Rantz, *Suggested Criteria for Hydrologic Design of Storm Drainage Facilities in the San Francisco Bay Region, California,* U.S. Geological Survey, Menlo Park, Calif., 1971.
11. U.S. Department of Agriculture, Soil Conservation Service, *National Engineering Handbook, Section 4, Hydrology,* GPO, Washington, D.C., 1972.
12. U.S. Department of the Interior, Bureau of Reclamation, *Design of Small Dams,* GPO, Washington, D.C., 1973.
13. Virginia Soil and Water Conservation Commission, *Virginia Erosion and Sediment Control Handbook,* VSWCC, Richmond, Va., 1980.
14. Wright-McLaughlin Engineers, *Urban Storm Drainage Criteria Manual,* Denver Regional Council of Governments, Denver, Colo. 1969.

chapter 5

Estimating Soil Loss with the Universal Soil Loss Equation

Soil conditions are a principal factor in determining the erosion potential at a site. Yet although estimating runoff is a common practice, estimating soil loss is not, particularly in urban areas.

Soil loss estimates have three important applications for erosion control planning:

1. To identify erosion-prone areas on a site
2. To compare the effectiveness of different control measures
3. To estimate the volume of sediment storage needed in a sediment basin

Thus, by estimating soil loss, the erosion control planner will be able to avoid disturbing highly erodible areas, select the most effective control measures for a site, and avoid costly oversizing or undersizing of sediment basins.

A number of methods for assessing soil loss have been developed. They range from simple, qualitative models to elaborate watershed simulations. Qualitative models rely on subjective evaluation of a series of criteria. Watershed simulation models are often very theoretical. Several empirical models also are available. Most models are best suited to estimating erosion from very large areas (more than 1 mi^2) and lack the accuracy for use on small sites such as construction sites.

The universal soil loss equation (USLE) is an empirical model developed by the U.S. Dept. of Agriculture (USDA) (21) to estimate sheet and rill erosion from agricultural lands. The equation is a simple arithmetic relation composed of five factors that influence erosion. The equation has been tested nationwide and in certain areas overseas. (13) Some refinement of the factors has occurred, and research continues as use of the equation expands.

This chapter describes the use of the universal soil loss equation for estimating soil loss on construction sites. The more recent USDA manual, *Predicting Rainfall Erosion Losses—A Guide to Conservation Planning,* includes some information about use of the equation on construction sites. (20) In addition, the Transportation Research Board published a report, *Erosion Control During Highway Construction, Manual on Principles and Practices* (10), which presents detailed procedures for use of the soil loss equation for erosion control planning during highway construction.

5.1 BASIC SOIL CHARACTERISTICS

Soil engineers and agricultural scientists describe the properties of soils differently because their interests are substantially different. Both soil and civil engineers are familiar with the unified and AASHTO systems that focus on the engineering properties of soils. These classifications are based on the physical properties of the soil. Initially, soils are described as either coarse- or fine-grained. Coarse-grained soils are further described by degree of sorting of particle sizes. Fine-textured soils are further distinguished by their liquid and plasticity limits. Particle size analysis is not usually performed. In contrast, the USDA system of soil classification used by the U.S. Soil Conservation Service (SCS) is directed at characteristics of soils important for agricultural uses, such as texture, organic matter, and nutrient content. A particle size analysis is necessary before a soil can be classified by using the USDA system. In the USLE, since it was originally developed for use in agricultural areas, the USDA system is used.

The following section describes soil characteristics that are important to the use of the universal soil loss equation. Four soil characteristics that affect erodibility are:

- Texture
- Organic matter content
- Structure
- Permeability

5.1a Soil Texture

Soil texture depends on the proportions, by weight, of sand, silt, and clay in a soil—often referred to as the particle size distribution. Table 5.1 lists the USDA

TABLE 5.1 USDA Particle Size Classes (14)

Particle name	Size, mm
Gravel	Greater than 2
Sand	2–0.1
Very fine sand	0.1–0.05
Silt	0.05–0.002
Clay	Less than 0.002

particle size classes. A triangle is used to present the soil texture names according to particle size content (Fig. 5.1). The percents of sand, silt, and clay in a soil add up to 100. By knowing any two components, one can find the texture name for the soil. For example, a soil with 40 percent sand and 40 percent silt is called a loam (Example 1, Fig. 5.1). A loam also contains 20 percent clay. A sample with 20 percent sand and 60 percent silt is called a silt loam (Example 2), whereas one with 60 percent sand and 30 percent silt is called a sandy loam (Example 3).

In the bottom part of Fig. 5.1 the USDA, unified, and AASHTO classification systems are compared. Note the size in each system that differentiates silt from sand; the engineering systems change the classification at 0.74 mm, the USDA system at 0.05 mm. This difference is important because the silt and very fine sand particles in this size range are most susceptible to erosion and are therefore of interest in erosion control planning. The particle size also is important because the ability of a sediment basin to trap soil is primarily related to particle size (see Sec. 8.2). The smaller the particle, the larger the basin must be to capture it. Each sediment basin should be designed to capture a certain size particle called the *design particle*. If a soils analysis is to be done on a site, the site planner should request that the design particle size be a threshold in the analysis (i.e., specify the percent, by weight, of particles larger or smaller than that size).

Sandy soils generally have a higher permeability than fine-textured soils have. The amount of runoff is lower; and since the particles are relatively large (and thus heavy), they are not carried far in any runoff that does occur. Sand particles will settle out of runoff at the bottom of a slope or in a channel with a gentle slope. Very fine sand particles, however, behave like silt particles.

Silt is the most important particle size class when soil erodibility is evaluated. The higher the silt content, the more erodible a soil is, because silt-sized particles are small enough to reduce the permeability of a soil and are also easily carried by runoff. Control measures should be designed to prevent the erosion of silt, or at least to contain it on-site.

Clay is the smallest particle size class. A soil with a high clay content tends to be quite cohesive—the particles stick together in clumps. Runoff does not pick up clay particles as easily as it does silt. However, once clays are suspended in runoff, they will not settle out until they reach a large, calm water body. These very small particles have so low a settling velocity that they will be carried long

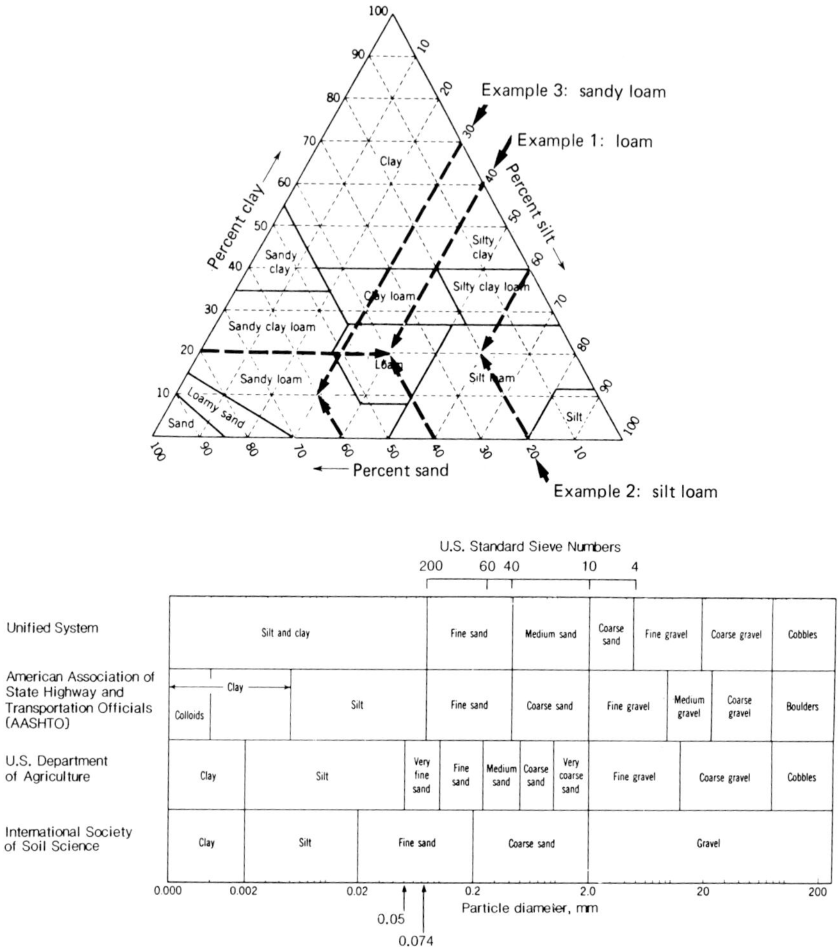

Fig. 5.1 Guide to textural classification of soils. (3)

distances until still water is reached or until salt water increases precipitation by causing them to clump together again in aggregates.

It is easiest to prevent erosion of sandy soils. Silts are most susceptible to erosion, but they can be recaptured on-site by applying the control measures described in Chap. 8. Clays are the most difficult to trap once erosion has occurred, so control measures must focus on preventing their erosion in the first place.

Although texture is a principal soil characteristic affecting erodibility, three other characteristics have a strong influence on erosion potential; they are organic matter, soil structure, and permeability.

5.1b Organic Matter

Organic matter *within a soil* is decomposed plant and animal litter. It consists of colloidal particles as small as and smaller than clay particles. This kind of organic matter helps bind the soil particles together, improves soil structure, and increases permeability and water-holding capacity. Soils with organic matter are less susceptible to erosion and more fertile than soils without organic matter. On a construction site, where extensive grading has removed the original topsoil and exposed layers of earth that have no plant roots growing in them, the organic matter content will be nil. Such subsoils are likely to be more erodible and less fertile than surface soils.

In another sense of the term, organic matter means plant residue, or other organic material that is applied to the soil surface. Surface-applied mulch reduces erosion by reducing the impact of raindrops and by absorbing water and reducing runoff. It provides a more hospitable environment for plant establishment, and it eventually decomposes and improves the structure and fertility of the soil. The uses of mulch in erosion control are described in Chap. 6.

5.1c Soil Structure

Soil structure refers to the arrangement of particles in a soil. In an undisturbed soil with established vegetation, organic matter binds the particles into aggregates. It thereby produces what is called a granular structure, which is desirable because permeability and water-holding capacity are increased and the clumped particles are more resistant to erosion. Grading and compaction of soils during construction destroy the natural structure, reduce permeability, and increase runoff and erodibility. The direct impact of raindrops on a soil unprotected by mulch or vegetation also breaks up soil aggregates and increases erodibility.

5.1d Soil Permeability

Soil permeability refers to the ability of the soil to allow air and water to move through it. Permeability classes are listed in Table 5.2. Soil texture, structure, and organic matter all contribute to permeability. Sites with highly permeable soils absorb more rainfall, produce less runoff, are less susceptible to erosion, and support plant growth more successfully.

One disadvantage of higher permeability relates to slope stability. Graded areas must meet certain standards of compaction to ensure a stable foundation surface. Compaction reduces the permeability of the soil. Infiltration of water into a large fill is *not* desirable because it may reduce the fill's stability. Yet by reducing infiltration, surface runoff and surface erosion are increased. When grass is planted on these fills and paved diversion ditches are installed midslope to carry away excess runoff, surface erosion is reduced.

TABLE 5.2 USDA Soil Permeability Classes (14)

Class	Estimated inches/hour through saturated undisturbed cores under ½-in head of water
Very slow	Less than 0.06
Slow	0.06–0.2
Moderately slow	0.2–0.6
Moderate	0.6–2.0
Moderately rapid	2.0–6.0
Rapid	6.0–20
Very rapid	More than 20

5.2 USING THE UNIVERSAL SOIL LOSS EQUATION TO ESTIMATE SOIL LOSS

5.2a The Equation

The general form of the universal soil loss equation is:

$$A = R \times K \times \text{LS} \times C \times P$$

where A = soil loss, tons/(acre) (year)
R = rainfall erosion index, in 100 ft · tons/acre × in/hr
K = soil erodibility factor, tons/acre per unit of R
LS = slope length and steepness factor, dimensionless
C = vegetative cover factor, dimensionless
P = erosion control practice factor, dimensionless

In metric units,

$$A \text{ (in t/ha)} = R \text{ (in J/ha)} \times K \text{ (in t/J)} \times \text{LS} \times C \times P$$

To convert the English units of R (100 ft · tons/acre × in/hr) to a convenient metric equivalent, multiply R by 1.70. This yields R in units of 10^7 J/ha × mm/hr. Multiply English units of K by 1.32. The product will then be in units of tonnes/hectare. Alternatively, to convert A from tons/acre to tonnes/hectare, multiply by 2.24.

The soil loss is an estimated annual average. The rainfall erosion index contains both an energy component and an intensity component. The LS, C, and P factors are ratios of soil loss from the site to soil loss from a unit area of a standard plot with these characteristics: 72.6 ft (22.1 m) long, 9 percent slope, tilled, bare soil.

To calculate soil loss, each of the factors is assigned a numerical value. The five factors are then multiplied together to produce an estimate of soil eroded from the site in an average year. When an estimate of required sediment storage volume is needed, the units of weight can be converted to volume by assuming a density for the sediment.

Careful evaluation of site characteristics is important to obtain reasonable soil loss estimates. Construction sites may vary from less than one acre to several hundred acres. Because soil erodibility varies greatly over large areas, a soil loss estimate is needed for each drainage area controlled by a sediment basin. A drainage area should be no larger than 100 acres (40 ha) and preferably much smaller. If a drainage area contains more than one soil type, it should be further subdivided into areas of homogeneous soil type. In addition, the length-slope LS, cover C, and surface condition P factors may differ with parts of a site. To produce the most accurate estimate, calculate soil loss separately for each homogeneous area (area of similar K, LS, C, and P) and add the estimates together.

Generally, a site can easily be divided into drainage areas served by a sediment basin or trap. Then, within each drainage area, homogeneous categories can be described, such as house pads, cut and fill slopes between pads, and unpaved road surfaces. Certain erosion control practices are associated with each one of these categories. House pads normally have a very gentle slope and a compacted surface which is not seeded, so LS, C, and P values should be the same for all of them. Slopes between pads are often at a 2:1 gradient, 20 to 40 ft (6 to 12 m) long, trackwalked and seeded. Examples of the use of these factors in the soil loss equation are provided in the following sections.

Note that the USLE assigns different meanings to two letters also used in the rational method runoff equation. The distinctions listed below must be kept clear to avoid miscalculations:

Symbol	Rational method	USLE
A	Acres (hectares)	Tons/acre (tonnes/hectare)
C	Runoff coefficient	Vegetative cover factor

5.2b Rainfall Erosion Index *R*

The rainfall erosion index R is a measure of the erosive force and intensity of rain in a normal year. The two components of the factor are the total energy E and the maximum 30-min intensity I_{30} of storms. The rainfall erosion index is the sum of the product $E \times I_{30}$ for all the major storms in an area during an average year. Values of R have been computed for the continental United States from rainfall records and probability statistics, so R should not be considered a precise factor for any given year or location. Its principal value, and that of the soil loss equation itself, is as a predictive tool and risk evaluator: Construction activities in areas with high R values will require greater attention to erosion

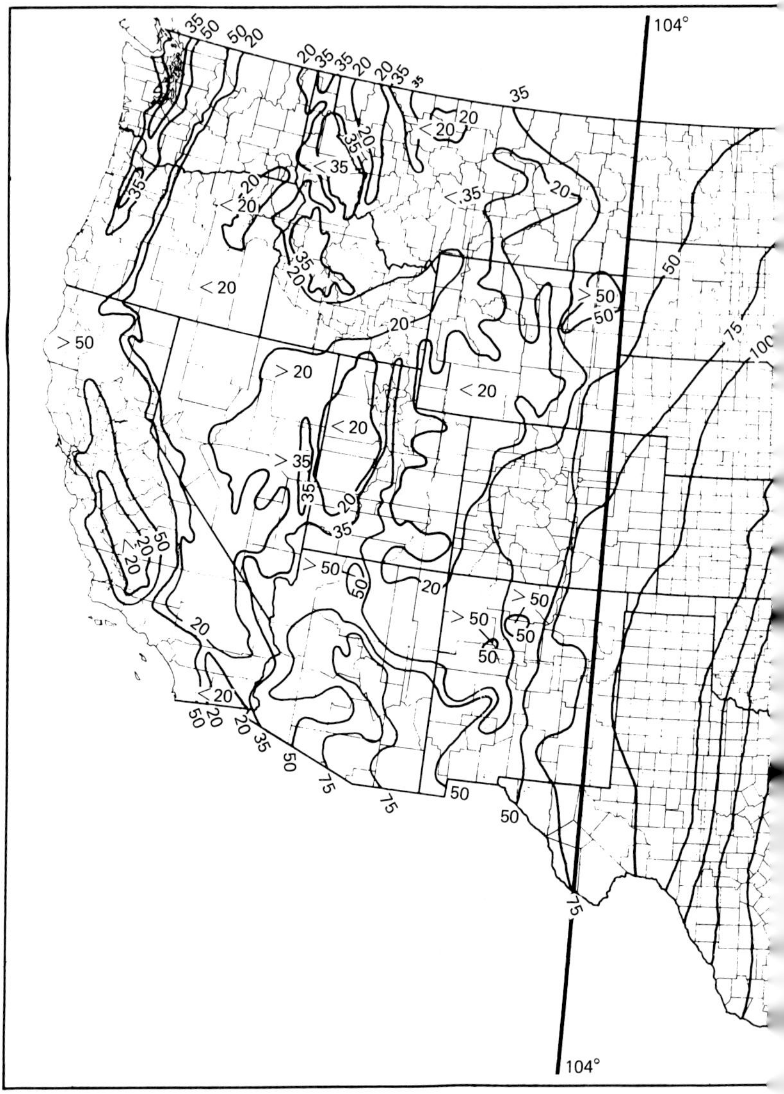

Fig. 5.2 *R* values for areas east of 104°. Because of irregular topography in the w
United States, calculate *R* values in this region by using local rainfall data. *R* is ir
of 100 ft · tons/acre per in/hr. To convert *R* to units of 10^7 J/ha per mm/hr, multi
1.70. (20) Scale is in miles.

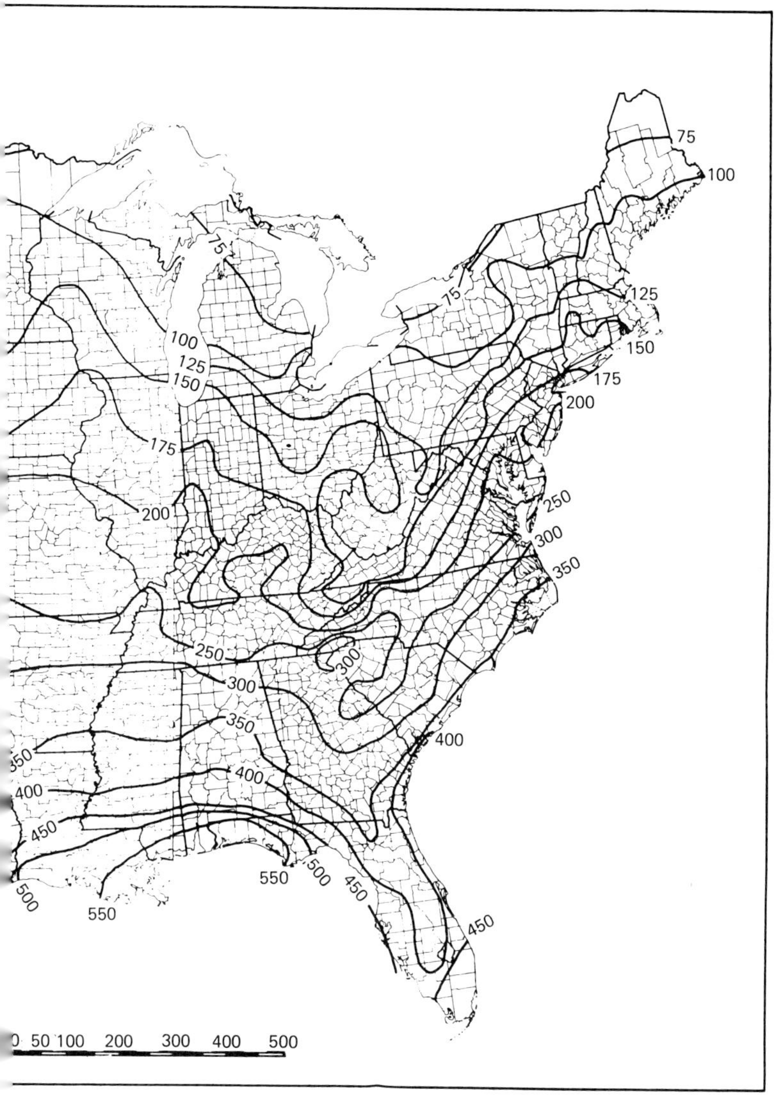
75
100
75
15
125
100
125
150
150
175
175
200
200
250
300
350
250
300
300
350
400
350
400
400
450
550
500
450
450
500
550
0 50 100 200 300 400 500

control practices than construction in areas with low R values. If a more precise value for R is needed, other references (10, 20, 21) that explain how to calculate R for individual storms and years from local data should be consulted.

An "isoerodent" map, prepared by Wischmeier for the USDA (20) and shown in Fig. 5.2, is used to find the R value for sites east of the Rocky Mountains (approximately 104° west longitude). R can be interpolated for points between the lines. Contact local soil conservation service offices for more detailed information on R values in areas covered by this map. West of the 104th west meridian, irregular topography makes use of a generalized map impractical. For the western states, R is calculated by using rainfall data. Results of investigations at

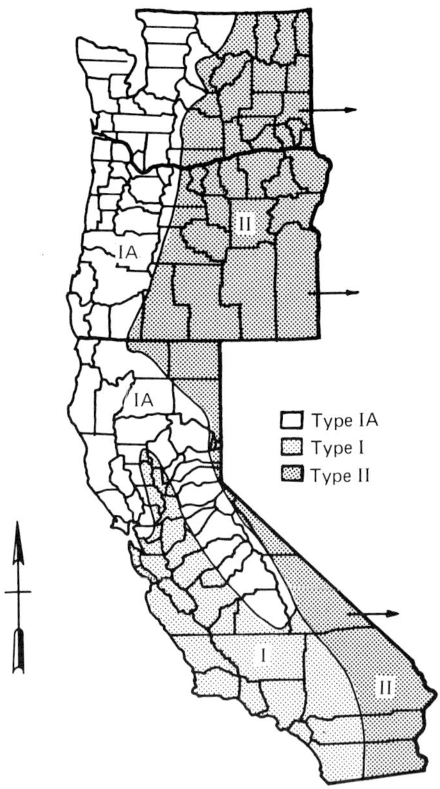

Fig. 5.3 Distribution of storm types in the western United States. (4) Type II storms occur in Arizona, Colorado, Idaho, Montana, Nevada, New Mexico, Utah, and Wyoming also.

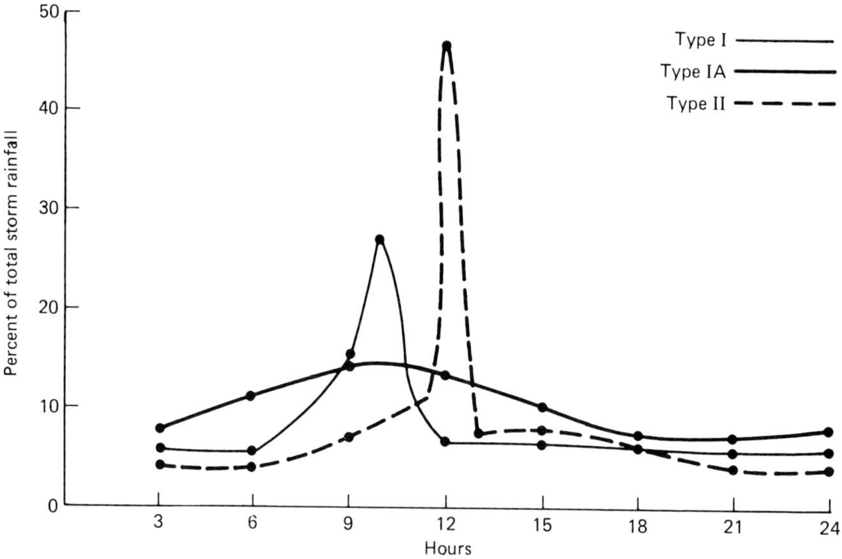

Fig. 5.4 Time distribution of rainfall within storm types. Adapted from unpublished data provided by Wendell Styner, U.S. Department of Agriculture, Soil Conservation Service, West Technical Service Center, Portland, Oregon, October 28, 1981.

the Runoff and Soil Loss Data Center at Purdue University showed that R values in the western states could be approximated with reasonable accuracy by using 2-year, 6-hr rainfall data. (20) Regression equations for three different storm types (I, IA, and II) are used to calculate R values. Figure 5.3 shows the distribution of type I, IA, and II storms throughout the western states.

A storm type is distinguished by the rainfall distribution within the storm. Figure 5.4 illustrates the time distributions of rainfall within the three types of storms. A type II storm is characterized by gradually increasing rainfall followed by a strong peak in rainfall intensity that tapers off to low-intensity rain. Type II storms occur in the following areas:

- The eastern parts of Washington, Oregon, and California (east of the Sierra Nevada)
- All of Idaho, Montana, Nevada, Utah, Wyoming, Arizona, and New Mexico

Type I and IA storms occur in a maritime climate. Type I is typical of storms that occur in southern and central California. These storms have a milder but definite peak similar to that of the type II storms. Type IA storms, which are characteristic of storms in coastal areas of northern California, Oregon, Washington, and the western slopes of the Sierra Nevada, have a low broad peak in the rainfall distribution.

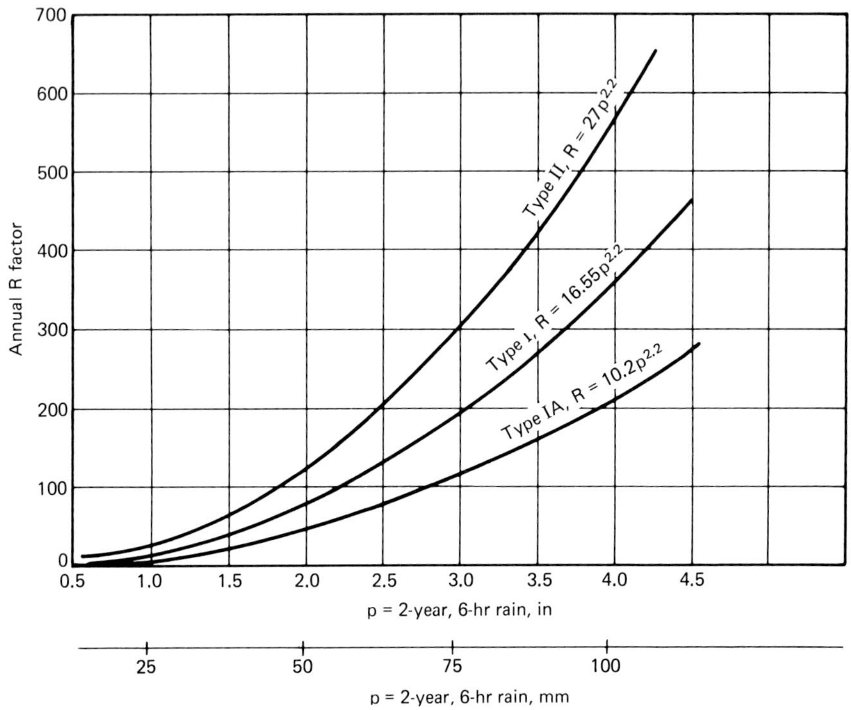

Fig. 5.5 Relations between average annual erosion index and 2-year, 6-hr rainfall in California. (14)

The differences in peak intensity are reflected in the coefficients of the equations for the rainfall factor. Figure 5.5 is a graphical representation of the equations. The equations, also shown on the curves for each individual storm type, are:

$$R = 27p^{2.2} \qquad \text{type II}$$
$$R = 16.55p^{2.2} \qquad \text{type I}$$
$$R = 10.2p^{2.2} \qquad \text{type IA}$$

where p is the 2-year, 6-hr rainfall in inches. (If p is in millimeters, the equations become: $R = 0.0219p^{2.2}$, type II; $R = 0.0134p^{2.2}$, type I; $R = 0.00828p^{2.2}$, type IA.)

The R value is rounded to the nearest whole number. When the rainfall time distribution curves (Fig. 5.4) and the corresponding R value equations are compared, it is evident that the stronger the peak intensity of the typical storm, the higher the rainfall erosion index.

EXAMPLE 5.1

Find: The average annual R value for Sacramento, California.

Given: The 2-year, 6-hr rainfall is 1.2 in (30.5 mm).

Solution: Sacramento is in the type I storm area. Thus

$$
\begin{aligned}
R &= 16.55p^{2.2} \qquad [0.0134 \times (p, \text{ in mm})^{2.2}] \\
\text{where } p &= 1.2 \text{ in } (30.5 \text{ mm}) \\
R &= 24.72, \text{ or } 25
\end{aligned}
$$

The rainfall erosion index does not account for erosion caused by snowmelt runoff. In any area where snow accumulates and the soil freezes, snowmelt runoff increases erosion losses. Until researchers develop a predictive method for this type of erosion, an addition component of the R value, termed R_s, should be added to the rainfall erosion index to determine a total R factor R_t. R_s is estimated by multiplying the average total winter precipitation (December through March) in inches (mm/25.4) of water by 1.5 [(mm/25.4) × 1.5 = 0.059 × mm].

EXAMPLE 5.2 Consider a site that has an R factor of 25 and receives 16 in (406 mm) of precipitation during the four winter months:

$$
\begin{aligned}
R_s &= 1.5(16 \text{ in}) = 24 \qquad [0.059(406 \text{ mm}) = 24] \\
R_t &= R + R_s \\
&= 25 + 24 \\
&= 49
\end{aligned}
$$

The R value is used to estimate the average annual soil loss. If erosion protection is required for less than one year, the soil loss for a portion of a year can be estimated by using a derivative of the R value. Since R is proportional to rainfall, the R value for a short time period can be calculated by multiplying the average rainfall during the shorter time period by the annual R value and dividing the product by the average annual rainfall. For example, suppose you wish to estimate soil loss in January. January rainfall averages 2 in (51 mm), and annual rainfall averages 20 in (510 mm). Then

$$
R_{\text{Jan.}} = \frac{2 \text{ in}}{20 \text{ in}} \times R_{\text{annual}} \qquad \left(\frac{51 \text{ mm}}{510 \text{ mm}} \times R_{\text{annual}}\right)
$$

EXAMPLE 5.3

Given: A site in California on the western slope of the Sierra Nevada where 2-year, 6-hr rainfall is 1.6 in (41 mm), December–March precipitation is 27.6 in (701 mm), and the storm type is IA.

Find: R, R_s, and R_t.

Solution: For type IA storm area,

$$R = 10.2p^{2.2} = 28.7$$
$$R_s = 1.5(27.6 \text{ in}) = 41.4 \qquad [0.059(701 \text{ mm}) = 41.4]$$
$$R_t = R_s + R = 28.7 + 41.4 = 70.1$$

5.2c Soil Erodibility Factor *K*

The soil erodibility factor K is a measure of the susceptibility of soil particles to detachment and transport by rainfall and runoff. Texture is the principal factor affecting K, but structure, organic matter, and permeability also contribute. K values range from 0.02 to 0.69.

Several methods can be used to estimate a K value for a site, but a nomograph method using analyses of site soils is the most reliable. If a recent soil survey for the area has been published and minimal soil disturbance is anticipated, the K value listed in the survey of the soil series found on the site can be used.

Nomograph Method

The preferred method for determining K values is the nomograph method. Use of the nomograph requires a particle size analysis to determine the percentages of sand, very fine sand, silt, and clay. The size range for each class is listed in Table 5.1. ASTM D-422 (1) is a standard hydrometer analysis for particle size distribution. (Specific particle sizes can be designated in the request for analysis. More typically, values are reported for specified size intervals, such as every 5 or 10 μm. The fee for a particle size analysis is normally only a small fraction of the total fee for a geotechnical report.)

The determination of the K value should be based on the soil exposed during the critical rainfall months. Subsoils exposed during grading will have K values different from the topsoil K value. On large sites, several samples should be taken and analyzed separately to ensure that differences in soil texture are detected. If fill is imported, this material also should be characterized.

The more carefully the site soils are characterized, the more accurate the K values will be. If analysis indicates significant variation in soil erodibility, it might be advisable to use different K values for different parts of the site and to focus erosion control efforts on the most susceptible areas. A simpler and more conservative approach is to use the highest value obtained by analysis for all parts of the site, since it may not be possible to know exactly what soils will be exposed or how varied the soils are.

A nomograph developed by Erickson of the SCS-Utah office (6), based on the original nomograph provided by Wischmeier (21), is reproduced in Fig. 5.6. To use the nomograph, enter the triangle with any two of the particle size percents: total sand and silt; silt and clay; or clay and total sand. Use whole numbers. Follow the dashed straight lines to their point of intersection. From that point, follow parallel to the dotted curves to the right side of the triangle, where the K values are listed.

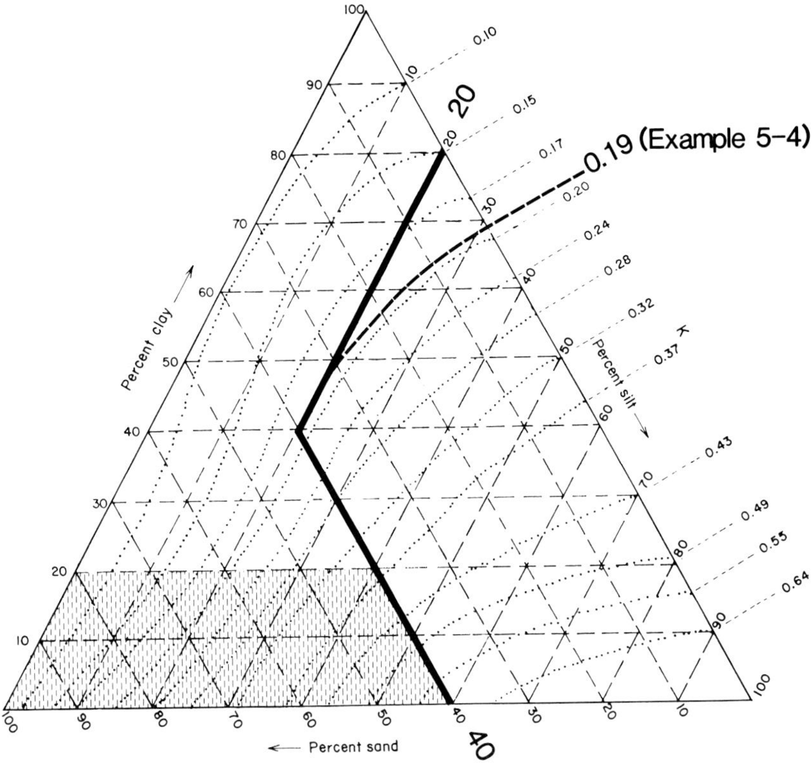

Fig. 5.6 Triangular nomograph for estimating K value. (6) See Table 5.3 for adjustments to K value under certain conditions.

EXAMPLE 5.4

Given: A soil with the following particle size distribution.

Component	Size, mm	Fraction, %
Sand	2.0–0.1	30
Very fine sand	0.1–0.05	10
Silt	0.05–0.002	20
Clay	Less than 0.002	40

Find: Texture and K value.

Solution: Entering Fig. 5.1 with 40 percent total sand and 20 percent silt, the texture is found to be on the border between clay and clay loam. Entering Fig. 5.6 with the same percents (see bold lines), the K value is found to be 0.19.

Table 5.3 describes adjustments to the K factor. Adjustment 1 is a correction for very

TABLE 5.3 Adjustments to K Value (6)

1. For soils with greater than 15% *very fine sand* (vfs) make the following adjustment:
   a. If texture is *coarser than loam* (shaded area in Fig. 5.6): Subtract 5% from the % vfs and add the difference to the silt content. Consider the remaining 5% vfs to be part of the % total sand.
   b. If texture is *loam and finer* (outside shaded area on Fig. 5.6), subtract 10% from the % vfs and add the difference to the silt content. Consider the remaining 10% vfs to be part of the % total sand.
   c. Find the K value by using the adjusted sand and silt contents.

2. The nomograph assumes 2% organic matter, structure other than granular, and 0–15% rock content. The correction factors are as follows.
   a. Organic matter: Add or subtract correction factor to K value as indicated in the following table.

K value	Correction factor when percent organic matter is				
	0	1	2	3	4
Greater than 0.40	+0.14	+0.07	0	−0.07	−0.14
0.20–0.40	+0.10	+0.05	0	−0.05	−0.10
Less than 0.20	+0.06	+0.03	0	−0.03	−0.06

   b. Rock content: Rock content is defined as the percent (by volume) of soil particles greater than 2 mm.

Unadjusted K value from Fig. 5.6	K values adjusted for rock content as follows		
	15–35%	35–60%	60–75%
0.10	0.05	0.05	0.02
0.15	0.10	0.05	0.02
0.17	0.10	0.05	0.02
0.20	0.10	0.05	0.02
0.24	0.15	0.10	0.05
0.28	0.15	0.10	0.05
0.32	0.17	0.10	0.05
0.37	0.20	0.10	0.05
0.43	0.24	0.15	0.10
0.49	0.28	0.15	0.10
0.55	0.32	0.17	0.10
0.64	0.37	0.20	0.15

   Add or subtract the correction factors at the right to the K value to correct for the following structures and permeabilities.

   c. Structure:

Very fine granular	−0.09
Fine granular	−0.06
Moderate or coarse granular	−0.03

   d. Permeability:

Compact soil or pH greater than 9.0	+0.03
Many medium or coarse pores	−0.03

fine sand content. Very fine sands (0.05 to 0.1 mm) are as easily eroded as silts; so if their content exceeds 15 percent, an adjustment to the silt content is made.

Adjustment 2 addresses situations in which soil conditions differ from those used to construct the nomograph. The nomograph assumes 2 percent organic matter, low rock content, and structure other than granular (e.g., undisturbed). Construction sites soils often do vary from those conditions. Organic matter content will be practically zero on a site where grading activities expose the subsoil. The adjustment for zero organic matter ranges from +0.06 to +0.14, depending on the *K* value (adjustment 2*a*, Table 5.3).

Adjustment 2*b* in Table 5.3 is a correction for rock content greater than 15 percent. For this correction, rock content is the percent of soil larger than 2 mm in diameter. (The particle size analysis should include this information.) Generally, the *K* value decreases with greater rock content.

The corrections for structure (2*c*) and permeability (2*d*) are relatively minor—less than 0.1. Since there are no standard laboratory procedures for measuring these criteria, and since the *K* value is based on estimates of site conditions, these corrections may be omitted.

EXAMPLE 5.5

Given: A soil with the following particle size distribution.

Particle	Size, mm	Fraction, %
Sand	2.0–0.1	20
Very fine sand	0.1–0.05	25
Silt	0.05–0.002	20
Clay	Less than 0.002	35

Find: *K* value adjusted for very fine sand content. Compare with *K* found without adjustment.

Solution: Using Fig. 5.6, the intersection of the 45 percent total sand line and 20 percent silt line lies outside the shaded area. Since the very fine sand content is greater than 15 percent, follow adjustment 1*b*: subtract 10 percent from very fine sand and add 15 percent (the difference) to silt. Adjusted particle size contents are:

Particle	Fraction, %
Total sand	20 + 10 = 30
Adjusted silt	20 + 15 = 35
Clay (unchanged)	35

Entering Fig. 5.6 with 30 percent sand and 35 percent silt, *K* is 0.30. Without the adjustment, *K* would have been 0.19.

Soil Survey Method

County soil surveys, published by the Soil Conservation Service, contain soil maps superimposed on aerial photographs. The maps permit easy location of sites and tentative determination of soil series. Recent surveys list the *K* value

TABLE 5.4 Example of SCS Table Listing K Values by Soil Series (17)
Alameda County, California, Western Part

Soil name and map symbol	Depth, in	Permeability, in/hr	Available water capacity, in/in	Soil reaction, pH	Salinity, mmhos/cm	Shrink-swell potential	Erosion factors	
							K	T
122, 123, 124								
Los Osos	0–8	0.2–0.6	0.17–0.19	5.6–6.5	<2	Moderate	0.32	2
	8–30	0.06–0.2	0.12–0.16	5.6–7.3	<2	High	0.28	
Millsholm	0–20	0.6–2.0	0.14–0.17	5.6–6.5	<2	Low	0.43	1
125, Marvin	0–4	0.06–0.2	0.15–0.17	6.6–7.8	<4	Moderate	0.43	5
	4–36	0.06–0.2	0.15–0.17	7.9–8.4	4–8	High	0.24	
	36–60	0.06–0.2	0.19–0.21	7.9–8.4	4–8	High	0.37	
126, Maymen	0–19	0.6–2.0	0.12–0.14	4.5–6.5	<2	Low	0.17	1
127, Maymen	0–19	0.6–2.0	0.12–0.14	4.5–6.5	<2	Low	0.17	1
Los Gatos	0–19	0.6–2.0	0.15–0.20	5.6–7.3	<2	Moderate	0.32	2
	19–40	0.2–0.6	0.14–0.20	5.6–7.3	<2	Moderate	0.37	
128, 129, Millsholm	0–20	0.6–2.0	0.14–0.17	5.6–6.5	<2	Low	0.43	1
130, Montara	0–14	0.2–0.6	0.17–0.02	6.6–8.4	<2	Moderate	0.32	1
131, Omni	0–6	0.2–0.6	0.14–0.19	7.4–8.4	<2	Moderate	0.43	5
	6–52	0.06–0.2	0.12–0.16	7.9–8.4	<2	High	0.37	
	52–60	0.06–0.2	0.12–0.18	7.9–8.4	<2	High	0.37	

for the soil series in a table of physical and chemical properties. A portion of one of these tables is reproduced as Table 5.4. This method of finding K should be used only if minimal soil disturbance is anticipated and a site analysis of soils is not available.

5.2d Length-Slope Factor LS

The slope length-gradient factor LS describes the combined effect of slope length and slope gradient. It is the ratio of soil loss per unit area on a site to the corresponding loss from a 72.6-ft- (22.1-m-) long experimental plot with a 9 percent slope.

Table 5.5 lists values of the LS factor for slopes 0.5 to 100 percent and lengths 10 to 1000 ft (3 to 305 m). The values are derived from Wischmeier's empirical equation (21), which is shown in the table. Israelson (10) has verified the LS values for steep slopes by using a rainfall simulator.

Determining LS

Since the LS factor has a considerable effect on predicted erosion (except for R, it is the only factor substantially greater than 1), care in figuring values for the factor is warranted. In particular, results of the soil loss calculation will be more accurate if the USLE is individually applied to portions of a site with similar slopes (similar gradient and length) and summing the individual soil loss estimates. Do not average slope steepness and length for an entire site. If an estimate for complex slopes is desired, use the method described by Foster and Wischmeier (8) or Israelson. (10)

Slope gradient can be expressed in either percent or as a ratio of horizontal to vertical height. Table 5.4 lists slope gradients both ways.

Slope length is the distance of overland flow to the nearest diversion or channel. For a long slope with several midslope diversions, use the slope length equal to the distance from the top of the slope to the first bench or the distance between benches, whichever is greater. To use Table 5.5, start with slope gradient and then move across the row to the right until you reach the column for the appropriate slope length. For example, a 10 percent slope that is 100 ft (30.5 m) long has an LS value of 1.37.

Discussion of LS

Slope gradient and slope length strongly influence the transport of soil particles once the particles are dislodged by raindrop impact or by runoff. In flat areas, runoff is slow and soil particles are not moved far from the point of raindrop impact. Thus, LS is small—less than 1.00 for slopes less than 9 percent with lengths less than 70 ft (21 m). On steep slopes, soil movement increases dramatically. For example, on a 100-ft- (30.5-m-) long slope, doubling the gradient from 3:1 to 1.5:1 triples the LS factor (i.e., triples the soil loss):

TABLE 5.5 LS Values* (10)

Slope ratio	Slope gradient s, %	10 (3.0)	20 (6.1)	30 (9.1)	40 (12.2)	50 (15.2)	60 (18.3)	70 (21.3)	80 (24.4)	90 (27.4)	100 (30.5)
		LS values for following slope lengths l, ft (m)									
	0.5	0.06	0.07	0.07	0.08	0.08	0.09	0.09	0.09	0.09	0.10
100:1	1	0.08	0.09	0.10	0.10	0.11	0.11	0.12	0.12	0.12	0.12
	2	0.10	0.12	0.14	0.15	0.16	0.17	0.18	0.19	0.19	0.20
	3	0.14	0.18	0.20	0.22	0.23	0.25	0.26	0.27	0.28	0.29
	4	0.16	0.21	0.25	0.28	0.30	0.33	0.35	0.37	0.38	0.40
20:1	5	0.17	0.24	0.29	0.34	0.38	0.41	0.45	0.48	0.51	0.53
	6	0.21	0.30	0.37	0.43	0.48	0.52	0.56	0.60	0.64	0.67
	7	0.26	0.37	0.45	0.52	0.58	0.64	0.69	0.74	0.78	0.82
12½:1	8	0.31	0.44	0.54	0.63	0.70	0.77	0.83	0.89	0.94	0.99
	9	0.37	0.52	0.64	0.74	0.83	0.91	0.98	1.05	1.11	1.17
10:1	10	0.43	0.61	0.75	0.87	0.97	1.06	1.15	1.22	1.30	1.37
	11	0.50	0.71	0.86	1.00	1.12	1.22	1.32	1.41	1.50	1.58
8:1	12.5	0.61	0.86	1.05	1.22	1.36	1.49	1.61	1.72	1.82	1.92
	15	0.81	1.14	1.40	1.62	1.81	1.98	2.14	2.29	2.43	2.56
6:1	16.7	0.96	1.36	1.67	1.92	2.15	2.36	2.54	2.72	2.88	3.04
5:1	20	1.29	1.82	2.23	2.58	2.88	3.16	3.41	3.65	3.87	4.08
4½:1	22	1.51	2.13	2.61	3.02	3.37	3.69	3.99	4.27	4.53	4.77
4:1	25	1.86	2.63	3.23	3.73	4.16	4.56	4.93	5.27	5.59	5.89
	30	2.51	3.56	4.36	5.03	5.62	6.16	6.65	7.11	7.54	7.95
3:1	33.3	2.98	4.22	5.17	5.96	6.67	7.30	7.89	8.43	8.95	9.43
	35	3.23	4.57	5.60	6.46	7.23	7.92	8.55	9.14	9.70	10.22
2½:1	40	4.00	5.66	6.93	8.00	8.95	9.80	10.59	11.32	12.00	12.65
	45	4.81	6.80	8.33	9.61	10.75	11.77	12.72	13.60	14.42	15.20
2:1	50	5.64	7.97	9.76	11.27	12.60	13.81	14.91	15.94	16.91	17.82
	55	6.48	9.16	11.22	12.96	14.48	15.87	17.14	18.32	19.43	20.48
1¾:1	57	6.82	9.64	11.80	13.63	15.24	16.69	18.03	19.28	20.45	21.55
	60	7.32	10.35	12.68	14.64	16.37	17.93	19.37	20.71	21.96	23.15
1½:1	66.7	8.44	11.93	14.61	16.88	18.87	20.67	22.32	23.87	25.31	26.68
	70	8.98	12.70	15.55	17.96	20.08	21.99	23.75	25.39	26.93	28.39
	75	9.78	13.83	16.94	19.56	21.87	23.95	25.87	27.66	29.34	30.92
1¼:1	80	10.55	14.93	18.28	21.11	23.60	25.85	27.93	29.85	31.66	33.38
	85	11.30	15.98	19.58	22.61	25.27	27.69	29.90	31.97	33.91	35.74
	90	12.02	17.00	20.82	24.04	26.88	29.44	31.80	34.00	36.06	38.01
	95	12.71	17.97	22.01	25.41	28.41	31.12	33.62	35.94	38.12	40.18
1:1	100	13.36	18.89	23.14	26.72	29.87	32.72	35.34	37.78	40.08	42.24

*Calculated from

$$\mathrm{LS} = \left(\frac{65.41 \times s^2}{s^2 + 10{,}000} + \frac{4.56 \times s}{\sqrt{s^2 + 10{,}000}} + 0.065\right)\left(\frac{l}{72.5}\right)^m$$

LS = topographic factor
l = slope length, ft (m × 0.3048)
s = slope steepness,
m = exponent dependent upon slope steepne
(0.2 for slopes $< 1\%$, 0.3 for slopes 1 to
0.4 for slopes 3.5 to 4.5%, and
0.5 for slopes $> 5\%$)

LS values for following slope lengths l, ft (m)												
150 (46)	200 (61)	250 (76)	300 (91)	350 (107)	400 (122)	450 (137)	500 (152)	600 (183)	700 (213)	800 (244)	900 (274)	1000 (305)
0.10	0.11	0.11	0.12	0.12	0.13	0.13	0.13	0.14	0.14	0.14	0.15	0.15
0.14	0.14	0.15	0.16	0.16	0.16	0.17	0.17	0.18	0.18	0.19	0.19	0.20
0.23	0.25	0.26	0.28	0.29	0.30	0.32	0.33	0.34	0.36	0.37	0.39	0.40
0.32	0.35	0.38	0.40	0.42	0.43	0.45	0.46	0.49	0.51	0.54	0.55	0.57
0.47	0.53	0.58	0.62	0.66	0.70	0.73	0.76	0.82	0.87	0.92	0.96	1.00
0.66	0.76	0.85	0.93	1.00	1.07	1.13	1.20	1.31	1.42	1.51	1.60	1.69
0.82	0.95	1.06	1.16	1.26	1.34	1.43	1.50	1.65	1.78	1.90	2.02	2.13
1.01	1.17	1.30	1.43	1.54	1.65	1.75	1.84	2.02	2.18	2.33	2.47	2.61
1.21	1.40	1.57	1.72	1.85	1.98	2.10	2.22	2.43	2.62	2.80	2.97	3.13
1.44	1.66	1.85	2.03	2.19	2.35	2.49	2.62	2.87	3.10	3.32	3.52	3.71
1.68	1.94	2.16	2.37	2.56	2.74	2.90	3.06	3.35	3.62	3.87	4.11	4.33
1.93	2.23	2.50	2.74	2.95	3.16	3.35	3.53	3.87	4.18	4.47	4.74	4.99
2.35	2.72	3.04	3.33	3.59	3.84	4.08	4.30	4.71	5.08	5.43	5.76	6.08
3.13	3.62	4.05	4.43	4.79	5.12	5.43	5.72	6.27	6.77	7.24	7.68	8.09
3.72	4.30	4.81	5.27	5.69	6.08	6.45	6.80	7.45	8.04	8.60	9.12	9.62
5.00	5.77	6.45	7.06	7.63	8.16	8.65	9.12	9.99	10.79	11.54	12.24	12.90
5.84	6.75	7.54	8.26	8.92	9.54	10.12	10.67	11.68	12.62	13.49	14.31	15.08
7.21	8.33	9.31	10.20	11.02	11.78	12.49	13.17	14.43	15.58	16.66	17.67	18.63
9.74	11.25	12.57	13.77	14.88	15.91	16.87	17.78	19.48	21.04	22.49	23.86	25.15
11.55	13.34	14.91	16.33	17.64	18.86	20.00	21.09	23.10	24.95	26.67	28.29	29.82
12.52	14.46	16.16	17.70	19.12	20.44	21.68	22.86	25.04	27.04	28.91	30.67	32.32
15.50	17.89	20.01	21.91	23.67	25.30	26.84	28.29	30.99	33.48	35.79	37.96	40.01
18.62	21.50	24.03	26.33	28.44	30.40	32.24	33.99	37.23	40.22	42.99	45.60	48.07
21.83	25.21	28.18	30.87	33.34	35.65	37.81	39.85	43.66	47.16	50.41	53.47	56.36
25.09	28.97	32.39	35.48	38.32	40.97	43.45	45.80	50.18	54.20	57.94	61.45	64.78
26.40	30.48	34.08	37.33	40.32	43.10	45.72	48.19	52.79	57.02	60.96	64.66	68.15
28.35	32.74	36.60	40.10	43.31	46.30	49.11	51.77	56.71	61.25	65.48	69.45	73.21
32.68	37.74	42.19	46.22	49.92	53.37	56.60	59.66	65.36	70.60	75.47	80.05	84.38
34.77	40.15	44.89	49.17	53.11	56.78	60.23	63.48	69.54	75.12	80.30	85.17	89.78
37.87	43.73	48.89	53.56	57.85	61.85	65.60	69.15	75.75	81.82	87.46	92.77	97.79
40.88	47.20	52.77	57.81	62.44	66.75	70.80	74.63	81.76	88.31	94.41	100.13	105.55
43.78	50.55	56.51	61.91	66.87	71.48	75.82	79.92	87.55	94.57	101.09	107.23	113.03
46.55	53.76	60.10	65.84	71.11	76.02	80.63	84.99	93.11	100.57	107.51	114.03	120.20
49.21	56.82	63.53	69.59	75.17	80.36	85.23	89.84	98.42	106.30	113.64	120.54	127.06
51.74	59.74	66.79	73.17	79.03	84.49	89.61	94.46	103.48	111.77	119.48	126.73	133.59

Length	100 ft (30.5 m)	100 ft (30.5 m)	100 ft (30.5 m)
Slope	3:1	2:1	1.5:1
LS	9.43	17.82	26.68
Factor increase	1	1.9	2.8

The effect of length is not as great as the effect of slope angle: LS increases 30 to 50 percent for each doubling of length. For example, on a 2:1 slope, LS doubles when L is quadrupled:

Slope	2:1	2:1	2:1
Length	30 ft (9.1 m)	60 ft (18.3 m)	120 ft (36.6 m)
LS	9.76	13.81	19.42
Factor increase	1	1.4	2

Thus, very long slopes and especially, long, steep slopes, should not be constructed. Those that already exist should not be disturbed.

Slope length can be shortened by installing midslope diversions. Local building codes often require terraces or drainage ditches at specified intervals. Chapter 70 of the *Uniform Building Code* specifies a 30-ft (9.1-m) interval. (9) Several erosion control manuals recommend 15-ft (4.6-m) intervals between terraces. (2, 18). Because these intervals are defined as vertical rise, the slope length would be somewhat longer.

Decreasing steepness will require use of more land and so must be incorporated early in the project design. To ensure slope stability, a maximum gradient is frequently recommended by the soils engineer.

5.2e Cover Factor *C*

The cover factor C is defined as the ratio of soil loss from land under specified crop or mulch conditions to the corresponding loss from tilled, bare soil. The C is *not* the same as the runoff coefficient C used in the rational method.

In the USLE, the C factor reduces the soil loss estimate according to the effectiveness of vegetation and mulch at preventing detachment and transport of soil particles. On construction sites, recommended control practices include the seeding of grasses and the use of mulches. These measures are often considered "temporary"—they are designed to control erosion primarily during the construction period. Permanent landscaping may be added later, or temporary erosion control plants may be left as a permanent cover. Any product that reduces the amount of soil exposed to raindrop impact will reduce erosion. Table 5.6 lists C factors for various ground covers. The C values for vegetation were obtained from USDA publications (14, 20); those for mulch were obtained from Burgess Kay at the University of California, Davis, who tested materials on experimental plots under a rainfall simulator. (11)

When the soil surface is bare, C is 1.0. At the other end of the scale, undisturbed native vegetation is assigned a value of 0.01; hence the advantage of retaining as much existing vegetation as possible is clear. A C value of 0.1 is used

TABLE 5.6 *C* Values for Soil Loss Equation*

Type of cover	*C* factor	Soil loss reduction, %
None	1.0	0
Native vegetation (undisturbed)	0.01	99
Temporary seedings:		
90% cover, annual grasses, no mulch	0.1	90
Wood fiber mulch, ¾ ton/acre (1.7 t/ha), with seed†	0.5	50
Excelsior mat, jute†	0.3	70
Straw mulch†		
1.5 tons/acre (3.4 t/ha), tacked down	0.2	80
4 tons/acre (9.0 t/ha), tacked down	0.05	95

*Adapted from Refs. 11, 15, and 20

†For slopes up to 2:1.

if a complete cover of newly seeded annual grasses is well established before the onset of rains.

In many areas, seed and wood fiber mulch are applied hydraulically shortly before the rainy season. The early rains cause the seeds to germinate, but a complete grass cover is not established until at least 4 weeks later. During the germination and early growth period, the wood fiber mulch provides only marginal protection. A *C* value of 0.5 is an appropriate average representing little protection initially and more thorough protection when the grass is well established.

On bare soils mulch can provide immediate reduction in soil loss, and it performs better than temporary seedings in some cases. Straw mulch is more effective than wood fiber mulch; it reduces loss about 80 percent (*C* value, 0.2) when it is applied at the rate of 3000 lb/acre (3.4 t/ha) and tacked down. Additional reduction is obtained with 8000 lb/acre (90 t/ha) of straw, but this rate may not be cost-effective.

Wood fiber mulch alone (without seed) provides very little soil loss reduction; it primarily helps seeds to become established so that the new grass can provide the erosion control. Other products, such as jute, excelsior, and paper matting, provide an intermediate level of protection; the *C* value equals approximately 0.3. Test results of various mulch treatments are presented in Chap. 6.

5.2f Erosion Control Practice Factor *P*

The erosion control practice factor *P* is defined as the ratio of soil loss with a given surface condition to soil loss with up-and-down-hill plowing. Practices that reduce the velocity of runoff and the tendency of runoff to flow directly downslope reduce the *P* factor. In agricultural uses of the USLE, *P* is used to describe plowing and tillage practices. In construction site applications, *P* reflects the roughening of the soil surface by tractor treads or by rough grading, raking, or disking.

TABLE 5.7 *P* Factors for Construction Sites (Adapted from Ref. 15)

Surface condition	*P* value
Compacted and smooth	1.3
Trackwalked along contour*	1.2
Trackwalked up and down slope†	0.9
Punched straw	0.9
Rough, irregular cut	0.9
Loose to 12-in (30-cm) depth	0.8

*Tread marks oriented up and down slope.

†Tread marks oriented parallel to contours, as in Figs. 6.9 and 6.10.

P values appropriate for construction sites are listed in Table 5.7.

- A surface that is compacted and smoothed by grading equipment is highly susceptible to sheet runoff and is assigned a *P* value of 1.3.
- Trackwalking is given a value of 1.2 if the vehicle traverses along the contour. The *P* value is relatively high because the depressions left by cross-slope tracking resemble up-and-down furrows and worsen runoff conditions.
- Trackwalking up and down slope reduces *P* to 0.9. The tread marks act as slope benches; they reduce runoff velocity and trap soil particles (see Fig. 6.10).
- Punched straw is assigned a *P* value of 0.9 because the action of punching the straw into the soil roughens the surface and creates a trackwalking effect.
- When the soil surface is disked or otherwise loosened to a depth of 1 ft, a slightly lower *P* value of 0.8 may be used. This condition is unlikely to occur on a construction site because compaction, not loosening, is required when fill slopes are constructed.

Clearly, changing the surface condition does not provide much direct reduction in soil loss; all the *P* values are close to 1.0. However, roughening the soil surface is essential before seeding because it greatly increases plant establishment (see Chap. 6) and thus also reduces the *C* factor. Vegetation, mulch, slope length, and gradient have far more significant effects on the erosion process and provide greater opportunities to reduce soil loss.

5.2g Combined Effects of LS, *C*, and *P*

Of the five factors in the USLE, the *R*, LS, and *C* factors have the widest range. Although *R* for a site is constant and *K* is essentially a constant, slope length and gradient, cover, and, to a limited extent, surface condition can be manipulated. Slope length and vegetative cover are the most effective and easily implemented measures.

Table 5.8 compares the effect on the soil loss estimates of varying LS, *C*, and *P*. For example, a building pad with a 1 percent slope, smooth surface, and no cover has a fractional soil loss potential. A 2:1 slope, common between terraced

TABLE 5.8 Combined Effects of LS, C, and P on Soil Loss

Slope gradient	Slope length	Surface condition	Ground cover	LS	C	P	Product LS × C × P	Reduction, %[a]
2:1	40 ft (12.2 m)	Smooth	Bare	11.27	1.0	1.3	14.65	—
2:1	40 ft (12.2 m)	Tracked; tread marks oriented up and down slope	Seed and wood fiber mulch, ¾ ton/acre (1.7 t/ha)	11.27	0.5	1.2	6.76	54
2:1	40 ft (12.2 m)	Tracked; tread marks parallel to contours	Seed and wood fiber mulch, ¾ ton/acre (1.7 t/ha)	11.27	0.5	0.9	5.07	65
2:1	20 ft (6.1 m)	Tracked; tread marks parallel to contours	Seed and wood fiber mulch, ¾ ton/acre (1.7 t/ha)	7.97	0.5	0.9	3.95	73
2:1	40 ft (12.2 m)	Tracked; tread marks parallel to contours	Punched straw, 1.5 tons/acre (3.4 t/ha)	11.27	0.2	0.9	2.03	86
2:1	20 ft (6.1 m)	Tracked; tread marks parallel to contours	Punched straw, 1.5 tons/acre (3.4 t/ha)	7.97	0.2	0.9	1.43	90
1%	100 ft (30.5 m)	Smooth	Bare	0.12	1.0	1.3	0.16	—

[a]Compared to the example on the first line (LS × C × P = 14.65).

housepads on hillside lots, with a length of 40 ft (12.2 m) and without cover has an LS $\times$ C $\times$ P product of 14.65. Table 5.8 compares other conditions by using this example as a reference. Hydraulic seeding and mulching reduce the erosion potential 54 percent if trackwalked across the slope (tread marks pointing up and down the slope) and 65 percent if trackwalked perpendicular to the slope (tread marks parallel to contours).* Cutting slope length in half by installing a midslope diversion produces an additional 8 percent reduction to 73 percent lower erosion potential.

A midslope diversion has little effect on soil loss from the slope if punched straw is used because straw reduces the C value dramatically. At a rate of 1.5 tons/acre (3.4 t/ha) of straw, the soil loss product is reduced 86 percent for a 40-ft (12.2-m) slope length and 90 percent for a 20-ft (6.1-m) slope length.

5.2h Limitations of the USLE

Procedures for assigning numerical values to the five factors in the USLE have been discussed. Care should be taken in choosing those values, since significantly different soil loss estimates will be obtained. The R and K factors are constants at most sites, but the C and LS factors vary substantially with the control measures used. C can vary from almost 0 to 1; LS ranges from almost 0 to more than 100. There are limitations to the USLE that should be recognized, but they do not reduce the USLE's usefulness for estimating soil loss from temporarily disturbed areas such as construction sites. The following observations are pertinent to USLE use.

1. *It is empirical.* The equation is empirical and does not mathematically represent the actual erosion process. Predictive errors can be overcome by the use of empirical coefficients. Several authors (7, 12) have proposed additional factors, exponents, and other ways to calculate the R factor. However, the process of fitting the equation to a particular set of data makes the equation less general and more difficult to apply.
2. *It predicts average annual soil loss.* The equation was developed to predict soil loss on an average annual basis. Unusual rainfall seasons, especially higher than normal rainfall and atypically heavy storms, may produce more sediment than estimated. Sediment basins whose storage volume is based on a USLE estimate of soil loss should be inspected after each storm to ensure that adequate storage volume remains.
3. *It estimates sheet and rill erosion, NOT gully erosion.* The USLE produces an estimate of sheet and rill erosion. Gully erosion, caused by concentrated flows of water, is not accounted for by the equation and yet can produce large volumes of eroded soil. Major causes of gully erosion include:

*Trackwalking perpendicular to the slope also increases the rate of plant establishment (see Figs. 6.10 and 6.11). When this combined effect is considered, the soil loss reduction for this condition will probably be greater than 65 percent.

- Improperly designed or installed channels. Channels must be designed and installed to assure adequate capacity and nonerosive velocities in unlined sections.
- Too-long slope lengths. Shortening slope lengths on steep slopes greatly reduces the erosive potential of rainfall.
- Lack of diversion structure at the top of exposed slopes. Runoff from an area above a disturbed slope should never be allowed to cross the disturbed slope.
- Discharge of concentrated runoff onto an undisturbed, vegetated slope. Most slopes, and particularly steep ones, cannot accommodate concentrated flows even if they are completely covered with vegetation.

Prevention of gullies and repair of any gullies that do occur is preferable to trying to accommodate soil loss from a gullied site in a sediment basin.

4. *It does not calculate sediment deposition.* The USLE estimates soil loss, not soil deposition. Generally, sediment deposition at the bottom of a watershed will be less than the total soil loss in that watershed. As soon as runoff from a slope area reaches the toe of the slope or a channel of lesser slope, the larger particles will be deposited. The farther the runoff travels, the smaller the percentage of eroded soil being carried.

Nevertheless, the USLE can be used to calculate the sediment storage volume required in retention structures. Sediment basins on construction sites typically serve small areas of 5 to 100 acres (2 to 40 ha). The runoff has not traveled far, and the basin is intended to serve as the settling area. The USLE soil loss estimate can be used as a conservative measure of potential sediment storage needs. This assumption saves the time that would be required for more detailed soil analysis and sediment transport calculations and leaves a margin of safety for storage of additional material eroded from unexpected sources such as gullies or unlined channels. However, if drainage on a site is improperly controlled and gully erosion is extensive, the USLE will greatly underestimate the sediment storage requirements.

5.3i Using the USLE

Sample Soil Loss Calculation; Step-by-Step Procedure

1. Determine the R factor.
2. Based on soil sample particle size analysis, determine the K value from the nomograph (Fig. 5.6). Repeat if you have more than one soil sample.
3. Divide the site into sections of uniform slope gradient and length. Assign an LS value to each section (Table 5.5).
4. Choose the C value(s) to represent a seasonal average of the effect of mulch and vegetation (Table 5.6).

5. Set the P factor based on the final grading practice applied to the slopes (Table 5.7).
6. Multiply the five factors together to obtain per acre soil loss.
7. Multiply soil loss per acre by the acreage to find the total volume of sediment. If the soil loss prediction shows excessive volume lost from the site, consider (*a*) working only a portion of the site at one time, (*b*) altering the slope length and gradient, or (*c*) increasing mulch application rate or seeding.

EXAMPLE 5.6

Given: A 4-acre (1.6-ha) site on a 400-ft- (122-m-) long, 20 percent, smooth-graded slope. No mulching or seeding is planned, but the slope will be trackwalked by driving a tractor up and down the slope. A sediment trap will be built at the lower edge of the slope. R is 34.

A representative sample of the site soil contains:

Particle and size	Fraction, %
Sand, greater than or equal to 0.1 mm	51.0
Very fine sand, 0.05–0.1 mm	18.6
Silt, 0.002–0.05 mm	20.9
Clay, less than 0.002 mm	9.5

Find: The estimated soil loss from the site in one year.

Solution:

STEP 1. R is given; it is 34.

STEP 2. The texture of the site soil is called sandy loam. Since the very fine sand fraction is more than 15 percent and the texture is coarser than loam, we use adjustment 1*a* (Table 5.3).

Subtract 5 percent from the very fine sand and add it to the 51 percent sand fraction. Add the remaining very fine sand (13.6 percent) to the silt content. We now have the percents of sand, silt, and clay to use in the nomograph (Fig. 5.6):

Particle	Original percent	Percent change	Adjusted percent
Sand	51.0	+5	56.0
Silt	20.9	+13.6	34.5
Very fine sand	18.6	−18.6	0.0

On the nomograph, find 56 percent on the "percent sand" side of the triangle and 35 percent on the "percent silt" side. Two lines drawn parallel to the dashed lines intersect near the base of the triangle. The point of intersection is below the 0.32 curve (dotted line), so we find a K value of 0.34.

STEP 3. Table 5.5 gives an LS value of 8.16 for a 400-ft- (122-m-) long, 20 percent slope.

STEP 4. The C factor for no mulch or ground cover is 1.0 (Table 5.6).

STEP 5. Driving a tractor up and down a slope will create tread marks parallel to the slope contours. The P factor from Table 5.7 is thus 0.9.

STEP 6. Compute the soil loss:

$$\begin{aligned} A &= R \times K \times \text{LS} \times C \times P \\ &= 34(0.34)(8.16)(1.0)(0.9) \\ &= 84.9 \text{ tons/(acre)(year)} \qquad [190.5 \text{ t/(ha)(year)}] \end{aligned}$$

STEP 7. 84.9(4 acres) = 340 tons [(190.5 t/ha)(1.6 ha) = 305 t] of soil loss in one season.

Mulch is a recommended practice for controlling erosion. Determine the soil loss if 1.5 tons/acre (3.4 t/ha) of straw mulch is tacked down on the soil surface. The new C is 0.20.

$$\begin{aligned} A &= R \times K \times \text{LS} \times C \times P \\ &= 34(0.34)(8.16)(0.2)(0.9) \\ &= 17 \text{ tons/(acre) [38 t/(ha)(year) or 68 tons (61 t) lost from the 4-acre (1.6-ha) site} \end{aligned}$$

Mulching with straw is a very effective way to reduce erosion; soil loss is reduced 80 percent.

Estimating Erosion from Undeveloped Areas

The USLE can be used to estimate erosion from undeveloped areas. The East Bay Regional Park District, near Oakland, California, provides a useful example of how the USLE can be used for planning. The district operates a recreational lake, Lake Temescal, in the hills above the city. Because of highway construction and residential development in the Lake Temescal watershed, the lake had to be dredged three times in 15 years at considerable expense. To assess the problem and support the district's recommendation that erosion control be required of any future developments, the district used the USLE to estimate the maximum potential soil loss. (5)

Potential building sites were inventoried to determine the developable acreage. The R value was determined to be 35. The K value was estimated from the recent soil survey of Alameda County. (19) Length and slope were selected for each lot. Cover and surface condition factors were assigned maximum values; in other words, no erosion control and full disturbance were assumed. The soil loss per lot was calculated and summed throughout the entire watershed, giving a total of 137,613 yd^3 (105, 213 m^3). Table 5.9 shows the format used for the soil loss calculations.

If development of the potential building sites were spread over 30 years, about 4600 yd^3 (3500 m^3) of soil per year (137,613 yd^3/30 years or 105,213 m^3/30 years) would be eroded from the watershed. If development were to occur more rapidly, over 20 years, the soil loss rate would be nearly 7000 yd^3 (5350 m^3) per year. These soil loss estimates were then compared to the amounts of sediment dredged from Lake Temescal in past years. Between 1963 and 1978 an average of 5500 yd^3/year (4200 m^3/year) of sediment was removed from the lake. Thus, the actual soil loss rate over a 15-year period was found to be close to the maximum predicted soil loss rate for the next 20 to 30 years. These calculations illustrate that unless erosion caused by development is controlled, costly dredging operations will continue to be necessary to maintain the lake for recreation.

TABLE 5.9 Example: Calculation of Soil Loss from Developable Land in Lake Temescal Watershed (5)

Lot no.	Soil Type[a]	Area, acres	Average slope length, ft: Actual	Average slope length, ft: Rounded[b]	Average slope, %: Actual	Average slope, %: Rounded[b]	R	K	LS	C	P	A, yd^3/(acre)(year)[c]	Soil loss, yd^3/year[d]
1	126	0.90	235	250	43	45	35	0.17	24.03	1.0	1.3	185.9	167.3
2	126	1.37	160	150	66	66.7	35	0.17	32.68	1.0	1.3	252.8	346.3
3	126	2.75	244	250	64	66.7	35	0.17	36.60	1.0	1.3	283.1	778.5
4	126	0.90	122	100	43	45	35	0.17	15.20	1.0	1.3	117.6	105.8
5	126	3.67	141	150	44	45	35	0.17	18.62	1.0	1.3	144.0	528.6
6	126	1.37	160	150	44	45	35	0.17	18.62	1.0	1.3	144.0	197.3
7	126	2.85	119	100	64	66.7	35	0.17	26.68	1.0	1.3	206.4	588.2
8	126	6.88	254	250	48	50	35	0.17	28.18	1.0	1.3	218.0	1,499.6
9	126	2.85	150	150	53	55	35	0.17	25.09	1.0	1.3	194.1	174.7
10	160	0.90	132	150	53	55	35	0.32–0.43	25.09	1.0	1.3	490.9	441.8
11	126	40.40	346	350	46	50	35	0.17	33.34	1.0	1.3	257.9	10,418.6
12	126	90.91	816	800	47	50	35	0.17	50.41	1.0	1.3	390.0	35,447.8
13[e]	126: 80%	73.46 } 91.82	2308	2000	28	30	35	0.17	50.30	1.0	1.3	389.1 } 1373.2	28,583.3 } 46,651.4
	160: 20%	18.36						0.32–0.43	50.30			984.1	18,068.1
14	126: 70%	17.35 } 24.79	458	500	34	33.3	35	0.17	21.09	1.0	1.3	163.1 } 470.1	2,830.3 } 5,115.1
	158: 30%	7.44						0.32				307.1	2,284.8

[a]126: Maymen loam
127: Maymen–Los Gatos Complex
158: Xerorthents, Altamont Complex
160: Xerorthents, Millsholm Complex

[b]Rounded values were used in computing LS.

[c]Soil loss converted from tons to cubic yards by assuming a density of 1 ton/yd^3.

[d]Soil loss equals A multiplied by acreage.

[e]60% of area is designated for a sports complex. This alters the following values:
Area = 55.09 acres
A = 1167.2 yd^3/(acre)(year)
Soil loss = 27,990.8 yd^3/year

Evaluating Erosion Control Measures

Each factor of the USLE either provides a clue to site susceptibility to erosion or suggests a means of effectively reducing erosion. High R and K values point out potentially troublesome sites. Slope length and gradient (LS), cover C, and surface condition P are factors that can be altered, and the LS and C values are most significant. By computing the partial product of LS, C, and P (see Sec. 5.2g), the site planner can compare the effectiveness of different erosion control measures.

REVIEW QUESTIONS

1. What kinds of information about a construction site does a soil loss estimate provide?
2. What is the most important distinction between the USDA and unified soil classification systems?
3. What particle size class is most susceptible to erosion?
4. What particle size class is most difficult to settle out of suspension?
5. Name several advantages and disadvantages of high permeability.
6. Define the terms of the universal soil loss equation.
7. How do you determine the R factor in areas *without* snowmelt runoff?
8. What additional component of the R factor applies to areas *with* snowmelt runoff?
9. What areas require the use of the R_s component?
10. What information about a soil is needed to find a K value by using the nomograph?
11. Give an example of a particle size analysis for which adjustment 1*a* from Table 5.3 must be applied.
12. What portion of the erosion process is reflected in the LS factor?
13. What is the simplest way to reduce LS on a long, steep slope?
14. Discuss ways to reduce the cover factor C and compare the effectiveness of each method.
15. What are the two surface preparation practices that have a P value greater than 1?
16. Discuss the limitations of the USLE and the implications of these limitations for erosion control efforts.
17. ***Given:*** A 15-acre (6-ha) site on a moderately sloping hillside just west of Philadelphia, Pennsylvania, has been terraced for homesites. Soil has been identified as silt loam with the following characteristics:

	Size, mm	%
Total sand	2–0.5	11.3
Very fine sand	0.10–0.05	7.2
Total silt	0.05–0.002	65.9
Fine silt	0.02–0.002	41.7
Clay	Less than 0.002	22.8

(Note *overlap* between total sand and very fine sand size ranges and between total silt and fine silt size ranges.)

Ten acres (4 ha) will be housepads 120 ft (36.6 m) long with 2 percent slope, graded smooth. The remaining 5 acres (2 ha) will be 3:1 slopes, each 20 ft (6.1 m) in length. After rough grading is done, the slopes will be track-walked and hydroseeded with annual ryegrass; the housepads will not be seeded or mulched. Perimeter dikes will divert housepad runoff from the seeded slopes.

Find: Soil loss for one season.

REFERENCES

1. American Society for Testing and Materials, *Annual Book of ASTM Standards,* Philadelphia, 1982.
2. Association of Bay Area Governments, *Manual of Standards for Erosion and Sediment Control Measures,* Oakland, Calif. 1981.
3. C. A. Black, *Soil Plant Relationships,* John Wiley & Sons, Inc., New York, 1968.
4. F. L. Brooks and J. W. Turelle, *Technical Notes,* Conservation Agronomy No. 32, U.S. Department of Agriculture, Soil Conservation Service, West Technical Service Center, Portland, Oreg. 1975.
5. East Bay Regional Park District, *Assessment of Impact of Development of Vacant Land in the Lake Temescal Watershed,* EBRPD, Oakland, Calif. 1981.
6. A. J. Erickson, *Aids for Estimating Soil Erodibility—"K" Value Class and Soil Loss Tolerance,* U.S. Department of Agriculture, Soil Conservation Service, Salt Lake City, Utah, 1977.
7. E. M. Flaxman, "Predicting Sediment Yield in the Western United States," *Journal, Hydraulics Division, ASCE,* vol. 98, 1972, pp. 2073–2085.
8. G. R. Foster and W. H. Wischmeier, "Evaluating Irregular Slopes for Soil Loss Prediction," *American Society of Agricultural Engineers Transactions,* vol. 17, 1974, pp. 305–309.
9. International Conference of Building Officials, *Uniform Building Code,* ICBO, Whittier, Calif., 1982.
10. E. Israelson, *Erosion Control During Highway Construction—Manual on Principles and Practices,* Transportation Research Board, National Cooperative Highway Research Program Report No. 221, TRB, Washington, D.C., 1980.
11. B. L. Kay, "Straw as an Erosion Control Mulch," *Agronomy Progress Report No. 140,* University of California, Davis, Agricultural Experiment Station Cooperative Extension, Davis, Calif. 1983.

12. D. B. Simons, R. M. Li, and T. J. Ward, "Estimation of Sediment Yield for a Proposed Roadway Design," *Proceedings, International Symposium on Urban Stormwater Management,* Lexington, Ky., 1978.

13. F. R. Troeh, J. A. Hobbs, and R. L. Donahue, *Soil and Water Conservation for Productivity and Environmental Protection,* Prentice-Hall, Inc., Englewood Cliffs, N.J., 1980.

14. U.S. Bureau of Plant Industry, Soils and Agricultural Engineering, *Soil Survey Manual,* U.S. Department of Agriculture, Agriculture Handbook No. 18, GPO, Washington, D.C., 1951.

15. U.S. Department of Agriculture, Soil Conservation Service, *Guides for Erosion and Sediment Control in California,* U.S.D.A., SCS, Davis, Calif., 1977.

16. U.S. Department of Agriculture, Soil Conservation Service, *Procedure for Computing Sheet and Rill Erosion on Project Areas,* Technical Release No. 51, U.S.D.A., SCS, Engineering Division, Washington, D.C., 1975.

17. U.S. Department of Agriculture, Soil Conservation Service, *Soil Survey Laboratory Data and Descriptions for Some Soils of California,* Soil Survey Investigations Report No. 24, U.S.D.A., SCS, Washington, D.C., 1973.

18. U.S. Department of Agriculture, Soil Conservation Service, *Standards and Specifications for Erosion and Sediment Control in Developing Areas,* U.S.D.A., SCS, College Park, Md., 1975.

19. L. E. Welch, *Soil Survey of Alameda County, California, Western Part,* U.S. Department of Agriculture, Soil Conservation Service, Washington, D.C., 1981.

20. W. H. Wischmeier and D. D. Smith, *Predicting Rainfall Erosion Losses—A Guide to Conservation Planning,* Agriculture Handbook No. 537, U.S. Department of Agriculture, Science and Education Administration, Washington, D.C., 1978.

21. W. H. Wischmeier and D. D. Smith, *Predicting Rainfall Erosion Losses from Cropland East of the Rocky Mountains,* Agriculture Handbook No. 282, U.S. Department of Agriculture, Washington, D.C., 1965.

chapter 6

Vegetative Soil Stabilization Methods

Vegetation is the most cost-effective form of erosion control. It is also self-healing and attractive. Since vegetation *prevents* erosion, it is a much more desirable control measure than straw bale dikes, silt fences, and sediment traps and basins. Those devices can only trap sediment after soil has already eroded. Only a fraction of the eroded soil can be recaptured on the site, and the land is often left gullied and scarred. Vegetation reduces erosion by:

- Absorbing the impact of raindrops
- Reducing the velocity of runoff
- Reducing runoff volumes by increasing water percolation into the soil
- Binding soil with roots
- Protecting soil from wind

This chapter focuses on methods for quickly revegetating exposed soils. The primary purpose is to prevent *surficial erosion* in the first few years after the original ground cover has been removed. Use of vegetation for protecting man-made waterways is discussed in Sec. 6.6. The use of vegetation for slope stabilization, landslide prevention, and permanent landscaping is beyond the scope of this handbook. A number of good references on those topics are available. (5, 12)

Grass is the most effective type of plant for erosion control in most areas. It germinates and grows quickly and provides a complete ground cover. Once a site is stabilized with grass, other species, such as native shrubs and flowers, may invade the site and replace the grass. This replacement process occurs naturally and typically takes about 5 years.

Planting trees and shrubs is not effective for initial erosion control because not enough of the land surface is protected. In time, however, the deep, strong roots of the woody plants will help to stabilize large soil masses and leaf litter will protect the surface. Flowers* are more effective for erosion control than trees or shrubs but less effective than grass. Besides being slow-growing, many flowers are tap-rooted and do not produce a fibrous root mass. Legumes and other herbaceous plants, however, may be seeded along with grasses to improve soil fertility and to add color to the landscape.

It must be noted, however, that vegetative soil stabilization will be effective only if the site is geologically stable and is engineered properly. Thus, slopes should not be overly steep and diversion devices such as dikes and ditches should be installed to prevent runoff from crossing newly planted slopes.

6.1 PLANTS FOR EROSION CONTROL

Plants for erosion control must be well suited to the local environment. They must be vigorous and troublefree and adapted to the soils and climate of a site. While it is important to select plants that are suited to the local area, it is also desirable to select plants that will survive best under construction site conditions. Plant species that require irrigation, fertilization, high maintenance, and "just right" conditions are not ideal. Easy-to-grow, troublefree plants are best. Developers are more likely to use vegetation for erosion control if they have good success with their first attempts at revegetation.

6.1a Criteria for Plant Selection

Plant selection should be based on the effectiveness of the plants for erosion control and on a number of other criteria:

- *Complete soil protection.* Plants should have dense growth and fibrous roots that provide a *continuous* soil cover with no soil exposed between individual plants.
- *Fast growing.* Plants should be able to germinate and grow rapidly to a size and areal extent that can provide good erosion protection.
- *Easy to plant.* Plants should be suitable for seeding by hand broadcasting (using a hand-held, mechanical broadcaster), by drilling, or by hydraulic jet.

*The term "flowers," as used in this handbook, refers to herbaceous plants other than grasses; examples are poppies, lupines, and daisies. Many flowers are legumes.

- *Adapted to poor soils.* Plants should grow adequately in low-fertility, rocky, acid, or alkaline soils or in soils with poor drainage.
- *Adapted to local environment.* Plants should grow well in the local climate and be competitive with undesirable local plants.
- *Regrowth in subsequent years.* Annual plants should reseed well, and perennials should provide adequate regrowth after dormancy.
- *Available commercially.* Seeds should be readily available from seed supply companies.
- *Low maintenance.* Plants should require little or no irrigation, fertilization (after the first year), or mowing.
- *Low cost.* The cost for seed, application, maintenance, and slope repair should be minimal.
- *Low fire hazard.* Plants should not produce excessive fuel volumes, particularly in dry climates such as in the western United States.
- *Drought tolerance.* Plants should be able to survive dry periods without irrigation, particularly in climates with a dry season.

The erosion control planner should try to select plants that meet as many of the above criteria as possible. The first two, complete soil protection and fast growth, are particularly important for erosion control in the first few years after grading.

A mix of plant species should be used when revegetating. Having a variety of species will increase the chances for success. A single plant type is always more susceptible to failure because of disease, unusual weather, lack of a key nutrient, or some factor peculiar to it.

6.1b Plant Types

Annual Grasses

Many annual grasses, including cereals, meet most of the above erosion control criteria. They are quick to germinate, and they provide the fastest ground cover of any group of plants. They provide a fibrous root mat which protects the entire soil surface. Many species of annual grass are readily available commercially, some at very low cost. Common examples are annual ryegrass *(Lolium multiflorum)*, oats *(Avena sativa)*, and barley *(Hordeum vulgare)*. These species are used for erosion control on both the west and east coasts of the United States. (1, 10, 16) Tables 6.1 and 6.2 provide descriptions of these and other plants that are commonly used for erosion control in California and Virginia.

Perennial Grasses

Perennial grasses develop slower than annuals and require more moisture and soil. Many species require irrigation, particularly in areas with a dry season.

TABLE 6.1 Characteristics of Erosion Control Plants Used in California

Plant species	Height, ft(′) or in (″) (cm)	Erosion rating		Fuel volume	Typical seeding rates[a] lb/acre (kg/ha)	Comments
		First year	Following years			
Annual grasses						
Brome, "Blando"	1–2½′ (30–16)	Exc.	Good	Med.	40–60 (45–67)	Establishes on compact soils better than barley.
Fescue, "Zorro" annual	10–12″ (25–30)	Exc.	Good	Low	10–20 (11–22)	Best adapted to shallow soils.
Ryegrass, Italian or annual	1–3′ (30–91)	Exc.	Poor	Med.	40–60 (45–67)	Much fertilizer required. Turns gray-black and inhibits reseeding of other species.
Ryegrass, "Wimmera 62"	1–3′ (30–91)	Exc.	Poor	Med	40–60 (45–67)	Pros and cons similar to those of annual ryegrass; matures earlier.
Barley	2–3′ (61–91)	Exc.	Poor	Med.	300 (337)	Good for early erosion control, but requires high seeding rate.
Oats	3–4′ (91–122)	Good	Poor	High	300 (337)	Requires high seeding rate; blends with range landscape.

Perennial grasses						
Fescue, creeping red	1–1½′ (30–46)	Good	Good	Low	15–30 (17–34)	Irrigation required.
Fescue, tall	3–6′ (91–183)	Good	Good	Low	15–30 (17–34)	Irrigation required; widely adaptable, grows very tall unless mowed.
Hardinggrass	3–4′ (91–122)	Fair	Good	Low	10–20 (11–22)	Min. 20 in rainfall; bunchy growth can block view.
Orchardgrass, "Berber" or "Palestine"	2–4′ (61–122)	Good	Good	Low	15–30 (17–34)	Min. 20 in rainfall; compatible with wildflowers. "Berber" has better winter growth, better adapted than wheatgrass.
Perlagrass ("Perla" koleagrass)	3–4′ (91–122)	Good	Good	Low	10–20 (11–22)	Min. 20 in rainfall; bunchy growth can block view.
Ryegrass, perennial	1–3′ (30–91)	Good	Poor	Low	15–30 (17–34)	Cannot tolerate drought; may be useful short-term cover.
Wheatgrass, "Luna" pubescent	24–40″ (61–102)	Fair	Good	Low	10–40 (11–45)	Best-performing wheatgrass variety; plant on fill slopes only.

TABLE 6.1 *(Continued)*

Plant species	Height, ft(′) or in (″) (cm)	Erosion rating: First year	Erosion rating: Following years	Fuel volume	Typical seeding rates[a] lb/acre (kg/ha)	Comments
Annual legumes						
Clover, bur	½–1′ (15–30)	Fair	Good	Low	5–20 (6–22)	Basic soils only.
Clover, crimson	1–2′ (30–61)	Fair	Poor	Low	5–20 (6–22)	Compatible with grasses; has colorful red flowers.
Clover, rose	1–1½′ (30–46)	Fair	Good	Low	5–20 (6–22)	Compatible with grasses; has colorful pink flowers.
Subclover	½–1′ (15–30)	Poor	Fair	Low	3–10 (3–11)	Poor grass competitor; requires grazing or mowing to maintain stand.
Vetch, "Lana" woolypod	1–3′ (30–91)	Good	Fair	High	10–75 (11–84)	Vigorous growth but difficult to mow.
Perennial legumes						
Clover, strawberry	1–2′ (30–61)	Poor	Poor	Low	5–20 (6–22)	Irrigation required; better performance in moist climate.
Trefoil, broadleaf	1–2′ (30–61)	Poor	Poor	Low	5–20 (6–22)	Irrigation preferred; does best in wet areas.

TABLE 6.1 *(Continued)*

Plant species	Height, ft(′) or in (″) (cm)	Erosion rating: First year	Erosion rating: Following years	Fuel volume	Typical seeding rates[a] lb/acre (kg/ha)	Comments
Trefoil, narrowleaf	1–2′ (30–61)	Poor	Poor	Low	5–20 (6–22)	Similar to broadleaf; tolerates slightly drier conditions.
			Flowers			
California poppy	1–2′ (30–61)	Fair	Poor	Low	2–10 (2–11)	Calif. native with colorful orange flowers; does not persist well with grasses or where heavily fertilized.
Lupine, valley	1–3′ (30–91)	Poor	Poor	Low	5–15 (6–17)	Calif. native with purple flowers.
Lupine, "Gedling" golden	1–3′ (30–91)				1–5 (1–6)	Calif. native with showy golden flowers.
Lupine, spider	1–3′ (30–91)	Poor	Poor	Low	1–5 (1–6)	Calif. native with light to deep blue flowers.
Lupine, foothill	1–3′ (30–91)	Poor	Poor	Low	1–5 (1–6)	Calif. native with purple flowers.
			Shrubs			
Australian saltbush	1′ (30)	Fair	Good	Low	6–12 (7–13)	Blue-green-colored shrubs; good on fill slopes, competes poorly with grasses.
California buckwheat	1–3′ (30–91)	Poor	Fair	Low	5–10 (6–11)	Calif. native with brown flowers; competes poorly with grasses.

[a]Varies according to number of seeds per pound and germination.

TABLE 6.2 Virginia Specifications for Seeding Temporary Erosion Control Plants (16)

	Seeding rate			North[a]			South[b]			
Species	lb/acre	lb/1000 ft^2	kg/ha	3–1 4–30	5–1 8–15	8–15 11–1	2–15 4–30	5–1 9–1	9–1 11–15	Plant characteristics
OATS *(Avena sativa)*	100	2	112	×	—	—	×	—	—	Use spring oats.
RYE *(Secale cereale)*	170	3	191	×	—	×	×	—	×	Use for fall seedings, winter cover. Tolerates cold and drought.
GERMAN MILLET *(Setaria italica)*	60	1.5	67	—	×	—	—	×	—	Warm season annual. Dies at first frost.
ANNUAL RYEGRASS *(Lolium multiflorum)*	60	1.5	67	×	—	×	×	—	×	Do not use where volunteers would be a problem later.
WEEPING LOVEGRASS *(Eragrostis curvula)*	3	0.06	3	—	×	—	—	×	—	Short-lived perennial, 2–3 years. Tolerates hot, dry slopes and acid, infertile soils.
KOREAN LESPEDEZA[c] *(Lespedeza stipulacea)*	20	0.5	22	×	×	—	×	×	—	Warm season annual legume. Tolerates acid soil.
CRIMSON CLOVER[d] *(Trifolium incarnatum)*	15	0.4	17	—	—	×	—	—	×	Cool season annual legume; begins growth in fall and dies in late spring.

[a] Northern Piedmont and mountain region.

[b] Southern Piedmont and coastal plain.

[c] May be used, with a grass or grain, at half the seeding rate of any spring seeding.

[d] May be used, with a grass or grain, at half the seeding rate of any fall seeding.

× *May* be planted between these dates.

— May *not* be planted between these dates.

Seedlings do not compete well in mixtures with annual grasses. Fire hazard is generally lower because perennials green-up earlier in the fall and stay green longer in the spring than annuals; they shorten the fire season. In areas with summer rainfall, they may be green the entire growing season.

Some commonly used perennial grasses are red fescue *(Festuca rubra)*, tall fescue *(Festuca arundinacea)*, and perennial ryegrass *(Lolium perenne)*.

Legumes

The ability of legumes to make their own nitrogen makes them a valuable addition to infertile soils. Legumes are often planted alone or in combination with grasses. The legumes commonly used for erosion control include red clover *(Trifolium pratense)*, rose clover *(Trifolium hirtum)*, and birdsfoot trefoil *(Lotus corniculatus* and *Lotus tenuis)*.

Legume seed must be coated (innoculated) with the correct bacteria immediately before seeding (Fig. 6.1). The bacteria form colonies on the plants' roots as the legumes develop. This symbiotic relationship is essential for normal, healthy growth. Figure 6.2 shows a young, healthy legume seedling whose seed was properly innoculated. Figure 6.3 shows an uninnoculated seedling that was planted at the same time. This seedling will never grow into a healthy plant.

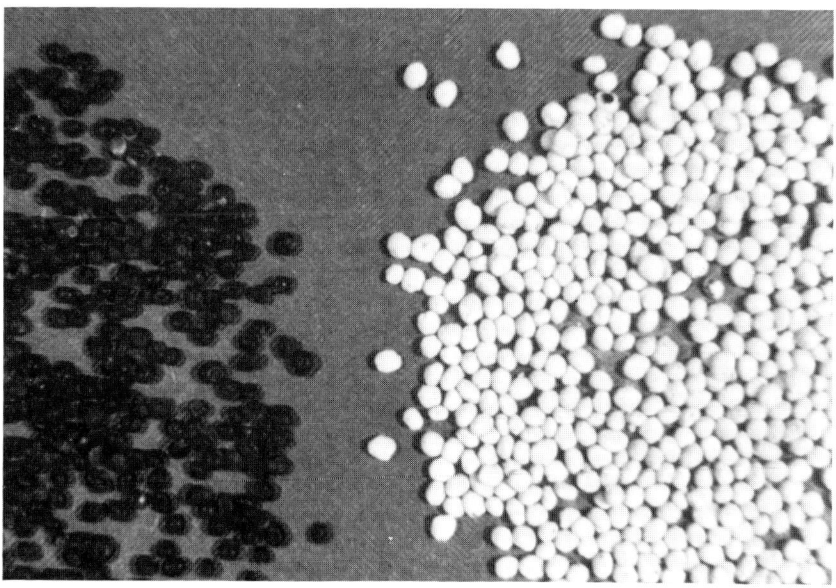

Fig. 6.1 Two examples of innoculated legume seeds. (Courtesy of Burgess Kay)

Fig. 6.2 Healthy, innoculated legume seedling. (Courtesy of Burgess Kay)

Flowers

Flowers are often desired in erosion control seed mixes because they add color to a landscape. However, they do not develop as fast or provide as much soil protection as grasses. Because flowers are poor competitors, they may be best adapted to the poorest soils because of the reduced weed growth on such soils. They should not be seeded where grass seed and fertilizer application is heavy as this will promote strong grass competition.

Flower seed tends to be the most expensive part of erosion control seed mixes, yet the flowers are often completely obscured by the grasses after the first year or two. Figures 6.4 and 6.5 show an area along Highway 113 near Sacramento, California, that was seeded with a mix of flowers and grasses. When the first photograph was taken (Fig. 6.4) the site was ablaze with bright orange California poppies. In the second photograph (Fig. 6.5), taken at the same spot exactly 2 years later, no flowers are visible. The money that was spent for flower seed on this site might better have been spent for some other purpose. Flowers can be used more effectively on less erodible areas, where grasses can be omitted from the seed mix. Some flowers, such as lupines, are legumes and should be innoculated with the correct bacteria.

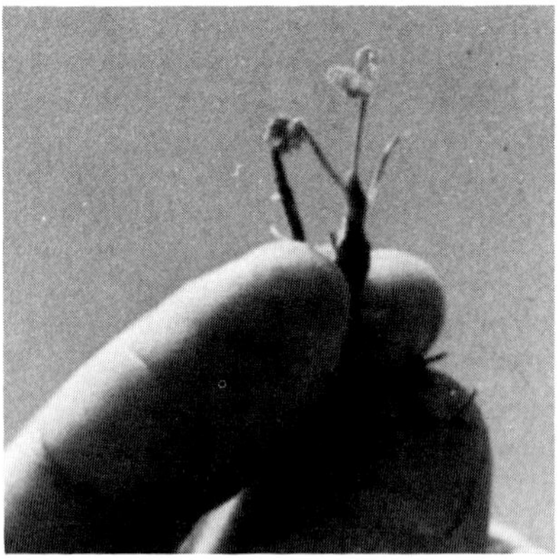

Fig. 6.3 Stunted growth of uninnoculated legume seedling. (Courtesy of Burgess Kay)

Fig. 6.4 California poppies blooming on erosion control site near Davis, California, April 1978. (Courtesy of Burgess Kay)

Shrubs

Shrubs are less effective for erosion control because of their slow growth and limited ground coverage. Most require transplanting from containers and irrigation. Deep-rooted, woody shrubs can, in time, aid in stabilizing large soil masses, and they may be desirable as permanent landscaping. Shrubs are frequently used in combination with other ground covers, such as wood chips, to provide both long-term erosion protection and landscaping. On high-elevation and desert sites, because of the harshness of the environment, native shrubs are valuable for erosion control because they may be the only plants that can survive in such locations. Gray and Leiser (5) and Schiechtl (12) provide extensive information on the use of shrubs for erosion control. Several of the other references

Fig. 6.5 Davis, California, site, April 1980; flowers have been replaced by grasses. (Courtesy of Burgess Kay)

listed at the end of this chapter discuss revegetation techniques on mountain, desert, and other difficult sites. (3, 9, 11, 14)

Trees

Trees, like shrubs, are less effective for initial erosion because of their slow growth. In the long term, however, they can provide both slope stability and erosion protection. A well-developed leaf canopy and leaf or needle droppings protect the soil from raindrop impact. Tree roots that extend deep into the soil help prevent soil slippage and landslides.

Trees are not well adapted to the shallow soil of cut slopes, but they may be used successfully for erosion control on fill slopes. Transplanting is the most common planting method. Staking bundles of willow cuttings into soils on slopes has been an effective erosion control practice. Several references describe this technique, called willow wattling, in detail. (3, 5, 9, 12) Willows are also easily established from cuttings. Gray and Leiser (5) and Schiechtl (12) provide extensive information on the use of trees for slope protection.

6.1c Plant Lists

Many state and regional agencies and local governments publish handbooks of erosion control specifications, which may be either requirements or recommendations for erosion control practices within the state or local jurisdiction. For example, Maryland's handbook, *1983 Maryland Standards and Specifications for Soil Erosion and Sediment Control,* contains detailed planting specifications for revegetating exposed soils in Maryland. (10) It includes a series of tables covering temporary and permanent vegetation using grasses, legumes, vines, shrubs, and trees. Virginia publishes a similar handbook (16); Table 6.2 illustrates the specifications for temporary seeding in that state. The Association of Bay Area Government's *Manual of Standards for Erosion and Sediment Control Measures* contains vegetation recommendations and plant lists suitable for the valleys, foothills, and coastal areas of California. (1) Before starting a revegetation project, check with the local jurisdiction to see if there are local specifications for plant materials.

In areas where there are no specific requirements for selecting plants for erosion control, which is true in most of the United States and the rest of the world, plant lists may be helpful. Lists of plants suitable for erosion control have been compiled in numerous sources. A few of these sources are briefly described below. Gray and Leiser (5) present an annotated bibliography of lists of a wide variety of plants for erosion control. Schiechtl (12) presents an extensive, worldwide list of plants for erosion control, including plants suitable for a variety of difficult sites—shorelines, dunes, streambanks, and mine tailings. *Reclamation of Drastically Disturbed Lands* (11) presents a series of papers on revegetating denuded soils in both dry and humid regions of the United States. *Methods of Quickly*

Vegetating Soils of Low Productivity, Construction Activities (3) gives case histories of revegetation efforts in 10 areas in the United States. The book contains extensive lists of plants, literature, and sources of information about plants and revegetation. Thornburg (14) presents a lengthy list and description of plants suitable for the arid and semiarid regions of the western United States.

6.1d Other Sources of Information on Plants for Erosion Control

Because of the great variability of environmental conditions from place to place and the value of local experience, it is usually advisable to consult a local expert. The U.S. Soil Conservation Service's plant materials centers are a good source of information. These centers, which are located in 22 states, have staffs who are continually studying the performance of a wide variety of plant types for both agriculture and erosion control. Table 6.3 gives the addresses of these plant materials centers. Other good sources of information are the local farm adviser or county agent (listed in the telephone book under County Offices—Cooperative Extension), state colleges, or universities, and the local office of the U.S. Department of Agriculture, Soil Conservation Service.

6.2 SITE PREPARATION

6.2a Construction Site Conditions

Construction practices, such as extensive clearing and earth moving, produce large areas of exposed soil with a high potential for erosion. Grading creates cut and fill areas that have soil conditions substantially different from the original ones. Cut slopes are typically stripped down to the subsoil or parent material (rock). The exposed slope is often hard, rocky, and low in fertility. Initial plant regrowth on such slopes will probably be scraggly and thin. Fill areas generally contain a mixture of topsoil, subsoil, rock, and anything else that had to be disposed of. A common fill material may be the grading spoils from other construction projects. Thus, a fill area may contain the right types of soil for plant growth, but they are often mixed up or inverted or may not have originated at that site. Because of the probable presence of some topsoil in the fill, plant seeds may be present to recolonize the area. However, the unpredictable soil composition and potential for differential compaction and settling make fill areas particularly susceptible to erosion.

Before a disturbed site is revegetated, the earth should be shaped and roughened to create a favorable environment for plants to grow in. It is a common practice to grade slopes until they have a hard, smooth surface (Fig. 6.6). Such slopes give a false impression of "finished grading" and a job well done. Seedling

TABLE 6.3 U.S. Soil Conservation Service Plant Materials Centers

ALASKA: Star Route B, Box 7440, Palmer, AK 99645
ARIZONA: 3241 North Romero Road, Tucson, AZ 85705
CALIFORNIA: P.O. Box 68, Lockeford, CA 95237
COLORADO: P.O. Box 448, Meeker, CO 81641
FLORIDA: 14119 Broad Street, Brooksville, FL 33512
GEORGIA: Route 3, Patton Drive, Americus, GA 31709
HAWAII: P.O. Box 236, Hoolehua, Molokai, HI 96729
IDAHO: P.O. Box AA, Aberdeen, ID 83210
KANSAS: Route 2, Box 314, Manhattan, KS 66502
KENTUCKY: Plant Materials Center, Quicksand, KY 41363
MARYLAND: Building 509, Agricultural Research Center, Beltsville, MD 20705
MICHIGAN: Route 7, 7472 Stoll Road, East Lansing, MI 48823-9807
MISSISSIPPI: Route 3, Box 215A, Coffeeville, MS 38922
MISSOURI: R.R. 1, Box 9, Elsberry, MO 63343
MONTANA: Route 1, Box 1189, Bridger, MT 59014
NEW JERSEY: Route 1, Box 236A, Cape May Courthouse, NJ 08210
NEW MEXICO: 1036 Miller Street, S.W., Los Lunas, NM 87031
NEW YORK: Box 360A, RD 1, Route, 352, Corning, NY 14830
NORTH DAKOTA: P.O. Box 1458, Bismarck, ND 58501
OREGON: 3420 Northeast Granger Avenue, Corvallis, OR 97330
TEXAS: Route 1, Box 155, Knox City, TX 79529
EAST TEXAS: Agriculture Building, SFASU, P.O. Box 13000, SFA Station, Nacodoches, TX 75962
SOUTH TEXAS: Caesar Kleberg Wildlife Research Institute, Texas A&I University, P.O. Box 218, Kingsville, TX 78363
WASHINGTON: Room 257, Johnson Hall, Washington State University, Pullman, WA 99164-6428

roots have a difficult time penetrating these surfaces. A light rain or wind can easily carry away the seeds.

6.2b Grading and Shaping

Graded slopes that are to be vegetated should not be too steep—preferably 2:1 or flatter. The chance for successful revegetation will be greater on gentler slopes. If a slope is steeper than 3:1, stair-stepping it with benches will help vegetation to become established (Fig. 6.7). The benches will also trap soil eroded from the slopes above. Benching is particularly desirable on steep cut slopes where the rock is soft, but not so soft that the benches may collapse.

Fig. 6.6 A smooth-graded slope makes a poor environment for establishing plants.

Both cuts and fills should be rounded in shape to reduce erosion potential and blend with the natural landscape (Figs. 1.7*c* and 6.8). Rounding is particularly important at the top of cuts. There should never be a sharp break in slope, as in Fig. 1.7*b*. Slopes should be left in a rough condition. Breaking up clods and smoothing the surface is not only unnecessary but will reduce the chances for successful revegetation.

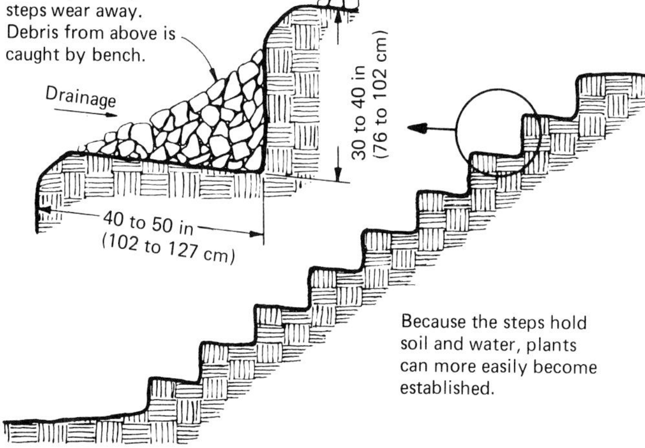

Fig. 6.7 Stair-stepping a cut slope will increase the chances for successful revegetation. (Adapted from 16)

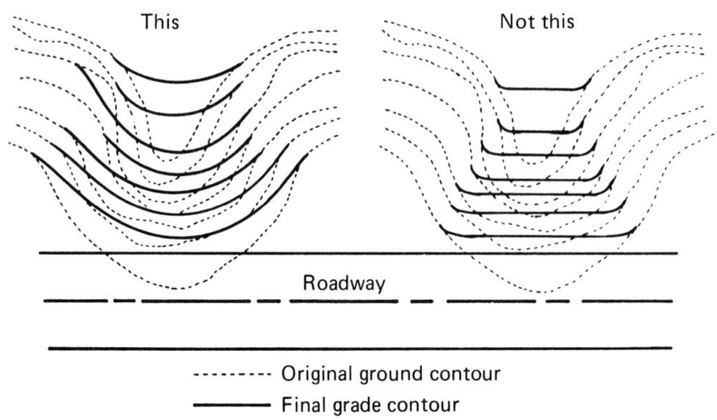

Fig. 6.8 Man-made slopes should be rounded, not squared off. (5)

6.2c Surface Roughening

Graded areas should be roughened before seeding. Soil may be roughened in a variety of ways. One of the most popular ways is to drive a bulldozer or crawler tractor up and down the slope (Fig. 6.9). This process, called trackwalking, leaves a pattern of tread imprints parallel to slope contours (Fig. 6.10). The tread

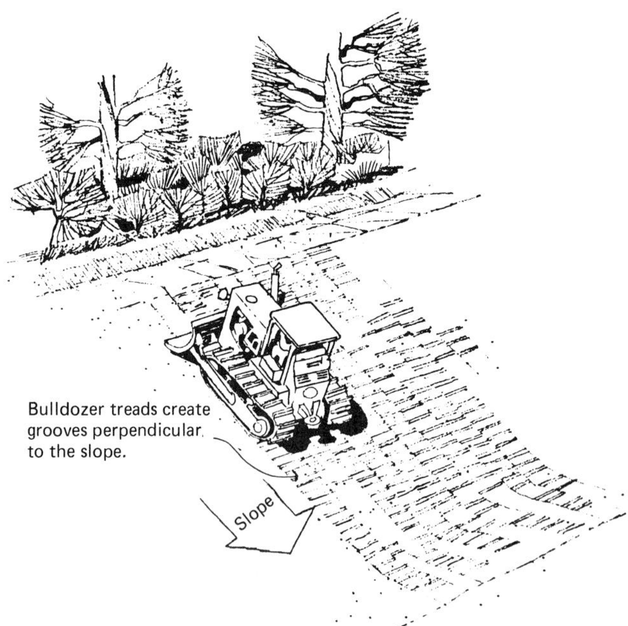

Fig. 6.9 Bulldozer treads create grooves perpendicular to the slope. (16)

Fig. 6.10 Trackwalked slope before seeding.

indentations are ideal places to trap seeds and encourage plants to become established (Fig. 6.11). The tracks also slow the velocity of runoff and thus lower the *P* factor in the universal soil loss equation (see Sec. 5.2f).

A serrated wing blade attached to the side of a bulldozer (Fig. 6.12) is designed to roughen small cut and fill slopes. In Fig. 6.13 a spike-toothed harrow, a common piece of farm equipment, is shown being used to break up the soil on a cut slope. The soil should be loosened to a depth of 2 to 4 in (5 to 10 cm).

Fig. 6.11 Trackwalked slope several weeks after seeding. Seedlings are growing only in the tread depressions. (Courtesy of Burgess Kay)

Fig. 6.12 Serrated wing blade on bulldozer used for roughening slopes. (Courtesy of Burgess Kay)

6.3 SEEDING

This section describes procedures for planting grasses, legumes, flowers, and other herbaceous plants. These plants provide quick ground cover, which is needed for initial erosion control. Use of trees and shrubs for long-term erosion control was discussed in Sec. 6.2.

6.3a Time of Planting

The time of planting is important and is determined by local conditions. The best time to plant is when temperatures, moisture, and sunlight are adequate. In the eastern United States, planting may be done from February or March until November, varying somewhat with location and plant type. In Virginia, for example, the planting season runs from February 15 to November 15 in the southern and coastal portions of the state and from March 1 to November 1 in the northern piedmont and mountain regions (Table 6.2).

On the west coast, the planting time is affected by the long summer dry season. In the valleys, foothills, and coastal areas of California, the best time to seed is early fall. The light autumn rains typical in this region will cause the seeds to germinate. About 30 days after germination the grass seedlings will have grown to a size that can provide erosion protection from the heavier winter rains. In

Fig. 6.13 Spike-toothed harrow breaks up soil on cut slope. (Courtesy of Burgess Kay)

northern California, planting by October 1 provides a 90 percent probability that seeds will be in the ground before the first rainfall great enough to cause germination [assumed to be a 1-in (2.5-cm) storm] and a 90 percent probability that the first highly erosive rain [2 in (5 cm) or greater] will not occur for over 30 days. (4)

In the deserts of the western United States, the best time to plant is midwinter or early spring; in the mountains, it may be early fall or spring. Because of the great variability in optimum planting times from region to region, consult a local expert before seeding (see Sec. 6.1d for places to contact).

6.3b Seed Application

The key factor in seeding is to get the seeds evenly distributed and in contact with the soil. Best results are obtained when seeds are covered with a shallow layer of soil. Seed should be covered to a depth of ¼ to ½ in (0.6 to 1.3 cm). It should not have a soil cover greater than 1 in (2.5 cm). Seeding can be done by hand, by machine (seed drill), by hydraulic jet, or by aircraft. The steepness of slope and size of area determine the proper equipment to use. Various seeding methods are briefly discussed below.

Fig. 6-14 Breast seeder. (Courtesy of Burgess Kay)

Hand Broadcast Seeding

Small, gently sloping or flat areas can be effectively seeded by hand. Breast seeders (or "belly-grinders") are inexpensive (Fig. 6.14). Labor is 1 to 3 hr/acre (2.5 to 7.5 hr/ha). Seed can be covered by lightly raking the soil or by dragging a chain over the surface. The seed may become covered naturally on a rough, loose seedbed.

Hydraulic Seeding and Mulching

Hydraulic seeding and mulching (hydroseeding) is the most efficient means for seeding steep slopes (typically slopes 2:1 or steeper). It is a one-step process for spraying a slurry of seed, fertilizer, wood fiber mulch, and water (Fig. 6.15).

Fig. 6.15 Hydraulic seeding and mulching.

Fig. 6.16 Seed drill. (Courtesy of Burgess Kay)

The critical factor in hydroseeding is the ability of the fiber to adhere to the soil and hold the seed in place during rainfall and wind. The effectiveness of various seeding and mulching practices is discussed in Sec. 6.5.

Seed Drilling

Seed drills are commonly used in agriculture (Fig. 6.16); they can be used only on flat or gently sloping sites. Because seed drills plant seeds at the proper depth, they provide the best stands of grass. In addition, when seed is drilled, seed and fertilizer rates may be cut in half.

6.4 FERTILIZING

Fertilizer is essential to the establishment of vegetation, particularly on cut slopes. The key elements are nitrogen, phosphorus, potassium, and sulfur. If one of these elements is not present in a soil, plant growth will be severely stunted.

Figure 6.17 shows four pots that were planted with the same type of grass at the same time. The pots contained a California soil that was not deficient in potassium. The left-rear pot was fertilized with a mix of nitrogen, phosphorus, and sulfur. The left-front pot was fertilized with nitrogen and sulfur only. The right-front pot was given phosphorus and sulfur, and the right-rear pot was given nitrogen and phosphorus. Only the pot that was given all three nutrients had healthy plant growth.

Fig. 6.17 Effect of nutrient omission on plant growth. (Courtesy of Burgess Kay)

Soil acidity, or pH, is another important variable for plant growth. Some plants require acid soils, and others require alkaline soils. If a soil is too acid for the desired plants, lime can be added to the fertilizer mix to raise the pH to the desired range. The lime may be in the form of pulverized dolomite or limestone.

Soil salinity is also a concern, particularly in arid climates and along roads where deicing salts are used. One solution to salinity problems is to choose salt-tolerant plants. An alternative is to improve drainage so that the salt will be leached from the soil.

The amount and composition of fertilizer to use depends on local soil conditions. The local farm adviser or soil conservation service office (Sec. 6.1d) should know the fertilizer requirements in your area. Chemical soil tests can also be used to determine nutrient needs on-site, although the tests can be costly and time-consuming. Because the chemical composition of a soil can vary significantly at different locations and depths on a single small site, a soil test can produce erroneous or misleading results.

Many state, regional, and local agencies have developed fertilizing specifications for use in their jurisdictions. In the San Francisco Bay Area, the recommended nutrient mix for temporary and permanent grass seeding is 500 lb/acre (568 kg/ha) of 16-20-0 (16 percent nitrogen, 20 percent phosphorus and 0 percent potassium) with 15 percent sulfur. (1) In Virginia, the recommended fertilizer for temporary erosion control seeding is 10-20-20 at 450 lb/acre (511 kg/ha). (16) Maryland's specifications call for 600 lb/acre (682 kg/ha) of 10-10-10. (10)

Fig. 6.18 Straw mulch (left) and wood fiber mulch (right). (Courtesy of Burgess Kay)

As was stated earlier, it is preferable to use plants that can survive without repeated fertilization. Legumes can be planted to increase soil nitrogen. Fertilizing is costly, and on most sites you cannot count on having someone there to do it year after year. Excessive fertilization can cause water pollution. However, on sites with sandy or coarse-grained soils, such as decomposed granite, reapplication of fertilizer may be necessary for continued plant cover.

6.5 MULCHING

6.5a Introduction

Application of mulch protects a disturbed site from erosive forces until plants are large enough to do the job. Mulches also increase plant establishment by conserving moisture and moderating temperatures. Mulch helps hold fertilizer, seed, and topsoil in place in the presence of wind, rain, and runoff and maintains moisture near the soil surface. Mulch materials include straw, wood fiber, wood chips, bark, fabric or plastic mats, soil, and gravel.

The choice of mulch is determined by site characteristics, product availability, cost, and effectiveness. The two most common mulches used on construction sites are straw and wood fiber (Fig. 6.18). These are discussed in detail in the following sections. In general, straw mulch offers the best results for both site protection and encouragement of plant growth. Wood fiber mulch offers a weed-free mulch of low fire hazard, with possible laborsaving in application methods, but it is seldom as effective as straw. Test results on the effectiveness of a variety of mulch treatments are presented in Sec. 6.5e.

Fig. 6.19 Straw blower. (Courtesy of Burgess Kay)

6.5b Straw Mulch

Straw is an excellent mulch material. Because of its length and bulk, it is highly effective at absorbing raindrop impact and moderating the climate on the soil surface. Straw pieces tend to interweave with each other on the ground; thus they trap soil and reduce the possibility of the straw washing or blowing away. Because straw is a by-product of agriculture, it is generally readily available near agricultural areas, often at modest cost.

Application Procedure

Straw should be applied at a rate of 3000 to 8000 lb/acre (3.4 to 9.0 t/ha). The rate depends on the anchoring method, as described below. Mulch should not be applied too heavily. Soil should be visible through the straw mat. Nonmechanically anchored straw that is applied at a rate higher than 4000 lb/acre (4.5 t/ha) may produce a mat too dense for seedlings to penetrate.

Straw can be applied either by blower or by hand. Straw blowers (Fig. 6.19) have a range of about 50 ft (15 m). Some commercial straw blowers advertise a capability of up to 15 tons/hr (13.6 t/hr) and distances up to 85 ft (26 m). It is difficult to apply straw uphill or under windy conditions.

Holding Straw in Place

Straw *must be anchored* to keep it from blowing away. Washing away is usually not a problem because wet straw is heavy and is not easily moved. Common methods of holding straw in place include:

- Crimping, disking, rolling, or punching into the soil
- Covering with netting
- Spraying with a chemical or fiber binder (tackifier)

Crimping, disking, rolling, and punching are mechanical methods for anchoring straw. When straw is mechanically anchored, a high application rate is required because some of the straw is incorporated into the soil and some is punched into a vertical position and no longer serves as a mulch cover. Long straw, preferably at least 6 in (15 cm) in length, is desirable. Unfortunately, new agricultural practices are resulting in shorter lengths. The flails used in straw blowers will further shorten straw, although they can be adjusted to reduce the effect.

Crimping is accomplished with commercial machines which use blunt, notched disks which are forced into the soil by a weighted, tractor-drawn carriage. They will not penetrate hard soils, and they cannot be pulled on steep slopes. When incorporated by crimping, straw is applied at a rate of 4000 to 8000 lb/acre (4.5 to 9.0 t/ha).

Rolling or punching is done with a specially designed roller (Fig. 6-20). A sheepsfoot roller, commonly used in soil compaction, is *not* satisfactory for incorporating straw. Specifications of the California Department of Transportation contain the following provisions: (7)

Fig. 6.20 Straw roller. (Courtesy of Burgess Kay)

Fig. 6.21 Straw roller being lowered and raised by using a winch on the back of a truck. (Courtesy of Burgess Kay)

> Roller shall be equipped with straight studs, made of approximately ⅞ inch (2.2 cm) steel plate, placed approximately 8 inches (20 cm) apart, and staggered. The studs shall not be less than 6 inches (15 cm) long nor more than 6 inches (15 cm) wide and shall be rounded to prevent withdrawing the straw from the soil. The roller shall be of such weight as to incorporate the straw sufficiently into the soil so that the straw will not support combustion, and will have a uniform surface.

The roller may be tractor-drawn on flat areas or gentle slopes, and on steeper slopes it may be lowered by gravity and raised by a winch in yo-yo fashion, commonly from a flat-bed truck (Fig. 6.21). A roller can thus be used on much steeper slopes than a crimper. Requirements for the roller-on-a-winch procedure are (1) soil soft enough for the roller teeth to penetrate and (2) vehicle access to the top of the slope. This is a common treatment on highway fill slopes in California, where 4000 lb/acre (4.5 t/ha) of straw is applied. Then the slope is rolled, another 4000 lb/acre (4.5 t/ha) is applied, and the slope is rolled again.

A variety of nets can be used to hold straw in place: twisted and woven kraft paper, plastic fabric, poultry netting, concrete reinforcing wire, and even jute. Price and the required length of service should determine the product to use. Netting should be anchored at enough points to prevent the net from whipping in the wind, which would rearrange the straw. If a slope is difficult to penetrate with punching or rolling equipment, straw should be applied at a rate of 3000 to 4000 lb/acre (3.4 to 4.5 t/ha) and held in place with netting. The netting must be so installed that it contacts the mulch or soil uniformly. Methods for anchoring netting are described in Secs. 6.5d and 6.6d.

Perhaps the most common method of holding straw, particularly in the eastern United States, is with a gluey substance called a *tackifier*. This method may be used on relatively steep slopes which have limited access and soil too hard for crimping or punching. Asphalt emulsion, the tackifier used most commonly, is applied at 200 to 500 gal/acre (1.9 to 4.7 kL/ha) either over the top of the straw or simultaneously with the straw-blowing operation. Tests have shown that 600 gal/acre (5.7 kL/ha) is superior to 400 gal/acre (3.8 kL/ha) and that 200 gal/acre (1.9 kL/ha) is marginally satisfactory. Wood fiber, and other products used in combination with wood fiber, have been demonstrated to be equally effective, similar in cost, and environmentally more acceptable (Table 6.4). Plantago gum is made from plantain *(Plantago insularis)*. The remaining products are emulsions used in making adhesives, paints, and other products. Though wood fiber alone is effective as a short-term tackifier, glue must be added to give protection beyond a few weeks. Increasing the application rate of any of the materials will increase its effectiveness. (7)

Table 6.4 shows that straw bound together with a tackifier is highly resistant to wind. No more than 50 percent of the straw was removed in short-duration winds greater than 84 mi/hr (135 km/hr) when tackifiers were applied at the levels shown.

TABLE 6.4 Effect of Tackifier Products on Wind Stability of Barley Straw Broadcast at 2000 lb/acre (2.24 t/ha) (7)

	Application rates			
Treatment	Binder, gal/acre (L/ha)*	Wood fiber, lb/acre (kg/ha)	Water, gal/acre (L/ha)	Wind speed at which 50% of straw was blown away, mi/hr (km/hr)†
None				9 (14)
SS-1 asphalt	200 (1871)			40 (64)
SS-1 asphalt	400 (3741)			80 (129)
SS-1 asphalt	600 (5612)			84 (135)
Fiber only		484 (544)		47 (76)
Fiber only		736 (827)		84 (135)
Fiber only		986 (1107)		84 (135)
Plantago gum and fiber	100 (112)*	150 (168)	700 (6548)	84 (135)
Styrene butadiene copolymer emulsion	60 (561)	75 (84)	400 (3741)	84 (135)
Polyvinyl acetate	100 (935)	250 (281)	1000 (9354)	54 (87)
Copolymer of methacrylates and acrylates	100 (935)	250 (281)	1000 (9354)	76 (122)

*Plantago gum is in units of lb/acre (kg/ha).

†84 mi/hr (135 km/hr) was the maximum wind speed that could be measured with the equipment used in this test.

Disadvantages of Straw Mulch

Although straw mulch is a highly effective and inexpensive erosion control measure, it does have a few drawbacks:

- Weed growth is common (less common with rice straw, but rice straw is difficult both to obtain and to spread).
- Fire hazard is greater than with wood fiber mulch.
- Cleanup cost may be high if the mulch is applied when it is windy.
- It is difficult to apply more than 50 ft (15 m) from the nearest access point (blowers have a shorter range than hydraulic jets).

6.5c Wood Fiber Mulch

Wood fiber mulch consists of fine wood fibers (similar to the fibers used in paper making) to which a green dye is added to make the mulch visible on the ground when it is being applied. It is relatively inexpensive, readily available commercially, and easy to apply with a hydroseeder. Wood fiber mulch, because of the small size and weight of the wood fibers, does not by itself provide much erosion protection. Though it provides a nearly complete ground cover, it does not have enough mass to absorb the energy of raindrops and flowing water. Its primary function is to assist plant establishment, mostly by holding seeds to the slope.

Wood fiber is applied hydraulically. Typically, wood fiber, seed, fertilizer, and water are combined in a tank and applied together as a slurry. Wood fiber should be applied at a rate of at least 1000 lb/acre (1.1 t/ha). However, to produce a true

Fig. 6.22 Jute netting used to stabilize a steep slope below a building pad.

mulch effect with wood fiber (i.e., to hold moisture near the soil surface and moderate temperature), the material must be applied at a rate of at least 2000 lb/acre (2.2 t/ha). Increasing the rate to 3000 lb/acre (3.4 t/ha) will increase the effectiveness. However, the resulting increase in effectiveness may not be great enough to justify the added cost (see Sec. 6.5e).

Though wood fiber mulch is not as effective as straw, it is preferable to straw mulch under these conditions:

- On steep cut slopes, 2:1 or steeper. (It is difficult to hold straw in place on such slopes, but wood fiber can hold seed in place long enough for the seed to germinate.)
- Where access by vehicles is not possible within 50 ft (15 m).
- Where weed growth or fire hazard is a major problem.
- If the mulch must be applied on a windy day.

Wood fiber can also be used very effectively as a tackifier for straw mulch (see Table 6.4). When it is used for that purpose, it should be applied at a rate of 750 lb/acre (0.84 t/ha).

6.5d Nettings

Heavy-weight netting materials such as jute (woven fibers) and excelsior (a material also used for evaporative cooler pads) can be effective mulches. Not only do these dense nettings hold soil in place; they also absorb water and hold moisture near the soil surface. Less dense nettings such as chicken wire and plastic mesh may be used to hold other mulch materials in place, but by themselves they provide little if any soil protection.

Jute netting is particularly desirable for quickly stabilizing critical slopes and small sites in urban areas (Fig. 6.22). It must be so applied that it is in complete contact with the soil. If it is not, erosion will occur beneath it. Contact is improved by applying the netting over a layer of straw. Netting should be securely anchored to the soil with No. 11 gauge wire staples at least 6 in (15 cm) long. On very hard or rocky soils, the netting can be anchored by using large nails and washers.

One drawback to using materials such as jute and excelsior for erosion control is high cost. The installed cost for jute netting ranges from 14 to 36 cents/ft^2 (\$1.50 to 3.90/m^2). Comparative costs of various seed and mulch treatments are presented in Table 6.6.

6.5e Comparison of Costs and Effectiveness of Various Mulch Treatments

Test Procedures

Burgess Kay has performed extensive tests at the University of California, Davis, on the effectiveness of straw, wood fiber, and nettings in controlling erosion. (8)

Fig. 6.23 Mulch effectiveness test on loam soil with a 2:1 slope. From left to right: straw at 3000 lb/acre; wood fiber at 2500 lb/acre; wood fiber at 3000 lb/acre, bare soil. (Courtesy of Burgess Kay)

Mulches were tested on 2- by 4-ft (0.6- by 1.2-m) boxes of soil inclined at 5:1 and 2:1 slopes (horizontal to vertical ratio) (Fig. 6.23). Rainfall was simulated by using 3-mm-diameter drops falling 15 ft (4.5 m) at the rate of 6 in/hr (152 mm/hr) for periods of 2 to 6 hr. The soil boxes were designed to allow rapid drainage if water moved through the top 6 in (15 cm) of soil. Soil washed from each box was collected, dried, and weighed.

Eight soils were used in the tests; they were taken from construction sites and road cuts throughout northern California. Some samples were subsoils or mixtures of profiles from throughout a site. The composition of the eight soils varied widely; it ranged from a clay loam to an uncemented fine sand (Table 6.5).

Test Results

Figure 6.24 shows the rate of erosion from each of the soil types when no protection was used. Soil loss was greater from all soil types when inclined at 2:1 than at 5:1. The most erodible soils were those with high percentages of fine sand and

TABLE 6.5 Textures and Particle Size Percentages of Soils Used in Mulch Tests (8)

Texture	Name	Clay	Silt	Sand	Gravel
Uncemented fine sand	Arnold	2	3	95	0
Very gravelly coarse sand	Decomposed granite	3	4	41	52
Gravelly sandy loam	Cieneba	9	9	49	33
Sandy clay loam	Dibble	21	18	61	0
Loam	Los Osos	17	48	35	0
Loam	Yolo	22	45	33	0
Loam	Auburn-Sobrante	21	43	36	0
Clay loam	Altamont	29	45	26	0

silt, including the three loam soils. The uncemented fine sand did not erode initially, but it would liquefy when saturated and flow downslope. On the 2:1 slope, this liquefied sand flowed rapidly and thus produced the very high erosion rate. The least erodible soils were the coarse, gravelly, decomposed granite and the soil with the highest clay content.

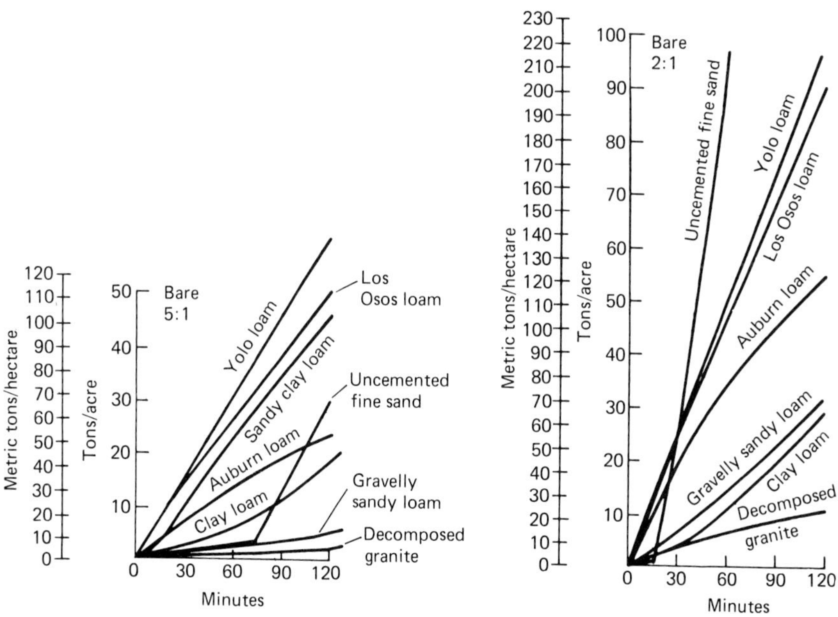

Fig. 6.24 Erosion rates from unprotected soil surfaces inclined at 5:1 and 2:1 slopes. (8)

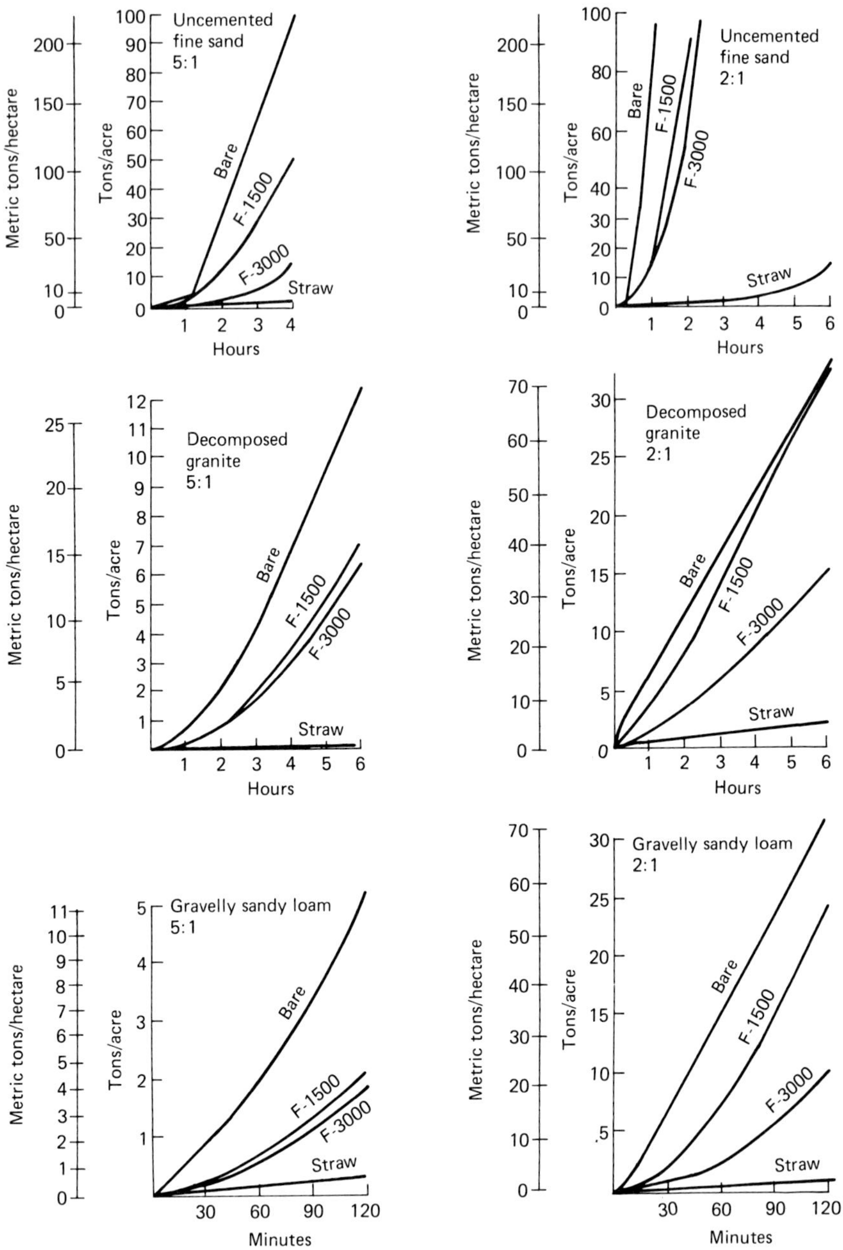

Fig. 6.25 Effectiveness of wood fiber mulches and straw mulch on seven soils at 5:1 and 2:1 slopes. (8)

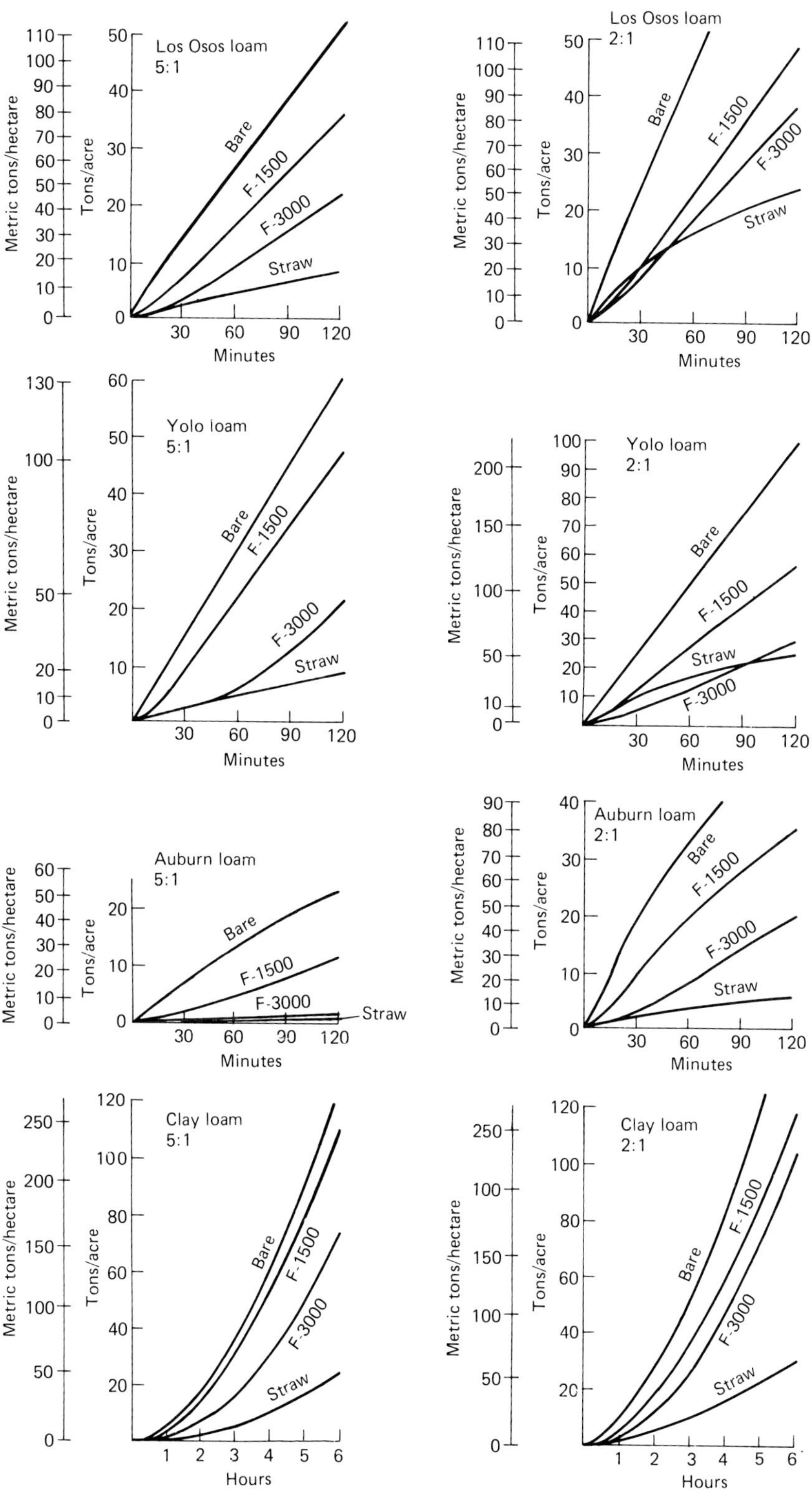
Los Osos loam
5:1
Bare
F-1500
F-3000
Straw
Metric tons/hectare
Tons/acre
Minutes
Los Osos loam
2:1
Yolo loam
5:1
Yolo loam
2:1
Auburn loam
5:1
Auburn loam
2:1
Clay loam
5:1
Clay loam
2:1
Hours

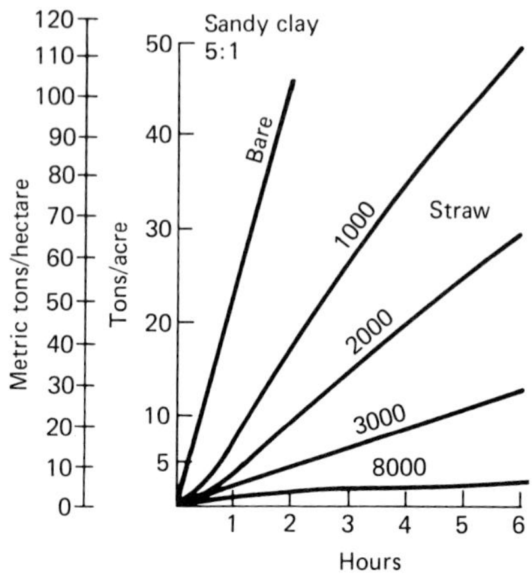

Fig. 6.26 Effect of varying straw application rates on dibble sandy loam with a 5:1 slope. (8)

Figure 6.25 compares the effectiveness of wood fiber mulch and straw mulch on seven of the soil types at 5:1 and 2:1 slopes. The wood fiber* was applied hydraulically at rates of 1500 and 3000 lb/acre (1.7 and 3.4 t/ha). Barley straw was applied at 3000 lb/acre (3.4 t/ha) and tacked down with asphalt emulsion at 200 gal/acre (1870 L/ha). Straw mulch provided much greater protection on all soils, but particularly on the uncemented fine sand, the decomposed granite, and the clay loam. The mulch treatments were similar in effectiveness on both 5:1 and 2:1 slopes. The wood fiber mulch, though less effective than straw, did offer some protection. Increasing the commonly used rate of 1500 lb/acre (1.7 t/ha) of fiber to 3000 lb/acre (3.4 t/ha) provided additional protection, but not in all cases.

Varying the application rate of straw had a significant effect on soil loss. Application rates of 1000, 2000, 3000, and 8000 lb/acre (1.1, 2.2, 3.4, and 9.0 t/ha) were tested on a sandy clay loam at 5:1 slope (Fig. 6.26). The 8000 lb/acre (9.0 t/ha) treatment was punched into the soil by using a shovel. At 1000 lb/acre (1.1 t/ha) of straw, soil loss was reduced by two-thirds relative to the bare soil. At 2000 lb/acre (2.2 t/ha), soil loss was cut in half again, relative to the 1000 lb/acre (1.1 t/ha) test. Each successive increase in the application rate cut the previous erosion rate roughly in half.

Several commonly available fabrics—jute, excelsior, and a paper strip–synthetic yarn product (Hold/Gro®)†—were also tested on the same sandy clay loam soil as above (Fig. 6.27). Straw and jute were the most effective treatments, and the difference in effectiveness of the two was not significant. Excelsior was less effective but was better than the paper product. The most effective of all treatments tested was jute on top of 3000 lb/acre (3.4 t/ha) of straw. This treatment was even more effective than 8000 lb/acre (9.0 t/ha) of punched straw. Although jute alone was very effective on the 5:1 slope, its effectiveness diminished over time. It was as effective as 8000 lb/acre (9.0 t/ha) of straw for over 2 hr, but then its performance deteriorated as soil washed from beneath the fabric. This last test result demonstrated the critical importance of good soil contact when fabrics are used. A layer of straw underneath the fabric will improve the contact (Fig. 6.27, bottom curve at right).

*Silvafiber®, a virgin wood fiber, was used in the tests. Silvafiber® is a product of Weyerhaeuser Co., Tacoma, Washington.

†Hold/Gro® is a product of Gulf States Paper Corp., Tuscaloosa, Alabama.

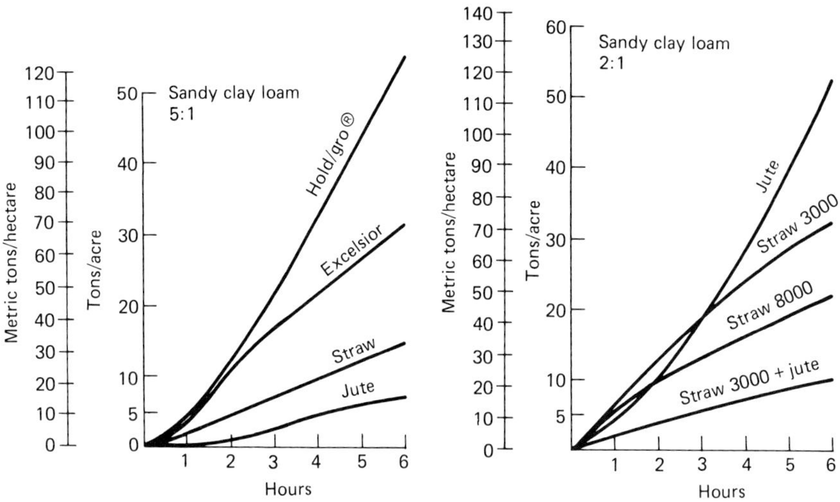

Fig. 6.27 Effectiveness of erosion control fabric and straw on dibble sandy clay loam at 5:1 and 2:1 slopes. (8)

Summary

Table 6.6 compares the cost and effectiveness of a wide variety of seeding and mulching treatments, including seeding without mulching, seeding and wood fiber mulching, seeding and straw mulching, and seeding and fabric mulching. The treatments were rated both for their initial erosion control effectiveness (before germination) and for their long-term effectiveness after vegetation has been established. The ratings were based on a combination of test results and field experience.

The first three treatments listed in the table offer little or no soil protection until grass is established. The effectiveness of treatment 1 is also low after plant establishment because many of the seeds are washed or blown away or eaten before they have a chance to germinate.

The cost-effectiveness ratings (last two columns of Table 6.6) were developed as a means of comparing the different treatments. The formula used to calculate this rating is given at the bottom of the table. When an erosion control treatment is to be selected, it is necessary to consider both short- and long-term effectiveness. The best practices are those that have high ratings in both categories (treatments 5 and 6). Treatment 2 (seed and fertilizer drilled into the soil, with no mulch) has the highest combined short- and long-term cost-effectiveness but the lowest short-term cost-effectiveness. If erosion control during the first months after seeding is critical, this treatment should not be used.

The most cost-effective technique in both categories in the table is number 5 [straw blown on at a rate of 3000 lb/acre (3.4 t/ha) and glued down]. This technique, at a cost of about $1000/acre ($2500/ha), offers moderately high preger-

TABLE 6.6 Summary of Erosion Control Practices, Effectiveness, and Cost[a]

Treatment	Comments	Effectiveness before plant establishment[b]	Effectiveness after plant establishment[b]	Approx. cost,[c] \$/acre (\$/ha)	Short-term cost-effectiveness[d]	Combined short- and long-term cost effectiveness[d]
1. Seed and fertilizer broadcast on the surface; seed not covered with soil; no mulch.	Inexpensive and fast. Effective only on rough seedbeds with minimal slope and erodibility where seed will be covered naturally with soil. Suitable for remote or noncritical areas where machinery cannot be taken.	0	1–4	330 (820)	0	6
2. Seed and fertilizer drilled.	Lowest-seed-mortality method, but limited to friable areas no steeper than 3:1. Soil must be loose before drilling. Plants establish only in rows. Rates of seed and fertilizer may be reduced by 50%.	0	6–8	210 (520)	0	33
3. Seed, fertilizer, and 1500 lb/acre (1.7 t/ha) wood fiber applied hydraulically.	Common hydromulch mix in California. Advantages include holding seed and fertilizer in place on steep and smooth slopes where there may not be an alternative method. Only a minimal mulch effect. Seed is not covered with soil.	2	3–5	790 (1950)	3	8
4. Seed, fertilizer, and 3000 lb/acre (3.4 t/ha) wood fiber applied hydraulically.	More effective than treatment 3 in some cases. Provides more of a true mulch effect than treatment 3 provides.	4	4–6	1280 (3160)	3	7
5. Seed and fertilizer broadcast with hydroseeder. Straw	Very effective as energy absorber and in encouraging plant	5–7	7–9	1000 (2470)	6	14

applied with blower at 3000 lb/acre (3.4 t/ha) and anchored with 300 lb/acre (337 kg/ha) wood fiber and 60 lb/acre (67 kg/ha) organic binder.	establishment. Straw forms small dams to hold some soil. May be weedy depending on straw source. Not for cut slopes steeper than 2:1 or longer than 50 ft (15 m). Cost increases significantly when slopes are over 50 ft (15 m) from access or application is uphill. Mobilization costs are high.					
6. Seed and fertilizer broadcast with hydroseeder. Straw broadcast at 4000 lb/acre (4.5 t/ha), rolled to incorporate and then broadcast again at 4000 lb/acre (4.5 t/ha) and rolled again.	Common on highway fill slopes in California. Very effective. Not possible on most cut slopes. Top-of-slope access is required for rolling equipment. High mobilization costs.	6–8	8–10	1540 (3810)	5	10
7. Jute or excelsior mats held in place with wire staples. Seed and fertilizer as in treatment 1.	Good on small sites and critical slopes. Very expensive. Weed-free. Not recommended on rocky soils. Loses effectiveness if not entirely in contact with soil. More effective if applied over straw.	7–9	8–10	15,800[e] (39,040)	1	1

[a]Source: Burgess Kay, University of California, Davis, and Robert Crowell, Cagwin & Dorward, Landscape Contractors and Engineers, San Rafael, California.

[b]1 = minimal, 10 = excellent. Ratings assume treatments are properly applied.

[c]1984 west coast contract prices.

[d]Cost-effectiveness = 1000/cost per acre/effectiveness rating
The higher the number, the more cost-effective the treatment is. The short-term cost-effectiveness was computed by using the effectiveness rating for "before plant establishment." The combined short- and long-term cost-effectiveness was computed by using the sum of the effectiveness ratings for before and after plant establishment. When the effectiveness rating was a range, an average rating was used in the calculation.

[e]East coast costs for this treatment are considerably lower [$6050 to $13,310/acre ($14,950 to $32,890/ha)]. (6) See Tables 3.6 and 3.7 for additional cost data.

mination erosion protection and very high protection after plant establishment. Hydraulic seeding and mulching with wood fiber does provide adequate erosion protection if plants are established, but its pregermination effectiveness is low to moderate. Jute netting and jute over straw are highly effective but very costly treatments. They are best for small sites and critical areas.

The initial cost of applying and anchoring straw mulch is somewhat higher than the cost of hydraulic mulching with wood fiber. However, since wood fiber mulch provides far less erosion protection prior to plant establishment, reapplication will probably be required more often than if straw were used. In addition, sediment removal, slope repair, and reseeding may be required if wood fiber mulch fails to hold the soil. Thus the ultimate cost of using wood fiber mulch may well be higher than the cost of using straw.

6.6 GRASSED WATERWAYS

6.6a Purpose and Applications

Grass can be used to protect both temporary and permanent waterways from erosion. When the flow velocity in a channel exceeds the limit of the soil material (Sec. 7.3c), erosion will occur unless a channel lining is used. A grass lining reduces erosion by lowering water velocity over the soil surface and by binding soil particles with roots. Grass-lined channels are also aesthetically pleasing. The main disadvantages of grassed waterways are that such waterways require regular maintenance (primarily mowing) and occupy more space than paved or rock-lined channels. Grassed waterways should not be used on slopes greater than 15 percent or on slippage-prone slopes.

Permanent grassed waterways are best used where there is a relatively constant year-round flow of water. Their use is much more common in the eastern United States than in the west, where flows are much more irregular. Grass may also be used to protect temporary drainageways, such as dikes and swales.

6.6b Design Criteria

Design Storm

Unless otherwise specified in local or state regulations, the waterway should be designed to carry, as a minimum, the peak discharge from a 10-year, 24-hr storm. Where the consequences of overflow are likely to be serious, the capacity of the waterway should be increased correspondingly.

Channel Shape

Grassed waterways may be V-shaped, parabolic, or trapezoidal. V-shaped channels are generally used where flows will be small, as along roadsides. A channel with a parabolic cross section resembles a natural waterway. Because it has

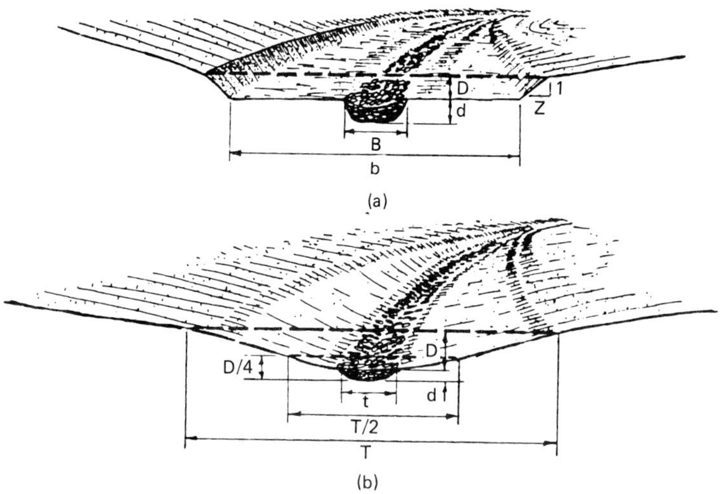

Fig. 6.28 Grassed waterway with stone center. (*a*) Trapezoidal cross section. (*b*) Parabolic cross section. (1, 10)

gentle side slopes, it is easily mowed. Trapezoidal channels should have stable side slopes that are not steeper than 2:1. Flatter side slopes are easier to plant and also to maintain.

Channel shape is often selected for the type of excavation equipment that is available. V-shaped channels are easily constructed with blade-type equipment. Bulldozers and scrapers are well suited to the construction of parabolic channels. They can also be used to construct trapezoidal channels *if* the bottom width of the channel is wider than the blade or bucket.

When base flows of extended duration and of other than immediate storm runoff origin are expected, structural protection should be installed. Three alternative types of base flow protection measures are:

- Stone center (Fig. 6.28)
- Subsurface drain
- Gabion mattress

Channel Dimensions

Channel cross sections, dimensional relations, and formulas are presented in Fig. 6.28 and in Appendix B. Use the methods described in Chap. 4 to calculate whether the initial channel design will be adequate to handle the expected flows. Also check whether the peak velocity will exceed the limit for the type of vegetation desired (see below). If the velocity will be too high, making the channel broader and shallower may solve the problem.

TABLE 6.7 Characteristics of Grasses Suitable for Lining Waterways (1, 13, 15)

Common name	Persistence/growth form	Description	Rating[a]	Gradient, %	Max. velocity ft/sec (m/sec)
Annual ryegrass or Italian ryegrass	Annual/bunchgrass	Well adapted to the Pacific coast states; commonly used for erosion control on bare soils; establishes rapidly but does not reseed well; best for first-year protection only.	***	0–5 5–10	5 (1.5) 4 (1.2)
Bermuda grass	Perennial/sod-forming	Well adapted to the southeastern, central, and southwestern United States; tough, vigorous grass that tolerates wet and dry, acid and alkaline soils; develops slowly from seed; may be used without other grass species.	****	0–5 5–10 Over 10	6 (1.8) 5 (1.5) 4 (1.2)
Kentucky bluegrass	Perennial/sod-forming	Grows throughout the humid regions of the United States; requires irrigation if planted elsewhere; a common turf grass.	***	0–5 5–10 Over 10	5 (1.5) 4 (1.2) 3 (0.9)
Redtop	Perennial/sod-forming	Grows well in the northeastern United States and northern Great Plains on poorly drained, acidic, and low-fertility soils; becomes coarse and stemmy.	***	0–5	3 (0.9)

Reed canarygrass	Perennial/sod-forming	Grows throughout the United States on both acid and alkaline soils; tolerates flooding and standing water; requires irrigation in dry climates.	***	0–5 5–10 Over 10	5 (1.5) 4 (1.2) 3 (0.9)
Smooth bromegrass	Perennial/sod-forming	"Northern" type is adapted to western Canada and the northern Great Plains; "southern" type is adapted to the central Great Plains; grows best on high-fertility soils.	***	0–5 5–10 Over 10	5 (1.5) 4 (1.2) 3 (0.9)
Tall fescue	Perennial/bunchgrass	Grows in humid areas of the United States and throughout the United States if irrigated; common turfgrass; may be used without other grass species.	****	0–5 5–10 Over 10	5 (1.5) 4 (1.2) 3 (0.9)
Western wheatgrass	Perennial/sod-forming	Grows in the northern and central Great Plains and western United States; tolerates drought, high alkalinity, and high sodium.	***	0–5 5–10	5 (1.5) 4 (1.2)

[a]Erosion protection ratings are as follows: * = fair; ** = good; *** = excellent; **** = superior.

6.6c Selection of Vegetation

Base the selection of vegetation for lining waterways on the local climate and soil conditions and on the capacity of the grass to resist erosion when handling the maximum expected flow velocities. Table 6.7 summarizes the characteristics of grasses recommended for planting in waterways in various regions of the United

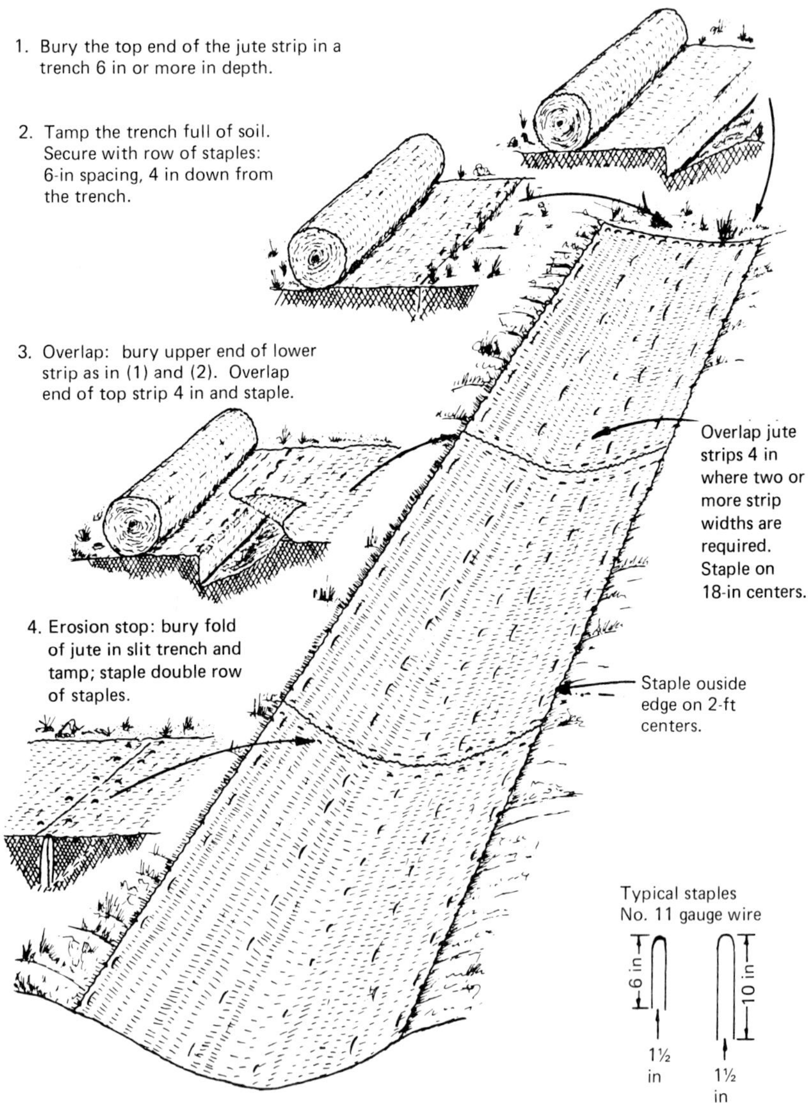
1. Bury the top end of the jute strip in a trench 6 in or more in depth.
2. Tamp the trench full of soil. Secure with row of staples: 6-in spacing, 4 in down from the trench.
3. Overlap: bury upper end of lower strip as in (1) and (2). Overlap end of top strip 4 in and staple.
4. Erosion stop: bury fold of jute in slit trench and tamp; staple double row of staples.
Overlap jute strips 4 in where two or more strip widths are required. Staple on 18-in centers.
Staple ouside edge on 2-ft centers.
Typical staples No. 11 gauge wire
6 in
10 in
1½ in
1½ in

Fig. 6.29 Detail for stabilizing grassed waterways with jute mat. (10)

TABLE 6.8 Grass Establishment Methods (Adapted from Ref. 16)

Conditions	Method
• Slopes less than 5% • Velocity less than 3 ft/sec (0.9 m/sec) • Drainage can be diverted during grass establishment period • Erosion-resistant soils	• Seed and straw mulch; anchor straw mechanically or with a glue. • Establish Bermuda grass by sprigging.
• Slopes less than 5% • Velocity less than 5 ft/sec (1.5 m/sec) • Drainage cannot be diverted during grass establishment period • Moderately erodible soil	• Seed and straw mulch; anchor straw with jute netting or a synthetic erosion liner.
• Slopes greater than 5% • Velocity 5 to 6 ft/sec (1.5 to 1.8 m/sec) • Drainage cannot be diverted during grass establishment period • Highly erodible soil	• Sodding.

States. The right-hand column of the table gives the maximum permissible velocity for each grass type. Use the Manning and continuity equations to calculate flow velocity in the channel (see Chap. 4).

6.6d Grass Establishment

The proper technique for establishing grass in a waterway depends on the severity of planting conditions. It is best to seed when there is no water flowing in the channel. On the west coast, grass should be seeded during the dry season and established with irrigation before October 1. In other areas, flows should be diverted during the grass establishment period, if possible. Where flows cannot be diverted, seed and mulch should be anchored in place with jute matting or another suitable lining material. Figure 6.29 illustrates the procedure for installing a jute lining. Table 6.8 presents three alternative grass establishment methods and the conditions under which each method should be used. They are listed in order of increasing severity and cost.

6.6e Maintenance

Grassed waterways should be inspected after seeding and after each major storm. Eroded areas and areas damaged by vehicles or rodents or otherwise damaged should be reseeded and mulched immediately. Channels must be mowed or headed periodically to maintain the drainage capacity of the waterway.

REVIEW QUESTIONS

1. In what ways does vegetation help prevent erosion?
2. Why is grass a better erosion control plant than flowers, shrubs, or trees?
3. Name four beneficial effects of mulching.
4. What are the advantages of straw over other types of mulch materials?
5. Why is it important to roughen a slope before seeding?
6. Why is the time of seeding important?
7. Why is it important to seed by October 1 in the valleys, foothills, and coastal portions of California?
8. What are the possible causes of erosion on a slope that has been properly seeded and mulched?
9. If erosion control during the initial months after seeding is a critical concern, what extra precautionary measures (in addition to seeding) are recommended?
10. How can too much mulch inhibit plant growth?

REFERENCES

1. Association of Bay Area Governments, *Manual of Standards for Erosion and Sediment Control Measures,* ABAG, Oakland, Calif., 1981.
2. R. P. Beasley, *Erosion and Sediment Pollution Control,* The Iowa State University Press, Ames, Iowa, 1972.
3. F. W. Bennett and R. L. Donahue, *Methods of Quickly Vegetating Soils of Low Productivity, Construction Activities,* U.S. Environmental Protection Agency, Washington, D.C., 1975.
4. D. Farber, "Recommended Planting Time for Vegetative Stabilization of Construction Sites," Technical Memorandum No. 51, in *San Francisco Bay Area Environmental Management Plan, Appendix J,* Association of Bay Area Governments, Oakland, Calif., 1981.
5. D. H. Gray and A. T. Leiser, *Biotechnical Slope Protection and Erosion Control,* Van Nostrand Reinhold, New York, 1982.
6. K. R. Kaumeyer, and R. E. Benner, "Comparative Costs of Implementing Erosion and Sediment Control Practices in Maryland," unpublished paper, State of Maryland Water Resources Administration, Annapolis, Md., 1983.
7. B. L. Kay, "Mulches for Erosion Control and Plant Establishment on Disturbed Sites," *Agronomy Progress Report No. 87,* University of California, Davis, Agricultural Experiment Station Cooperative Extension, Davis, Calif., 1978.

8. B. L. Kay, "Straw as an Erosion Control Mulch," *Agronomy Progress Report No. 140,* University of California, Davis, Agricultural Experiment Station Cooperative Extension, Davis, Calif., 1983.
9. A. T. Leiser et al., Departments of Environmental Horticulture and Agronomy and Range Sciences, University of California, Davis, *Revegetation of Disturbed Soils in the Tahoe Basin,* California Department of Transportation, Sacramento, Calif., 1974.
10. Maryland Water Resources Administration, U.S. Soil Conservation Service, and State Soil Conservation Committee, *1983 Maryland Standards and Specifications for Soil Erosion and Sediment Control,* MWRA, Annapolis, Md., 1983.
11. F. W. Schaller and P. Sutton (eds.), *Reclamation of Drastically Disturbed Lands,* American Society of Agronomy, Crop Science Society of America, Soil Science Society of America, Madison, Wis., 1978.
12. H. Schiechtl, *Bioengineering for Land Reclamation and Conservation,* The University of Alberta Press, Edmonton, Alberta, 1980.
13. G. O. Schwab, *Elementary Soil and Water Engineering,* 2d ed., John Wiley & Sons, Inc., New York, 1971.
14. A. Thornburg, *Plant Materials for Use on Surface Mined Lands in Arid and Semiarid Regions,* U.S. Department of Agriculture, Soil Conservation Service, GPO, Washington, D.C., 1982.
15. F. R. Troeh et al., *Soil and Water Conservation for Productivity and Environmental Protection,* Prentice-Hall, Inc., Englewood Cliffs, N.J., 1980.
16. Virginia Soil and Water Conservation Commission, *Virginia Erosion and Sediment Control Handbook,* 2d ed., VSWCC, Richmond, Va., 1980.

chapter 7

Design and Installation of Water Conveyance and Energy Dissipation Structures

Chapters 7 and 8 provide design guidelines for structural control measures that reduce erosion and retain sediment on construction sites. Structural measures fall into three general categories: conveyance facilities, energy dissipation structures, and sediment retention structures. The conveyance facilities described in this chapter are designed to carry runoff in a nonerosive manner into a natural stream or a permanent storm drain system. Energy dissipation structures are designed to reduce concentrated flow velocities to nonerosive rates. Sediment retention structures, presented in Chap. 8, are designed to remove sediment carried in runoff.

Structural erosion and sediment control measures are more effective when they are combined with vegetative measures and good grading practices as part of a comprehensive erosion control plan. Control measures may be either permanent or temporary depending on whether they will remain in use after development is complete. In some cases, permanent measures can be designed to also serve temporary purposes during construction. Using permanent facilities for temporary erosion control during development reduces the total cost of erosion control.

Sediment basins and traps can be located around storm drain inlets. The riser on a sediment basin can discharge to the permanent storm drain system. Temporary drainageways and sediment basins can become permanent open space and function as stormwater management systems. Diversions and subsurface drainage on fill slopes are essential for prevention of slope failure. Both temporary and permanent measures are discussed in the following sections, but the emphasis is on temporary structures.

The design of temporary measures is simplified by carefully predetermining certain design factors. Maximum drainage area and slope restrictions are noted throughout the discussion of these measures. These limitations are important; if they are not observed, the simplest measures will fail and erosion control will not be achieved. Perhaps the most common cause of failure is lack of recognition of the size of peak storm flows and the erosive forces that these flows generate. Permanent measures require more careful design and installation. Permanent waterways and energy dissipators require engineered designs, just as storm drain systems do. All of them must be capable of carrying the design peak flow specified by the local jurisdiction.

Although design and construction guidelines are presented in some detail, this chapter is not a substitute for training in hydraulic and structural engineering. The materials presented are guidelines to assist in the design of erosion control measures. Measures which require engineering must be designed by or under the supervision of a registered professional engineer. The standards and specifications provided should not be considered rigid requirements, particularly where local experience has shown that a smaller structure will work or that a larger one is needed. The authors encourage site planners and engineers to continuously seek new, more reliable solutions for controlling erosion and sediment.

The focus of this chapter is on the design and stabilization of *man-made* water conveyance facilities on construction sites. However, it is likely that natural stream channels will cross or be adjacent to such sites and that the natural streams may themselves have erosion problems. Streambank stabilization is briefly discussed in Sec. 7.9, and references to comprehensive sources of information on this subject are given at the end of the chapter.

7.1 PRACTICAL CONSIDERATIONS IN CONTROL MEASURE DESIGN

7.1a Control Measures Should Be Economical

Most erosion control measures are temporary. Use of costly materials and elaborate designs is not cost-effective.

Use Materials Available On-Site

Construct control measures with materials present on the site whenever possible. Use earth from grading to build dikes, rocks for riprap and outlet protection, and logs for check dams.

Keep Structures Simple

Simple structures will do the job in most cases. They are easier to design, are less prone to improper installation, and cost less. Examples of simple structures are riprap aprons as energy dissipators, gravel-lined swales, and excavated sediment traps around storm drain inlets.

Take Advantage of Permanent Facilities

Regardless of erosion control needs, developers must build permanent storm drain systems for projects. By making these permanent facilities double as temporary erosion control measures, overall costs are reduced. Install the storm drain system early. Use storm drain inlets as the outlets for sediment traps. A riser pipe installed over a storm drain can serve as the outlet for a sediment basin.

7.1b Install the Most Important Control Measures First

Don't wait for the first rains to arrive before implementing the plan. If one large sediment basin is to treat runoff from a project, put that basin in first. As each fill slope is finished, build dikes along the top of the slope and have the slope seeded.

7.1c Install Control Measures Correctly

Control measures won't work if they are not installed properly. Follow the standards and specifications in approved reference manuals (1, 6, 19). Visit the site during the early rains to be sure the measures are properly located and constructed. Make sure that erosion is not beginning to occur around the structural measures or anywhere else on the site.

7.1d Time Installation of Control Measures to Minimize Land Disturbance

If a sediment basin is to be located at the bottom of a fill, build the basin before you construct the fill. Once the fill is in place, you may not be able to get heavy equipment and materials down to the basin site. Similarly, install midslope diversions and backfill utility and drainage trenches before seeding to avoid disturbing the seedbed. Reseed by hand if necessary.

7.1e Don't Block a Natural Drainageway

Don't install control measures across a creek or stream bed. Runoff from adjacent undisturbed areas does not need treatment. The additional runoff volumes from undisturbed areas will require larger structures with unnecessary extra costs. Higher than expected flow reaching a structure in a drainageway is one of the most common causes of failure. Catch and treat the runoff from the disturbed area *before* it reaches a creek.

7.1f Place Control Measures Out of the Way of Construction Operations

If you put a barrier across a vehicle route, it will get run over or pushed aside. Try to place measures out of the way. For example, locate a sediment basin in a future park site, green belt, or cul-de-sac—a place where it does not have to be removed to make way for future construction. Another good location is the last lot in a tract scheduled for construction.

7.1g Make Field Modifications Where Necessary

The inspector and superintendent should compare measures shown on the plans with the field conditions and ask themselves: What is this measure supposed to do? Does the placement of the measure make sense? Is it going to work as well as it is intended to? Would something else be more effective and easier to maintain or less likely to be disturbed by operations?

In many jurisdictions, suitable modifications that are agreed upon and signed off by both the inspector and superintendent can be made. If you are changing a plan, make sure that any drainage diversions are accounted for in the design of erosion and sediment control measures. *Note:* In some jurisdictions, the director of public works or another responsible official may have to approve a plan modification.

7.1h Provide Access for Maintenance

Sediment basins should have an access road nearby so that heavy equipment can be used to clean out the sediment. Waterways require access for streambank repair, debris removal, or mowing. Most control measures require some form of regular maintenance.

7.1i Be Pragmatic

Suppose the site conditions require a certain storage capacity in a sediment basin. Suppose also that a basin in the most convenient location for it can easily meet 90 percent of that demand but meeting the last 10 percent will require excavating into a steep hill. Rather than go to the extra expense—and the increased erosion and slope stability hazard associated with disturbing the hill—look for other ways to solve the problem. Perhaps some additional mulch or vegetative cover would reduce sediment delivery. Alternatively, you can schedule more frequent inspections and cleanings.

Every site is unique. The methods and measures remain the same, but the combination used differs with the site. Evaluate each situation and use common sense.

7.2 TYPICAL APPLICATIONS OF WATER CONVEYANCE STRUCTURES

Temporary and permanent drainageways offer effective means of reducing slope erosion and preventing damage to both construction areas and off-site properties. When a drainageway is placed above a disturbed slope, it reduces the *volume* of water reaching the disturbed area by intercepting runoff from above. When a drainageway is placed horizontally across a disturbed slope, it reduces the *velocity* of runoff flowing down the slope by shortening the distance that the runoff can flow directly downhill. The size and type of structure may vary with the drainage area and the degree of slope; a temporary furrow may suffice for a small area. The basic principle is to intercept water flowing across or toward a disturbed area and convey it at a 1 to 5 percent grade toward a sediment basin, an erosion-resistant drainage channel, or a fairly level, well-vegetated area such as a meadow.

Swales, dikes, and waterways convert sheet flow to concentrated flow. The energy thus accumulated can erode the channel surface unless some means is provided to reduce the velocity of flow. Check dams, energy dissipators, and channel linings perform this function.

All water conveyance structures must have adequate outlets. An outlet may be a man-made or a natural waterway, but it is likely that some kind of outlet protection will be necessary. Riprap aprons and energy dissipators are used to convert pipe flow to channel flow and reduce the velocity of channel flow in the transition from a paved channel to a natural waterway. A variety of conveyance structures and channel and outlet protection are described in the following sections.

7.2a Dikes and Swales

Dikes and swales are temporary, nonengineered channels. A *dike* is a ridge of compacted soil. A *swale* is a ditch cut into the soil. Dikes and swales are the simplest ways to convert sheet flow to channel flow and convey the runoff to a permanent channel, a storm drain, or a sediment retention structure. A combination dike and swale is easily constructed by a single pass of a bulldozer followed by wheel-walking the ridge to compact it. The most common uses of dikes and swales are:

- To intercept runoff and divert it from an exposed or newly seeded slope toward an acceptable outlet (Fig. 7.1).
- To intercept runoff from undisturbed areas and prevent it from entering a disturbed area such as a house pad.

Fig. 7.1 Dike prevents road runoff from discharging onto newly seeded slope.

Fig. 7.2 Cross-road swale.

- To reduce the velocity of runoff flowing down an unpaved street surface or access road (Fig. 7.2). Dikes and swales so used are often called waterbars or cross-road drains.
- To collect the runoff from a gently sloping, disturbed area and carry it to a sediment retention structure before releasing it to a natural drainageway or into the storm drain system (Fig. 7.3).

Figure 7.4 illustrates the use of a swale and dike system both to *divert* off-site runoff away from the disturbed area and to *collect* on-site runoff and direct it to a sediment retention structure.

7.2b Pipe Slope Drains and Paved Chutes

The terms "pipe slope drain" and "chute" refer to structures that carry concentrated runoff down steep slopes without causing erosion. Sometimes they are referred to as grade stabilization structures. A pipe slope drain is constructed with flexible or rigid pipe and can be temporary or permanent. Chutes, also

Fig. 7.3 V ditch.

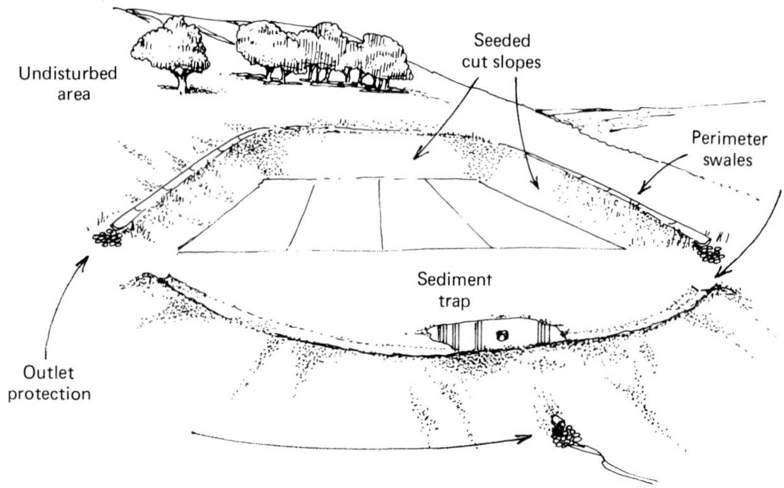

Fig. 7.4 Perimeter swales used to control runoff around a disturbed area.

Fig. 7.5 Pipe slope drain.

called flumes or spillways, are paved, open, permanent stormwater conveyance structures. They are commonly used to:

- Temporarily convey runoff down the face of a cut or fill slope until permanent facilities are installed and the slope is stabilized with vegetation (Fig. 7.5)
- Prevent further cutting of a gully (Fig. 7.6)
- Serve as a permanent drainageway down a steep slope where no other, less visible solution is possible
- Serve as an emergency spillway for a sediment basin

7.2c Permanent Waterways

A permanent waterway is a man-made drainage channel designed, shaped, and lined to provide for safe disposal of surface runoff over many years. In U.S. Soil Conservation Service publications, the term "diversion" is sometimes used to describe such a waterway. A permanent waterway can be lined with concrete,

Fig. 7.6 Temporary pipe slope drain installed to prevent further cutting of a gully.

Fig. 7.7 Midslope diversion.

asphalt, gravel, grass, or a combination of grass and rock. The choice depends on design factors discussed later. A permanent waterway can be used to:

- Divert runoff away from cut or fill slopes
- Shorten long slopes and thereby reduce runoff velocity and soil loss (Fig. 7.7)
- Intercept shallow subsurface drainage and help reduce soil slump
- Convey runoff to the permanent storm drain system
- Provide the entire stormwater conveyance system (Fig. 7.8)

Fig. 7.8 Grass- and gravel-lined waterway used as a storm drain.

Figure 2.5 depicts a large fill slope with concrete-lined diversions at regular intervals. The diversions are required by the Uniform Building Code for drainage control (5). The soil surface is stabilized with a grass cover.

Figure 7.8 shows part of the drainage system at the Village Homes subdivision in Davis, California. Rather than construct a conventional storm drain system composed of pipes and drop inlets, the developer built a network of vegetated waterways resembling natural creeks. The cost of building this natural-looking drainage system, including its landscaping, was less than the estimated cost of constructing a conventional storm drain system. The system has functioned well. During heavy rains, when adjacent subdivisions with pipe storm drain systems were experiencing flooding of streets and sidewalks, the waterways of the Village Homes did not overflow. But although grass-lined permanent waterways offer several advantages over paved waterways, they do require more space and maintenance and are not suitable on steep hillsides.

7.2d Check Dams

A check dam is a small temporary dam built across a swale or drainage ditch. Its function is to reduce the velocity of flow in the channel and thus reduce erosion of the channel bed. A check dam can be used:

- As a substitute for channel lining in a temporary swale. (This is not the preferred approach.)
- To protect a grass-lined channel during initial establishment of the vegetation.

Check dams are closely related to sediment barriers; the materials and construction are similar. Sandbags, logs, rock, silt fences, gravel filters, and straw bale dikes can function as check dams or sediment barriers depending on placement. Check dams are located in flow paths; their primary purpose is to slow velocity and reduce channel bed erosion. Sediment capture is a secondary function. Sediment barriers, discussed in Chap. 8, intercept runoff before it enters a channel and serve principally to capture sediment.

7.2e Channel Linings

The concentrated flows carried by drainageways frequently develop velocities sufficient to erode the channels. Temporary channels with slope gradients less than 3 percent may remain unlined if velocities are sufficiently low, but channels with greater slopes generally require linings to prevent erosion of the channel beds. Permanent channels should always be lined or vegetated regardless of slope. Channel linings have several secondary functions that influence the choice of lining material; they are:

- To slow velocity in channels and reduce peak flow. A lining that slows velocity not only reduces erosion but also reduces peak flow by spreading it over a longer time period. Thus it can be used as part of a storm water management system. However, since lowering velocity reduces the flow rate, a larger channel may be required.

- To encourage infiltration into the soil. If a lining is permeable, infiltration of water occurs, and that encourages plant growth.
- To prevent infiltration into the soil where slope stability is a major concern. These linings must be impermeable.

Permeable lining materials include:

- Gravel or rock
- Grass
- Grass-gravel combinations
- Jute or other fabrics

The choice of lining is governed by velocity of flow, cost, need for permanence, aesthetics, desirability of infiltration, and many other factors. Gravel and fabrics can be used as temporary linings. Gravel is simplest: it is used in small swales; it withstands gentle flows; and it is a common building material (Fig. 7.9). Linings such as jute netting, excelsior, or synthetic filter fabric can provide protection for up to 15 percent slopes. However, there is always a danger of the water running underneath the fabric.

Grass and grass-gravel combinations are practical as permanent linings. Grass linings can be installed on slopes up to 10 percent. A gravel low-flow channel in the center of a grassed waterway is desirable when a constant minimum flow exists, since it is difficult to maintain grass under continuous water flow conditions (Fig. 7.8). Large rock may be required to withstand higher velocities (Fig. 7.10).

Fig. 7.9 Gravel-lined swale.

Fig. 7.10 Rock-lined waterway.

Fig. 7.11 Plastic-lined ditch.

When infiltration is not desirable, an impermeable lining is required. It can be one of the following:

- Grouted riprap
- Concrete
- Asphalt
- Pipe
- Plastic sheeting

Plastic sheets or half sections of plastic or metal pipe provide quick, temporary impermeable linings (Fig. 7.11), but care must be taken to prevent the flow from getting under the lining. Such linings generally work for only a short time, and they should not be used under extreme conditions (e.g., steep slopes and high flows). Grouted riprap is often too expensive and impractical for use in slope diversions, but it can, and often does, serve to protect stream banks (Fig. 7.12). Grouted rock also serves to slow velocity in the waterway.

Permanent top-of-slope and midslope diversions on constructed fill are examples of situations in which an impermeable lining *must* be used. Excessive water infiltration into a fill slope may threaten the slope's stability. Asphalt and concrete are the two most commonly used impermeable linings (Fig. 7-13). Since these linings increase flow velocity rather than decrease it, higher peak flows may

Fig. 7.12 Grouted riprap channel protection.

result, and they should be accounted for when downstream facilities are designed.

7.2f Outlet Protection

The purpose of outlet protection is to prevent erosion around outlets and in the downstream reaches of natural drainage channels. The outlets of pipes, lined channels, and other structures that concentrate flow are points of critical erosion potential. Storm water which is transported through man-made conveyance systems often reaches a velocity which causes erosion in the receiving channel. To prevent scour at outlets, a flow transition structure is needed to dissipate the flow's high energy and reduce the flow velocity to a level which will not erode the receiving channel.

Energy dissipation devices are usually permanent structures and thus require engineered design. The most commonly used device for outlet protection is a lined apron. The aprons can be lined with riprap (Fig. 7.14), grouted riprap, or concrete. Two typical concrete energy dissipators are illustrated in Figs. 7.15 and 7.16. The energy dissipator shown in Fig. 7.15 is not a good design because it

Fig. 7.13 Asphalt-lined channel.

Fig. 7.14 Riprap used for outlet protection.

discharges at a steep angle, which causes the water to accelerate after it hits each obstacle. In addition, the last row of obstacles is located too far above the base of the chute. This structure could have been improved by locating it at the bottom of the slope, where the pipe could discharge through obstacles on a more level slope. In most cases, smaller, less visually obtrusive structures such as rock aprons (Fig. 7.14) can do the job.

7.3 DESIGN AND INSTALLATION OF DIKES AND SWALES

Because dikes and swales are small-scale temporary channels, they do not require formal design. They may be reconstructed daily to protect exposed areas as grading proceeds. However, they do convert sheet flow to channel flow, and thus they increase erosion potential in the channel. Certain design factors must be observed to ensure proper function.

Fig. 7.15 Poorly designed energy dissipator.

Fig. 7.16 Impact-type energy dissipator.

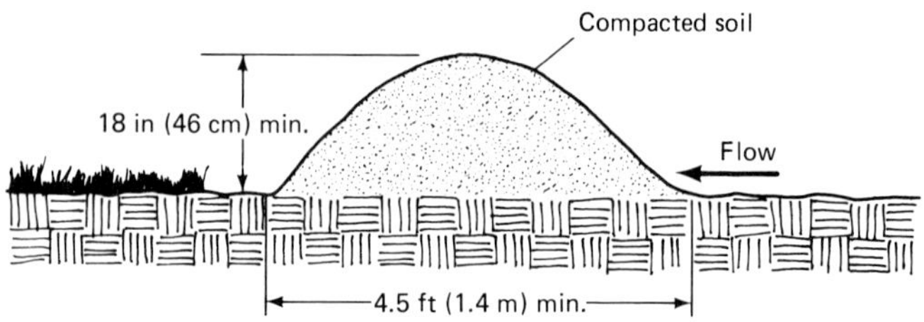

Fig. 7.17 Example of temporary diversion dike. (19)

7.3a Drainage Area

The drainage area contributing runoff to a temporary earth berm or swale should be less than 5 acres (2 ha). A number of publications (1, 6, 19) present standard designs and specifications for dikes and swales. The standard designs are typically sized for a 5-acre (2-ha) maximum drainage area. If a larger area is served, channel design must include runoff calculations and follow the specifications for waterways (Sec. 7.5). A larger cross section, and often a channel lining, will be required to handle the expected flows.

7.3b Channel Shape and Size

Temporary dikes and swales do not have precise cross sections; they are simply shaped by grading equipment to a round, V, or trapezoidal contour.

Dikes

The minimum height for a dike is 18 in (46 cm) of compacted material. The side slopes should have a 2:1 or flatter slope and be stabilized by seeding with grass if the dike is to serve for more than 30 days. The top width should be at least 2 ft (0.6 m) (Fig. 7.17).

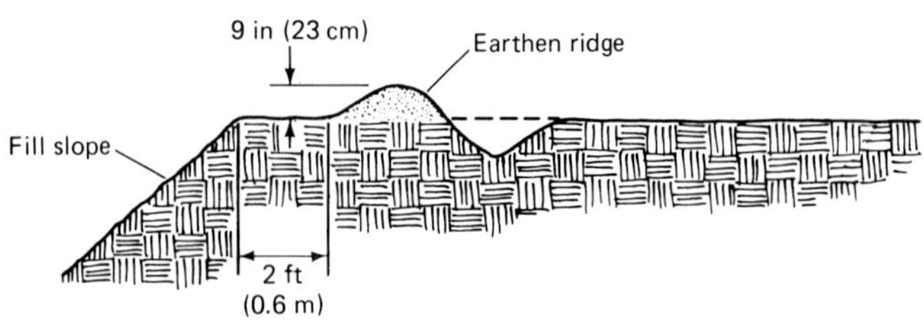

Fig. 7.18 Temporary top-of-slope diversion: earth berm and swale. (19)

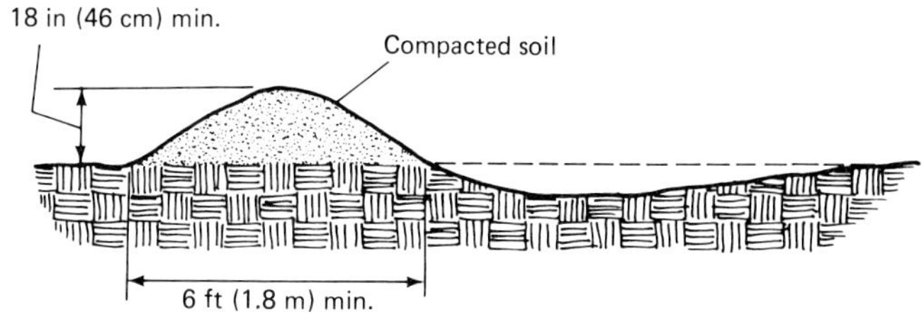

Fig. 7.19 Example of a typical earthen structure for temporary right-of-way diversion. (19)

Swales

A swale is a ditch cut into the soil. The ditch should be at least 9 to 12 in (23 to 30 cm) deep and have 2:1 or flatter side slopes. The width should be up to 7 ft (2.1 m), depending on flow conditions. The width should be calculated by an engineer using the rational or another accepted method.

Dike and Swale Combinations

A swale with a dike on the downslope side of the channel is easily constructed by running a grader or small bulldozer with its blade tilted to one side near the edge of an exposed slope (Figs. 7.18 and 7.19). Unless the berm is compacted, this diversion should not remain in place for more than 1 week.

Cross-Road Drain

A swale or dike built diagonally across a right-of-way serves to reduce roadway erosion; it is called a cross-road drain, waterbar, or right-of-way diversion. Compacted earth dikes can sustain limited amounts of vehicle traffic, but they are susceptible to breaching by wheels. If a breach occurs, severe gullying in the roadbed may result. A preferred design is a cross-road swale or dike and swale combination (Fig. 7.19). Swales are less prone to breaching by vehicle wheels, but they must be broad and shallow to prevent jarring of vehicles. Filling the swale with gravel may prolong its life.

Spacing of cross-road drains is determined by the slope of the road surface as listed in the accompanying table. (19)* Caution should be exercised in using cross-road drains on roads with slopes greater than 25 percent. Cross-road drains should *not* discharge onto sensitive areas such as steep, bare slopes.

Slope of Road	Spacing, ft (m)
Less than 7%	100 (30)
7–25%	75 (23)
25–40%	50 (15)
Greater than 40%	25 (8)

*Reference 13 provides more detailed information on cross-road drain spacing, which takes into account soil type as well as the slope of the road.

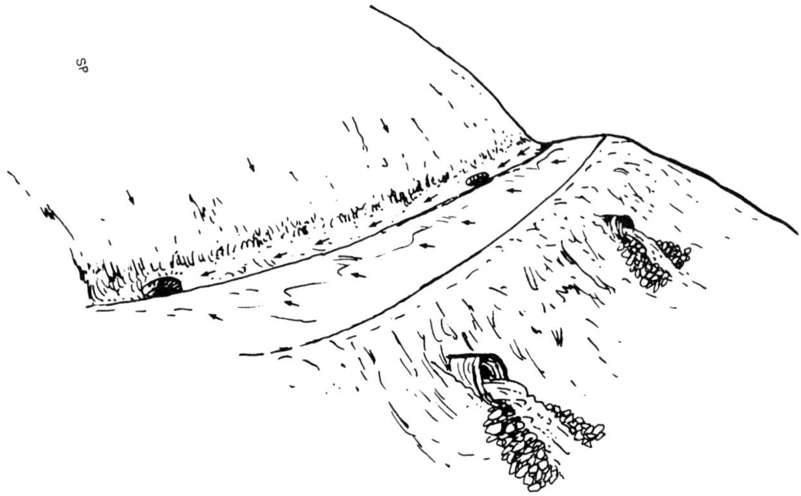

Fig. 7.20 Culverts under roads are a more permanent and more costly solution to controlling road erosion.

Two alternatives to cross-road swales are illustrated in Figs. 7.20 and 7.21. Installing culverts underneath a road at regular intervals, as an alternative to open channels at grade, eliminates the possibility of damage by vehicles. Disadvantages of that approach are high installation cost and a higher cost to remove the drains when temporary roads are no longer needed. [Many public agencies, such as the University of California Institute of Transportation Studies (8), publish guidelines for sizing and installing culverts.] An open-top box culvert made of wood (Fig. 7.21) is a lower-cost solution which has been found to work well. It is built by using lumber commonly available on construction sites.

7.3c Flow Velocity

Dikes and swales must have positive drainage (i.e., a minimum slope of 1 to 3 percent). Low points in a channel will collect water, which may overtop the dike and cause erosion of the slope below (Fig. 7.22). Slopes steeper than 3 percent are not desirable in unlined channels, because the flow velocity may cause erosion of the channel bed. Table 7.1 lists the maximum velocities allowable for different soil types. Because unlined channels at a construction site are unlikely to be aged, as assumed in this table, it is strongly recommended that the *lower velocities* (the column for clear water) be used as the *upper limits* for permissible flow. Expected flow velocity is calculated by first determining peak flow by the rational method and then determining velocity by using Manning's equation, as explained in Chap. 4.

If the maximum velocity will be exceeded, widening the channel may lower the velocity enough to avoid the necessity of a channel lining. All other condi-

(a)

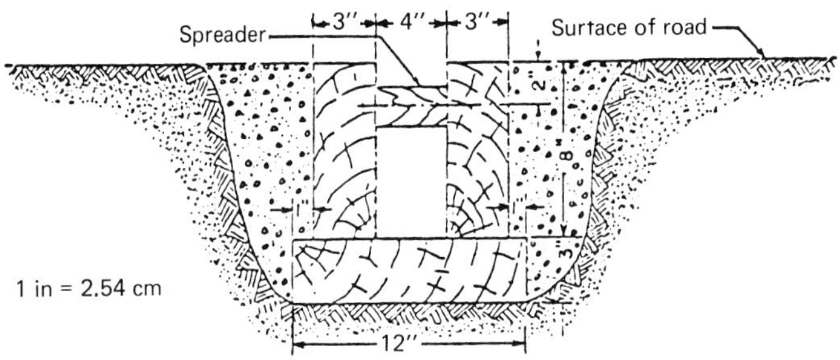

(b)

Fig. 7.21 Open-top wooden box culvert. (*a*) General view; (*b*) cross section. (13)

Fig. 7.22 Improper swale installation: water ponds at low point.

TABLE 7.1 Maximum Permissible Velocities for Unlined Channels (Adapted from Ref. 15)*

Material	Mean velocity	
	Clear water, ft/sec (m/sec)	Silty water, ft/sec (m/sec)
Fine sand, colloidal	1.50 (0.457)	2.50 (0.762)
Sandy loam, noncolloidal	1.75 (0.533)	2.50 (0.762)
Silt loam, noncolloidal	2.00 (0.610)	3.00 (0.914)
Alluvial silts, noncolloidal	2.00 (0.610)	3.50 (1.067)
Ordinary firm loam	2.50 (0.762)	3.50 (1.067)
Volcanic ash	2.50 (0.762)	3.50 (1.067)
Stiff clay, very colloidal	3.75 (1.143)	5.00 (1.524)
Alluvial silts, colloidal	3.75 (1.143)	5.00 (1.524)
Shales and hardpans	6.00 (1.829)	6.00 (1.829)
Fine gravel	2.50 (0.762)	5.00 (1.524)
Graded loam to cobbles, noncolloidal	3.75 (1.143)	5.00 (1.524)
Graded silts to cobbles, colloidal	4.00 (1.220)	5.50 (1.676)
Coarse gravel, noncolloidal	4.00 (1.220)	6.00 (1.829)
Cobbles and shingles	5.00 (1.524)	5.50 (1.676)

*As recommended by Special Committee on Irrigation Research, ASCE, 1926. Based on uniform flow in continuously wet, aged channels.

tions being equal, shallower channels have lower flow velocities. Alternatively, a channel lining may be installed.

7.3d Channel Lining

If velocity exceeds the maximum for the soil type, a lining is required. In temporary swales, gravel is the easiest lining material to use. Grass or a fabric such as jute also may be used. See Sec. 7.7 for further details.

7.3e Stabilization

If a dike or swale must serve for more than 30 days, it should be seeded with grass (see Chap. 6).

7.3f Outlet

The diverted runoff, if free of sediment, must be released to a stabilized outlet or channel. Sediment-laden runoff must be diverted to a sediment-retention structure (see Chap. 8).

7.3g Construction Specifications

It is recommended that the following construction specifications be incorporated in erosion control plans or grading plans.

Dikes

- Whenever feasible, dikes shall be built before construction begins.
- Dikes shall be compacted.
- Dikes shall be seeded and mulched within 15 days of construction.
- Dikes shall be located to minimize damage by construction operations and traffic.
- In areas where freezing temperatures occur, frozen materials shall not be incorporated in dikes or fill nor shall a dike or fill be placed on a frozen foundation. (As the material thaws, it may settle and shift.)

Dike and Swale Diversions above Fill Slopes

- Diversions shall be constructed at the top of each fill at the end of each work day as needed.
- Diversions shall be located at least 2 ft (0.6 m) uphill from the top edge of each fill (Fig. 7.18).
- The dike portion of the diversion shall be constructed with a uniform height along its entire length.

Temporary Right-of-Way Diversions

- Diversions shall be installed as soon as the right-of-way has been cleared or graded.
- All earthen diversions shall be machine- or hand-compacted in 8-in (20-cm) lifts.
- Outlets of diversions shall be located on undisturbed or stabilized areas when possible. Field locations of diversions should be adjusted as needed to make use of stabilized outlets.

7.4 DESIGN AND INSTALLATION OF PIPE SLOPE DRAINS AND PAVED CHUTES

7.4a Pipe Slope Drains

Pipe slope drains are used to convey storm water from the top to the bottom of a slope. They are usually used in conjunction with top-of-slope diversion dikes or swales. It is very important that these temporary structures be installed properly, since their failure will often result in severe gully erosion. The entrance section must be securely entrenched; all connections must be watertight; and the conduit must be staked securely to the slope. Pipes are preferred to open chutes because the water cannot spill out and cause erosion of the slope.

Drainage Area

The maximum drainage area per pipe slope drain is 5 acres (2 ha). For larger areas, a paved chute should be installed.

Pipe Conduit

Heavy-duty flexible pipe or corrugated metal pipe can be used. The diameter of the pipe should be uniform throughout the pipe's length. Reinforced hold-down grommets should be spaced at 10-ft (3-m) intervals or closer.

The accompanying table was developed by the U.S. Soil Conservation Service as a guide for sizing pipe slope drains in Maryland, where the annual rainfall is approximately 40 in (1020 mm). (6, 14) The pipe dimensions for such drains on a project site should, however, be calculated by a qualified engineer and be based on local conditions.

Pipe diameter, in (cm)	Maximum drainage area, acres (ha)
12 (30)	0.5 (0.2)
18 (46)	1.5 (0.6)
21 (53)	2.5 (1.0)
24 (61)	3.5 (1.4)
30 (76)	5.0 (2.0)

Entrance Section

The entrance consists of a standard flared end section for culverts with a minimum 6-in (15-cm) metal toe plate (Fig. 7.23). This toe plate, also called a cutoff wall, prevents runoff from undercutting the pipe inlet. The slope of the entrance should be at least 3 percent (Fig. 7.24).

Dikes

An earth dike is used to direct runoff into a slope drain. The height of the diversion dike at the inlet should be the diameter of the pipe plus 1 ft (30 cm) of freeboard. Where the dike height is more than 18 in (46 cm) at the inlet, slope the top of the dike at the inlet at a gradient of 3:1 or flatter to connect with the remainder of the dike. Maximum side slope steepness should be 2:1.

Outlet Protection

Since slope drains carry concentrated runoff, outlet protection is essential. The sample drawing for a pipe slope drain (Fig. 7.24) includes a riprap apron. For a

Fig. 7.23 Flared entrance with toe plate at pipe slope drain inlet.

given pipe diameter D the apron should be $3D$ wide and $6D$ long, and it should consist of 6-in- (15-cm-) diameter stone placed at least 12 in (30 cm) deep. The apron should have sides sloping inward, and it should be at least as deep as the pipe diameter. Note that the slope of the outlet section of the pipe slope drain is nearly flat before the drain discharges onto the riprap apron. Thus the outflow is not accelerating as it enters or leaves the apron. See Sec. 7.8 for further details on design of outlet protection.

If the runoff is sediment-laden, a sediment trap should be installed *below* the energy dissipator at the pipe outlet. If the area draining to a pipe slope drain is larger than about 5 acres (2 ha), flow at the receiving channel below the outlet protection should be calculated to check for erosive velocity.

Construction Specifications

It is recommended that the following construction specifications be incorporated in erosion control plans or grading plans.

- The inlet pipe shall have a slope of 3 percent or steeper.
- The top of the earth dike over the inlet pipe, and dikes carrying water to the pipe, shall be at least 1 ft (30 cm) higher at all points than the top of the inlet pipe.

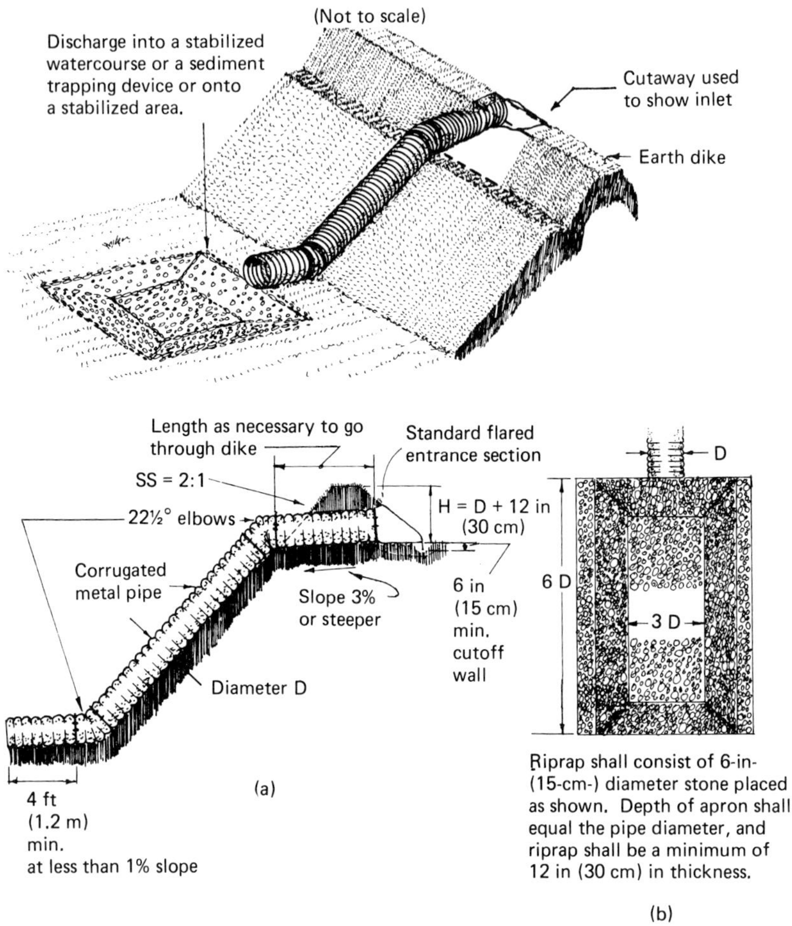

Fig. 7.24 Pipe slope drain. (*a*) Profile; (*b*) riprap apron plan. Drainage area must not exceed 5 acres. (Adapted from 14)

- The soil around and under the inlet pipe and entrance section shall be hand-tamped in 4-in (10-cm) lifts to the top of the earth dike.
- The inlet pipe shall be corrugated metal pipe with watertight connecting bands.
- If flexible tubing is used, it shall be the same diameter as the inlet pipe and shall be constructed of durable material with hold-down grommets spaced no more than 10 ft (3 m) apart. Flexible tubing shall be securely fastened to corrugated metal pipe with metal strapping or watertight collars. Flexible tubing shall be securely anchored to the slope.

- If pipe is plastic or corrugated metal, slope drain sections shall be securely fastened together and have watertight fittings. Stakes shall be driven into the soil on both sides of the pipe at 10-ft (3-m) intervals to prevent sideways movement of the pipe. At a minimum, stakes shall be placed at the top and bottom of the slope where the pipe bends to conform to the slope. A riprap apron shall be provided at the outlet. This apron shall consist of 6-in- (15-cm-) diameter stone placed as shown in the sample drawing (Fig. 7.24). Follow-up inspection and any needed maintenance shall be performed after each storm.

7.4b Paved Chutes

A concrete- or asphalt-lined chute can be installed to convey storm water runoff down a slope on a permanent basis. Proper installation is essential. Runoff calculation *must* be used to size the channel. In particular, sufficient freeboard must be assured, since damage from overflow may be severe.

Drainage Area

The specifications in this section apply only to paved chutes with drainage areas up to 36 acres (15 ha). Larger chutes may be designed where additional flow capacity is required.

Chute Channel

The chute channel should be designed to carry the peak flow from at least a 10-year-frequency storm, unless otherwise specified. Various references (1, 14, 19) provide sizing specifications for paved chutes. The table below gives paved chute sizing guidelines developed for Maryland. (14) Group A chutes are for drainage areas up to 18 acres (7.3 ha). They have these additional sizing specifications:

Group A		Group B	
Bottom width, ft (m)	Maximum drainage area, Acres (ha)	Bottom width, ft (m)	Maximum drainage area, Acres (ha)
2 (0.6)	5 (2)	4 (1.2)	14 (5.7)
4 (1.2)	8 (3)	6 (1.8)	20 (8.1)
6 (1.8)	11 (5)	8 (2.4)	25 (10.1)
8 (2.4)	14 (6)	10 (3.0)	31 (12.5)
10 (3.0)	18 (7)	12 (3.7)	36 (14.6)

- The height of the dike at the entrance is at least 1.5 ft (0.46 m).
- The depth of the chute down the slope is at least 8 in (20 cm).
- The length of the inlet and outlet sections is 5 ft (1.5 m).

Group B chutes are for drainage areas from 14 to 36 acres (5.7 to 14.6 ha). They have these additional sizing specifications:

- The height of the dike at the entrance is at least 2 ft (0.6 m).
- The depth of the chute down the slope is at least 10 in (25 cm).
- The length of the inlet and outlet sections is 6 ft (1.8 m).

Although the above criteria could be used as a guide for sizing chutes in many parts of the United States, chute dimensions should be calculated by a qualified engineer and be based on local conditions. The chute most frequently has a trapezoidal cross section and is constructed of reinforced concrete. The slope of the chute should be no greater than 1.5:1. Bends in the chute are not desirable, since overtopping is likely at high flows.

Entrance Apron

A paved approach apron is needed to prevent erosion at the entrance to the chute (Fig. 7.25). A cutoff wall at the upstream edge of the apron should be continuous with the apron and extend at least 18 in (46 cm) into the soil below the channel. The cutoff wall should be constructed of reinforced concrete at least 6 in (15 cm) in thickness. The apron should slope toward the chute at a minimum 2 percent gradient.

Dike

The earth dike at the chute inlet should have sufficient freeboard to prevent overtopping due to ponding at the entrance section from discharges in excess of design flow. The minimum freeboard should be 12 in (30 cm) or 1.5 times the computed depth of flow, whichever is greater.

Outlet Protection

The velocity at the base of all paved chutes should be calculated to determine what kind of outlet protection is needed. On most small chutes, a riprap apron should provide adequate outlet protection. Sec. 7.8b provides guidelines for designing outlet protection structures.

Construction Specifications

It is recommended that the following construction specifications for paved chutes be incorporated in erosion control plans or grading plans.

- The subgrade shall be constructed to the required elevations. All soft sections and unsuitable material shall be removed and replaced with suitable material. The subgrade shall be thoroughly compacted and shaped to a smooth, uniform surface. For portland cement concrete chutes, the subgrade shall be moist at the time the concrete is poured.

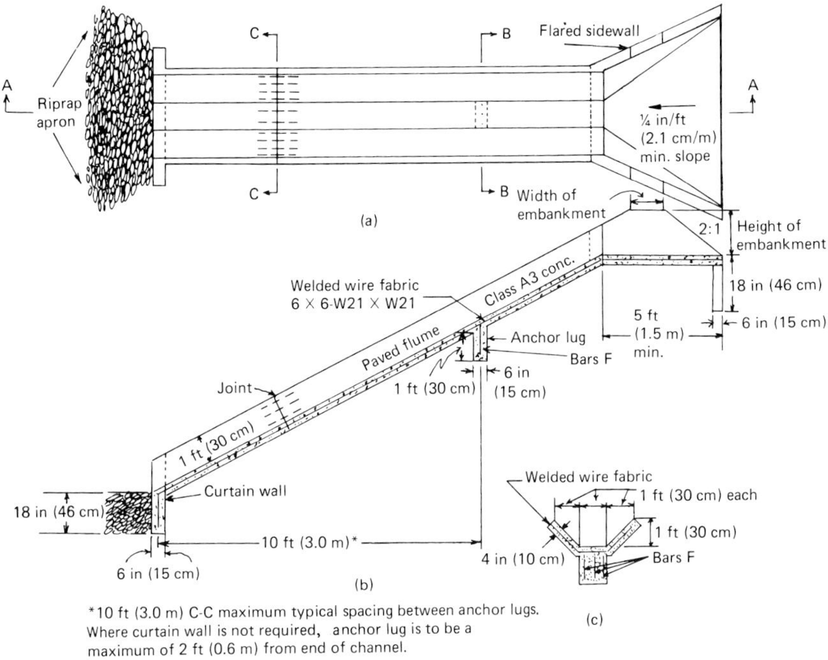

Schedule of Reinforcing Steel

Mark	No.	Length 2:1	Length 11/2:2	Size	Spacing C-C	Shape
F	3	1′2″	1′2″	4	4 in (10 cm)	Straight
G	9	1′5″	1′5″	4	6 in (15 cm)	Straight
Dowels	10	1′6″	1′6″	4	5 in (13 cm)	Straight

No. shown are for one anchor lug, curtain wall, and joint

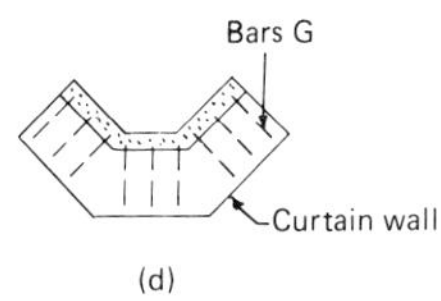

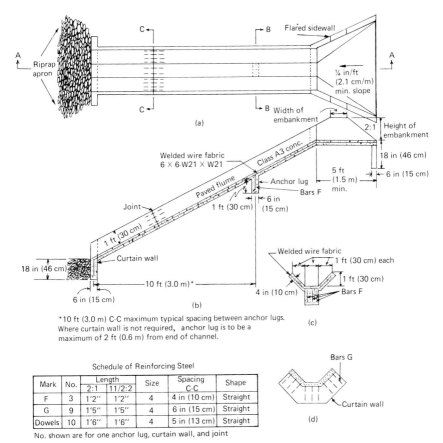

Fig. 7.25 Paved chute construction details. (*a*) Plan; (*b*) section A-A; (*c*) section B-B; (*d*) section C-C. (Adapted from 19)

- The cut or fill slope shall not be steeper than 1.5:1 and shall not be flatter than 20:1.
- The chute channel shall have at least a 4-in (10-cm) thickness of reinforced concrete.
- Anchor lugs shall be spaced at a maximum of 10 ft (3 m) on center for the length of the chute. Where no curtain wall is required, an anchor lug shall be constructed within 2 ft (0.6 m) of the end of the chute. Anchor lugs are to be as wide as the bottom of the chute, extend at least 1 ft (0.30 m) into the soil below the chute, and have a thickness of 6 in (15 cm). Anchor lugs shall be reinforced with No. 4 reinforcing steel bars placed on 4-in (10-cm) centers (Fig. 7.25).
- Expansion joints shall be provided approximately every 90 ft (27 m). Eighteen-inch (46-cm) dowels of No. 4 reinforcing steel placed on 5-in (13-cm) centers shall be located at all joints (Fig. 7.25).

- The top of the earth dike at the entrance, and the tops of the dikes carrying water to it, shall not be lower at any point than the top of the lining at the entrance of the structure.
- The lining shall be placed beginning at the lower end of the chute and proceeding up the slope to the upper end. The lining shall be well compacted and free of voids. The lining surface shall be reasonably smooth.
- Anchor lugs and curtain walls shall be formed to be continuous with the channel lining.
- Traverse joints for crack control shall be provided at approximately 20-ft (6-m) intervals and when more than 45 min elapses between consecutive concrete placements. All sections should be at least 6 ft (1.8 m) long. Crack control joints may be formed by using a ⅛-in- (0.3-cm-) thick removable template, by scoring or sawing to a depth of at least ¾ in (1.9 cm), or by an approved "leave-in" type of insert.
- The entrance floor at the upper end of the structure shall have a slope toward the outlet of 2 to 4 percent.
- The cutoff walls at the entrance and at the end of the discharge aprons shall be continuous with the lining.
- The lining shall consist of type 2 portland cement concrete [3000 lb/in^2 (211 kg/cm^2)], bituminous concrete, or a comparable nonerodible material.
- An energy dissipator of adequate design shall be used to prevent erosion at the outlet (Sec. 7.8).

7.5 DESIGN AND INSTALLATION OF PERMANENT WATERWAYS

Permanent waterways must be engineered in accordance with two primary design criteria. First, the channel must have sufficient capacity to pass the peak flow from at least the 10-year-frequency storm, depending upon local or state regulations and the risk to downstream facilities. Second, the channel lining must be resistant to erosion at the design peak flow. Both the capacity of the channel and the velocity of flow are functions of channel lining, cross-sectional area and shape, and slope. The task of the designer is to determine a channel geometry and lining which will have sufficient capacity and durability.

Open-channel hydraulics, the determination of appropriate channel cross sections and capacities, is a major element in hydraulic engineering. Channel design charts greatly facilitate the design of drainage channels. With such references as V. T. Chow's *Open Channel Hydraulics* (4), the Federal Highway Administration's *Design Charts for Open Channel Flow* (15) and *Design of Roadside Drainage Channels* (16), and other engineering manuals, an engineer can readily determine the hydraulic geometry of an adequate channel.

7.5a Asphalt- and Concrete-Paved Channels

The design and installation of permanent drainage swales and ditches on cut and fill slopes is governed by Chapter 70, Section 7012 of the Uniform Building Code (UBC) or by local ordinances. (5) Permanent drainage systems built to the requirements of the UBC or local ordinances generally do not need design calculations. The UBC specifies that swales or ditches on terraces shall have a minimum gradient of 5 percent and be paved with reinforced concrete not less than 3 in (7.6 cm) thick or an approved equal paving. They must be at least 1 ft (0.3 m) deep and have a minimum paved width of 5 ft (1.5 m). A single run of swale or ditch must not collect runoff from a tributary area exceeding 13,500 ft^2 [about 0.3 acre (0.13 ha)] without discharging into a down drain.

The UBC also requires that paved interceptor drains be installed along the tops of all cut slopes where the tributary area above a cut has a drainage path greater than 40 ft (12 m). These drains are to be paved with a minimum of 3 in (7.6 cm) of reinforced concrete or gunnite. The minimum depth is 1 ft (0.3 m), and the minimum paved width is 2½ ft (0.76 m). The slope of the drain must be approved by the building department official.

For the design of paved channels other than those required to protect cut and fill slopes, consult references such as those cited above.

7.5b Riprap-Lined Channels

Channels lined with riprap can be designed to withstand most flow velocities by choosing a stable stone size. A procedure for the design of rock linings is included in Sec. 7.7b. The procedure is based on the assumption that the size of the channel has been tentatively sized based on capacity requirements, and the appropriate rock size is determined by a trial-and-error process. References 1 and 19 provide design and construction specifications for permanent riprap.

7.5c Grass-Lined Waterways

The design of grassed waterways is complicated by the fact that the value of Manning's n changes with the height of the grass. The SCS has compiled tables for parabolic and trapezoidal waterways that include the effect of grass retardance on flow. These tables are reproduced in Appendix B. The Federal Highway Administration (15) provides several channel charts based on the SCS tables and includes instructions for devising new charts.

To design a grass-lined waterway, choose a suitable grass from Table 6.7. Grasses are grouped into four retardance classes, which are based on the heights shown in Table B-1 in Appendix B. Figure B-1 can then be used to determine n from the product of the velocity and the hydraulic radius.

If tables exist for the retardance class of the desired grass lining, the channel can easily be sized by using the SCS tables in Appendix B. Example 7.1 illustrates the procedure. If tables have not been prepared and no alternative choice of grass would allow use of the existing tables, Fig. B-1 is used in a trial-and-error process to find a channel size with adequate capacity in which velocity does not exceed the maximum permissible for the soil type. Example 7.2 illustrates use of Manning's equation to size a trapezoidal, grass-lined channel.

EXAMPLE 7.1 Design of a Parabolic Grass-Lined Channel

Given: A temporary parabolic waterway will be lined with annual ryegrass. The gradient is 10 percent, and the peak flow for a 10-year storm is 13.6 ft^3/sec (0.385 m^3/sec).

Find: The appropriate top width T and depth D of the channel.

Solution:

STEP 1. *Flow.* Q is given as 13.6 ft^3/sec (0.385 m^3/sec).

STEP 2. *Permissible Velocity.* The permissible velocity V_1 is a function of the grass type and slope. In Table 6.7 the permissible velocities for various grasses used for erosion control are listed. From Table 6.7, maximum permissible velocity for annual ryegrass is 4 ft/sec (1.2 m/sec) at 10 percent slope, so $V_1 = 4$ ft/sec (1.2 m/sec).

STEP 3 *Retardance Class.* Ryegrass is not listed in Table B.1. By following the instructions in the note on Table B.1, and with the knowledge that ryegrass can grow taller than 10 in (25 cm), retardance class B will be used to assure capacity when the grass is tall and velocity slower.

STEP 4. *Top Width and Depth.* By Table B.3 (sheet 14),

Flow = 13.6 ft^3/sec (0.385 m^3/sec)
Slope = 10 percent
Permissible velocity = 4 ft/sec (1.2 m/sec)

From the column for $V_1 = 4$ ft/sec (1.2 m/sec) and $Q = 15$ ft^3/sec (0.42 m^3/sec), since this is the lowest flow listed:

Top width = 15.7 ft (4.79 m)
Depth = 0.76 ft (0.23 m)

The column labeled V_2 represents the velocity when the grass is tall and unmowed. In this example, V_2 is 1.86 ft/sec (0.567 m/sec). The depth D in the table, 0.76 ft (0.23 m), is designed to accommodate the deeper, slower flow through unmowed grass.

The dimensions in the table represent the actual depth and width of water flow. Freeboard height must be added to the dike along the channel or to the depth of excavated channels.

EXAMPLE 7.2 Design of a Trapzoidal, Grass-Lined Channel by Using Manning's Equation

Given: $Q_{peak} = 20$ ft^3/sec (0.566 m^3/sec)
Slope = 5 percent
Side slopes = 2:1
Bermuda grass lining

Find: A trapezoid channel cross section for the given conditions.

Solution: Use the following three steps to solve this problem:

STEP 1. Use the continuity equation to determine a cross-sectional area; choose a base width (or depth); and solve for depth (or base width) and hydraulic radius. Use the given Q and V_{max} for the first estimate of area.

STEP 2. Multiply V_{max} by the hydraulic radius, and then use Table B.3 (sheet 14) to estimate values for Manning's n at the lower retardance level (with faster flow). Here V_{max} provides a conservative estimate of n. Experiment with channel size to obtain an initial V equal to V_{max}. This step assures that V_{max}, maximum permissible velocity, will not be exceeded at peak flow.

STEP 3. Solve Mannning's equation at higher retardance (lower velocity) to assure adequate capacity of the channel when the vegetation becomes taller. The cross-sectional area will have to be increased somewhat to accommodate this situation, usually by increasing the depth. Again, use a trial-and-error process.

Solution: Table 6.7 states that, at 5 percent grade, the maximum permissible velocity for Bermuda grass is 6 ft/sec (1.8 m/sec).

Q_{peak} = 20 ft^3/sec (0.566 m^3/sec)
S = 5 percent
z = 2 (2:1 side slopes)
V_{max} = 6 ft/sec (1.8 m/sec)

STEP 1. Assume V = 6 ft/sec (1.8 m/sec) and find the cross-sectional area:

$$A = \frac{Q}{V} = \frac{20 \text{ ft}^3/\text{sec}}{6 \text{ ft/sec}} = 3.33 \text{ ft}^2 \qquad \left(\frac{0.566 \text{ m}^3/\text{sec}}{1.8 \text{ m/sec}} = 0.314 \text{ m}^2\right)$$

Assuming d = 1 ft (0.3 m) for ease in calculations, solve for base width b:

$$A = db + zd^2$$

$$3.33 \text{ ft}^2 = b(1 \text{ ft}) + 2[(1 \text{ ft})^2] \qquad \{0.314 \text{ m}^2 = b(0.3 \text{ m}) + 2[(0.3 \text{ m})^2]\}$$

$$b = 1.33 \text{ ft} \qquad (0.45 \text{ m})$$

Find the hydraulic radius:

$$r = \frac{A}{\text{WP}} = \frac{3.33 \text{ ft}^2}{b + 2d\sqrt{2^2+1}} \qquad \left(\frac{0.314 \text{ m}^2}{b + 2d\sqrt{2^2+1}}\right)$$

$$= \frac{3.33 \text{ ft}^2}{1.33 \text{ ft} + (2 \text{ ft})(\sqrt{5})} \qquad \left[\frac{0.314 \text{ m}^2}{0.45 \text{ m} + (0.6 \text{ m})(\sqrt{5})}\right]$$

$$= 0.57 \text{ ft} \qquad (0.18 \text{ m})$$

STEP 2. Using V_{max} for V:

$$V \times r = (6 \text{ ft/sec})(0.57 \text{ ft}) = 3.42 \text{ ft}^2/\text{sec} \qquad [(1.8 \text{ m/sec})(0.18 \text{ m}) = 0.32 \text{ m}^2/\text{sec}]$$

From Table B.1, Bermuda grass mowed to a height of 2.5 in (6.4 cm) has a retardance D. From Fig. B.1 at $V \times r$ = 3.42 ft^2/sec (0.32 m^2/sec), the retardance D is n_D = 0.039. Now solve Manning's equation for V:

$$V = \frac{1.49}{n} r^{2/3} S^{1/2}$$

$$V = \frac{1.49}{0.039} (0.57 \text{ ft})^{2/3}(0.05^{1/2}) \qquad \left[\frac{(0.18 \text{ m})^{2/3}(0.05^{1/2})}{0.039}\right]$$

$$= 5.87 \text{ ft/sec} \qquad (1.8 \text{ m/sec})$$

Since 5.87 ft/sec is less than the 6 ft/sec maximum for Bermuda grass, we increase b slightly to try to get V closer to the actual V_{max}. We try b = 1.5 ft (0.46 m) and d = 1 ft (0.3 m). Then

$$r = \frac{bd + 2d^2}{b + 2d\sqrt{5}}$$

$$= \frac{(1.5\text{ ft})(1\text{ ft}) + 2[(1\text{ ft})^2]}{1.5\text{ ft} + 2(1\text{ ft})(\sqrt{5})} \qquad \left\{\frac{(0.46\text{ m})(0.3\text{ m}) + 2[(0.3\text{ m})^2]}{(0.46\text{ m}) + 2(0.3\text{ m})(\sqrt{5})}\right\}$$

$$= 0.59\text{ ft} \qquad (0.18\text{ m})$$

Again using Fig. B.1:

$$V_{max} \times r = (6\text{ ft/sec})(0.59\text{ ft}) = 3.54\text{ ft}^2/\text{sec} \qquad [(1.8\text{ m/sec})(0.18\text{ m}) = 0.324\text{ m}^2/\text{sec}]$$

$$n_D = 0.038$$

$$V = \frac{1.49}{0.038}(0.59\text{ ft})^{2/3}(0.05^{1/2}) = 6.2\text{ ft/sec} \qquad \left[\frac{(0.18\text{ m})^{2/3}(0.05^{1/2})}{0.038} = 1.9\text{ m/sec}\right]$$

$$A = bd + 2d^2 = 1.5\text{ ft}^2 + 2\text{ ft}^2 = 3.5\text{ ft}^2 \qquad (0.14\text{ m}^2 + 0.19\text{ m}^2 = 0.33\text{ m}^2)$$

The cross-sectional area at this velocity and bottom width is bigger than Q/V at V_{max} (3.5 > 3.33). In order for flow to actually fill this channel, peak flow would have to exceed the given Q_{peak}. Therefore, a channel bottom width of 1.5 ft (0.46 m) provides nonerosive flow at V_{max} and the given Q_{peak}.

STEP 3. Now we solve for V with tall grass. From Table B.1, tall Bermuda grass is in retardance class B.

Given: Q = 20 ft³/sec (0.57 m³/sec)
b = 1.5 ft (0.46 m) (from step 2)

Assume: d = 1.2 ft (0.37 m) (slightly deeper than before because velocity through tall grass will be slower)

$$r = \frac{bd + zd^2}{b + 2d\sqrt{z^2 + 1}}$$

$$= \frac{(1.5\text{ ft})(1.2\text{ ft}) + 2[(1.2\text{ ft})^2]}{1.5\text{ ft} + 2(1.2\text{ ft})(\sqrt{5})} \qquad \left\{\frac{(0.46\text{ m})(0.37\text{ m}) + 2[(0.37\text{ m})^2]}{0.46\text{ m} + 2(0.37\text{ m})(\sqrt{5})}\right\}$$

$$= \frac{4.68\text{ ft}^2}{6.87\text{ ft}} \qquad \frac{0.44\text{ m}^2}{2.11\text{ m}}$$

$$= 0.68\text{ ft} \qquad (0.21\text{ m})$$

To find an initial V at B retardance, solve the continuity equation:

$$A = bd + zd^2 = 4.68\text{ ft}^2 \qquad (0.43\text{ m}^2)$$

$$V = \frac{Q}{A} = \frac{20\text{ ft}^3/\text{sec}}{4.68\text{ ft}^2} = 4.27\text{ ft/sec} \qquad \left(\frac{0.57\text{ m}^3/\text{sec}}{0.43\text{ m}^2} = 1.3\text{ m/sec}\right)$$

$$V \times r = (4.27\text{ ft/sec})(0.68\text{ ft}) = 2.91\text{ ft}^2/\text{sec} \qquad [(1.3\text{ m/sec})(0.21\text{ m}) = 0.27\text{ m}^2/\text{sec}]$$

$$n_B = 0.08\text{ (Fig. B.1)}$$

$$V = \frac{1.49}{0.08}(0.68\ \text{ft})^{2/3}(0.05^{1/2}) \qquad \left[\frac{(0.21\ \text{m})^{2/3}(0.05^{1/2})}{0.08}\right]$$

$$= 3.23\ \text{ft/sec} \qquad (0.99\ \text{m/sec})$$

Since 3.23 is less than the 4.27 initial V we started with, we try again with a larger d. Assume d = 1.5 ft (0.46 cm); then

$$r = \frac{(1.5\ \text{ft})^2 + 2[(1.5\ \text{ft})^2]}{1.5\ \text{ft} + 2(1.5\ \text{ft})(\sqrt{5})} \qquad \left\{\frac{(0.46\ \text{m})^2 + 2[(0.46\ \text{m})^2]}{0.46\ \text{m} + 2(0.46\ \text{m})(\sqrt{5})}\right\}$$

$$= \frac{6.75\ \text{ft}^2}{8.21\ \text{ft}} \qquad \left(\frac{0.635\ \text{m}^2}{2.52\ \text{m}}\right)$$

$$= 0.82\ \text{ft} \qquad (0.25\ \text{m})$$

$$V = \frac{Q}{A} = \frac{20\ \text{ft}^3/\text{sec}}{6.75\ \text{ft}^2} = 2.96\ \text{ft/sec} \qquad \frac{0.57\ \text{m}^3/\text{sec}}{0.635\ \text{m}^2} = 0.90\ \text{m/sec}$$

$$V \times r = (2.96\ \text{ft/sec})(0.82\ \text{ft}) = 2.43\ \text{ft}^2/\text{sec} \qquad [(0.90\ \text{m/sec})(0.25\ \text{m}) = 0.225\ \text{m}^2/\text{sec}]$$

$$n_B = 0.09$$

$$V = \frac{1.49}{0.09}(0.82\ \text{ft})^{2/3}(0.22) \qquad \left[\frac{(0.25\ \text{m})^{2/3}(0.22)}{0.09}\right]$$

$$= 3.19\ [\text{somewhat larger than initial } V\text{, 2.96 ft/sec (0.97 m/sec)}]$$

Try d = 1.45 ft (0.44 m)

$$r = \frac{(1.5\ \text{ft})(1.45\ \text{ft}) + 2[(1.45\ \text{ft})^2]}{1.5\ \text{ft} + 2(1.45\ \text{ft})(\sqrt{5})} \qquad \left\{\frac{(0.46\ \text{m})(0.44\ \text{m}) + 2[(0.44\ \text{m})^2]}{(0.46\ \text{m}) + 2(0.44\ \text{m})(\sqrt{5})}\right\}$$

$$= \frac{6.38\ \text{ft}^2}{7.985\ \text{ft}} \qquad \left(\frac{0.59\ \text{m}^2}{2.428\ \text{m}}\right)$$

$$= 0.80\ \text{ft} \qquad (0.24\ \text{m})$$

$$A = 6.38\ \text{ft}^2 \qquad (0.59\ \text{m}^2)$$

$$V = \frac{Q}{A} = \frac{20\ \text{ft}^3/\text{sec}}{6.38\ \text{ft}^2} = 3.14\ \text{ft/sec} \qquad \left(\frac{0.57\ \text{m}^3/\text{sec}}{0.59\ \text{m}^2} = 0.966\ \text{m/sec}\right)$$

$$V \times r = (3.14\ \text{ft/sec})(0.80\ \text{ft}) = 2.51\ \text{ft}^2/\text{sec} \qquad [(0.97\ \text{m/sec})(0.24\ \text{m}) = 0.23\ \text{m}^2/\text{sec}]$$

$$n_B = 0.088$$

$$V = \frac{1.49}{0.088}(0.80\ \text{ft})^{2/3}(0.22) \qquad \left[\frac{(0.24\ \text{m})^{2/3}(0.22)}{0.088}\right]$$

$$= 3.21\ [\text{larger than but very close to } V = 3.14\ \text{ft/sec (0.97 m) above}]$$

Thus, depth for maximum flow is approximately 1.45 ft (0.44 m). Therefore, the channel dimensions should be:

b = 1.5 ft (0.46 m)
d = 1.45 ft (0.44 m) + freeboard

7.6 DESIGN AND INSTALLATION OF CHECK DAMS

No formal design is required for a check dam; however, the guidelines below should be observed.

7.6a Design and Installation Guidelines

The drainage area of the ditch or swale being protected should not exceed 10 acres (4 ha) except in low-rainfall areas. Straw bale check dams should have drainage areas less than 2 acres (0.8 ha). The maximum height of the check dam should be about 2 ft (0.6 m), and the center should be at least 6 in (15 cm) lower than the outer edges (Figs. 7.26 and 7.27). The maximum spacing between the dams should be such that the toe of the upstream dam is at the same elevation as the top of the downstream dam (Fig. 7.26).

Check dams can be constructed by using any materials on the site which can withstand the flow of water. Stone and log check dams are the sturdiest. When installing a straw bale check dam, make sure it is:

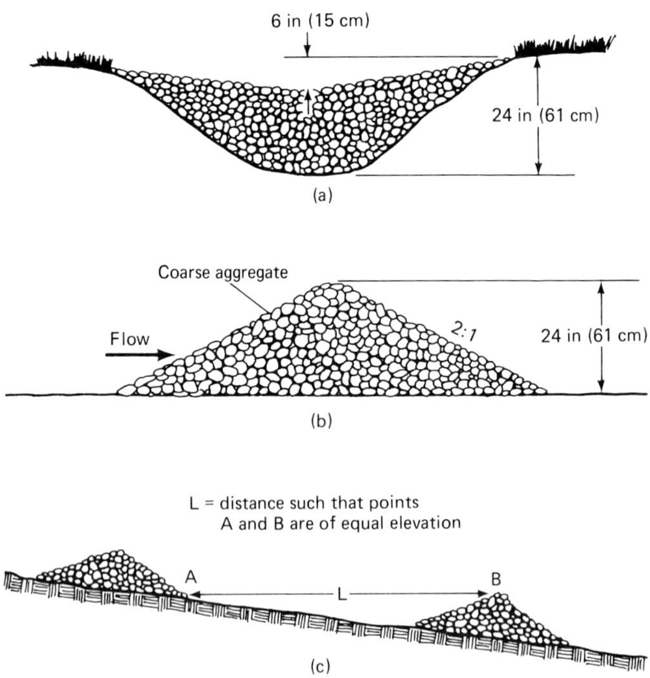

Fig. 7.26 Rock check dam design details. (*a*) Elevation view; (*b*) cross section; (*c*) spacing between check dams. (19)

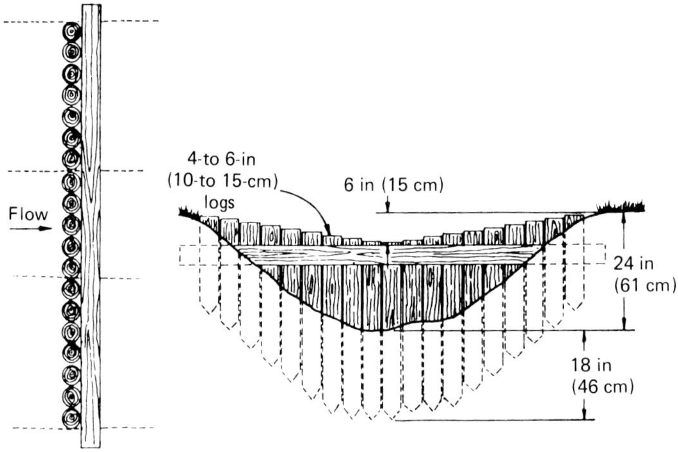

Fig. 7.27 Log check dam design details. (19)

- Centered in the drainageway so that flow will not go around the ends
- Staked
- Entrenched

For more information on straw bale installation, see Sec. 8.5a.

7.6b Maintenance

Check dams are not intended as sediment-trapping devices, but larger-sized particles will accumulate behind them. The sediment should be removed when it accumulates to one-half of the original height of the dam. Be sure to place the spoils where they will not be washed into a drainage system.

7.6c Removal

Check dams should be removed when they are no longer needed. In permanent drainageways, check dams should be removed when a permanent lining is installed. If the permanent lining is grass, the check dam should be left in place until the grass has matured sufficiently to protect the waterway.

7.7 DESIGN AND INSTALLATION OF CHANNEL LININGS

Erosion within waterways can be reduced through the selection of appropriate lining materials. The choice of lining depends on expected velocity, soil type, cost, and need for permanence. Calculate velocity and depth of flow as described in Chap. 4. Common lining materials include:

- Earth
- Rock
- Grass
- Grass and rock combination
- Fabric
- Pavement

7.7a Unlined Channels

Table 7.1 lists the maximum permissible velocities in unlined channels according to soil type. Generally, sandy, noncohesive soils tend to be very erodible, mixtures of sand, clay, and colloids are moderately erodible, and large-grained gravel, clay, and silt mixtures are erosion-resistant.

Channels with slopes less than 3 percent may remain unlined; but unless the channel is a small swale, a lining is advisable if the channel is expected to serve throughout an entire season. Figure 7.28 shows an unlined diversion that became a deep channel in the course of an average rainy season.

7.7b Rock Linings

Gravel or rock is the simplest kind of lining. Rock linings can be made to withstand most velocities if the proper size of rock is selected. Basically, the sequence of construction is to place a filter layer on the soil and then place a layer of riprap on top of the filter layer. The filter layer is important to prevent soil movement out through the riprap, which would result in the settling and eventual failure of the lining. The filter may be a special filter cloth or properly graded layers of sand and gravel.

Sample Design Procedure to Determine Stone Size for Riprap-Lined Channels (14)

The design procedure for riprap-lined channels is adapted from the National Cooperative Highway Research Program Report No. 108, entitled *Tentative Design Procedure for Riprap-Lined Channels* (14). It is based on the tractive force method and covers the design of riprap in two basic channel shapes: trapezoidal and triangular.

Note: This procedure is for uniform flow in channels and is *not* to be used for design of riprap energy dissipators. See Sec. 7.8 for design guidelines for outlet protection and energy dissipators.

The procedure is based on the assumption that the channel is already designed and the remaining problem is to determine the riprap size that would

Fig. 7.28 Eroding swale/dike; lining is needed.

be stable in the channel. The designer first determines the channel dimensions by the use of Manning's equation. The n value for use in Manning's equation is determined by first estimating a riprap size and then determining the corresponding n value for the riprapped channel from Fig. 7.29.

TRAPEZOIDAL CHANNELS

1. Calculate the bottom width-to-depth (b/d) ratio and enter Fig. 7.30 to find the ratio P/R of the wetted perimeter to the hydraulic radius.
2. Enter Fig. 7.31 with the channel bottom slope S_b, Q, and P/R to find the median riprap diameter d_{50} for straight channels. (By definition, 50 percent by weight of a rock mixture is greater than or less than the d_{50} size.)
3. Enter Fig. 7.29 to find the actual n value corresponding to the d_{50} from step 2. If the estimated and actual n values are not in reasonable agreement, another trial must be made. Recalculate the channel size by assuming a larger or smaller riprap, as appropriate, assign a new n value from Fig. 7.29, and return to step 1.

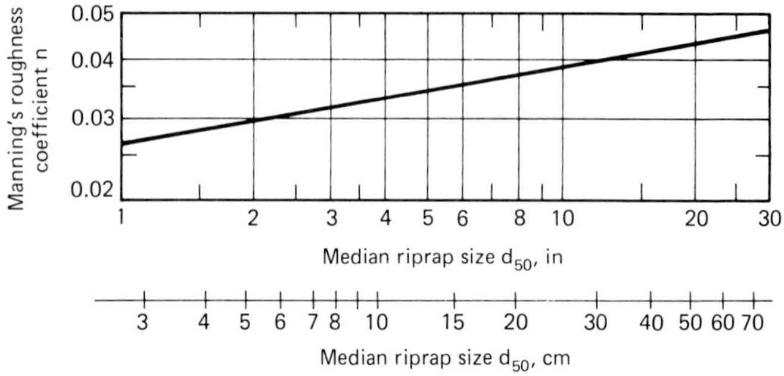

Fig. 7.29 Manning's *n* value for riprap-lined channels. (14)

4. For channels with bends, calculate the ratio B_s/R_o, where B_s is the channel surface width and R_o is the radius of the bend. Enter Fig. 7.32 and find the bend factor F_B. Multiply the d_{50} for straight channels by the bend factor to determine the riprap size to be used in bends. If the d_{50} for the bend is less than 1.1 times the d_{50} for the straight channel, then the size for the straight channel may be used in the bend. Otherwise, the larger stone size calculated for the bend should be used. The riprap should extend across the full channel section and extend upstream and downstream from the ends of the bend a distance equal to 5 times the bottom width.
5. Enter Fig. 7.33 to determine the maximum stable side slope of the riprap surface.

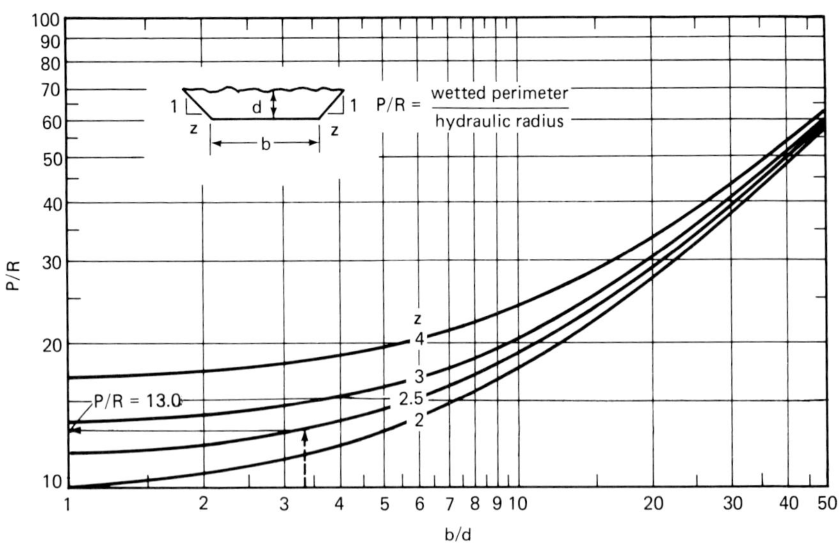

Fig. 7.30 Relation of *P/R* to *B/D* in trapezoidal channels. (14)

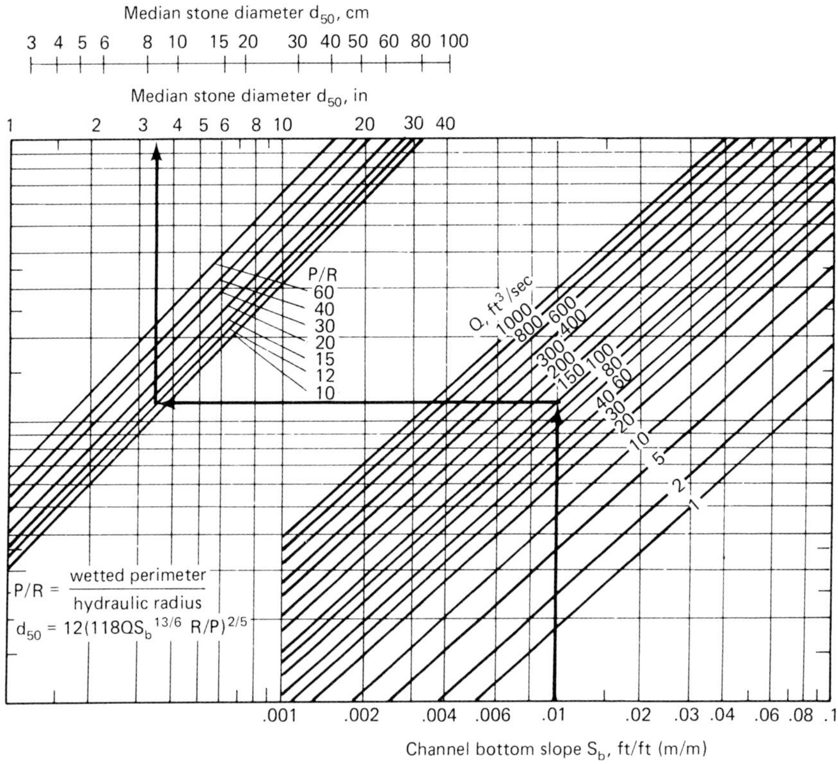

Fig. 7.31 Median riprap diameter for straight trapezoidal channels. (14)

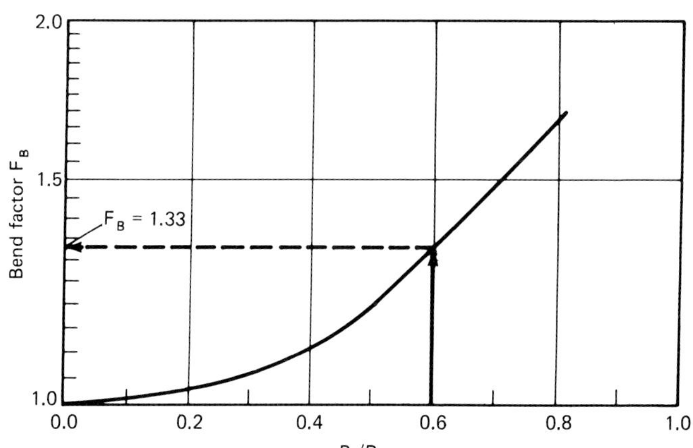

Fig. 7.32 Riprap size correction factor for flow in channel bends.
d_{50} (for bend) = d_{50} (for straight) $\times F_B$
B_s = channel surface width
R_o = mean radius of bend

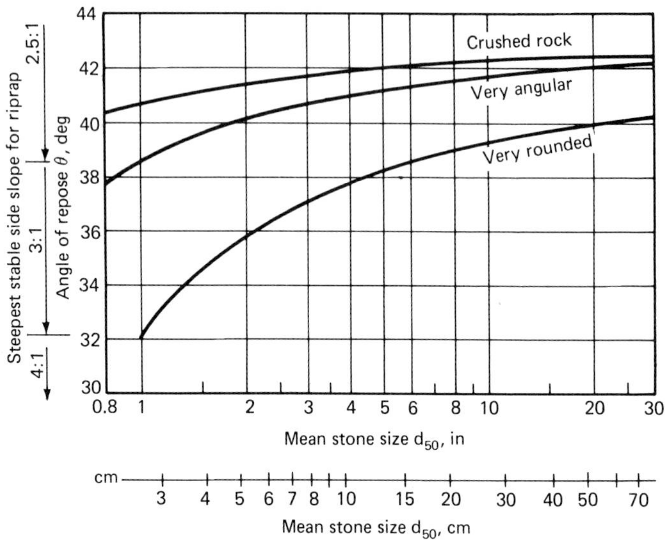

Fig. 7.33 Maximum riprap side slope for given riprap size. (14)

TRIANGULAR CHANNELS

1. Enter Fig. 7.34 with S_b, Q, and Z and find the median riprap diameter d_{50} for straight channels.
2. Enter Fig. 7.29 to find the actual n value. If the estimated and actual n values are not in reasonable agreement, another trial must be made.
3. For channels with bends, see step 4 under "Trapezoidal Channels."

Since temporary drainage channels on construction sites are generally small ditches with flows of less than 1 ft^3/sec (0.028 m^3/sec), Fig. 7.35 was developed as a handy riprap sizing chart for them. Figure 7.35 was derived from Fig. 7.34.

When writing plan specifications, specify stones 1.5 times the d_{50} as the maximum stone size in the mixture. The thickness of the riprap layer should be 1.5 times the maximum stone size, but not less than 6 in (15 cm). Freeboard should be added to the channel depth and should not be less than 0.2 times the depth of flow or 0.3 ft (9 cm), whichever is greater.

EXAMPLE 7.3 Riprap Channel Lining Design Calculation (14)

Given: Trapezoidal channel with

Q = 75 ft^3/sec (1.2 m^3/sec)

S_b = 0.01 ft/ft (m/m)

z = 2.5:1

Mean bend radius R_o = 25 ft (7.6 m)

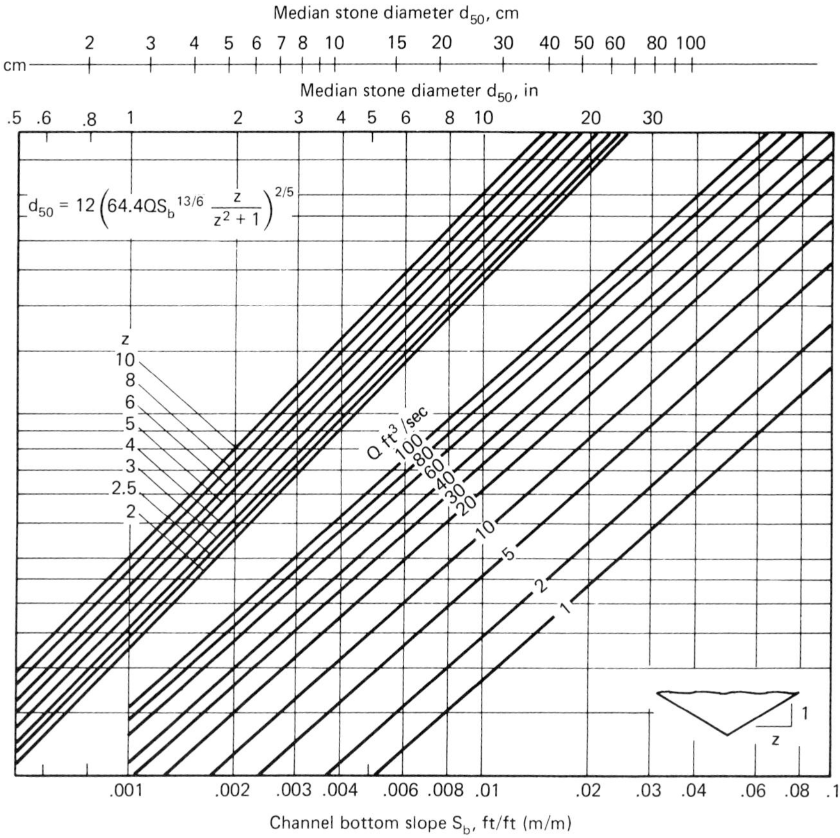

Fig. 7.34 Median riprap diameter for straight triangular channels. (14)

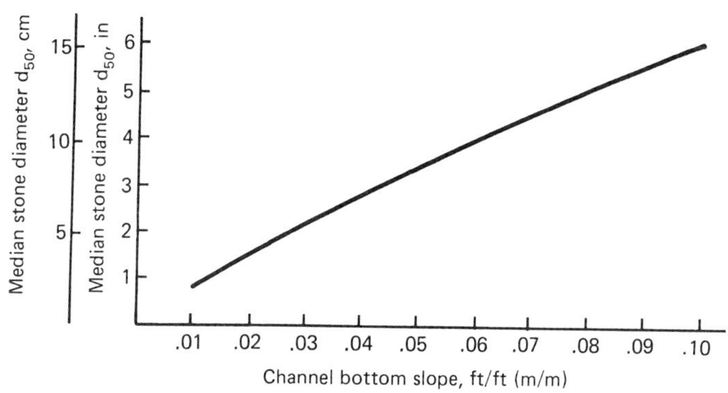

Fig. 7.35 Riprap sizing guide for lining triangular channels with flows of 1 ft³/sec (0.028 m³/sec) or less; assumes 2:1 side slopes.

n = 0.033 estimated and used to design the channel to find that b = 6 ft (1.8 m) and d = 1.8 ft (0.55 m)]

The type of rock available is crushed stone.

Find: Riprap size for straight and bend sections of channel.

Solution: *Straight channel reach.*

b/d = 6 ft/1.8 ft = 3.3 (1.8 m/0.55 m = 3.3)

From Fig. 7.30, P/R = 13.0

From Fig. 7.31, d_{50} = 3.4 in (8.6 cm)

From Fig. 7.29, n (actual) = 0.032, which is reasonably close to the estimated n of 0.033

Maximum riprap size = 1.5(3.4 in) = 5.1 in [1.5(8.6 cm) = 12.9 cm]

Riprap thickness = 1.5(5.1 in) = 7.7 in [1.5(12.9 cm) = 19.4 cm]

Use 5 ain (13 cm) as riprap size and 8 in (20 cm) as riprap layer thickness.

Channel bend

$$B_s = b + 2\,zd$$

$$= 6\text{ ft} + 2(2.5)(1.8\text{ ft}) \qquad [1.8\text{ m} + 2(2.5)(0.55\text{ m})]$$

$$= 15\text{ ft} \qquad (4.6\text{ m})$$

$$B_s/R_o = 15\text{ ft}/25\text{ ft} = 0.6 \qquad (4.6\text{ m}/7.6\text{ m} = 0.6)$$

From Fig. 7.32, F_B = 1.33. Since F_B is greater than 1.1, the bend factor must be used.

The riprap size in bend d_{50} = (3.4 in)(1.33) = 4.52 in [(8.6 cm)(1.33) = 11.4 cm]

Maximum riprap size in bend = 1.5(4.52 in) = 6.78 in [1.5(11.4 cm) = 17.1 cm] = 18.9 cm)

Riprap thickness = 10.2 in (25.9 cm)

Use 7.8 in (18 cm) for riprap size and 10 in (25 cm) for riprap layer thickness.

The heavier riprap for the bend should extend upstream and downstream from the ends of the bend a distance of 5 times the bottom width, or 5 × 6 ft = 30 ft (9.1 m). Figure 7.33 shows that the riprap for both d_{50} = 3.4 in (8.6 cm) and d_{50} = 4.52 in (11.4 cm) will be stable on a 3:1 side slope.

Freeboard = 0.2(1.8 ft) = 0.36 ft [0.2(0.55 m) = 0.11 m]; therefore, minimum freeboard is 0.36 ft (0.11 m). *Use 0.4 ft* (0.12 m).

Channel depth should be 1.8 ft + 0.4 ft = 2.2 ft (0.55 m + 0.12 m = 0.67 m)

Note: The cross-sectional area for this channel is 19 ft^2 (1.8 m^2) at a full flow of 100 ft^3/sec (2.8 m^3/sec). This produces a calculated water velocity of 5.2 ft/sec (1.6 m/sec). By Table 7.1, this velocity is permissible for *unlined,* aged channels of certain soil types. Why do we need the riprap? Because experimental data as reported in Fig. 7.36 shows that unlined channels have a maximum permissible depth of flow, which is exceeded in this example.

Temporary Rock Lining

The use of riprap larger than drain rock as a temporary erosion control measure is seldom seen because of the high cost of transporting rock. Riprap is more often used as outlet protection for the permanent storm drain system.

To save on cost, riprap can be used to line only critical sections of a waterway such as bends. Caution should be exercised to ensure that erosion will not occur

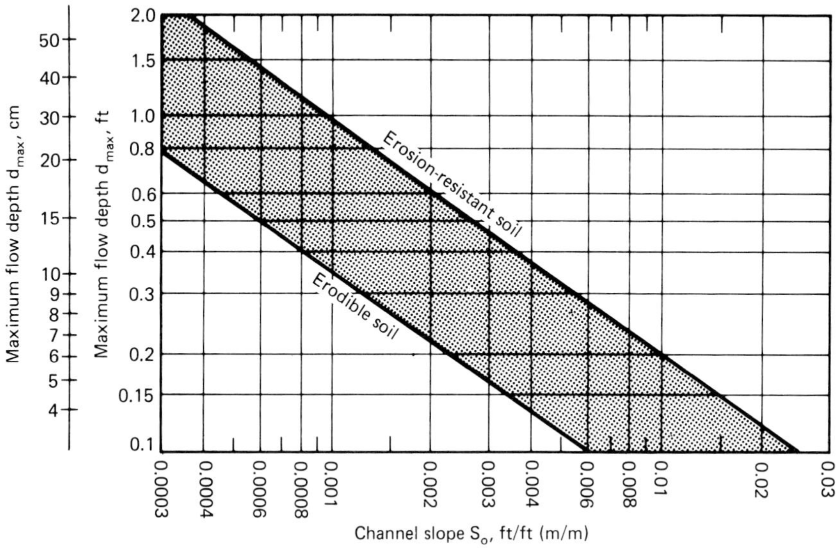

Fig. 7.36 Maximum permissible depth of flow for unlined (bare soil) channels. (17)

at the boundary between lined and unlined sections of channel. Proper installation, with rock laid flush with the unlined channel bottom, is required.

As a lining for temporary swales, dikes, and waterbars on roadways, gravel without a filter layer can be used. The following guidelines should apply. Where the channel slope is 1 to 5 percent and the maximum flow velocity from the 10-year-frequency storm exceeds the maximum permissible velocity for the soil type (see Table 7.1), line the flow area with coarse aggregate [2- to 3-in (5- to 8-cm) stone] pressed into the soil in a layer at least 6 in (15 cm) thick. Extend the lining up the sides of the dike or swale to a height of at least 8 in (20 cm) measured vertically. The lining should cover the entire flow area. If the drainageway is to serve more than 30 days, a filter layer should be used.

7.7c Vegetative Linings

Temporary and permanent waterways can be lined with grass; Sec. 7.5c contains channel design guidelines. Descriptions of various grass types and specifications for installing grass linings are provided in Chap. 6. Grassed waterways are best suited to level or gently sloping terrain. Because grassed waterways have lower velocities than paved channels, they generate lower peak flows. Thus, they can serve as an important part of a storm water management system. Grassed waterways may be designed to resemble natural creeks and thus enhance the appearance of a development. In addition, fewer outlet erosion problems are likely to occur from grassed waterways. In developments where land is available to meet their additional space requirements, grassed waterways should be seriously considered as an alternative to permanent paved channels or underground pipes.

Although permanent grassed waterways offer significant benefits, various problems arise from using grass-lined channels as a temporary measure. Flow should be kept out of the channel during the grass establishment period to prevent young seedlings from being washed away. Large amounts of eroded soil are often deposited in channels during construction. Removal of the sediment is necessary, but the removal process can damage the grass lining. In addition, the lining is easily damaged by construction traffic. Finally, mowing, and sometimes irrigation, is required.

7.7d Fabric Linings

Fabric linings can serve as temporary linings of small swales and sediment trap spillways. They can also protect a grassed waterway during the establishment period. Fabric linings can withstand somewhat higher velocities and greater slopes than grass. Design charts published by the Federal Highway Administration (17) are useful in selecting a material.

Undercutting is a serious problem in the use of any fabric lining. To help avoid the problem, the edges of the fabric should be buried in the ground (see Fig. 6.29). Where two pieces of fabric overlap, the upstream piece must lie *on top* of the downstream piece. Figure 7.37 illustrates what happens when the overlapping is done incorrectly. Note the gully forming under the downstream piece of mesh. Because fabric is permeable, some water will always flow beneath the lining; thus the fabric must be continuously in contact with the soil. Several references (1, 6, 19) provide detailed specifications for installing fabric channel linings (see Sec. 6.6d).

Jute Mesh

A design curve for jute mesh is shown in Fig. 7.38. Jute mesh is woven with jute yarn which varies from ⅛ to ¼ in (0.32 to 0.64 cm) in diameter and has openings about ⅜ by ¾ in (0.95 by 1.9 cm). Steel pins or staples are used to hold the jute mesh in place.

Excelsior Mat

A design curve for excelsior mat is shown in Fig. 7.39. Excelsior mat is composed of dried, shredded wood [0.8 lb/yd^2 (0.43 kg/m^2)] covered with a fine paper net. The paper net, reinforced along the edges, has an opening size of approximately ½ by 2 in (1.3 by 5.1 cm). The mat is held in place by steel pins or staples with five staples per 6 ft (1.8 m) of mat. Three rows of staples are used, one along each side of the mat and one row in the middle. At the start of each roll, four or five staples are spaced approximately 1 ft (0.3 m) apart. Where more than one mat is required, the mats are butt-joined and securely stapled.

Other Materials

Filter fabric similar to the material used to construct silt fences can be used as a channel liner beneath riprap. It should not, however, be used as an exclusive

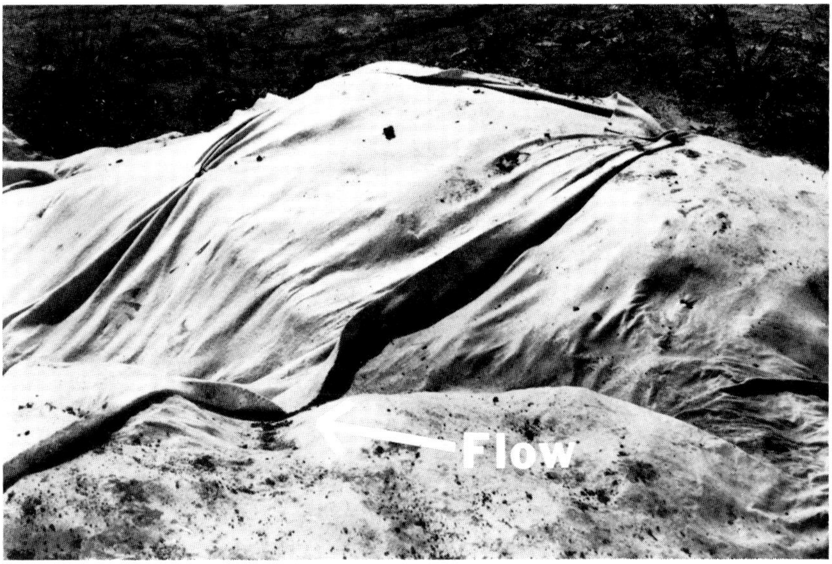

Fig. 7.37 Improper overlap of a fabric lining.

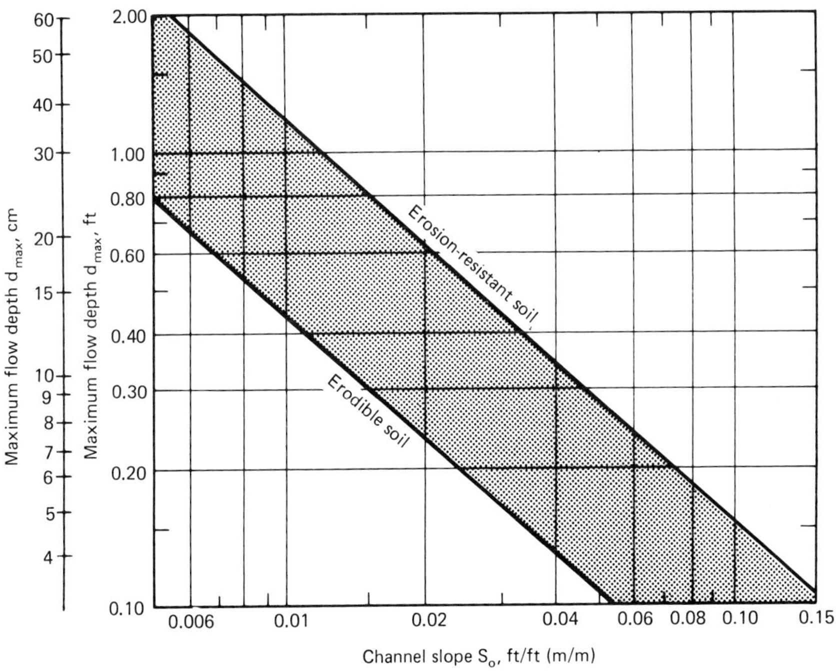

Fig. 7.38 Jute mesh lining design graph. (17)

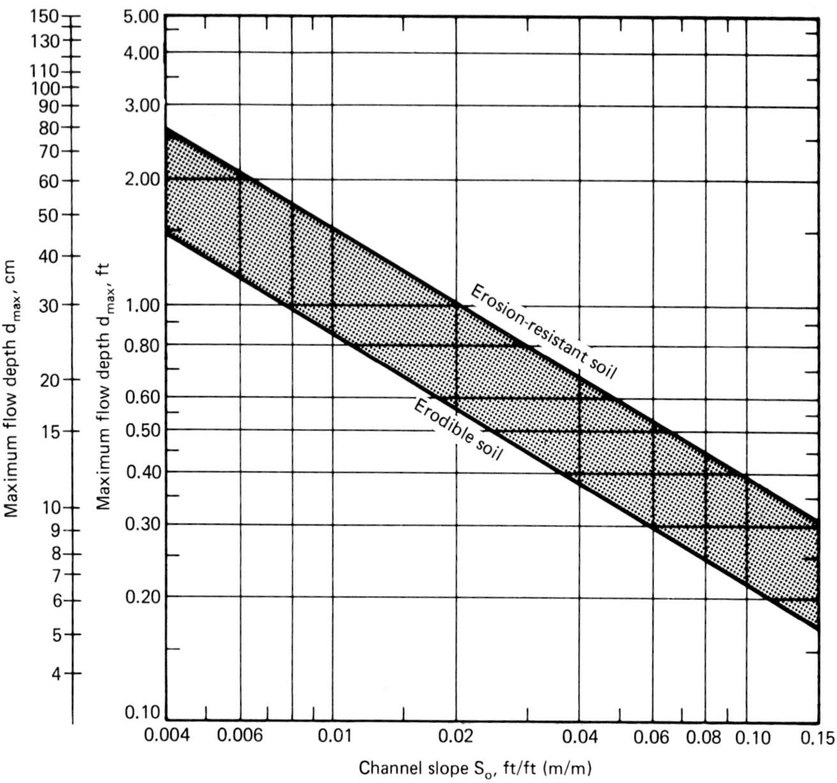

Fig. 7.39 Excelsior mat lining design graph. (17)

lining material. The fabric must be designed to be continuously wet. Check with the manufacturer for the proper material to use. The installation procedure for a filter lining is similar to the procedure for installing jute mesh, described above.

7.7e Pavement

Rigid pavements, constructed of asphalt or concrete, are permanent linings. Concrete linings tend to be more durable than asphalt linings and require less maintenance. The specifics of design and construction are left to the experienced engineer and contractor.

Although pavement does protect the channel bed, its smooth surface causes flows to increase in velocity. When paved channels discharge into natural waterways, higher velocities can cause erosion. Site designers should therefore look at the entire drainage system, both above and below a project, when designing channels and channel protection measures.

Because pavement is impermeable, water cannot infiltrate into the channel bed. This characteristic is important on steep or unstable slopes, where the subsoil must be kept dry. On more level sites, infiltration is desirable because it reduces flow volumes and encourages plant growth.

Like all drainage channels, paved channels must have positive drainage and adequate freeboard. Failure to provide either feature may result in overflows and serious erosion damage (Figs. 7.22 and 10.7).

7.8 DESIGN AND INSTALLATION OF OUTLET PROTECTION STRUCTURES (ENERGY DISSIPATORS)

The outlets of pipes, paved channels, and other structures that concentrate flow are points of critical erosion potential. Storm water which is transported through man-made conveyance systems sometimes reaches velocities high enough to erode the receiving channel. The condition is likely to occur where a pipe or paved channel discharges to a natural waterway. To prevent scour and erosion at outlets, a transition structure is needed to dissipate the water's high energy and reduce its velocity. These structures are called outlet protection devices or energy dissipators. The storm water can then safely enter a natural or other erodible channel used for drainage.

Perhaps the simplest and most effective type of energy dissipator is a rock-lined apron at the end of a pipe or culvert. Riprap aprons, which are described in detail in Sec. 7.8b, also blend well with natural streams. Other common types of energy dissipators, such as concrete impact-type structures, are briefly discussed following the section on riprap aprons. References such as the Federal Highway Administration's *Hydraulic Design of Energy Dissipators for Culverts and Channels* (18), Bohan (2), and the Portland Cement Association's handbook (9) provide design specifications for a variety of energy dissipators.

Outlet protection devices are usually permanent structures. They require engineered designs to accommodate the velocity and volume of flows generated. They are preferably constructed on level grade for a distance which is related to the outlet flow rate and the tailwater level. The sill or transition to the natural channel should be level with and at the same slope as the receiving channel.

7.8a Outlet Erosion

In designing protection against erosion below pipe outlets, it is important to consider what type of erosion to expect. The two general types of channel instability that can develop downstream from a culvert or storm-drain outlet are general channel degradation (gullying) and a localized erosion referred to as a scour hole. Distinction between the two conditions of scour and prediction of the type to be anticipated for a given field condition can be made by comparing the original ground slope at the outlet to the slope required for stability (Fig. 7.40).

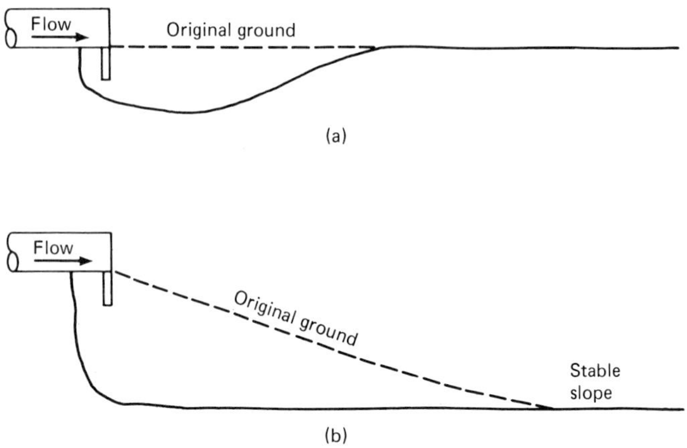

Fig. 7.40 Two types of outlet erosion: (*a*) scour hole; (*b*) gully scour. (2)

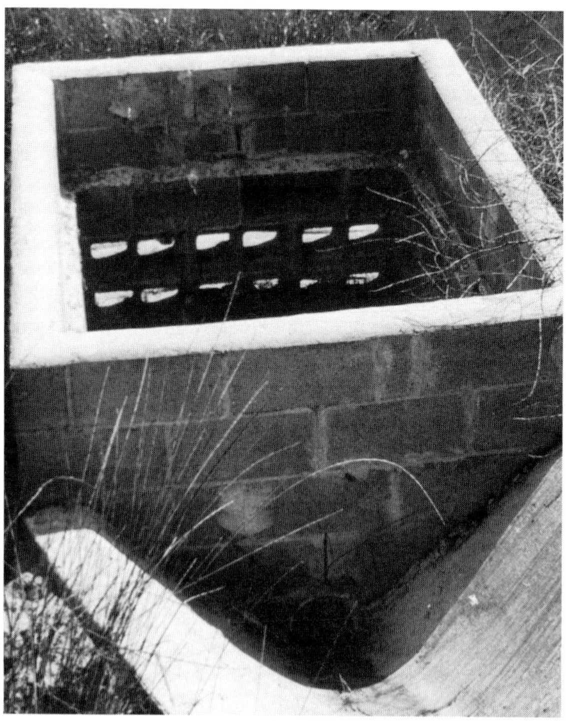

Fig. 7.41 Outlet with energy dissipator discharging onto grass-covered slope.

Gully Scour

Gully scour can occur where the velocity of flowing water exceeds the natural erosion resistance of the soil or other ground cover. It can begin to occur when any one of the following conditions exists:

- Concentrated flow is discharged onto a slope or into a small swale or canyon.
- Runoff to a natural drainageway is increased because of development.
- A sharp drop in elevation occurs at a pipe or channel outlet.

In the last case, "waterfall erosion" or "headcutting" occurs; it enlarges a gully upstream. This process is accentuated by slope failure at the gully headwall.

An effective way to prevent gully scour is to install a drop structure where there is a large change in the elevation of a channel. Another option is to line the receiving channel with riprap to a point downstream where the slope is gentle or the channel is significantly wider. It is not sufficient merely to provide protection at an outlet if the slope below the outlet is too steep. Figure 7.41 shows a small concrete swale discharging through an energy dissipator onto a grass-covered 10 percent slope. Figure 7.42 shows a gully forming a short distance below the dissipator. After only one season this gully was over 6 ft (1.8 m) deep.

Fig. 7.42 Gully erosion below energy dissipator in Fig. 7.41.

Scour Hole

A scour hole can occur below an outlet even if the receiving channel is nearly level. Natural channel velocities are almost universally less than culvert outlet velocities because the channel cross section is generally larger than the culvert flow area. Thus the flow exiting a culvert must rapidly adjust to a pattern controlled by the channel characteristics.

Erosion is caused by the impact of concentrated, high-velocity flow and by turbulence as the water loses energy and slows to flow in a larger channel. Figure 7.43 shows, in plan and cross section, the effect of scour at a pipe outlet. If a scour hole below a partcular outlet will not be objectionable, it may be desirable to allow it to form, since the hole will later act as an energy dissipator. However, pipe outlets still must be protected against undermining through the use of a

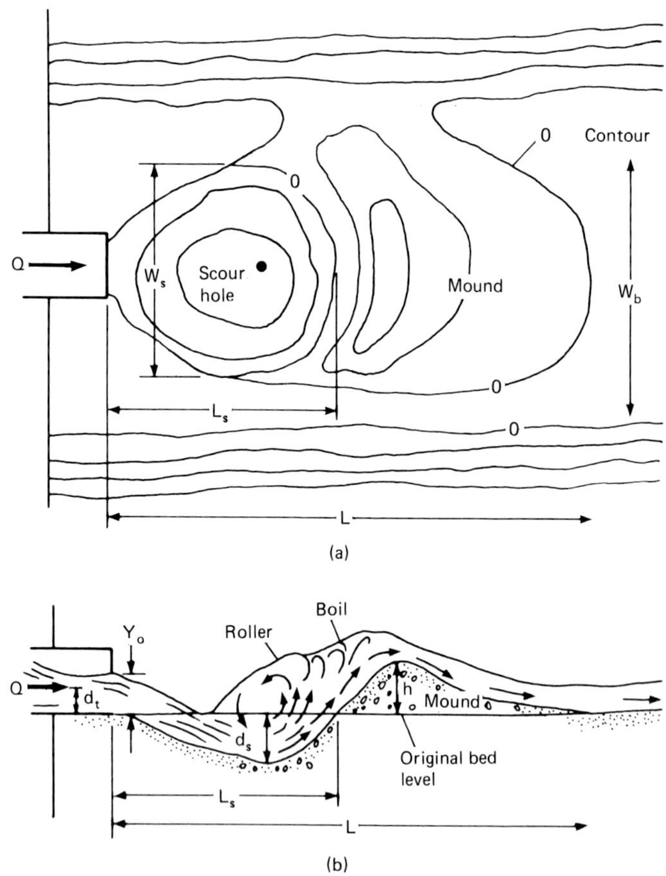

Fig. 7.43 (*a*) Plan and (*b*) cross section of a scour hole; see the reference for design details. (7)

cutoff wall. The proper depth for the wall can be determined from such references as Bohan (2) and FHWA (18).

7.8b Riprap Aprons

Proper apron design depends on whether minimum tailwater, maximum tailwater, or both conditions exist. Tailwater condition is the relation between the elevations of water surfaces in an outlet conduit and a receiving channel. Figure 7.44 illustrates minimum and maximum tailwater conditions for a round pipe flowing full.

Under minimum tailwater conditions, the water depth of the receiving channel, as calculated by Manning's equation, is less than one-half the discharge pipe diameter. The energy of the water discharging from the pipe will be dissipated by spreading on the apron and by turbulence from impact with the riprap of the apron.

Under maximum tailwater conditions, the water depth of the receiving channel is greater than one-half the pipe diameter. Energy will be dissipated by turbulence due to impact of the discharge stream with both the receiving tailwater and the rocks of the riprap apron.

Procedure for Design of Riprap Apron (14)

The following procedure is for the design of a level apron of length and flare such that the expanding flow (from pipe or conduit to channel) loses sufficient velocity and energy that it will not erode the downstream channel reach. The design curves are based on round pipes flowing full. The curves provide the apron size and the median diameter d_{50} for the riprap. There are two curves, one for the minimum tailwater condition (Fig. 7.45) and the other for the maximum tailwater condition (Fig. 7.46).

The first step in using this procedure is to determine the tailwater condition. Use Manning's equation, Q_{peak} in the receiving channel, and the channel dimensions to solve for cross-sectional area and then depth of flow in the receiving channel. Compare depth of flow to pipe diameter to determine tailwater condition. Then enter the appropriate chart with the discharge and the pipe diameter

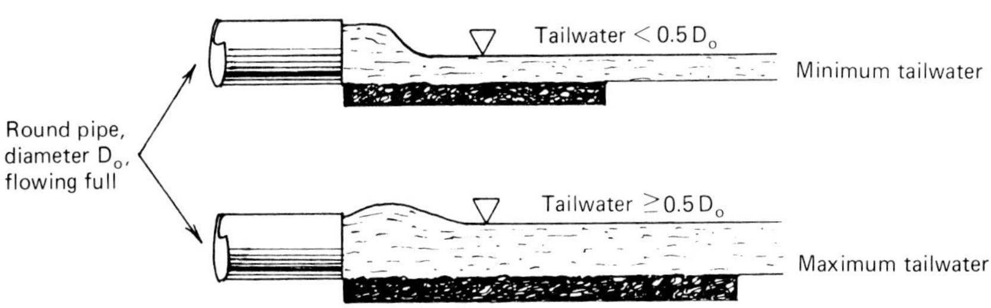

Fig. 7.44 Minimum and maximum tailwater conditions. (6, 14)

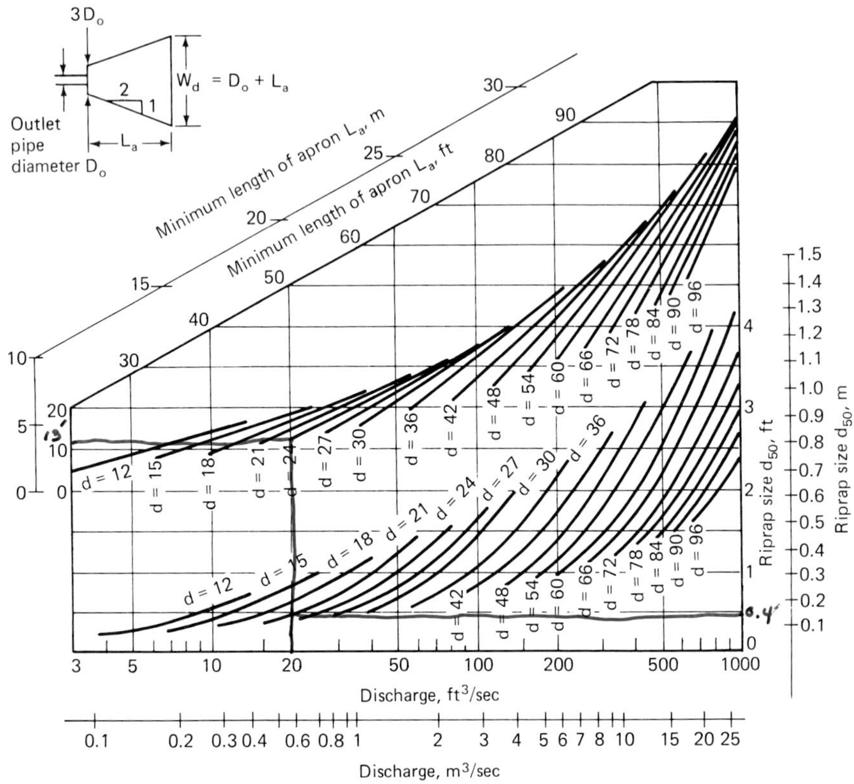

Fig. 7.45 Design of riprap outlet protection from a round pipe flowing full; minimum tailwater conditions. (6, 14)

to find the riprap size and apron length. The apron width at the pipe end should be 3 times the pipe diameter. Where there is a well-defined channel immediately downstream from the apron, the width of the downstream end of the apron should be equal to the width of the channel. Where there is no well-defined channel immediately downstream from the apron, minimum tailwater conditions apply and the width of the downstream end of the apron should be equal to the pipe diameter plus the length of the apron.

EXAMPLE 7.4 Riprap Outlet Protection Design Calculation for Minimum Tailwater Condition

Given: A flow of 6 ft^3/sec (0.17 m^3/sec) discharges from a 12-in (30-cm) pipe onto a 2 percent grassy slope with no defined channel.

Find: The required length, width, and median stone size d_{50} for a riprap apron.

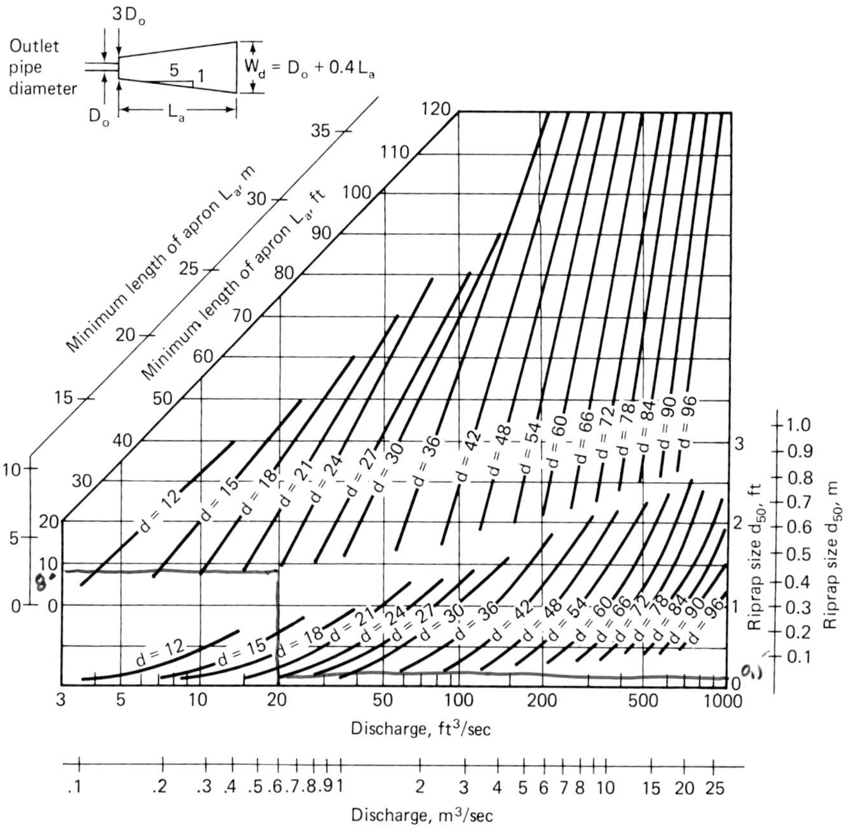

Fig. 7.46 Design of riprap outlet protection from a round pipe flowing full; maximum tailwater conditions. (6, 14)

Solution: Since the pipe discharges onto a flat area with no defined channel, a minimum tailwater condition can be assumed.

By Fig. 7.45, the apron length L_a and median stone size d_{50} are 10 ft (3 m) and 0.3 ft (9 cm), respectively. The upstream apron width W_u equals 3 times the pipe diameter D_o:

$$W_u = 3 \times D_o$$

$$= 3(1 \text{ ft}) = 3 \text{ ft} \qquad [3(0.3 \text{ m}) = 0.9 \text{ m}]$$

The downstream apron width W_d equals the apron length plus the pipe diameter:

$$W_d = D_o + L_a$$

$$= 1 \text{ ft} + 10 \text{ ft} = 11 \text{ ft} \qquad (0.3 \text{ m} + 3.0 \text{ m} = 3.3 \text{ m})$$

Note: When a concentrated flow is discharged onto a slope (as in this example), gullying can occur downhill from the outlet protection. The spreading of concentrated flow onto a slope has limited applications. It should be done only with very low, non-sediment-bearing flows and onto well-established vegetation on a relatively flat slope. One example might be to take the overland flow collected behind a dike above a small construction site, route this clean runoff around the site, and redistribute it onto a gently sloping area below the site.

EXAMPLE 7.5 Riprap Outlet Protection Design Calculation

Given: A 45-acre (18-ha) naturally vegetated site on a gentle slope. The site drains into an intermittent stream. The site is to be developed into a subdivision in which grading and street improvements will have been completed just before the rainy season. Natural peak drainage to the stream is 20 ft^3/sec (0.57 m^3/sec). After development, runoff from the site will discharge to the stream through a 24-in (61-cm) pipe.

The stream has a parabolic shape, a top width of 10 ft (3 m), a depth of 1 ft (0.3 m), a slope of 2 percent, and a roughness n of 0.045. The stream drains only the 45-acre (18-ha) site.

Problem: Design a riprap outlet to protect the streambed.

Solution:

STEP 1. By using Manning's equation, determine the capacity of the stream before development:

$$Q = V \times A = \frac{1.49}{n} \times R^{2/3} \times S^{1/2} \times A$$

From Appendix B, Table B.2:

A for a parabolic channel $= \frac{2}{3} \times d \times T$

R for a parabolic channel $= \dfrac{2 \times d \times T^2}{3T^2 + 8d^2}$

Therefore,

$$Q = \frac{1.49}{0.045}\left\{\frac{2(1\text{ ft})[(10\text{ ft})^2]}{3[(10\text{ ft})^2] + 8[(1\text{ ft})^2]}\right\}^{2/3}(0.02^{1/2})(\tfrac{2}{3})(1\text{ ft})(10\text{ ft})$$

$$\left(\frac{1}{0.045}\left\{\frac{2(0.3\text{ m})[(3\text{ m})^2)]}{3[(3\text{ m})^2] + 8[(0.3\text{ m})^2}\right\}^{2/3}(0.02^{1/2})(\tfrac{2}{3})(0.3\text{ m})(3\text{ m})\right)$$

$$= 33.11[(0.65\text{ ft})^{2/3}](0.14)(6.67\text{ ft}^2)\ \{22.22[(0.19\text{ m})^{2/3}](0.14)(0.6\text{ m}^2)\}$$

$$= 23.2\text{ ft}^3/\text{sec} \qquad (0.62\text{ m}^3/\text{sec})$$

This flow is the maximum capacity of the stream.

STEP 2. Estimate the peak flow in the stream after development. The stream will receive more water after development because of greater runoff from the site. By using Table 4.1, we estimate C values for the site before and after development. Let's assume that natural $C = 0.3$ and postdevelopment $C = 0.6$. Now

$$Q = C \times i \times A$$

A (watershed area) will remain the same, and i (rainfall intensity) will probably decrease a little because the time of concentration will likely be shorter after development. However, to simplify our calculations, we will assume that both i and A remain constant. Therefore, the postdevelopment runoff is

$$\frac{0.6}{0.3}\,(20\ \text{ft}^3/\text{sec}) = 40\ \text{ft}^3/\text{sec} \qquad \left(\frac{0.6}{0.3}\,(0.57\ \text{m}^3/\text{sec}) = 1.14\ \text{m}^3/\text{sec}\right)$$

This flow will exceed the natural capacity of the stream. It may erode the streambank and cause flooding problems.

STEP 3. Determine how to accommodate the postdevelopment flow in a nonerosive manner. There are several ways we could handle the increased flow. We could further divide the subdivision so that approximately one-half drains into the stream and one-half

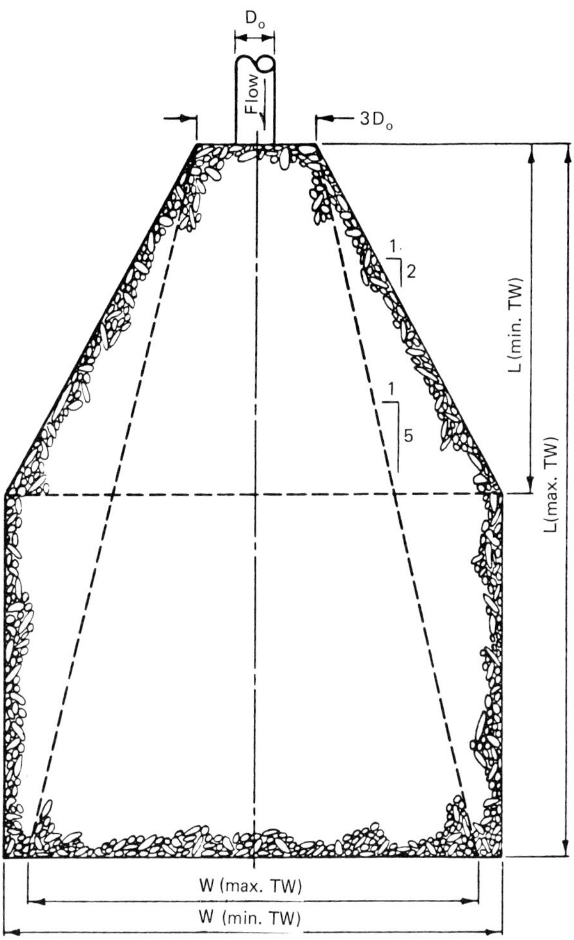

Fig. 7.47 Riprap blanket configuration for outlet protection; see the reference for design details. (2)

drains into a storm-drain network discharging into a larger volume of receiving water. The stream would thus discharge a flow equivalent to predevelopment conditions. Alternatively, we could construct on-site retention basins to limit peak site discharge to 20 ft^3/sec (0.57 m^3/sec). A poor third alternative is to widen the stream. We choose the first alternative.

STEP 4. Determine the tailwater condition. Depth of flow TW in the stream under a 20 ft^3/sec (0.57 m^3/sec) discharge would be slightly less than 1 ft (0.3 m), since at full flow Q is 22.2 ft^3/sec (0.63 m^3/sec) in the 1-ft (0.3-m) channel (step 1). Therefore, TW $<$ 1 ft (0.3 m) $<$ 0.5 pipe diameter, and we have a minimum tailwater condition.

STEP 5. Determine riprap size and apron dimensions. From Fig. 7.45:

d_{50} = 0.4 ft (0.12 m)

L_a = 12 ft (3.7 m)

STEP 6. Since the stream has a well-defined channel, the downstream apron width should be the width of the channel. The flare should be 1:2.

Note: If both minimum and maximum tailwater conditions will occur, the riprap apron should be designed to cover both conditions. Figure 7.47 illustrates how this can be done.

7.8c Other Types of Energy Dissipators

Figure 7.48 illustrates a wide variety of energy dissipator designs, from a simple T fitting on a CMP outlet to elaborate stilling basins. If the device is compact, it can be used in tight situations where there is not enough space for a riprap apron. Because these devices tend to be highly visible and permanent, they should be screened with landscaping to make them less obtrusive.

Most of the energy dissipators pictured in Fig. 7.48 use blocks or sills to impose resistance to flow. The Virginia Department of Highways design uses a single block and sill. The Colorado State University structure uses a row of blocks and a sill. The USBR Type IV basin, which uses staggered rows of blocks, is designed for moderate flows. The St. Anthony Falls stilling basin is designed for small culverts. The impact-type energy dissipator, USBR Type VI, is contained in a box-like structure and requires no tailwater for successful performance. The Contra Costa County energy dissipator is designed for small and medium-size culverts and also functions with no tailwater. A straight drop structure with blocks and a sill also functions as an energy dissipator.

Of these examples, the Contra Costa energy dissipator is best suited to the conditions on a construction site. It was developed at the University of California, Berkeley, in conjunction with Contra Costa County, California. The dissipator was developed to meet the following conditions:

1. To reestablish natural channel flow conditions downstream from a culvert outlet
2. To be self-cleaning and require minimum maintenance
3. To drain by gravity when not in operation
4. To be easily and economically constructed
5. To be applicable to a wide range of culvert sizes and operating conditions

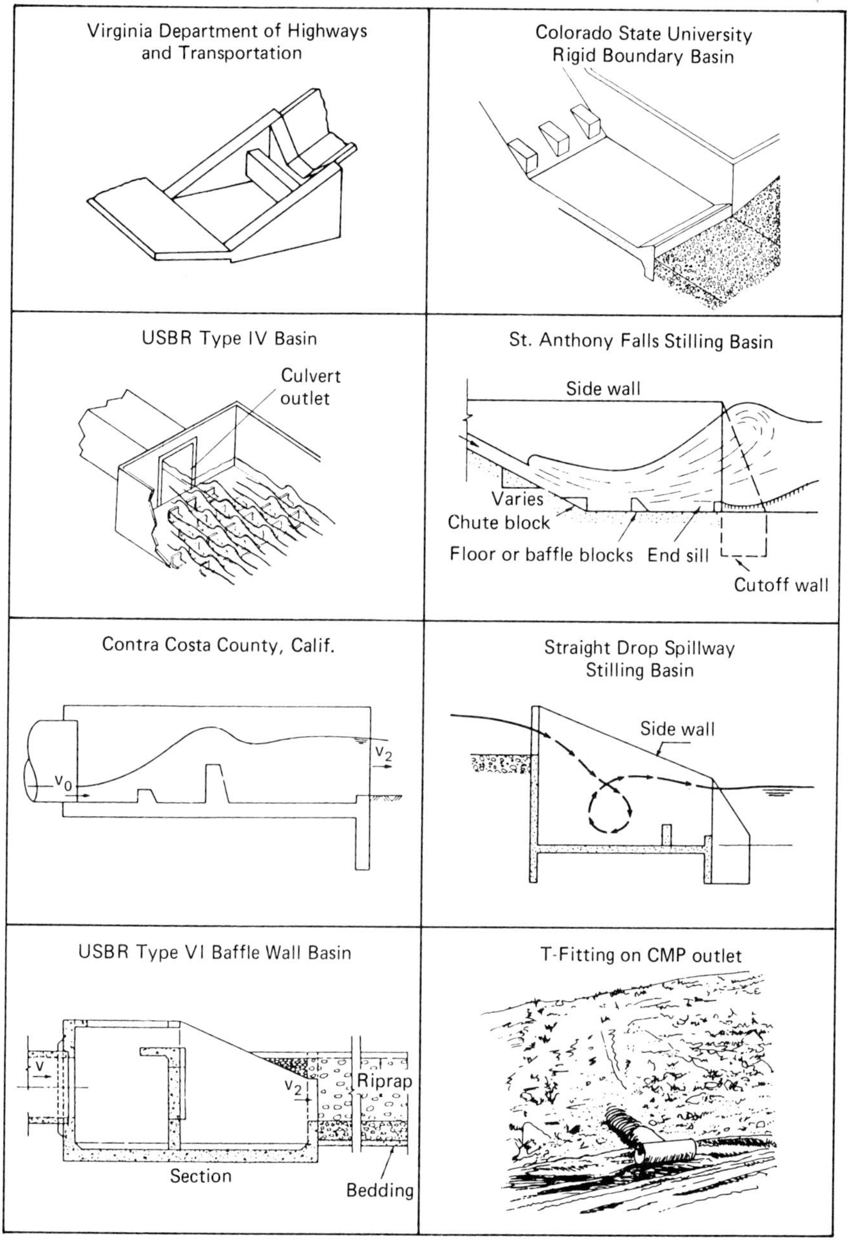

Fig. 7.48 Energy dissipators: eight sample designs. (18)

The dissipator is specifically designed for small and medium-size culverts of any cross section where the depth of flow at the outlet is less than the culvert height. It is applicable to medium- and high-velocity flows.

There are several good manuals on selecting and designing energy dissipators, including one published by the Federal Highway Administration (18). For design specifications pertaining to the devices illustrated in Fig. 7.48, consult this reference.

In some localities, energy dissipators have been regarded as a panacea for outlet erosion problems. Figures 7.41 and 7.42 graphically illustrate what can happen when an energy dissipator is used on too steep a slope. As with all erosion and sediment control measures, regular inspection and maintenance are required to ensure these devices are operating properly (see Chap. 11).

7.9 STREAMBANK STABILIZATION

Erosion of streambanks is a widespread problem and one which is difficult to control. A very large volume of material on the protection of streambanks is available. A detailed coverage of this subject, however, is beyond the scope of this book. The following section is a brief survey of the literature on streambank stabilization methods.

The U.S. Army Corps of Engineers is probably the foremost authority in the United States on streambank stabilization. Responding to a legislative mandate, the corps published in 1981 its *Final Report to Congress: The Streambank Erosion Control Evaluation and Demonstration Act of 1974* (10). This report contains an extensive survey of the types of measures being used for streambank protection. The document is composed of a main report supplemented with eight appendixes in separate volumes. Appendix A is a literature survey, and Appendix B is a survey of hydraulic research. For the past several years, the U.S. Army Engineers' Waterways Experiment Station in Vicksburg, Mississippi, has been conducting an annual week-long course on streambank protection. The course is accompanied by a large, loose-leaf binder of unpublished lecture notes, which provide an extensive amount of very practical information on streambank protection techniques (11). A more readily available document published by the Waterways Experiment Station is entitled *Streambank Protection Guidelines for Landowners and Local Governments* (12). This 60-page, paperback document describes the basic types of control measures, such as gabions (rock-filled wire baskets), and includes numerous sketches and color photographs. To order this publication, contact the local office of the U.S. Army Corps of Engineers.

Another useful book on streambank protection methods, entitled *Bank and Shore Protection in California Highway Practice* (3), is published by the California Department of Transportation. Highway departments in other states may publish similar documents.

REVIEW QUESTIONS

1. Describe and compare the major types of water conveyance structures. How do they differ in function from sediment retention structures?

2. Why is it not advisable to block a natural drainageway?
3. When should structural control measures be built?
4. List the design limitations for temporary dikes and swales with respect to drainage area, slope, and velocity.
5. Why is proper installation of pipe slope drains essential? Name several important installation details.
6. How is road erosion controlled if access must be maintained during the rainy season?
7. Why is channel erosion a common problem? What is the maximum permissible slope for unlined channels?
8. Compare channel linings with respect to (*a*) effect on velocity and peak flow, (*b*) permeability, and (*c*) usefulness as a temporary measure.
9. Under what conditions is a pavement channel lining necessary?
10. Give at least two reasons why grass-lined waterways are desirable. In what locations are they a reasonable alternative to pavement?
11. What is a "filter layer" under a riprap lining? Why is it important?
12. Discuss the important details of installing fabric linings.
13. Why is outlet protection so often necessary?
14. What are the two kinds of outlet erosion?
15. What kinds of structures can help prevent gully erosion? Outlet scour?
16. Design a parabolic grass-lined channel.
    Given: A parabolic waterway to be lined with Kentucky bluegrass. It will carry 50 ft^3/sec (1.4 m^3/sec) on a 3 percent slope.
    Find: The top width and depth of the channel and the velocity when the grass grows tall.
17. Design a riprap energy dissipator. For the waterway in Review Question 16, size a riprap apron to stabilize the discharge from a 24-in- (61-cm-) diameter pipe discharging 20 ft^3/sec (0.57 m^3/sec) into the head of the waterway. *Note:* The pipe is not carrying the full 50 ft^3/sec (1.4 m^3/sec) design capacity.

REFERENCES

1. Association of Bay Area Governments, *Manual of Standards for Erosion and Sediment Control Measures,* ABAG, Oakland, Calif., 1981.
2. J. P. Bohan, *Erosion and Riprap Requirements at Culvert and Storm-Drain Outlets,* Hydraulic Laboratory Investigation, U.S. Army Engineer Waterways Experiment Station, Vicksburg, Miss., 1970.
3. California Business and Transportation Agency, Department of Public Works, Division of Highways, *Bank and Shore Protection in California Highway Practice,* California Department of Transportation, 1970.
4. V. T. Chow, *Open Channel Hydraulics,* McGraw-Hill Book Company, New York, 1959.

5. International Conference of Building Officials, *Uniform Building Code,* ICBO, Whittier, Calif., 1982.
6. Maryland Water Resources Administration, U.S. Soil Conservation Service, and State Soil Conservation Committee, *1983 Maryland Standards and Specifications for Soil Erosion and Sediment Control,* MWRA, Annapolis, Md., 1983.
7. National Research Council, Highway Research Board, Group 2, *Design of Culverts, Energy Dissipators and Filter Systems,* Highway Research Record Number 373, NRC, HRB, Washington, D.C., 1971.
8. W. R. Naydo, J. W. Ross, and E. R. Rowe, *Street and Highway Drainage,* 2 vols., University of California, Berkeley, Institute of Transportation Studies, 1985.
9. Portland Cement Association, *Handbook of Concrete Culvert Pipe Hydraulics,* PCA, Skokie, Ill., 1964.
10. U.S. Army, Corps of Engineers, *Final Report to Congress: The Streambank Erosion Control Evaluation and Demonstration Act of 1974,* U.S. Army Engineer Waterways Experiment Station, Vicksburg, Miss., 1981.
11. U.S. Army Engineer Waterways Experiment Station, *Streambank Protection* (unpublished lecture notes), USAEWES, Vicksburg, Miss., 1983.
12. U.S. Army Engineer Waterways Experiment Station, *Streambank Protection Guidelines for Landowners and Local Governments,* USAEWES, Vicksburg, Miss., 1983.
13. U.S. Department of Agriculture, Forest Service, *Roads Handbook,* Handbook No. 2310, USDA, FS, Washington, D.C., 1968.
14. U.S. Department of Agriculture, Soil Conservation Service, *Standards and Specifications for Soil Erosion and Sediment Control in Developing Areas,* USDA, SCS, College Park, Md., 1975.
15. U.S. Department of Transportation, Federal Highway Administration, *Design Charts for Open Channel Flow,* Hydraulic Design Series No. 3, GPO, Washington, D.C., 1961, reprinted in 1979.
16. U.S. Department of Transportation, Federal Highway Administration, *Design of Roadside Drainage Channels,* Hydraulic Design Series No. 4, GPO, Washington, D.C., 1973.
17. U.S. Department of Transportation, Federal Highway Administration, *Design of Stable Channels with Flexible Linings,* Hydraulic Engineering Circular No. 15, GPO, Washington, D.C., 1975.
18. U.S. Department of Transportation, Federal Highway Administration, *Hydraulic Design of Energy Dissipators for Culverts and Channels,* Hydraulic Engineering Circular No. 14, GPO, Washington, D.C., 1975.
19. Virginia Soil and Water Conservation Commission, *Virginia Erosion and Sediment Control Handbook,* Richmond, Va., 1980.

chapter 8

Sediment Retention Structures

Earlier chapters described the use of vegetation to reduce erosion of soil from exposed areas (Chap. 6) and structural measures to carry runoff in a nonerosive manner to suitable outlets (Chap. 7). Despite the best efforts to reduce the quantity of eroded soil, however, some erosion is inevitable. This chapter discusses ways to capture and retain sediment once soil has been eroded.

Sediment retention structures do *not* stop erosion; they trap eroded soil before it can reach a water body or adjacent property. It is, of course, far better to prevent erosion than it is to trap sediment from eroded areas. Sediment retention devices should be used only as backup systems if preferred measures, such as vegetation and mulches, fail or if no other measures are feasible. Sediment retention structures are usually only temporary solutions until permanent measures, such as landscaping and paving, are in place.

Sediment retention structures are designed to remove and retain portions of the sediment being carried by runoff. In essence, they work by slowing the velocity of runoff and letting suspended soil particles settle by gravity. Typical structures include straw bale dikes, filter fabric fences, sediment traps, and sediment basins. There is very little performance data on sediment retention structures at construction sites. All too often these structures have been constructed with major flaws which prevent good performance. In other cases, retention structures have been properly constructed but not monitored or maintained. Fortunately, sediment retention structures for treatment of municipal wastewaters and mine drainage have been used and studied for many years. Sufficient information is available from these types of structures to provide a sound basis for sediment retention structure design for construction sites.

Sediment retention structures have serious weaknesses. The efficiency of sediment trapping is dependent upon watershed soil type. Clay and silt particles do not settle easily once they are suspended in water. Watersheds that contain soils high in clay and silt require large basins to capture the soil that has been eroded. Because of cost, space limitations on construction sites, and other practical problems, it is impossible to construct an ideal sediment basin. Therefore, a retention structure is typically designed with a removal efficiency of 50 to 75 percent. If such a structure performs as designed, as much as half the eroded soil will pass through the basin and be deposited in a water body such as a lake, bay, or stream.

Sediment retention structures require maintenance and cleaning at regular intervals. If too much sediment is allowed to accumulate in them, they will cease to function. Little or no settling will occur, and trapped sediment will be resuspended and washed away. Finally, sediment basins can pose a safety hazard to children when water is impounded in them. Therefore, these structures should be fenced or access to them should be restricted by other means.

The sediment retention structures described in this chapter are intended as temporary measures for use during the construction of a project. Structures similar to sediment basins can also be permanent devices. Permanent basins are sometimes used by public agencies to protect storm drain or water supply systems from debris and sediment from agricultural and other land outside the control of the agencies. They are also used to detain storm water peak flows for flood control purposes. For information on designing permanent structures, consult appropriate texts on hydraulics and flood control engineering. State and local regulations concerning embankment height, outlet design and location, fencing, design storm, and other criteria may also apply to both temporary and permanent sediment basins. Check with your city or county before designing such a structure.

8.1 TYPICAL APPLICATIONS OF SEDIMENT RETENTION STRUCTURES

8.1a Sediment Basins

The terms "sediment basins" and "silt basins," as used in this handbook, refer to engineered structures designed to treat runoff from drainage areas larger than 5 acres (2 ha) (Fig. 8.1). Typically, a basin will have:

- Compacted embankments.
- One or more inflow points carrying polluted runoff.
- Baffles to spread the flow throughout the basin.
- A securely anchored pipe riser as the principal spillway.

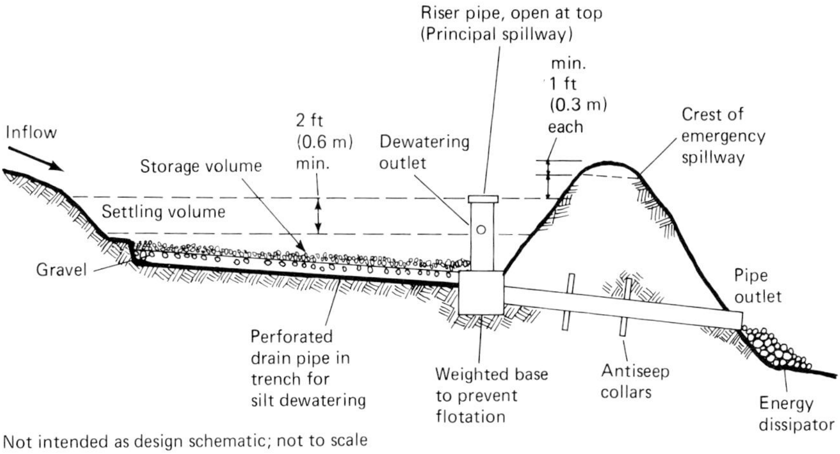

Fig. 8.1 Typical features of a sediment basin.

- An emergency spillway.
- A basin dewatering device.
- Outlet protection to reduce erosion at the pipe outlet if the basin discharges to a natural drainageway. Often, a sediment basin will discharge directly into the permanent storm drain system, in which case outlet protection is not a concern.

If sediment basins are to be used on a project, they should be among the first structures installed when grading begins, and they should remain in place until the drainage area is stabilized. The location of basins should be carefully thought out during the planning phase. For a basin to serve its intended lifetime, it should be placed out of the way of construction traffic, as in a cul-de-sac or on a house pad (Fig. 8.2). The basin in Fig. 8.3 occupies a future park site.

A sediment basin will usually be built near a low point on a site so it can trap a large amount of polluted runoff. It should *not,* however, be constructed where it will trap a substantial amount of clean runoff along with the dirty. A stream that drains a largely undisturbed watershed is thus a poor location for a basin. The sediment must be trapped before it reaches a natural stream. Figure 8.4 shows a basin placed at the bottom of a large fill slope. It treated slope runoff until the planted vegetation reduced the erosion rate. The basin discharges in the natural stream in the canyon.

8.1b Sediment Traps

Sediment traps may be designed in the same way as basins. The major difference from basins is that traps serve areas smaller than 5 acres (2 ha). They can also

Fig. 8.2 Sediment basin on a housepad.

Fig. 8.3 Sediment basin in a future park site.

Fig. 8.4 Sediment basin at base of seeded fill slope.

be predesigned (as to surface area and depth) for a geographical area by using the methods presented in Sec. 8.4. Sediment traps, being generally smaller than basins, are much easier to install and also more easily moved to keep up with grading activities. A sediment trap can be formed by excavation (Fig. 8.5) or by construction of embankments (Fig. 8.2).

The outlet may be an earth or stone spillway, a pipe riser (Fig. 8.2), or a storm drain inlet (Fig. 8.5). This last option is very commonly employed, since it utilizes permanent facilities at the site. In all other cases, outlet protection must be provided to prevent further erosion as the treated runoff leaves the trap.

8.1c Sediment Barriers

Sediment barriers are temporary, nonengineered retention structures. They treat runoff from small areas, and their installation requires less site disturbance than a sediment trap or basin would require. They are simple to construct; they can be repositioned as grading proceeds; and they are inexpensive. However, they are less durable than basins or traps and often require frequent maintenance. Sediment barriers can be built with many materials:

- Straw bales
- Filter fabric attached to a wire fence

Fig. 8.5 Sediment trap excavated around storm drain inlet.

- Filter fabric on straw bales
- Gravel and earth berms

Sediment barriers are most frequently placed below disturbed areas subject to sheet or rill erosion to capture sediment in runoff *before* the runoff reaches a channel. Typical locations for sediment barriers include the following:

- Along the contour of exposed slopes (Fig. 8.6)
- At the base of a slope (Fig. 8.7)
- Along a street or sidewalk to prevent silt from reaching the pavement
- At a storm drain inlet—sometimes called inlet protection (Figs. 8.8 and 8.9)

Sediment barriers can also be placed across *small* drainageways. Typical examples include:

- A straw bale–filter fabric barrier across a wide, gently sloping swale (Fig. 8.10)
- A silt fence across a swale (Fig. 8.11)

A sediment barrier should *not* be placed across a drainageway that carries a high-volume or high-velocity flow. A useful guideline is this: If the flow in a swale exceeds the capacity of a grass lining (Chap. 6), a sediment barrier should not be placed across the swale.

Fig. 8.6 Silt fences along slope contour.

Fig. 8.7 Straw bale dike at the base of a slope.

Fig. 8.8 Inlet protection: silt fence sediment barrier.

A sediment barrier in a swale is similar to a check dam, and both can be built with the same materials. However, the purpose of a sediment barrier is to trap sediment, whereas the purpose of a check dam is to prevent channel erosion by slowing the velocity of flow. In general, check dams are used under conditions of greater flow and larger drainage areas than sediment barriers. They must be more sturdily constructed. Only as a secondary function do check dams trap sediment. (See the discussion of check dams in Chap. 7.)

The last form of sediment barrier, a stabilized construction entrance, is somewhat different from the structures described above. It is a pad of crushed rock located where traffic enters or leaves a construction site (Fig. 8.53). The crushed rock reduces the tracking of mud and sediment onto a public right-of-way such as a paved street or parking lot.

8.2 DESIGN CONCEPTS FOR TEMPORARY SEDIMENT BASINS AND TRAPS

The most commonly used sediment retention structures are sediment basins and traps. Sediment basins are larger than sediment traps, and they are more precisely designed. The design theory is the same for all sediment retention struc-

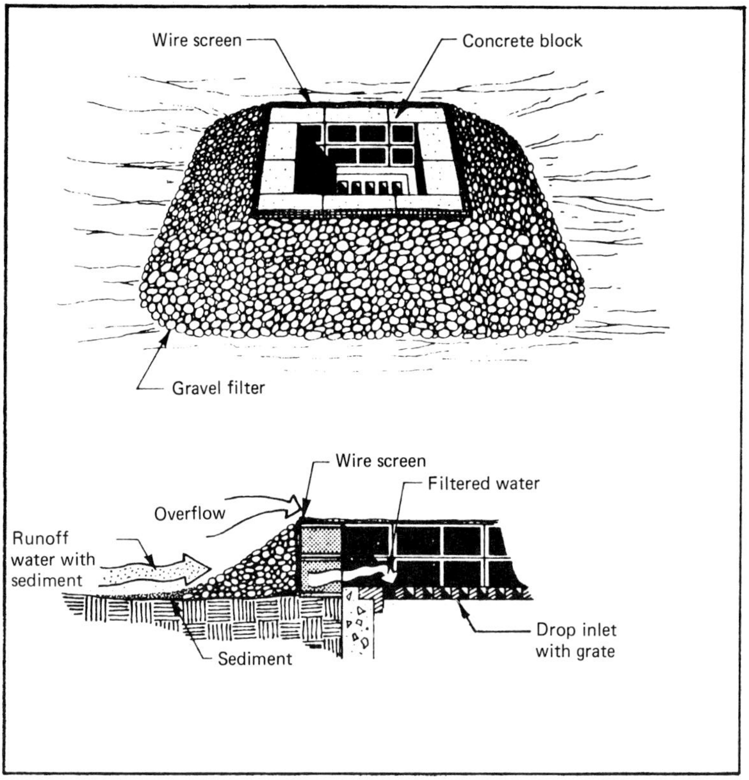

Fig. 8.9 Inlet protection: block and gravel drop inlet sediment barrier. (11)

Fig. 8.10 Straw bale–filter fabric sediment barrier across a swale.

Fig. 8.11 Silt fence across a swale.

tures, however, and it is presented below. In subsequent sections the differences between basins and other retention structures are described.

8.2a Design Theory for Surface Area Formula

Particles settle through water under the influence of gravity and follow one of three modes of settling:

1. Particles settle as separate elements with little or no interaction among them. This type of settling is usually found in waters with relatively low solids concentrations and is called free or ideal settling.
2. Independent particles coalesce or clump together during sedimentation. The larger resulting particles settle at a faster rate. This type of settling is often aided by the addition of chemicals which pull particles together.
3. At some concentration higher than in free settling, particles will start to interact and hinder settling. Instead of falling freely, the particles will settle as a group. This is called zone settling.

Once particles have settled to the bottom of a basin, they may be resting on other particles or be separated from them by electrostatic repulsion. Considerable water may be trapped among the particles. This water may be: (1) driven out as the weight of more particles is added to the top of the mass, (2) drained slowly at the bottom of the mass through capillary action and as particles shift and settle, or (3) evaporated when the overlying layer of water is removed.

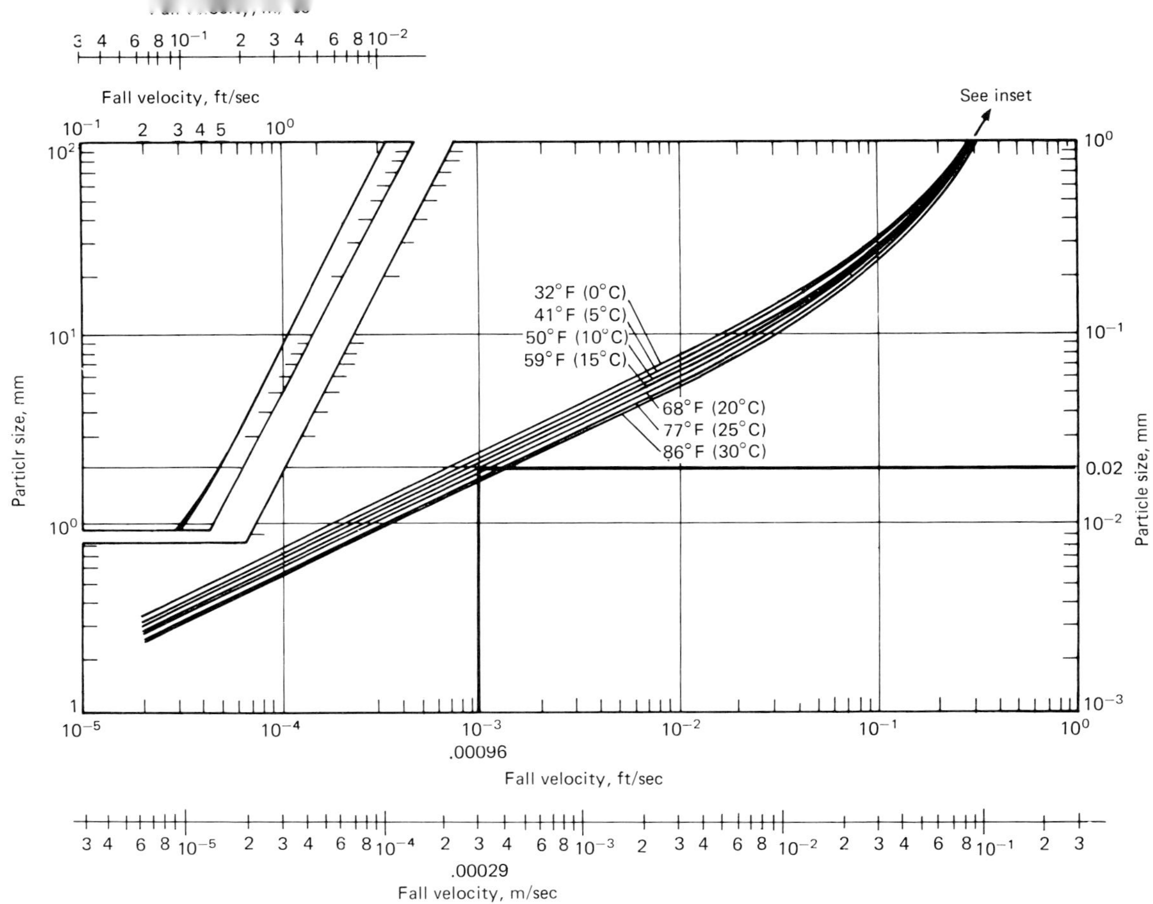

Fig. 8.12 Particle settling velocity curves. (7)

The design of sediment basins assumes free or ideal settling. It also assumes round soil particles and relatively uniform specific gravities. In reality, suspended particles are often rods, disks, or irregular lumps which settle more slowly than round particles. Variations in parent material can result in small particles with mass greater than that of larger particles. However, for purposes of sediment basin design at construction sites, the generalized assumptions are adequate. More information on sedimentation theory, including Stokes' and Newton's laws, can be found in other references. (2, 5)

Settling velocities of round soil particles have been calculated and plotted by Pemberton and Lara (7) as shown in Fig. 8.12. For a range of particle sizes and water temperatures, the settling velocities of the particles are plotted as a function of particle size. The curves assume free settling. These curves are essential in the following design procedure.

The Ideal Settling Basin

A simple model of an ideal sediment basin illustrates the fundamentals of basin design. For simplicity, it is assumed that soil particles have a uniform density (i.e., the smaller the particle the lower its weight). In Fig. 8.13, a flow Q (ft^3/sec) enters a basin of settling depth D, width W, and length L. It is assumed that we have "plug flow" in the basin, i.e., uniform flow in one direction.

A particle will travel horizontally with the water through the basin at a velocity of Q/WD. The particle will fall at a vertical velocity of V_s (specific for each particle size). The time T_Q for the particle to traverse the length of the basin will be

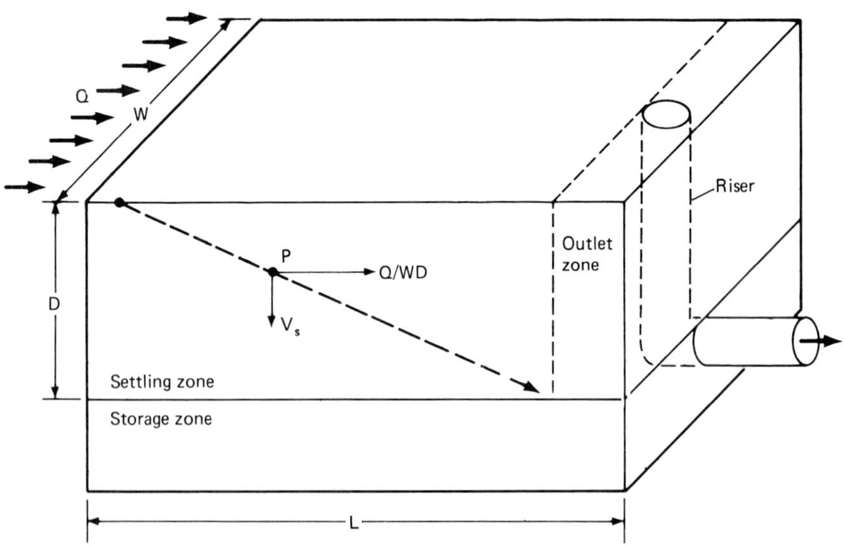

Fig. 8.13 Idealized sediment basin.

$$T_Q = \frac{L}{Q/WD}$$

The time for the particle to fall to the storage zone T_V will be

$$T_V = \frac{D}{V_s}$$

In a properly designed basin, the smallest particle to be captured will fall to the storage zone just before or as it reaches the outlet zone. Thus $T_Q = T_V$. Setting the transit and falling times equal gives us

$$T_Q = T_V = \frac{D}{V_s} = \frac{L}{Q/WD}$$

By simplifying the equation, we get

$$V_s L = \frac{DQ}{WD} \quad \text{or} \quad V_s = \frac{Q}{WL}$$

Note that WL is the suface area of the basin. Thus, by rearranging terms we have

$$WL = \text{basin surface area} = \frac{Q}{V_s}$$

We now have an ideal basin sized for removal of certain particles. The basin surface area has been established as a function of inflow Q and particle settling velocity. Note that basin depth and volume are not yet design factors!

Turbulence

The ideal basin is never constructed; it is only approached. Several factors affect performance; they include short circuiting, turbulence, bottom scour, riser design, temperature, and wind. First we must see how turbulence affects sizing of the ideal basin.

Turbulence in a basin is travel by water and particles in other than a straight line between inlet and outlet, i.e., travel in apparently random currents and swirls. Quiescent conditions, with little wasted motion of particles and laminar flow of the water, approximate the ideal sediment basin. *Turbulent conditions will lower basin efficiency.* Figure 8.14 was derived from work by Hazen (3) and the Bureau of Reclamation (7). It shows the change in sediment-trapping efficiency of a basin as the surface area of the basin is adjusted above and below the value computed with the formula Q/V_s. The graph also shows the loss of efficiency to be expected under turbulent flow conditions. Any well-designed basin should operate with an efficiency that lies somewhere between the two curves. To operate efficiently under turbulent conditions, a basin's surface area must be increased above Q/V_s.

Although there are equations which describe turbulent flow and its effect

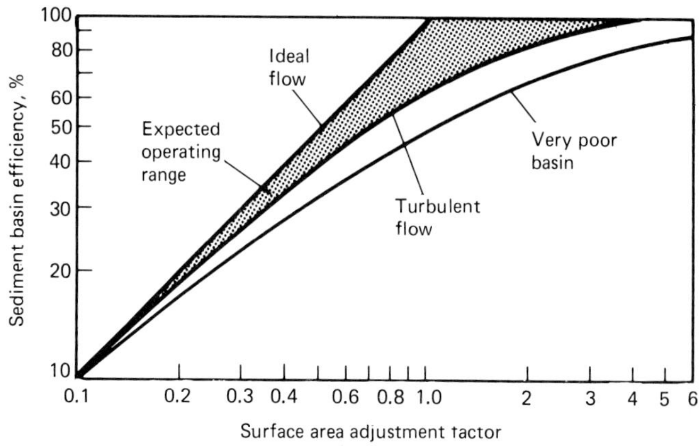

Fig. 8.14 Effect of turbulence on sediment basin efficiency. (3, 7)

upon performance, there is little guidance on reducing turbulence and increasing efficiency to that predicted for ideal conditions. More research needs to be performed on sediment basins before significant improvements in design can be proposed. However, the following practices can influence turbulence to some degree and reduce the surface area adjustment multiplier:

- Reduce water velocities through the basin. The ideal surface area is computed by using the equation $A = Q/V_s$. Increasing the surface area increases the cross-sectional area of the flow through the basin and decreases the horizontal flow velocity, and thus turbulence. An alternative way to reduce horizontal velocity is to increase basin settling depth. This has a small effect on reducing turbulence.
- Eliminate unnecessary angles or doglegs in the flow. When water flow changes direction, random currents that inhibit particle settling are set up. Long, straight basins are best.
- Reduce the effect of wind-induced turbulence. Large open water surfaces are affected by wind, which produces cross- and countercurrents that hinder settling and may resuspend bottom deposits. Using several smaller basins with a total capacity equal to the capacity of one large basin should improve capture efficiency. *Note: Multiple basins must be placed in parallel, not in series.* If small basins were in series, each one would be hydraulically overloaded and thus would perform poorly.

Surface Area Formula

The basin designer should select a surface area adjustment factor based on site conditions. A U.S. Environmental Protection Agency publication on erosion control for surface mining (6) proposes a surface area adjustment factor of 1.2. This factor is based on the assumption that the basin is well designed in all key aspects (shape, outlet location, riser design, etc.) but that turbulence and other nonideal conditions will reduce the basin's efficiency. The resulting sediment basin surface area sizing formula becomes:

$$A_s = \frac{1.2Q}{V_s}$$

where A_s is the appropriate surface area for trapping particles of a certain size and V_s is the settling velocity for that size particle.

8.2b Basin Efficiency

The trapping efficiency of a basin is a function of the particle size distribution of the inflowing sediment. Assuming ideal settling conditions, all particles with size and density equal to or larger than those of the design particle will be retained in the basin. In addition, some smaller particles will be captured while the basin is dewatering and the overflow rate has decreased. The additional capture ranges from 2 to 8 percent of the total sediment load. For our purposes the increase is not important enough to include in the calculations, particularly since the increase is offset by reduced capture efficiency of the basin when flow exceeds the design value.

Therefore, ideal basin efficiency corresponds to the percent of soil equal to or larger than the design particle size. For example, if a sediment basin on a site is designed to capture the 0.02-mm particle and 64 percent of the particles on this site are greater than or equal to 0.02 mm, the maximum efficiency of the basin is 64 percent. The only practical way to increase this efficiency is to increase the surface area of the basin.

8.2c Design Particle Size

The equation $A_s = 1.2Q/V_s$ defines the relation between size of particle to be captured and the surface area required for the basin. By applying this equation with the settling velocities V_s of various particle sizes from Fig. 8.12, Table 8.1 of surface areas per unit discharge is derived.

From Table 8.1 it is clear that the surface area required increases very rapidly as the particle size decreases. To capture the 0.02-mm particle, the area must be 6.5 times larger than the area required to capture the 0.05-mm particle. To capture the 0.01-mm particle, the basin area must be 4 times larger than for the 0.02-mm particle and 25 times larger than for the 0.05-mm particle. For particles smaller than 0.02 mm, the surface area requirement increases dramatically.

Where soils have a high content of clay or fine silt, increasing the size (and thus cost) of a basin will not bring about a proportional increase in basin efficiency. For example, a typical soil in the San Francisco Bay Area is 62 percent by weight composed of particles 0.02 mm and larger, but it is only 5 percent by

TABLE 8.1 Surface Area Requirements of Sediment Traps and Basins

Particle size, mm	Settling velocity, ft/sec (m/sec)	Surface area requirements, ft^2 per ft^3/sec discharge	(m^2 per m^3/sec discharge)
0.5 (coarse sand)	0.19 (0.058)	6.3	(20.7)
0.2 (medium sand)	0.067 (0.020)	17.9	(58.7)
0.1 (fine sand)	0.023 (0.0070)	52.2	(171.0)
0.05 (coarse silt)	0.0062 (0.0019)	193.6	(635.0)
0.02 (medium silt)	0.00096 (0.00029)	1,250.0	(4,101.0)
0.01 (fine silt)	0.00024 (0.000073)	5,000.0	(16,404.0)
0.005 (clay)	0.00006 (0.000018)	20,000.0	(65,617.0)

weight composed of particles in the 0.01- to 0.02-mm range. A surface area 4 times larger would be needed to capture 5 percent more of this soil.

A balance between the cost-effectiveness of a certain basin size and the desire to capture fine particles must be achieved. It is desirable to capture the very small soil particles (clays and fine silts) because they cause turbidity and other water quality problems. However, Table 8.1 shows that a basin would have to be very large to capture particles smaller than 0.02 mm, particularly clay particles 0.005 mm and smaller. Because of the high cost of trapping very small particles, the authors recommend 0.02 as the design particle size for sediment basins except in areas with coarse soils, where a larger design particle may be used. The 0.02-mm particle is classified as a medium silt by the AASHTO soil classification system.

8.2d Basin Discharge Rate

The peak discharge, calculated by the rational or another approved method, is used to size the basin riser. During any major storm, a sediment basin should fill with water to the top of its riser and then discharge at the rate of inflow to the basin. A sediment basin is not designed with a large water storage volume as is a reservoir. If the inflow exceeds the design peak flow used to size the riser, the overflow should discharge down an emergency spillway.

8.2e Design Runoff Rate

In the equation for surface area of a sediment basin, the discharge rate Q is a variable to be chosen by the designer. The above discussion of basin discharge rate shows that the discharge rate is, to a large extent, equal to the inflow. The riser is sized to handle the peak inflow to the basin. The authors suggest determining the surface area by the *average runoff of a 10-year, 6-hr storm* instead of the peak flow. A substantial savings in size, and therefore cost, is obtained, and basin efficiency is not significantly decreased.

Consider a basin designed to capture the 0.02-mm particle at the average runoff rate. The average rainfall per hour is 17 percent of the total rainfall in a 6-hr storm (Sec. 4.1f). On a site with soils with a moderately high clay content, under ideal settling conditions this basin would retain about 62 percent of the eroded soil (i.e., 62 percent of the soil, by weight, is composed of 0.02-mm or larger particles).

If the surface area of this basin were instead designed for the peak flow, it would be roughly 3 times larger. According to data from the U.S. Bureau of Reclamation (10), 25 percent of the total rainfall in a 6-hr storm falls in a ½-hr period (Fig. 4.2). Since the rainfall intensity i value is in units of inches (or millimeters) per hour, the peak flow can be calculated by using an i value of 50 percent of the 6-hr total. Since basin surface area is directly proportional to the discharge rate ($A = 1.2Q/V_s$) and the peak discharge rate in a 6-hr storm is 2.9 times the average rate ($50\% = 2.9 \times 17\%$), the surface area sized for the peak flow would be about 3 times the surface area sized for the average flow. The basin sized for the peak flow would capture, during most of the storm except the peak, particles with approximately one-third the settling velocity of the design particle. Since the 0.02-mm particle settles at 0.00096 ft/sec (0.00029 m/sec), particles with a settling velocity of 0.00032 ft/sec (0.000098 m/sec) would then be captured. These are approximately 0.01-mm particles.

Suppose a basin on a site with clayey soils were sized by using the peak runoff rate. For the purpose of illustration, suppose the soil composition were typical of the San Francisco Bay Area as in the preceding example (62 percent of particles, by weight, greater than 0.02 mm and 5 percent, by weight, from 0.01 to 0.02 mm). A basin with a large surface area based on the peak runoff would capture the 0.01- to 0.02-mm particles as well as particles greater than 0.02 mm, or 67 percent of the eroded material. The basin efficiency would be increased 8 percent (5/62) by tripling the surface area. Thus it is generally much more cost-effective to size a basin by using the average runoff rather than the peak, and basin efficiency will not be significantly lower.

8.2f Settling Depth

If a basin is too shallow, water flowing rapidly through the basin may resuspend settled particles and decrease efficiency of capture. A similar problem occurs in grit-settling chambers at sewage treatment plants, where velocity must be controlled to prevent particle resuspension. An equation that describes scour in a grit chamber (2) is:

$$V_{\text{scour}} = \frac{1.486}{n} \times r^{1/6}\left[k(S_s - 1) \times \frac{d}{304.8} \right]^{1/2}$$

where V_{scour} = horizontal water velocity that would just resuspend sediment, ft/sec
n = Manning's roughness coefficient, 0.016 to 0.020
r = hydraulic radius, ft
k = shape coefficient (0.04 for granular material)
S_s = specific gravity (2.65 for mineral particles)
d = design particle diameter, mm (0.02 for most sediment basins)

By using the figures in parentheses above,

$$V_{scour} = 0.0031 \frac{r^{1/6}}{n}$$

The horizontal water velocity in a basin is Q/WD. The horizontal velocity V_h must not exceed V_{scour}. Setting the velocities equal (or horizontal velocity less than scour limit) produces:

$$V_{scour} \geq V_h = \frac{Q}{WD}$$

For the 0.02-mm particle the settling velocity

$$V_s = \frac{Q}{WL} \approx 0.001 \text{ ft/sec} \qquad \text{and} \qquad \frac{Q}{W} = 0.001L$$

Therefore

$$V_{scour} = 0.0031 \frac{r^{1/6}}{n} \geq \frac{Q}{WD}$$

$$0.0031 \frac{r^{1/6}}{n} D \geq \frac{Q}{W} = 0.001L$$

Transposing and simplifying gives us:

$$\frac{L}{D} \leq 3.1 \frac{r^{1/6}}{n}$$

Substituting various values for r and n produces the accompanying table. It

Manning's coefficient	Values of length/depth when r = 1 ft (0.3 m)	r = 3 ft (0.9 m)	r = 10 ft (3 m)
0.016 (min)	193	232	284
0.018 (normal)	172	206	252
0.020 (max)	155	186	227

appears that the length-to-depth ratio should be less than the values in the table. It is difficult to predict the hydraulic radius of stray currents in a sediment basin, but in general, the radius should exceed 10 feet (3 m), as it certainly would if actual basin dimensions were used. For simplification, a conservative basin design criterion would be to keep the ratio of length to settling zone depth less than 200. The authors recommend that the settling zone of any sediment basin be maintained free of accumulated sediment for a depth of at least 2 ft (0.61 m).

8.2g Basin Length and Width

Figure 8.15 shows a jet of water entering a basin. This situation is comparable to that of a single inlet pipe at one end of the basin. Frictional forces slow the jet down, but the principle of conservation of momentum suggests that more fluid from the basin joins the jet. The length-to-width ratio of the jet, for water, is 8:1 (8). It can be seen in Fig. 8.15*a* that there are large dead spots in the basin which are not a part of the jet and which do not contribute to sedimentation. This is short-circuiting, and it *greatly* reduces a basin's performance.

In order to *approach* ideal design performance, the full surface area of the basin must be employed. The full employment can be secured by installing a baffle as in Fig. 8.15*b*. The baffle redistributes the flow across the entire width of

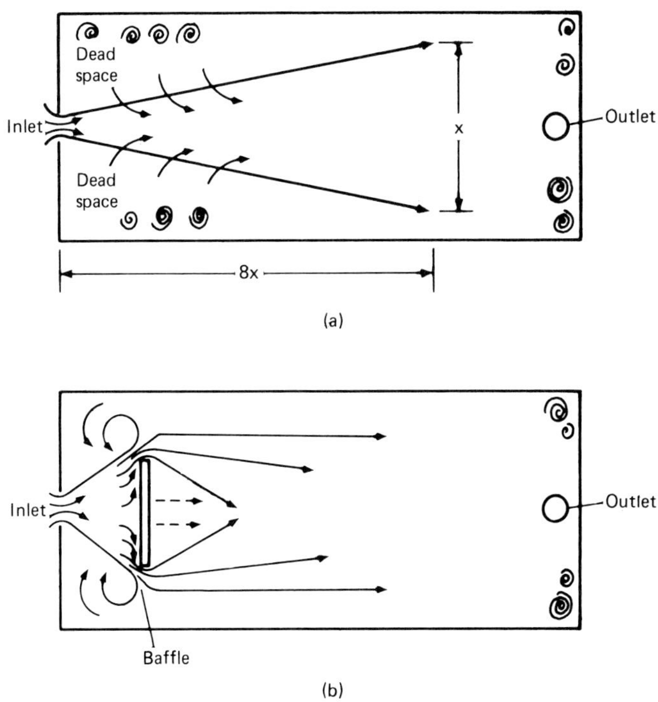

Fig. 8.15 Short-circuiting in a basin. (*a*) Velocity jet in a basin. (*b*) Redistribution of flow with a baffle.

the basin and makes maximum use of the settling area. A baffle located near the inlet is recommended for all basins, and it is *essential* for basins in which the length-to-width ratio is less than 10:1.

The exact design and placement of baffles have not been rigorously studied or developed. To a large extent these factors are at the discretion of the designer. Figure 8.16 illustrates some typical baffle placements. Placement of a baffle near the inflow causes the water to circulate around the baffle and traverse the basin in a more even distribution across the basin's width. Thus a baffle also serves to dissipate energy at the basin inlet (and thereby reduce scour and turbulence).

8.2h Outlet Scour

As water traveling through a basin approaches the riser, it will speed up. The area in which this speed becomes excessive is the outlet scour zone, and bottom deposits may be resuspended and discharged through the riser. This roughly cylindrical zone around the vertical riser has a circumference equal to the width W of the basin.

The proportion of the surface area of the basin lost to the outlet scour zone is approximately $W/4\pi L$. This proportion is usually less than the surface area adjustment factor of 1.2. Note that the longer and narrower basin will have a smaller outlet scour zone as L increases and W decreases. In *very* large basins, the scour zone can be reduced by multiple risers. The diameter of the scour zone at each riser can be reduced by increasing the number of risers, and it approaches the diameter of the riser as a limit.

8.2i Sediment Storage Depth

The volume of a basin consists of two portions: a settling volume and a storage volume. The settling volume requirement is 2 ft (0.6 m) times the required surface area. The settling depth of 2 ft (0.6 m) is the distance between the crest of the riser and the lowest dewatering hole on the riser.

The storage zone must be large enough to contain sediment deposits without decreasing the settling volume. The sediment yield to the basin can be estimated by using the universal soil loss equation (Chap. 5). The depth required for storage is determined by dividing the estimated annual sediment yield by the surface area of the basin. This depth will provide storage for one year's accumulated sediment. If a basin will be cleaned more frequently than once per year, the storage depth may be proportionately shallower.

8.2j Dewatering

A sediment basin should have a means for dewatering it after a storm. Dewatering is essential for three reasons. First, water ponded in a deep basin is a safety hazard, particularly to young children. Second, standing water is a potential

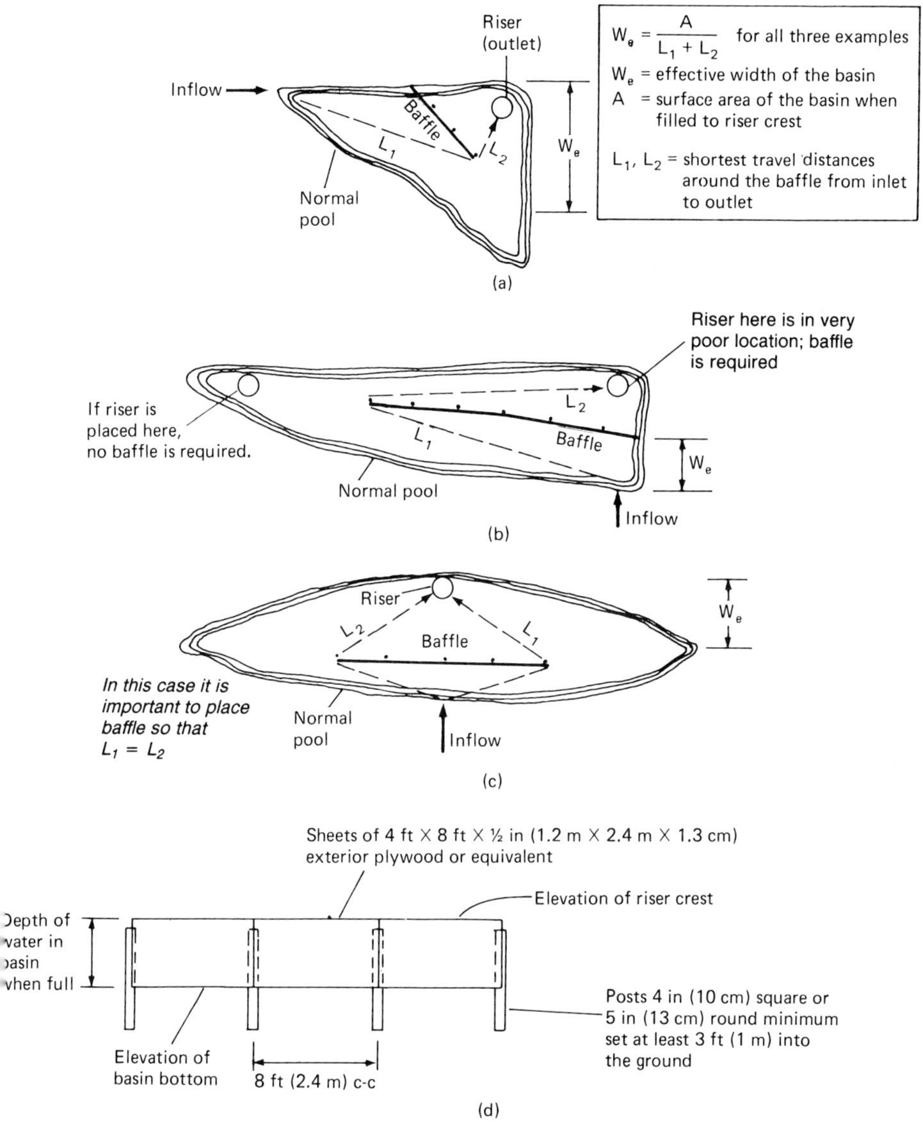

Fig. 8.16 (*a*) to (*c*) Three examples of a sediment basin baffle placement. (These plan views are not to scale.) (*d*) Baffle detail, elevation. (Adapted from 9)

mosquito breeding ground. Finally, accumulated sediment can be removed much more easily after the water is drained out. There are several aspects to dewatering that must be considered, and they are discussed below.

Time

One of the more important design variables available to the engineer is the time for dewatering. An informal survey of consultants and public works representatives in the San Francisco Bay Area uncovered a perception that 24 hr should be an appropriate basin dewatering period. As will be shown later, a basin designed to dewater within 24 hr will provide minimal if any treatment during commonly occurring light storms. Water would leave the basin as quickly as it came in and so would deposit no sediment.

Detention periods longer than 24 hr mandate fencing to restrict access to the basin, and a case could be made that fencing is needed even for basins that dewater within 24 hr. *Thus, as a safety measure, sediment basins should be fenced.* With this provision, dewatering periods can be extended to 2, 3, or more days. The longer dewatering period will allow the basin to trap sediment during light as well as heavy storms. To prevent creation of a mosquito breeding ground, sediment basins should be dewatered within 7 days.

Basin Condition

There are two extreme basin conditions that place limits on basin dewatering. When a basin is relatively new, or has recently been cleaned out, a storm will fill the settling zone and most of the sediment storage zone with water. The amount of water to be removed between storms is roughly the full volume of the basin. When a basin has been in service for some time, the sediment storage zone will be nearly full of sediment. Dewatering would consist of removing at least the volume in the settling zone and any free water in the sediment zone.

Wet sediment has the consistency of thick pea soup and can be thicker. Cleaning of the basin is greatly facilitated when this sediment is dewatered to a more manageable consistency. Therefore, a mechanism to dewater the sediment without losing sediment to the discharge is desirable. However, sediment need not be dewatered within 7 days.

Dewatering Techniques

Several techniques for dewatering basins are available. It should be understood at the outset that no single technique will work with all basins. Furthermore, an extensive evaluation of dewatering methods has not been conducted, nor is there a clearly preferred method. The following methods have been reported in the literature. When problems with these methods have been observed by the authors or reported by others, they are so noted.

The various methods for basin dewatering are shown in Fig. 8.17. Figure 8.17*a* presents a riser with a single hole placed approximately 2 ft (0.6 m) below the

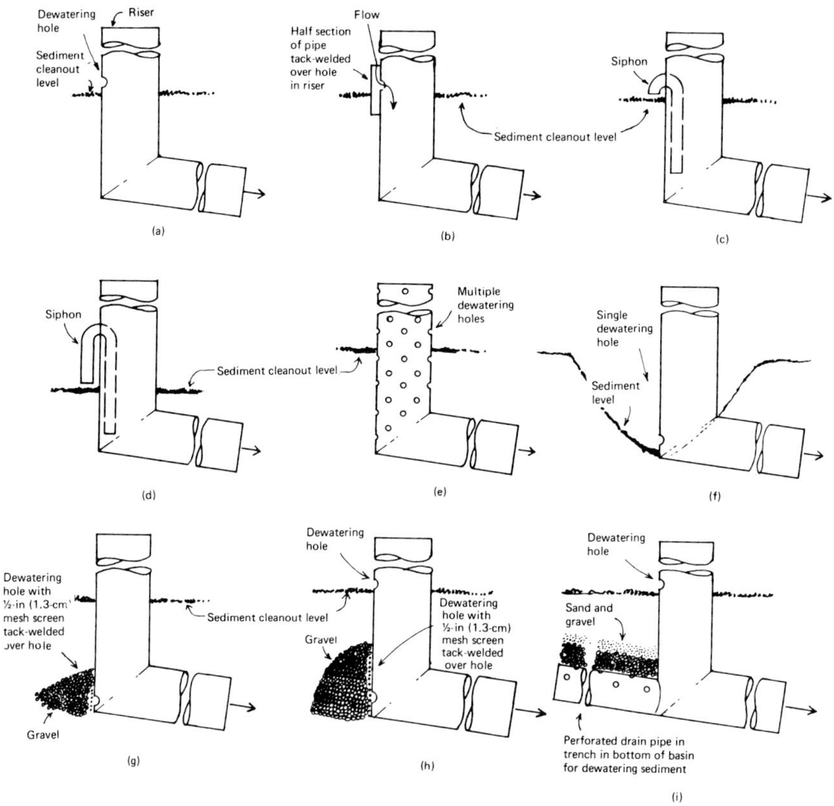

Fig. 8.17 Basin-dewatering methods. Methods (*e*) and (*f*) are poor; (*g*), (*h*), and (*i*) are best.

top. This hole will dewater the settling zone but *will not* dewater the sediment storage zone. The equation that describes flow through an orifice is:

$$Q = C_d \times A_o\sqrt{2gh}$$

where Q = flow, ft^3/sec (m^3/sec)
C_d = coefficient of contraction for an orifice, approximately 0.6 for a sharp-edged orifice
A_o = surface area of the orifice, ft^2 (m^2)
g = acceleration of gravity, 32.2 ft/sec^2 (9.81 m/sec^2)
h = head of water above the orifice, ft (m)

The time required for an orifice to dewater a basin can be obtained by integrating the change in basin volume over time as a function of depth of water. The resulting equation is

$$T = \frac{A_s\sqrt{2h}}{3600 \times A_o \times C_d\sqrt{g}}$$

where T = time, hr
A_s = surface area of basin, ft^2 (m^2)

Conversely, the size of orifice needed to dewater a basin within a specified time T is

$$A_o = \frac{A_s\sqrt{2h}}{3600 \times T \times C_d\sqrt{g}}$$

Thus a typical 10,000-ft^2 (930-m^2) basin would need a 0.068-ft^2 (0.0063-m^2) or 3.5-in- (9-cm-) diameter orifice to dewater a 2-ft (0.6-m) settling zone in 24 hr. Several orifices adding up to 0.068 ft^2 (0.0063 m^2) in surface area could be used.

Figure 8.17*b* presents the same riser with a shield tack-welded over the orifice. The shield greatly reduces clogging of the orifice, but it will somewhat lower the discharge rate through the orifice.

Figure 8.17*c* and *d* shows a siphon used to dewater a basin to the sediment level. This method, as presented in the SCS Maryland handbook (9), calls for a 4-in- (10-cm-) diameter siphon. With the system shown in Fig. 8.17*c*, runoff from small storms and base flow would flow through the siphon with no backing up and storage. In the system of Fig. 8.17*d* small flows would back up and discharge in a batch fill and drain process. This would allow some settling of the sediment. Neither system would dewater below the siphon.

The method shown in Fig. 8.17*e* has been frequently observed at California construction sites. It is *potentially the worst possible* to be employed. Although the basin would dewater rapidly, most runoff would flow straight across the basin and out the multiple holes. Not only would there be no backing up of the water, with associated particle settling, but the straight channel flow through the basin would *resuspend* previously deposited solids. The discharge from such basins has been found by the authors to contain more sediment than the runoff into the basins.

The dewatering hole shown in Fig. 8.17*f* can dewater the entire basin, including the sediment zone. However, this system is poor for two reasons. First, any orifice sized to dewater the basin quickly would directly pass low flows associated with the more common storms and provide either no treatment at all or negative treatment. Second, any sediment accumulating near the riser would be quickly washed through the orifice.

A number of consultants have suggested the systems shown in Figs. 8.17*g* and *h*. The gravel piled around the base of the riser acts as a coarse filter to keep sediment from escaping. These systems have not been monitored to assess their effectiveness. Additionally, the system shown in Fig. 8.17*i* has been suggested for the dewatering of sediment in large basins. Performance of this system also has not been verified. In all cases, the dewatering hole must be sized to prevent smaller storms from draining without any retention time in the basin.

Figure 4.2 shows that, within a typical 6-hr storm, for 4 hr the flow rate will be about 60 percent or less of the average rate. If, for example, the basin were designed for a 10-year interval storm that drops 2.6 in (66 mm) in 6 hr, a much more common storm may drop only 0.5 in (13 mm) in 6 hr. The average flow rate of the more common storm through the basin may be only 20 percent of the 10-year storm. For 4 of 6 hr of the common storm, the flow rate may be only 12 percent (60 percent of 20 percent) of the average flow rate from the 10-year storm. A dewatering hole sized to dewater a basin quickly (24 hr) would in all likelihood pass the small storm without ponding, and thus without trapping sediment. Dewatering holes sized to drain a basin over a longer period, such as 3 to 6 days, will back up the smaller flows and cause siltation of the basin.

EXAMPLE 8.1

Given: Basin surface area = 14,400 ft^2 (1338 m^2)
Basin depth D = 4.2 ft (1.3 m)
Q design = 2.64 ft^3/sec (0.0748 m^3/sec) [i = 0.44 in/hr (12 mm/hr)]

Find: Unobstructed orifice size at bottom of steel riser to dewater a full basin in 24 hr.

Solution:

$$A_o = \frac{A_s\sqrt{2h}}{3600 \times T \times C_d\sqrt{g}} = \frac{14{,}400(\sqrt{8.4})}{3600(24)(0.6)(\sqrt{32.2})} = \frac{41{,}735}{294{,}166}$$

$$= \left[\frac{(1338\text{ m}^2(\sqrt{2.6\text{ m}})}{3600(24)(0.6)(\sqrt{9.81\text{ m/sec}^2})} = \frac{2157}{162{,}368}\right]$$

$$= 0.14\text{ ft}^2\ (0.013\text{ m}^2)$$

$$= 20.4\text{ in}^2\ (133\text{ cm}^2)\text{, or five holes of 2.28 in (5.8 cm) diameter}$$

Now Find: Depth to which this basin will fill during a smaller storm precipitating 0.5 in (13 mm) in 6 hr.

Solution:

$$i_{\text{avg}} = \frac{0.5\text{ in}}{6\text{ hr}} = 0.083\text{ in/hr} \qquad \left(\frac{13\text{ mm}}{6\text{ hr}} = 2.2\text{ mm/hr}\right)$$

$$Q_{\text{avg}} = (2.64\text{ ft}^3\text{/sec})\left(\frac{0.083\text{ in/hr}}{0.44\text{ in/hr}}\right) = 0.5\text{ ft}^3\text{/sec}$$

$$\left[(0.0748\text{ m}^3\text{/sec})\left(\frac{2.2\text{ mm/hr}}{12\text{ mm/hr}}\right) = 0.014\text{ m}^3\text{/sec}\right]$$

For 4 of 6 hr,

$$Q_{\text{in}} = 0.6Q_{\text{avg}} = 0.3\text{ ft}^3\text{/sec} \qquad (0.0084\text{ m}^3\text{/sec})$$

$$= Q_{\text{out}}$$

Now $Q_{\text{out}} = C_d \times A_o\sqrt{2gh}$; therefore, the head of water h which would balance out Q_{in} and Q_{out} is

$$h = \left(\frac{Q_{\text{out}}}{C_d \times A_o\sqrt{2g}}\right)^2 = \left(\frac{0.3}{0.6(0.14)(\sqrt{64.4})}\right)^2 \left\{\left[\frac{0.0084\text{ m}^3\text{/sec}}{0.6(0.013\text{ m}^2)(\sqrt{19.6\text{ m/sec}^2})}\right]^2\right\}$$

$$= \frac{0.09}{0.45} = 0.2\text{ ft} \qquad (0.06\text{ m})$$

An empty basin would back up 0.2 ft (0.06 m) above the orifice before inflow and outflow equalized for the smaller storm. This would produce excessive horizontal water velocity in the basin and would scour existing bottom sediments. Therefore, the dewatering time should be increased. If the dewatering orifice were reduced to only one hole of 2.28 in (5.8 cm) diameter, the dewatering time would be increased to 5 days. The head that would then be needed to make Q_{out} equal to Q_{in} would be 5 ft (1.5 m), which means the basin would discharge properly over the top of the riser. In reality, the risk of plugging a single 2.28-in (5.8-cm) hole is high. Figure 8.17*g*, *h*, and *i* shows how to protect the dewatering hole or holes from clogging by using a wire mesh surrounded by gravel or by using a subsurface drain.

As a general rule, a basin designed to dewater quickly (short time T) will have a much lower trap efficiency during small storms. During small storms, runoff will drain from a basin which has a relatively large dewatering hole as fast as it enters the basin. Thus there is very little detention time for small particles to settle out. The benefits of rapid dewatering should therefore be weighed against the benefits of trapping sediment during small storms.

The following are some sample sediment basin design problems solved by using the surface area formula and the universal soil loss equation.

EXAMPLE 8.2 Calculation of Required Surface Area and Storage Volume

Given: A construction site has a drainage area of 12 acres (4.8 ha). The original plans call for smooth graded surfaces, without seeding or mulching. The site is on a hillside. House pads will be terraced up the slope with the following final gradients and slope lengths:

3 acres (1.2 ha): cut and fill embankments, 3:1 slopes, 35 ft (11 m) long
9 acres (3.6 ha): 10:1 slopes, 150 ft (46 m) long

Diversions will channel runoff away from the steep slopes and convey all site runoff to a single sediment basin. The particles of 79 percent of the soil, by weight, are equal to or larger than 0.02 mm. Average annual rainfall is 32 in (813 mm), and the 10-year, 6-hr rainfall is 2.62 in (67 mm). The rainfall erosion index (R factor) is 34, and the K factor is 0.32.

Find: (1) The required surface area to capture the 0.02-mm and larger particles, (2) the storage volume required to retain 1 year's soil loss, and (3) the basin depth required for once per year cleaning.

Solution:

STEP 1. Calculate average runoff by using the 10-year, 6-hr rainfall and the rational method.

$$Q_{avg} = C \times i \times A$$

C *Factor.* From Table 4.1, choose a value of 0.5 for bare, smooth earth and moderate slopes.

i *Factor.* The average intensity is 2.62 in (67 mm) in 6 hr, or 0.44 in (11 mm) in 1 hr.

A *Factor.* The area is 12 acres (4.8 ha).

$$Q_{avg} = C \times i \times A$$

$$= 0.5(0.44)(12) \qquad \left[\frac{0.5(11\ \text{mm/hr})(4.8\ \text{ha})}{360}\right]$$

$$= 2.64\ \text{ft}^3/\text{sec} \qquad (0.073\ \text{m}^3/\text{sec})$$

STEP 2. Sieve analysis indicates that 79 percent by weight of the soil is equal to or larger than 0.02 mm.

STEP 3. The settling velocity of the 0.02-mm particle V_s is 0.00096 ft/sec (0.00029 m/sec).

STEP 4. The minimum required surface area A_s is

$$A_s = \frac{1.2Q_{\text{avg}}}{V_s}$$

$$= \frac{1.2(2.64\ \text{ft}^3/\text{sec})}{0.00096\ \text{ft/sec}} \qquad \left[\frac{1.2(0.073\ \text{m}^3/\text{sec})}{0.00029}\right]$$

$$= 3300\ \text{ft}^2\ (302\ \text{m}^2)$$

STEP 5. The settling depth must be at least 2 ft (0.6 m) (see Sec. 8.2f).

STEP 6. The storage depth is estimated by using the USLE. Apply the USLE to each of the two length-slope categories on the site.

$$A = R \times K \times \text{LS} \times C \times P$$

R, K, C, and P will be the same for the two slope types; LS varies.

R *Factor.* A rainfall erosion index of 34 is given.

K *Factor.* A K factor of 0.32 is given.

C *Factor.* Since no vegetation or mulch is planned, C is 1.0.

P *Factor.* For smooth, compacted soil, P is 1.3.

LS Factor. By interpolation in Table 5.5:

3:1, 35 ft (11 m): LS is 5.57

10:1, 150 ft (46 m): LS is 0.81

Estimate Soil Loss for 1 Year. On 3 acres (1.2 ha), $A = 34(0.32)(5.57)(1.0)(1.3) = 79$ tons/acre (177 t/ha). Assuming 1 ton of soil = 1 yd^3 (1 t = 0.843 m^3), multiplying by 3 acres (1.2 ha) = 237 yd^3 (179 m^3).

On 9 acres (3.6 ha), $A = 34(0.32)(0.81)(1.0)(1.3)(9) = 11$ tons/acre (25 t/ha). Again assuming 1 ton of soil = 1 yd^3 (1 t = 0.843 m^3), multiplying by 9 acres (3.6 ha) = 99 yd^3 (76 m^3). Add the two together; multiply by 0.79 (the fraction of soil larger than 0.02 mm); and convert to cubic feet:

237 + 99 = 336 yd^3 (179 m^3 + 76 m^3 = 255 m^3) total soil loss

336(0.79)(27 ft^3/yd^3) = 7167 ft^3 [(225 m^3)(0.79) = 201 m^3] retained in basin (assuming ideal basin performance)

Now divide by minimum surface area:

$$\text{Storage depth} = \frac{7167\ \text{ft}^3}{3300\ \text{ft}^2} = 2.2\ \text{ft} \qquad \left(\frac{201\ \text{m}^3}{302\ \text{m}^2} = 0.67\ \text{m}\right)$$

EXAMPLE 8.3 Influence of Variable Basin Design Factors (Design Storm and Particle Size) on Required Basin Surface Area

Given: The same construction site as in Example 8.2, except the soil analysis is changed to read as follows:

Size range, μm	>425	250–425	180–250	125–180	63–125	36–63	23–36	14–23	10–14	7–10	4–7	1–4	≤1
Dry weight, %	35	5	3	4	7	4	2	4	3	3	2	5	22

Find:

1. The surface area and theoretical efficiency of the basin sized to capture the 0.02-mm particle
2. The surface area and theoretical efficiency of the basin sized to capture the 0.01-mm particle
3. The surface area and theoretical efficiency of the basin sized to capture the 0.06-mm particle
4. The efficiency of the basin sized to capture the 0.02-mm particle in the 10-year, 6-hr storm if the basin receives runoff from a 25-year, 6-hr storm with a rainfall of 2.91 in (74 mm) in 6 hr
5. The efficiency of the basin in (4) when the basin receives runoff from a 2-year, 6-hr storm with a rainfall of 1.81 in (46 mm) in 6 hr

Solution: (1) For a given particle capture size and geographic location, the surface area of a basin will not change as the soil analysis changes. The basin will still have a minimum surface area requirement of 3300 ft^2 (302 m^2) (see Example 8.2).

To find the theoretical efficiency, add together the weight percentages of particles greater than and equal to 0.02 mm. For the given soil the sum of particles from gravel size to 0.023 mm is 60 percent. Interpolating in the 0.023 to 0.014 size range adds approximately 2 percent. Therefore, the theoretical basin efficiency is 62 percent.

(2) From Fig. 8.12, the settling velocity of the 0.01-mm particle is 0.00022 ft/sec (0.000067 m/sec). A cool weather temperature of 50°F (10°C) is assumed.

The surface area will vary inversely as the ratio of settling velocities.

$$\frac{A_{0.01\ \text{mm}}}{A_{0.02\ \text{mm}}} = \frac{V_{0.02\ \text{mm}}}{V_{0.01\ \text{mm}}}$$

$$A_{0.01\ \text{mm}} = \frac{0.00096\ \text{ft/sec}}{0.00022\ \text{ft/sec}}(3300\ \text{ft}^2) = 14{,}400\ \text{ft}^2 \left[\frac{0.00029\ \text{m/sec}}{0.000067\ \text{m/sec}}(302\ \text{m}^2) = 1310\ \text{m}^2\right]$$

Again, summing particle sizes, this time to 0.01 mm, the basin's theoretical efficiency would be 67 percent.

(3) From Fig. 8.12, the settling velocity of the 0.06-mm particle is 0.0085 ft/sec (0.0026 m/sec). The surface area is

$$A_{0.06} = \frac{0.00096\ \text{ft/sec}}{0.0085\ \text{ft/sec}}(3300\ \text{ft}^2) = 373\ \text{ft}^2 \qquad \left[\frac{0.00029\ \text{m/sec}}{0.0026\ \text{m/sec}}(302\ \text{m}^2) = 34\ \text{m}^2\right]$$

Note the significant effect of design particle size on the size of the sediment basin. This smaller basin would capture 54 percent of incoming soil.

(4) The basin sized in Example 8.2 has a surface area of

$$A_s = \frac{1.2 \times Q_{avg}}{V_s} = 3300 \text{ ft}^2 \ (302 \text{ m}^2)$$

In the design, the basin's overflow rate was set equal to the particle settling velocity with an adjustment factor.

$$\frac{Q_{avg}}{A_s} = \frac{V_s}{1.2} = \frac{C \times i \times A}{A_s}$$

For a fixed basin size, the discharge rate will be proportional to the average storm intensity as follows:

$$\frac{Q_{avg10}}{Q_{avg25}} = \frac{V_{s10}}{V_{s25}} = \frac{i_{10}}{i_{25}}$$

where 10 and 25 refer to 10- and 25-year storm return periods and V_{s10} and V_{s25} refer to the settling velocities of the smallest particle that the basin can capture in a 10- or 25-year storm, respectively.

The settling velocity of the particle captured during the 25-year storm will be found from the ratio of average intensities [10-year, 6-hr, 2.62 in (67 mm) and 25-year, 6-hr, 2.91 in (74 mm)].

$$V_{s25} = \frac{V_{s10} \times i_{25}}{i_{10}} = \frac{0.00096 \text{ ft/sec}(0.48 \text{ in/hr})}{0.44 \text{ in/hr}} \left[\frac{(0.00029 \text{ m/sec})(12 \text{ mm/hr})}{11 \text{ mm/hr}}\right]$$

$$= 0.00105 \text{ ft/sec} \qquad (0.00032 \text{ m/sec})$$

The particle size corresponding to this velocity is 0.021 mm (Fig. 8.12).

From the sieve analysis we see that the 25-year storm would have a negligible impact on basin efficiency.

(5) Using a similar approach for the 2-year storm [i = 1.81 in/6 hr = 0.3 in/hr (7.6 mm/hr)],

$$V_{s2} = \frac{V_{s10} \times i_2}{i_{10}} = \frac{0.00096 \text{ ft/sec}(0.3 \text{ in/hr})}{0.44 \text{ in/hr}} \left[\frac{(0.00029 \text{ m/sec})(7.6 \text{ mm/hr})}{11 \text{ mm/hr}}\right]$$

$$= 0.00065 \text{ ft/sec} \qquad (0.00020 \text{ m/sec})$$

The corresponding particle size is 0.017 mm. From the sieve analysis we see that the basin gains in efficiency by approximately only 1 percent. However, we note that the basin can effectively trap sediment during both large and small storms.

8.3 DESIGN AND INSTALLATION OF SEDIMENT BASINS

The details of a sediment basin require an engineer's attention. This handbook provides guidance concerning the use of the surface area formula. Specifications for sizing the riser, installing antiseep collars, determining the length of pipe through the embankment, the embankment cross section, and more can be found in References 1, 9, and 11 and in civil engineering handbooks. This section highlights several key components of sediment basins and their proper installation. Some common faulty designs and installations are also illustrated.

Sediment basins can be costly to construct; but by using materials commonly found on construction sites and taking advantage of permanent storm drain facilities, costs can be minimized. Corrugated metal pipe can be used to construct the risers; rock and gravel can be used for outlet protection; and plywood can be used for baffles. Storm drain inlets can serve as risers if basins are excavated around them and thereby eliminate the need for a separate outlet system for the basin.

8.3a Principal Spillway Design and Installation

A pipe riser is the most common type of principal spillway. The minimum capacity of the riser should be equal to the peak flow expected from the design storm. For a basin with no additional emergency spillway, the riser must have the capacity to handle the peak flow from a rainfall event commensurate with the degree of hazard involved (e.g., if the hazard from overflow is great, a very large magnitude of rainfall should be used for design). In all cases the minimum diameter of a riser pipe should be 8 in (20 cm) when the pipe is used in combination with an emergency spillway. The elevation of the riser crest should be 1 ft (0.3 m) below the elevation of the spillway crest.

A concentric antivortex device and trash rack should be securely installed on top of the riser. The antivortex device allows the riser to carry the maximum possible flow without drawing air. The trash rack is essential for keeping this outlet of the basin clear and flowing. Figure 8.18 shows an effective design developed by the Soil Conservation Service (9). The riser design recommended in erosion control standards (1, 9, 11) calls for a single dewatering hole 2 ft (0.6 m) below the top of the riser, a weighted base to prevent flotation, and antiseep collars around the pipe through the embankment (Fig. 8.1). As shown in Fig. 8.19, the riser is firmly anchored in concrete, has a watertight connection, and is equipped with a trash rack. The only defect is that there are too many dewatering holes. It has become common practice in some areas to perforate the entire surface of a CMP riser (Fig. 8.20). The disadvantages of this practice, and suggested alternative arrangements for dewatering a basin, are discussed in Sec. 8.2j. Figure 8.17 illustrates several ways to dewater basins without putting multiple holes in the riser.

Weighted Base

The base of a pipe riser must be firmly anchored to prevent the riser from floating. At least one developer has seen a riser float and twist itself out of the embankment before the water level reached the dewatering hole. If the riser is more than 10 ft (3 m) in height, the forces acting on it must be calculated. As a minimum, a method of anchoring the pipe which provides a safety factor of 1.25 should be used (downward forces = 1.25 × upward forces).

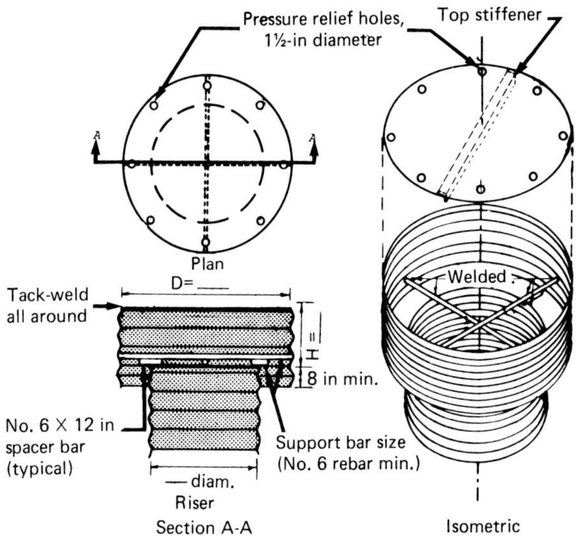

NOTES:

1. The cylinder must be firmly fastened to the top of the riser.
2. Support bars are welded to the top of the riser or attached by straps bolted to top of riser.
3. Pressure relief holes may be omitted, if ends of corrugations are left fully open when corrugated top is welded to cylinder.

Design Table

Riser diameter, in (cm)	Cylinder				Minimum top	
	Diam., in	Thickness gage	H, in (cm)	Minimum size support bar	Thickness	Stiffener
12 (30.5)	18 (45.7)	16	6 (15.2)	No. 6 rebar	16 gage	—
15 (38.1)	21 (53.3)	16	7 (17.8)	No. 6 rebar	16 gage	—
18 (45.7)	27 (68.6)	16	8 (20.3)	No. 6 rebar	16 gage	—
21 (53.3)	30 (76.2)	16	11(27.9)	No. 6 rebar	16 gage	—
24 (61)	36 (91.4)	16	13(33)	No. 6 rebar	14 gage	—
27 (68.6)	42 (106.7)	16	15(38.1)	No. 6 rebar	14 gage	—
36 (91.4)	54 (137.2)	14	17(43.2)	No. 8 rebar	12 gage	—
42 (106.7)	60 (152.4)	14	19(48.3)	No. 8 rebar	12 gage	—
48 (121.9)	72 (182.9)	12	21(53.3)	1¼-in (3.2-cm) pipe or 1¼ × 1¼ × ¼ angle	10 gage	—
54 (137.2)	78 (198.1)	12	25(63.5)	1¼-in (3.2-cm) pipe or 1¼ × 1¼ × ¼ angle	10 gage	—
60 (152.4)	90 (228.6)	12	29(73.7)	1½-in (3.8-cm) pipe or 1½ × 1½ × ¼ angle	8 gage	—

Fig. 8.18 Concentric trash rack and antivortex device for sediment basin riser pipe. (9)

Fig. 8.18 Design Table (*Continued*)

Riser diameter, in (cm)	Cylinder				Minimum top	
	Diam., in	Thickness gage	H, in (cm)	Minimum size support bar	Thickness	Stiffener
66 (167.6)	96 (243.8)	10	33(83.8)	2-in (5.1-cm) pipe or 2 × 2 × 3/16 angle	with 8 gage stiffener	2 × 2 × [illegible] angle
72 (182.9)	102(259.1)	10	36(91.4)	2-in (5.1-cm) pipe or 2 × 2 × 3/16 angle	with 8 gage stiffener	2½ × 2½ × angle
78 (198.1)	114(289.6)	10	39(99.1)	2½-in (6.4-cm) pipe or 2 × 2 × ¼ angle	with 8 gage stiffener	2½ × 2½ × angle
84(213.4)	120(304.8)	10	42(106.7)	2½-in (6.4-cm) pipe or 2½ × 2½ × ¼ angle	with 8 gage stiffener	2½ × 2½ × angle

NOTE: The criterion for sizing the cylinder is that the area between the inside of the cylinder and the outside of the riser is equal to or greater than the area inside the riser. Therefore, the above table is invalid for use with concrete pipe risers.

Fig. 8.19 Properly installed sediment basin riser pipe.

Fig. 8.20 Ineffective basin: riser perforated.

If the riser is 10 ft (3 m) or less in height, one of the two methods shown in Fig. 8.21 can be used to anchor it. One alternative is a concrete base 18 in (46 cm) thick with a width twice the diameter of the riser and with the pipe embedded 6 in (15 cm) into the concrete. A second alternative is a steel plate, at least ¼ in (6.4 mm) thick and having a width equal to twice the diameter of the riser, welded to the base of the riser. The plate must be covered with 2.5 ft (0.75 m) of stone, gravel, or compacted soil to prevent flotation.

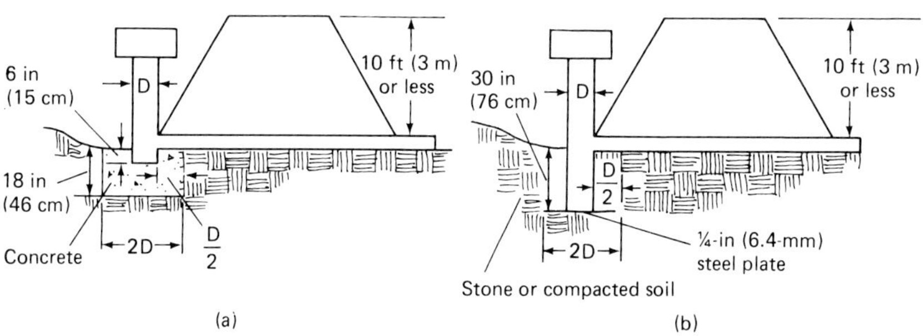

Fig. 8.21 Techniques for anchoring riser pipes for embankments less than 10 ft (3 m) high. (*a*) Concrete base; (*b*) steel base. (Adapted from 11)

Antiseep Collars

Antiseep collars are another necessary feature of larger basins. In the absence of such collars, water may seep along the pipe through the embankment and cause embankment failure. In general, antiseep collars are needed if either of the following two conditions exists:

1. The settled height of the embankment exceeds 10 ft (3 m).
2. The embankment has a low silt-clay content (unified soil class SM or GM), and the pipe through it is greater than 10 in (25 cm) in diameter.

The antiseep collars should be installed within the saturated zone. The maximum spacing between collars should be 14 times the projection of a collar above the barrel of the pipe (Fig. 8.22). A collar should not be closer than 2 ft (0.6 m) from a pipe joint. Collars should be placed sufficiently far apart to allow space for passage of hauling and compacting equipment during construction.

Watertight Connection

The connection between the pipe riser and the barrel of the pipe through the embankment must be watertight. If it is not, the basin will not fill and will not be effective. Figure 8.23 depicts a concrete riser that was simply placed in front of a metal culvert. Sandbags and straw bales were used to try to create a seal between the mismatched pipes—neither worked.

Embankment Failure

Improper installation of pipes and fill can cause embankment failure. Embankment fill must be compacted according to the specifications of the soils engineer. The bank should not be disturbed after construction. Figure 8.24 shows an embankment which was trenched to install the pipe and backfilled with loose earth. The uncompacted backfill simply washed away. The pipe should have been put in place before the embankment was built.

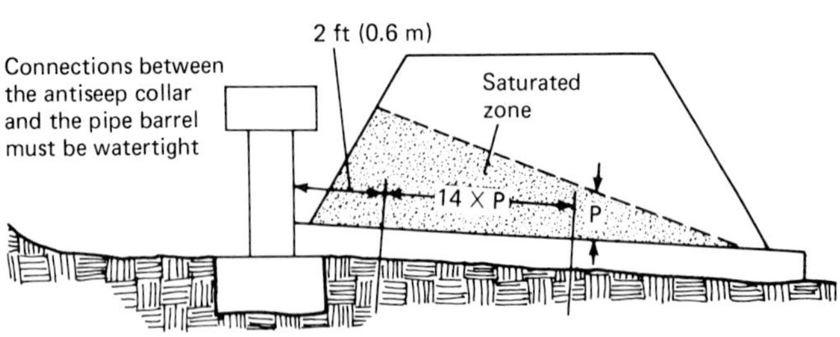

Fig. 8.22 Installation of antiseep collars. (11)

Temporary vegetative stabilization of the banks of sediment basins is advisable. The banks may be seeded at the same time other exposed soil on the site is seeded.

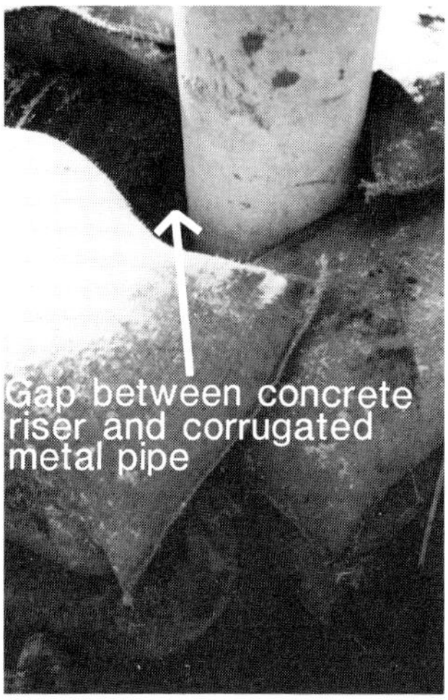

Fig. 8.23 Ineffective basin: riser not properly attached.

8.3b Emergency Spillways

Emergency spillways from a sediment basin are needed for larger basins when the blockage of the principal spillway (the riser) by trash or construction debris could cause the water level to rise over the basin embankment. Without an emergency spillway, the embankment would probably erode rapidly, which would lead to a sudden discharge of basin contents.

Emergency spillways should not be constructed on fill material. They should be wide enough to pass the peak rate of runoff from a 10-year-frequency storm or one commensurate with the degree of hazard involved. The freeboard (the difference between the high-water elevation in the spillway and the top of the embankment) should be at least 1 ft (0.3 m). Finally, emergency spillways require erosion protection; a suitable material is grass, riprap, asphalt, or concrete. Selection of the material should be based on the water velocities expected in the spillway. For more information on channel protection, see Sec. 7.7.

8.3c Cleaning

A sediment basin is sized to provide 2 ft (0.6 m) of settling depth and a certain storage depth. The storage depth to provide in a basin should depend on:

- Estimated sediment yield
- Depth that can easily be attained on a site (a function of subsoil conditions)
- Expected frequency of cleaning

Cleaning is required whenever sediment fills the storage zone. In the basin shown in Fig. 8.25 the entire settling and storage depth has been allowed to fill with sediment. Providing sufficient depth for storage of an entire season's sediment yield is the best solution. When site conditions or local regulations restrict storage depth, the basin must be cleaned out regularly if it is to function prop-

Fig. 8.24 Ineffective basin: embankment trenched to install outlet pipe after compaction.

erly. Lack of maintenance is a common reason for the failure of a basin to trap sediment. Schedules that call for frequent sediment removal are rarely adhered to.

Access can hamper a well-intentioned maintenance program. If a basin is located at the bottom of a fill slope, it may be impossible to clean it with heavy equipment (Fig. 8.4). If cleaning is required after every storm, machinery may not be able to reach a basin because of wet road conditions. Locating a basin near a paved road or placing gravel on an access road will make cleaning easier. Sediment removed from a basin should be placed where it will not reenter the basin or be washed into the storm drains (Fig. 8.26). Put it out of the path of flowing water. A dike can be built around a pile of dredge spoils to temporarily prevent runoff from carrying away the sediment. At the end of the season, the sediment must be permanently disposed of. Frequently, the silt is incorporated into fill material. The soils engineer can determine if the sediment can be used for other purposes.

Fig. 8.25 Ineffective basin: settling zone filled with sediment.

Fig. 8.26 Improper sediment disposal.

8.3d Outlet Protection

The outflow from a sediment basin may discharge into a storm drain system or into a natural drainageway. In the latter situation, outlet protection is required to ensure that erosion of the embankment and the natural channel does not occur. Figure 8.27 depicts a pipe protruding in midair; water falling out the end of the pipe eroded the embankment and completely filled the channel below with sediment.

The pipe outlet should be at the bottom of the embankment. The bottom of the pipe should be flush with the ground. Outlet protection, such as a riprap apron, should be provided (see Chap. 7).

8.4 DESIGN AND INSTALLATION OF SEDIMENT TRAPS

8.4a Design Factors

Surface Area

A sediment trap is a small sediment basin that drains an area of less than 5 acres (2 ha). It is sized by using a rule of thumb based on applying the surface area formula, $A = 1.2Q/V_s$, to a set of typical local conditions. To simplify the design process, a design storm and design particle size are preselected for a given geographical area. The rational method is applied to a hypothetical 1-acre (0.4-ha) site to find the Q to be used in the surface area formula. The design capacity is

Fig. 8.27 Improper installation: pipe extends beyond embankment.

then expressed in square feet (square meters) of surface area required per acre (hectare) of drainage area.

In the San Francisco Bay Area, for example, the authors designed the standard sediment trap on the basis of a moderately high rainfall of 30 in (762 mm) per year and a 0.02-mm design particle size. The 10-year, 6-hr storm at a site in the Bay Area with 30 in (762 mm) annual rainfall is 2.5 in (64 mm), or 0.42 in/hr (11 mm/hr). A runoff coefficient C of 0.5 was chosen to represent a smooth, graded area with no vegetation (Table 4.1). Applying the rational method, we have

$$Q = C \times i \times A = 0.5(0.42 \text{ in/hr})(1 \text{ acre}) = 0.21 \text{ ft}^3\text{/sec}$$

Using the surface area formula and the 0.02-mm particle's settling velocity gives us

$$A = \frac{1.2Q}{V_s} = \frac{1.2(0.21 \text{ ft}^3\text{/sec})}{0.00096 \text{ ft/sec}} = 263 \text{ ft}^2\text{/acre } (60 \text{ m}^2\text{/ha})$$

This formula means that there should be 263 ft^2 of sediment trap surface area (when the trap is full of water) for each acre of drainage area to the trap. For areas with significantly different rainfalls or soil textures, trap sizes can be adjusted by reapplying the formula.

Determining a standard trap size per acre of drainage area makes design simpler. Because the drainage area of traps is small, precise sizing is normally not necessary. If, however, the downstream impacts would be substantial were the structure to fail or a different design storm or design particle size is desired, the trap should be sized by applying the sediment basin sizing procedures.

Depth

If a sediment trap is to be effective, sufficient settling depth must be provided and must be supplemented with a certain amount of storage depth. In the trap designed for the San Francisco Bay Area, a minimum depth of 2 ft (0.6 m) was chosen; this provides 1 ft (0.3 m) of settling and 1 ft (0.3 m) of storage. That is equivalent to 19.4 yd^3/acre (36.7 m^3/ha) of drainage area, of which 9.7 yd^3 (8.4 m^3) is intended for sediment storage. For many sites this minimum depth may not provide storage capacity for an entire season's sediment yield.

To plan for a season's storage capacity, calculate the sediment yield and find the depth required on the basis of the surface area of the trap. If the soils in an area are relatively uniform, a standard depth per acre could be calculated by making assumptions about the factors in the USLE in much the same way as the standard surface area was determined by using the rational method and surface area formulas.

Cleaning

If depth for one season's sediment yield *cannot* be provided, either because of the site conditions or because a maximum depth limit is imposed by a local jurisdiction, periodic cleaning will have to be done. Since cleaning is difficult to guarantee, it is worthwhile to look for other ways to reduce the required depth (e.g., reduce sediment yield).

Additional midslope diversions to shorten slope length or the use of more sediment barriers may help. Installing several traps instead of one will provide more storage volume while minimizing the need to excavate. Example 8.4 illustrates the calculation of sediment storage volume, and some possible trade-offs are discussed.

Length-to-Width Ratio

The minimum length of flow through the trap should be 10 ft (3 m) where that is feasible. For traps draining less than 1 acre (0.4 ha), a minimum L/W ratio of 2:1 is suggested. Traps handling runoff from 1 to 5 acres (0.4 to 2 ha) should have an L/W ratio greater than 2:1.

Siting

A sediment trap should be built as close as possible to the source of sediment. It should be sited to impound runoff from the disturbed area only. In most cases, *the trap should not be built in a watercourse.* A sediment basin or trap located in a stream channel will needlessly impound clean runoff from undisturbed areas and necessitate a larger and more costly structure.

By using the natural depressions and the existing topography for storage areas and treating only the on-site runoff, it is often possible to construct several small traps and avoid construction of larger, more expensive basins. A trap can be built across a small drainageway as long as the drainage area does not exceed 5 acres (2 ha). Make sure, however, that the trap discharge structure can handle the peak flows from the area.

Never build basins or traps in series. A sediment basin or trap should *never* discharge into another basin or trap. A basin or trap is sized to remove suspended sediment from a certain flow. Placing several small basins in series overloads each one with the total flow from the entire drainage area above it. Also, the load may cause failure of the embankments.

EXAMPLE 8.4 Calculation of Sediment Storage in a Sediment Trap

Given: The 4-acre (1.6-ha) site in Example 5.6. A sediment trap will be constructed to capture sediment eroded from the entire site.

Find: The annual soil loss from the site, the volume of sediment that the trap should capture in 1 year, and the required frequency of cleaning.

Solution: The trap will be designed to capture particles 0.02 mm and larger by using the formula 263 ft^2/acre (60 m^2/ha) of drainage (see Sec. 8.4a). The trap will be 2 ft (0.6 m) deep.

STEP 1. *Soil Loss.* In the example in Sec. 5.2i, we calculated the soil loss as follows:

$$\text{Soil loss} = R \times K \times \text{LS} \times C \times P = 34(0.34)(8.16)(1.0)(0.9)$$

$$= 84.9 \text{ tons/(acre)(year)} \qquad 190.5 \text{ t/(ha)(yr)}$$

We assume that 1 ton of sediment deposited in a trap will occupy approximately 1 yd^3

(assume 1 t = 0.84 m^3), so the volume of eroded soil is estimated to be 85 yd^3/acre (160 m^3/ha). Multiplying by the area of the site gives us

$$(85\ yd^3/acre)(4\ acres) = 340\ yd^3\ [(160\ m^3/ha)(1.6\ ha) = 256\ m^3]\text{ soil loss per year}$$

STEP 2. *Sediment Capture.* The volume of soil captured in the trap is estimated by multiplying the soil loss by the trap efficiency. Trap efficiency is defined as the percent by weight of soil particles larger than or equal to the design particle size. Because 79.1 percent of this soil is larger than or equal to 0.02 mm, trap efficiency for this soil type is, ideally, about 79 percent.

$$(340\ yd^3)(0.79) = 269\ yd^3\ [(256\ m^3)(0.79) = 202\ m^3]$$

STEP 3. *Cleaning Frequency.* The available storage in a sediment trap designed with 263 ft^2/acre (60 m^2/ha) of drainage with a 1-ft (0.3-m) settling depth and a 1-ft (0.3-m) storage is:

$$(263\ ft^2)(1\ ft)(4\ acres) = 1050\ ft^3\text{ storage}$$

$$[(60\ m^2/ha)(0.3\ m)(1.6\ ha) = 29\ m^3\text{ storage}]$$

Convert cubic yards of sediment captured to cubic feet and compare with the storage volume:

$$\frac{(269\ yd^3)(27\ ft^3/yd^3)}{1050\ ft^3} = 6.9 \qquad \left(\frac{202\ m^3}{29\ m^3} = 6.6\right)$$

Thus, sediment would have to be cleaned out of the trap at least 6 times in a normal year.

Note: There are several ways to reduce soil loss. *Straw mulch* is very effective at reducing erosion. If 1.5 tons/acre (3.4 t/ha) of straw mulch were applied and tacked into the soil, C would decrease from 1.0 to 0.2 (see Table 5.6). Thus the soil loss would be reduced by 80 percent:

New $C = 0.2$

Soil loss = $R \times K \times LS \times C \times P = 85(0.2) = 17$ tons/acre = approximately 17 yd^3/acre (32 m^3/ha)

Multiplying by site acreage and trap efficiency and comparing with the storage volume gives us

$$\frac{(17\ yd^3)(4\ acres)(0.79)(27\ ft^3/yd^3)}{1050\ ft^3} = 1.4\text{ times per season}$$

$$\frac{[(32\ m^3)(1.6\ ha)(0.79)}{29\ m^3} = 1.4$$

With 1.5 tons/acre (3.4 t/ha) of straw mulch punched into the soil, soil loss is extremely small. This amount provides complete surface coverage, so no raindrop impact occurs and infiltration of water is maximized.

8.4b Construction Considerations

Sediment traps are constructed by:

- Excavating a hole in the ground
- Creating an impoundment with a low-head dam

Sediment traps should be located outside the area being graded and should be built prior to the start of grading activities or removal of existing vegetation. Constructing the traps first will provide protection from the first erosion. To minimize the area disturbed by them, the sediment traps should be located in natural depressions or in small swales or drainageways. Traps should be dimensioned to fit the site conditions and be so located as to facilitate periodic cleaning and not interfere with construction operations.

Embankments

The embankments can be up to 5 ft (1.5 m) high and should be constructed and compacted in 8-in (20-cm) lifts. Minimum top widths for various embankment heights are listed in Fig. 8.28. Side slopes should not be steeper than 2:1. The embankment should be seeded with temporary vegetation.

Outlets

The outlet can be a spillway in the embankment, a gravel section of the embankment, or a pipe (Fig. 8.29). The width, in feet, of earth or stone outlets should be roughly equal to 2 to 3 times the number of acres draining to the trap. The outlet crest should be at least 1 ft (0.3 m) below the top of the embankment. The outlet should be free of any restriction to flow.

The portion of the embankment below a stone outlet must be relatively impervious (e.g., timber, concrete block, or straw bales) to cause ponding. This impervious core should be covered by 6 in (15 cm) of stone. The crushed stone or gravel used in the outlet should meet American Association of State Highway Transportation Officials (AASHTO) M43, size No. 2 or 24, or its equivalent (such as MSHA No. 2).

A pipe outlet is commonly used in a sediment trap. Either plastic or corrugated pipe is installed as the embankment is built. The fill material around the pipe should be compacted in 4-in (10-cm) lifts. A minimum of 1.5 ft (0.5 m) of

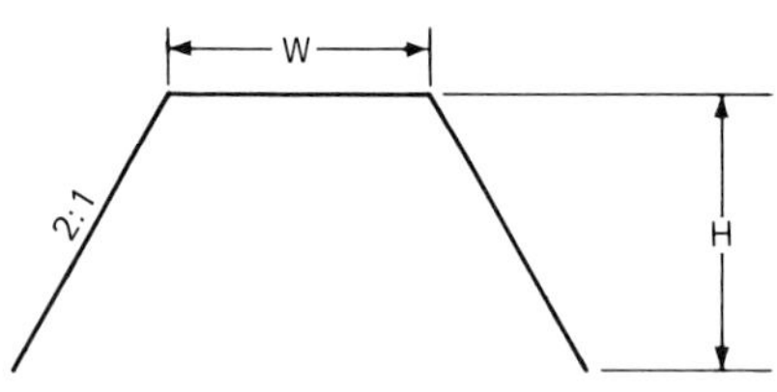

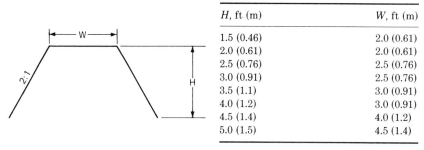

H, ft (m)	*W*, ft (m)
1.5 (0.46)	2.0 (0.61)
2.0 (0.61)	2.0 (0.61)
2.5 (0.76)	2.5 (0.76)
3.0 (0.91)	2.5 (0.76)
3.5 (1.1)	3.0 (0.91)
4.0 (1.2)	3.0 (0.91)
4.5 (1.4)	4.0 (1.2)
5.0 (1.5)	4.5 (1.4)

Spillway elevation should be 1 ft (0.3 m) below the top of the embankment. Pipe riser elevation should be at least 1.5 ft (0.46 m) below embankment top.

Fig. 8.28 Minimum top widths for sediment trap embankments of various weights. (11)

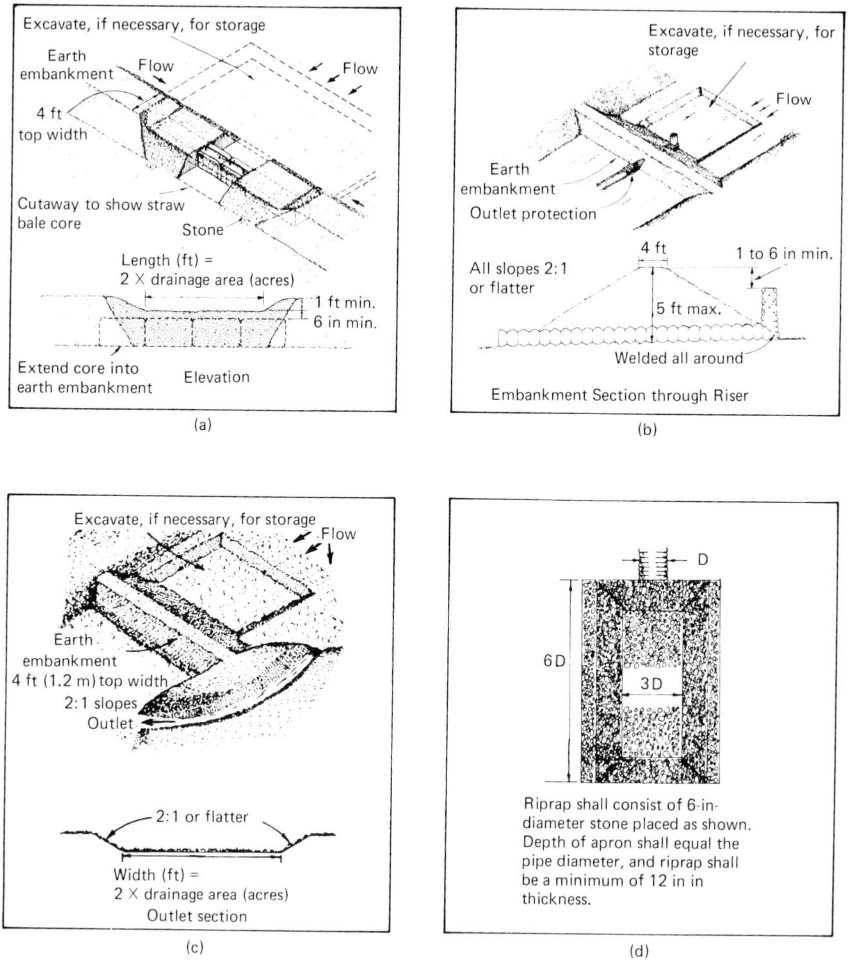

Fig. 8.29 Example of sediment trap outlets for drainage areas of less than 5 acres: (*a*) stone outlet; (*b*) pipe outlet; (*c*) earth outlet; (*d*) riprap apron for pipe outlets. (Adapted from 9)

fill should cover the pipe—2 ft (0.6 m) if equipment will be crossing over the embankment.

The pipe outlet should be constructed with a riser so that the trap fills to a depth of at least 1 ft (0.3 m) for storage and 1 ft (0.3 m) for settling. Figure 8.30 shows a sediment trap without a riser. This trap will not capture much sediment. The riser pipe can be of the same type as the pipe through the embankment. The diameter of the riser may be equal to or greater than the diameter of the pipe through the embankment, but the connection between the two pipes must be watertight. The top of the embankment should be at least 1.5 ft (0.5 m) above

Fig. 8.30 Ineffective sediment trap: no settling or storage depth.

the crest of the riser. Perforations in the riser should be kept to a *minimum.* A gravel base may be used to reduce flotation of the riser. (See the discussion of sediment basin design for further details.) Pipe diameter can be selected from the following table (9), but it should be checked by an engineer to ensure that the pipe has the capacity to carry peak flows:

Min. pipe diameter, in (cm)	Max. drainage area, acres (ha)
12 (30)	1 (0.4)
18 (46)	2 (0.8)
21 (53)	3 (1.2)
24 (61)	4 (1.6)
30 (76)	5 (2.0)

Outlet Protection

Whenever a flow of water is channeled or concentrated, protection from erosion at the outlet is usually needed. A pipe outlet should have a riprap apron below it. Figure 8.29 includes a sample drawing of a riprap apron for a pipe outlet. The apron should be 3 times as wide and 6 times as long as and equal in depth to the diameter of the pipe. The stones should be 6 in (15 cm) in diameter and be placed at least 12 in (20 cm) deep. The soil beneath the apron must be excavated so that the top of the stones will be roughly level with the bottom of the pipe. This apron is sized for flows from a drainage area of 5 acres (2 ha) or less. Outlet protection is discussed in more detail in Sec. 7.8b. If a sediment trap discharges into a paved street or a lined channel, additional outlet protection is probably unnecessary.

Excavated Sediment Traps

An excavated trap is simpler to build than an impoundment: A hole of the proper size is dug, and an outlet is provided. Excavated sediment traps can have the same kinds of outlets as are illustrated in Fig. 8.29. More frequently, though, excavated traps are constructed around storm drain inlets. Use of permanent storm drain inlets lowers the cost of the erosion control measures by avoiding the need to construct separate temporary structures. Outlet and channel bed erosion, a frequent and difficult problem on many sites, does not occur when storm drains serve as sediment trap outlets. Figure 8.31 is a sample drawing of an excavated trap around a storm drain inlet. The trap should be 2 ft (0.6 m) deep and have a surface area calculated by using the surface area formula. The shape can be suited to the location, but long, narrow shapes work best. One or two weep holes in the inlet will allow dewatering. Cleaning is required when the depth is reduced to 1 ft (0.3 m).

An excavated trap can be built in a small swale. This type of trap is similar to a check dam. The primary difference is the greater volume of sediment storage

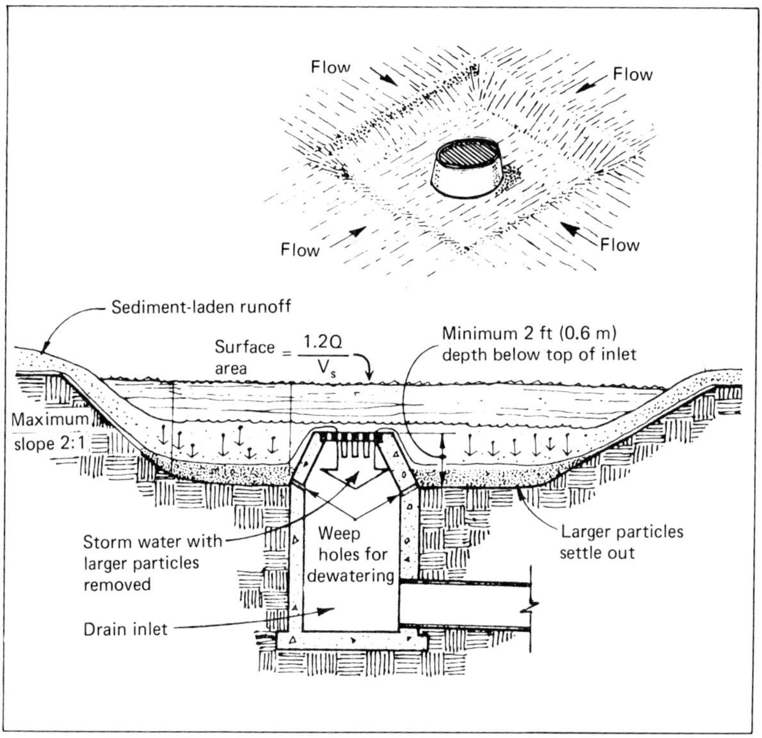

Fig. 8.31 Sample drawing: excavated sediment trap with storm drain inlet as outlet. (Adapted from 11)

obtained by excavation. The check dam itself serves as an outlet structure. (See Chap. 7 for further discussion of check dams.)

EXAMPLE 8.5 Excavated Sediment Trap

Given: A 4-acre (1.6-ha) project site is located near a major stream. The site is drained by a man-made swale that discharges into the stream. A local authority, as a condition of permit approval, requires all sediment traps to remove 80 percent of particles 0.062 mm and larger during the 10-year, 6-hr storm.

The average site slope is 5 percent. The water temperature is estimated to be 50°F (10°C) during winter storms. The mean annual rainfall is 20 in (51 cm), and the 10-year, 6-hr storm has a rainfall of 1.90 in (4.8 cm).

Find: The dimensions of a sediment trap which will meet the local authority's requirements. Prepare a sketch or sketches (plan, elevation, cross section, perspective, etc.) of the recommended trap. Show all pertinent dimensions and specifications for trap construction and outlet protection.

Solution: Because this problem specifies a particular particle size to capture, the surface area formula will be used to size the trap.

STEP 1. Apply the rational formula.

$C = 0.5$

Average storm intensity

$$i = \frac{1.90 \text{ in}}{6 \text{ hr}} = 0.32 \text{ in/hr} \qquad \left(\frac{48 \text{ mm}}{6 \text{ hr}} = 8 \text{ mm/hr}\right)$$

$$Q_{\text{avg}} = C \times i_{\text{avg}} \times A = 0.5(0.32 \text{ in/hr})(4 \text{ acres})$$

$$= 0.64 \text{ ft}^3\text{/sec} \quad \left[\frac{10.5(8 \text{ mm/hr})(1.6 \text{ ha})}{360} = 0.018 \text{ m}^3\text{/sec}\right]$$

STEP 2. Determine particle settling velocity V_s. From Fig. 8.12, the V_s for a 0.062-mm particle is 0.009 ft/sec (0.0027 m/sec).

STEP 3. Find the required surface area A of the sediment trap.

$$A = \frac{1.2Q_{\text{avg}}}{V_s} = \frac{1.2(0.64 \text{ ft}^3\text{/sec})}{0.009 \text{ ft/sec}} = 85 \text{ ft}^2 \qquad \left[\frac{1.2(0.018 \text{ m}^2\text{/sec})}{0.0027 \text{ m/sec}} = 8.0 \text{ m}^2\right]$$

STEP 4. Since we need to capture only 80 percent of the particles 0.062 mm and larger:

$$A = (85 \text{ ft}^2)(0.80) = 68 \text{ ft}^2 \qquad [(8.0 \text{ m}^2)(0.80) = 6.4 \text{ m}^2]$$

The standard depth for a sediment trap is 2 ft (0.6 m). An L/W of 2:1 is advisable. Figure 8.32 illustrates a possible design.

Note: The surface area is reduced only because less than 100 percent capture of the specified class of particles is required. Normally, the surface area formula is used to compute an area that will attain, ideally, 100 percent capture of the design particle size. The trap efficiency is then a function of the particle size distribution of the soils on the site and does *not* enter into the sizing equation.

If we had used the simplified formula in Sec. 8.4a, 263 ft²/acre (60 m²/ha) of drainage, the surface area would have been 263(4) = 1052 ft² (96 m²). This trap would capture particles much finer than 0.062 mm, but it would be substantially larger than required in this example.

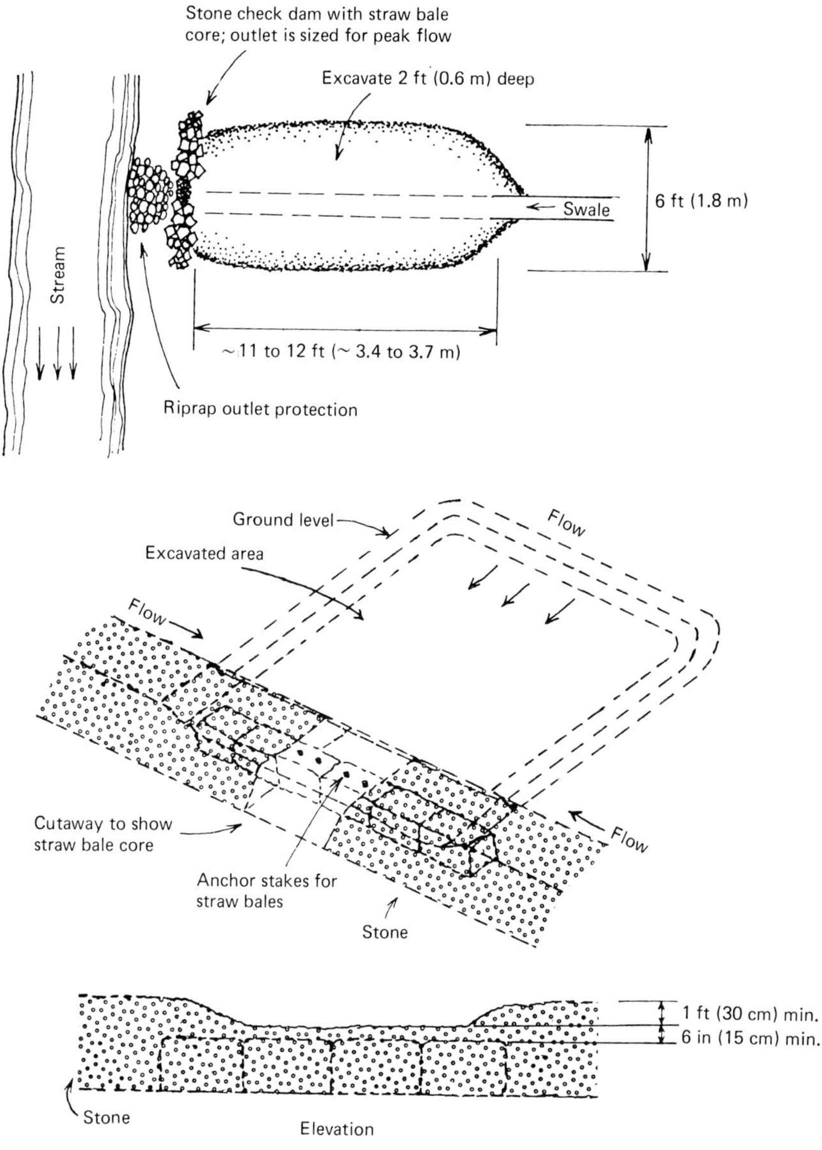

Fig. 8.32 Excavated sediment trap in Example 8.5.

8.5 DESIGN AND INSTALLATION OF SEDIMENT BARRIERS

Sediment basins and traps are designed to impound relatively large volumes of runoff or concentrated runoff from disturbed areas. Sediment barriers, in con-

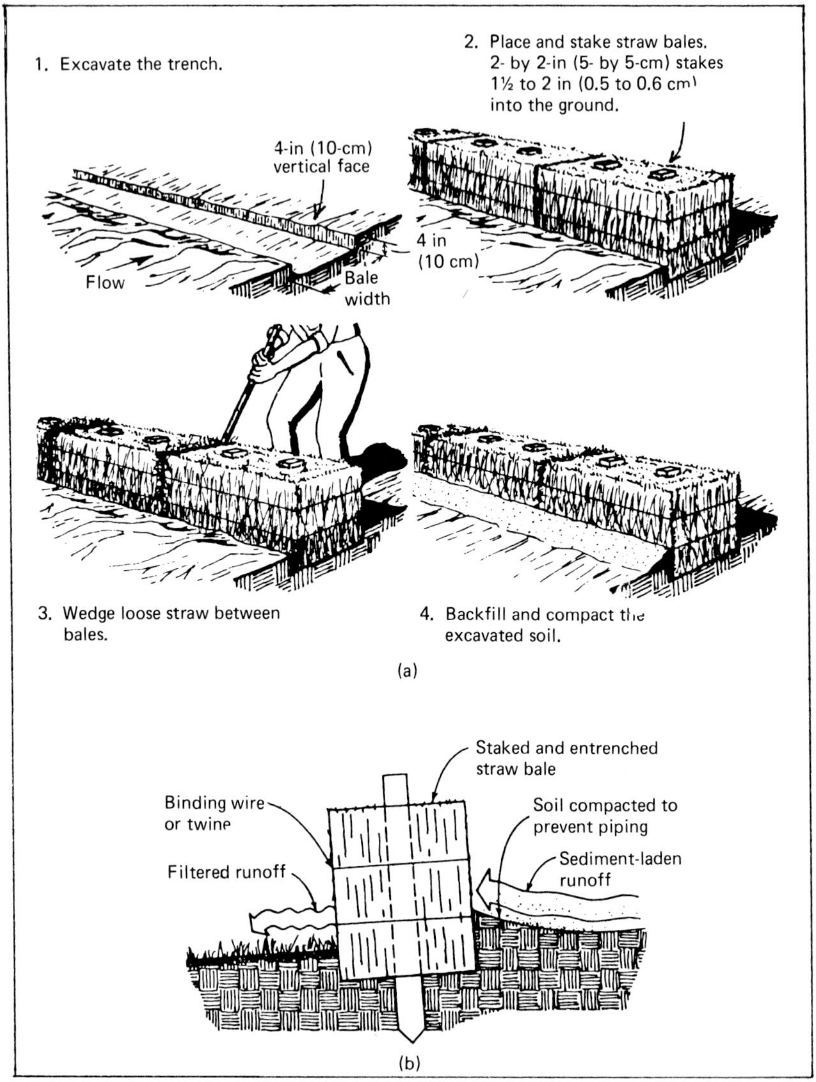

Fig. 8.33 Construction of a straw bale barrier. (*a*) Installation sequence. (*b*) Cross section of a properly installed straw bale. (Adapted from 11)

trast, are designed primarily to intercept and filter small volumes of "sheet flowing" runoff. Sediment barriers may be used when:

- The area draining to the barrier is 1 acre or less.
- The maximum slope gradient behind the barrier is 2:1.
- The maximum slope length behind the barrier is 100 ft (30 m).

In addition, when placed across a small swale, the barrier should *not* receive more than 1 ft^3/sec (0.028 m^3/sec) flow.

Sediment barriers have a useful life of 3 to 6 months depending on materials. Straw bales last only 3 months; silt fences can function for 6 months or longer if sediment accumulations are removed. Silt fences also trap a higher percentage of sediment than straw bales trap. (11)

Sediment barriers are perceived as inexpensive, easy solutions to erosion control. However, all too often they have been placed in poor locations and have been improperly installed and poorly maintained. Improper installation of straw bale barriers, in particular, has been a major problem. When a barrier fails, there is frequently more damage than if no barrier had been installed. The limitations listed above should be strictly observed. The need for proper installation cannot be overemphasized.

8.5a Straw Bale Dikes

Installation Procedure for Sheet Flow Applications

1. Excavate a 4-in- (10-cm-) deep trench the width of a bale and the length of the proposed barrier (Fig. 8.33). The barrier should follow the slope contour. If the barrier is at the toe of a slope, place it 5 to 6 ft (1.5 to 1.8 m) away from the slope if possible (Fig. 8.34). This placement will provide access for maintenance and allow coarse sediment to drop out of suspension before it reaches the barrier.
2. Place bales in the trench with their ends tightly abutting (Fig. 8.35). Corner abutment is not acceptable. A tight fit is important to prevent sediment from escaping through the spaces between bales.
3. All bales must be either wire-bound or string-tied. Install bales so that bindings are oriented around the sides rather than along the tops and bottoms of the bales (Fig. 8.33). If the binding is placed in contact with the soil, it will soon disintegrate and cause the bale to fall apart.

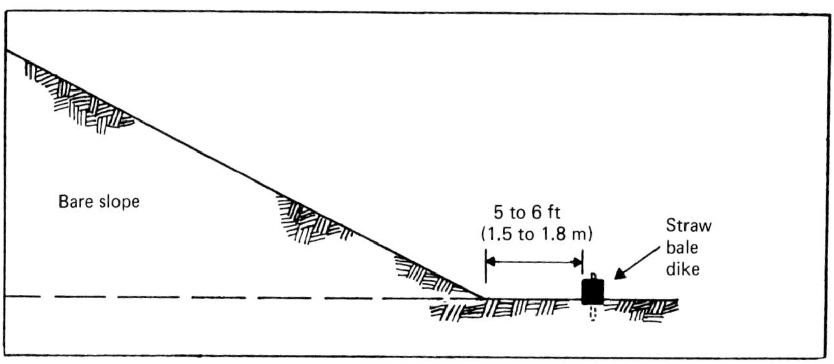

Fig. 8.34 Setback of barrier from toe of slope.

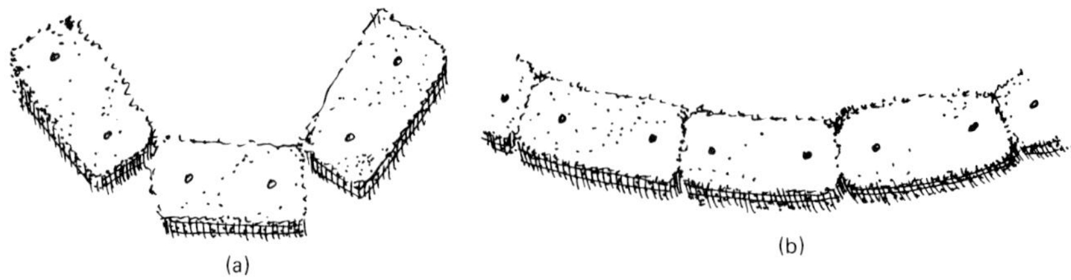

Fig. 8.35 (*a*) Incorrect and (*b*) correct abutment of straw bales in a trench.

Straw bales should be used, *not* hay bales. Hay bales cost more than straw because they contain the edible portion of the grain. Also, they rot faster, and so they require more frequent replacement.

4. Securely anchor each bale by driving at least two stakes through the bale. Drive the first stake in each bale toward the previously laid bale to force the bales together. Drive the stakes at least 1½ feet (0.5 m) into the ground. Wood stakes, 2 by 2 in by 4 ft (5 by 5 cm by 1.3 m), are best. Rebars also can be used as stakes, but they are not recommended because they can pose a hazard to equipment when bales disintegrate.
5. Fill any gaps between bales by wedging loose straw between the bales. Loose straw scattered over the area immediately uphill from a straw bale barrier tends to increase barrier efficiency. It is picked up by runoff and transported to holes in the barrier, which it tends to seal.
6. Backfill the trench with the excavated soil and compact it. The backfill soil should conform to the ground level on the downhill side of the barrier and should be built up to 4 in (10 cm) above the ground on the uphill side of the bales (bottom of Fig. 8.33).
7. Inspect and repair or replace damaged bales promptly. Straw bales typically deteriorate within 3 months when wet. Remove the straw bales when the upslope areas have been permanently stabilized.

Installation Procedure for Channel Flow Applications

Install straw bales as described above, with the following exceptions:

- Place bales in a single row, lengthwise, oriented *perpendicular to* the flow, and with ends of adjacent bales tightly abutting one another.
- Extend the barrier to such a length that the bottoms of the end bales are at a higher elevation than the top of the lowest middle bale to assure that sediment-laden runoff will flow either through or over the barrier but not around it (Fig. 8.36). Rock placed below the middle bale will dissipate the energy of the falling water and reduce downstream erosion.

"End runs" are a common cause of barrier failure. Often the problem is

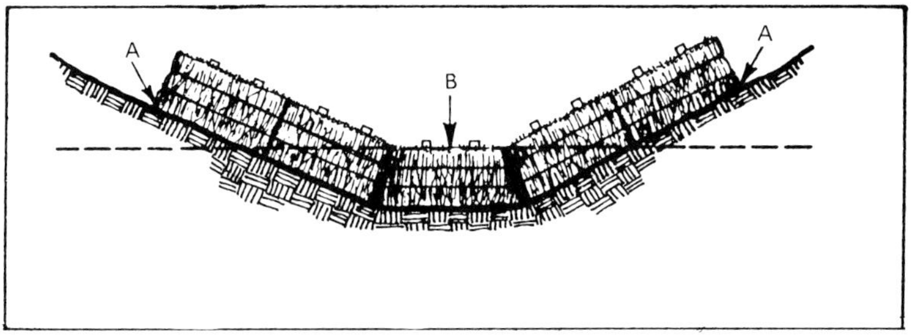

Fig. 8.36 Proper placement of a sediment barrier in a swale. Points *A* should be higher than point *B*. (11)

caused by not recognizing the center of the flow path during construction, with the result that the barrier is built off-center (Fig. 8.37). Staking is always important, but it is particularly crucial in channel flow applications because, until they are thoroughly wet, the bales will float (Fig. 8.38).

Other Applications

Straw bales can be used to construct check dams (see Chap. 7) and inlet protection (see Sec. 8.5d). They can also be used to form a sediment trap. Double- or triple-bale-height barriers can be used to construct silt traps that can do a good

Fig. 8.37 Straw bale dike not centered across flow path.

Fig. 8.38 Untrenched, unstaked bales dislodged by flow.

job if properly built (Fig. 8.39). However, 3 to 4 acres (1.2 to 1.6 ha) is the absolute maximum drainage area for such a barrier. The straw bales in a sediment trap *must* be supported by a 12- to 14-gauge wire fence. The bottom row of bales must be entrenched and backfilled. An outlet section should be provided to relieve water pressure.

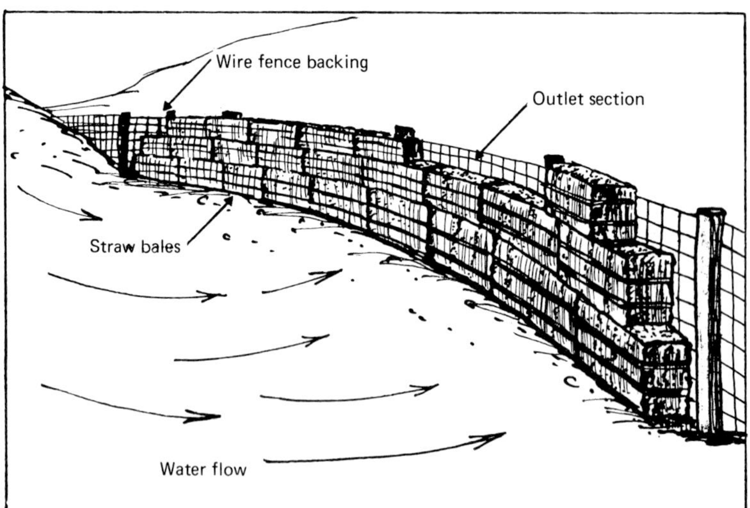

Fig. 8.39 Straw bale sediment trap.

Causes of Failure

Straw bale dikes have been widely misused. Rows of bales along a hilltop serve no useful function. An unstaked row of bales will not remain in place when water flows against it. Most straw bale dike failures are related to faulty installation:

- Bales not staked firmly into the ground
- Bales not trenched
- Bales not abutted tightly end-to-end
- Sufficient space for sediment entrapment and access for cleaning not provided
- Bales displaced by equipment not restored to their original position at the end of the day
- Barrier not centered in the flow path

Straw bale barriers should be inspected after each rain for displacement, undercutting, and end runs. Repairs should be performed immediately. See Chap. 10 for maintenance hints.

8.5b Silt Fences

A silt fence is a temporary structure of wood or steel fence posts, wire mesh fencing, and a suitable permeable filter fabric (Fig. 8.40). It has two functions: retention of the soil on the site and reduction of the runoff velocity across areas below

Fig. 8.40 Silt fence sediment barrier.

it. Although the fabric retains some soil particles by filtration at its surface, the portion of eroded soil that contacts the fabric is only a small portion of the total volume of retained solids. The reduction in runoff velocity at the fence causes suspended soil particles to settle.

Design Guidelines

A silt fence has the same design limitations as a straw bale dike:

- Drainage area 1 acre or less
- Maximum slope steepness 2:1
- Maximum flow path length to the fence 100 ft (30 m)
- No concentrated flows greater than 1 ft^3/sec

Figure 8.41 illustrates what can happen when a silt fence is placed on a slope that is too long and too steep.

A silt fence can last up to 6 months or longer, about twice as long as a straw bale dike. A properly installed silt fence is more effective than a straw bale dike and also more costly. The greater effectiveness of the silt fence is due to stronger construction, greater depth of ponding, and better installation practices. In addition, filter fabric allows fewer soil particles to pass through it.

Table 8.2 lists various commercially produced filter fabrics and some of their engineering characteristics; these fabrics are called *geotextiles* in the trade. The products are listed in alphabetical order by manufacturer, and no ranking or rat-

Fig. 8.41 Silt fence collapsing at base of slope that was too long and too steep.

ing is implied. The addresses and phone numbers of the manufacturers are listed in Table 8.3. Because of the need to match the product to the job, and because product availability changes from year to year, it is best to contact the manufacturer when deciding which product to use for a particular application. For example, a fabric suitable for a silt fence is often unsuitable for a riprap lining, and vice versa. Manufacturers will also advise on local suppliers of their products.

Selection of a filter fabric is based on soil conditions at the construction site [which affect the equivalent opening size (EOS) selection] and characteristics of the support fence (which affect the choice of tensile strength). The designer should specify a filter fabric that retains the soil found on the construction site yet will have openings large enough to permit drainage and prevent clogging. The U.S. Army Corps of Engineers, in its Civil Works Construction Guide Specification for Plastic Filter Fabric, Specification CW-02215 (4), recommends the following criteria for selection of the equivalent opening size:

1. If 50 percent or less of the soil, by weight, is fine particles smaller than the U.S. standard sieve No. 200, the EOS should be equal to or smaller than the sieve size that 85 percent of the soil can pass through.
2. For all other soil types, the EOS should be no larger than the openings in the U.S. Standard Sieve No. 70 [0.0083 in (0.21 mm)].

To reduce the chance of clogging, it is preferable to specify a fabric with openings as large as allowed by the criteria. No fabric should be specified with an EOS smaller than the openings of a U.S. Standard Sieve No. 100 [0.0059 in (0.15 mm)]. If 85 percent or more of a soil, by weight, is fine particles smaller than the openings in a No. 200 sieve [0.0029-in (0.074-mm)], filter fabric should not be used. Most of the particles in such a soil would not be retained if the EOS were too large, and they would clog the fabric quickly if the EOS were small enough to capture the soil.

Selection of fabric tensile strength and bursting strength characteristics depends on the support fence. Fabric attached to chain-link fence need not possess the same strength as one attached to a fence of 6- by 6-in (15- by 15-cm) reinforcing wire. Selection is thus based on standard engineering principles. Recommended fabric tensile strengths for various filter fence designs are listed in Table 8.4.

Other fabric characteristics also are important, such as retained strength after exposure to many hours of ultraviolet light. Many of the available fabrics meet a standard of better than 90 percent retained strength after exposure to 500 hr of light from a carbon arc. When comparing characteristics of fabrics made by different manufacturers, check to see if the fabrics were tested by using the same test standards.

Installation Procedure

As with straw bales, proper installation is important. Trenching, firmly setting posts, and securely stapling wire and fabric are key construction details. Figure 8.42 illustrates the basic steps outlined below.

TABLE 8.2 Filter Fabric Characteristics*

Manufacturer	Fabric name	Material	Equiv. opening size (U.S. std. sieve size)	Permeability coefficient, cm/sec	Tensile strength, lb (kg)	Burst strength, lb/in² (kg/cm²)
American Enka	Stabilenka T-80	Polyester	230–270	0.065	64 (29)	100 (7)
	Stabilenka T-100	Polyester	100	0.124	90 (41)	140 (10)
	Stabilenka T-140 N	Polyester	80–100	0.097	125 (57)	150 (11)
Amoco Fabrics	Propex 1199	Polypropylene	70–100	0.02	230 × 350 (106 × 159)	510 (36)
	Propex Silt Stop	Polypropylene	30–50	0.02	175 (80)	300 (21)
	Propex 4551	Polypropylene	70	0.02	120 (55)	300 (21)
Bradley Materials	Filterweave SF II	Polypropylene	40	0.01	150 (68)	300 (21)
	Filterweave 40	Polypropylene	40	0.01	300 × 225 (136 × 102)	500 (35)
	Filterweave 70	Polypropylene	70	0.02	380 × 280 (173 × 127)	540 (38)
	Polyfelt TS 500	Polypropylene	70–100	0.03	140 (64)	Unknown
	Polyfelt TS 600	Polypropylene	70–100	0.03	200 × 185 (91 × 84)	Unknown
	Polyfelt TS 700	Polypropylene	70–100	0.03	320 × 260 (145 × 118)	Unknown
	Polyfelt TS 750	Polypropylene	70–100	0.03	330 × 325 (150 × 148)	Unknown
	Polyfelt TS 800	Polypropylene	70–100	0.03	400 × 380 (182 × 173)	Unknown
Carthage Mills	Polyfilter X	Polypropylene	70	0.033–0.038	380 × 220 (173 × 100)	540 (38)
	Polyfilter GB	Polypropylene	40	0.2+	200 × 200 (91 × 91)	600 (42)
	Fabric 11	Polypropylene	40	0.005	120 (55)	200 (14)

Dupont	Typar 3201	Polypropylene	30	0.027	67 (30)	
	Typar 3341	Polypropylene	50	0.032	125 (57)	
	Typar 3401	Polypropylene	70–100	0.02	135 (61)	
	Typar 3471	Polypropylene	100	0.02	200 (91)	220 (16)
	Typar 3601	Polypropylene	140–170	0.014	203 (92)	263 (19)
Exxon	GTF 100S	Polypropylene	50–100	—	100 (45)	235 (17)
	GTF 400E	Polypropylene	70–100	0.01	390 × 250 (177 × 114)	525 (37)
Foss	Geomat 400	Polyester	100	7.6	185 (84)	337 (24)
	Geomat 600	Polyester	120	5.4	250 (114)	468 (33)
	Geomat 700	Polyester	120	4.9	320 (145)	564 (40)
Hoechst	Trevira Spunbond 1115	Polyester	70–100	0.3	130 × 110 (59 × 50)	220 (15)
	Trevira Spunbond 1120	Polyester	50–70	0.3	175 × 155 (80 × 70)	290 (20)
	Trevira Spunbond 1127	Polyester	70–100	0.3	260 × 225 (118 × 102)	380 (27)
Mirafi	Mirafi 100X	Polypropylene	40–70	0.04	120 (55)	200 (14)
	Mirafi 140S	Polypropylene	70–100	0.10	125 (57)	125 (9)
Nicolon	Nicolon 40/30A	Polypropylene	40	0.16	300 × 225 (136 × 102)	440 (31)
	Nicolon 70/06	Polypropylene	70	0.41	375 × 250 (170 × 114)	490 (35)
	Nicolon 100/08	Polypropylene	80–100	0.10	375 × 300 (170 × 136)	575 (41)
	Kontrol Fence	Polypropylene	70		150 (68)	180 (13)
Phillips	Supac 4½ (UV)	Polypropylene	70–100	0.2	140 (64)	260 (18)
	Supac 3WS (UV)	Polypropylene	40	0.01	125 (57)	230 (16)
	Supac 8NP	Polypropylene	70–100	0.22	260 (118)	450 (32)

*Based on manufacturers' data. Not intended to be a complete list.

TABLE 8.3 Filter Fabric Manufacturers*

American Enka Company
Enka, NC 28728
(704) 667-7713

Amoco Fabrics Company
550 Interstate North Parkway
Suite 150
Atlanta, GA 30099
(404) 955-0935

Bradley Materials Company
P.O. Box 368
Valparaiso, FL 32580
(904) 678-1105

Carthage Mills
1821 Summit Road
Cincinnati, OH 45237
(513) 242-2740

E. I. DuPont de Nemours and Co., Inc.
Explosives Products Division
1007 Market Street
Wilmington, DE 19898

Exxon Chemical Americas
380 Interstate North
Suite 375
Atlanta, GA 30339
(404) 955-2300

Foss Manufacturing Company
P.O. Box 277
Haverhill, MA 01830
(617) 374-0121

Hoechst Fibers Industries
Spunbond Business Group
P.O. Box 5887
Spartanburg, SC 29304
(800) 845-7597; from AK, HI, SC, and Canada,
(803) 579-5282

Mirafi, Inc.
P.O. Box 240967
Charlotte, NC 28224
(800) 438-1855; from NC, (704) 523-7477

Nicolon Corporation
3150 Holcomb Bridge Road
Suite 300
Norcross, GA 30071
(404) 447-6272

Phillips Fibers Corp.
Engineered Products Marketing
P.O. Box 66
Greenville, SC 29602
(803) 242-6600 or (800) 845-5737

*Not intended to be a complete list

TABLE 8.4 Recommended Tensile Strength for Filter Fabric (4)

Structure	Tensile strength,* lb	(kg)
3-ft (0.9-m) silt fence with reinforced backing of 6-in (15-cm) wire mesh; posts 10 ft (3 m) apart	120	(54)
3-ft (0.9-m) silt fence without reinforced backing; posts 6 ft (1.8 m) apart	200	(91)
18-in (0.5-m) silt fence without reinforced backing; posts 10 ft (3 m) apart	100	(45)
18-in (0.5-m) silt fence without reinforced backing; posts 3 ft (0.9 m) apart	30	(14)

*Tensile strength measured by test procedure ASTM D-1682G, as commonly reported in manufacturers' literature.

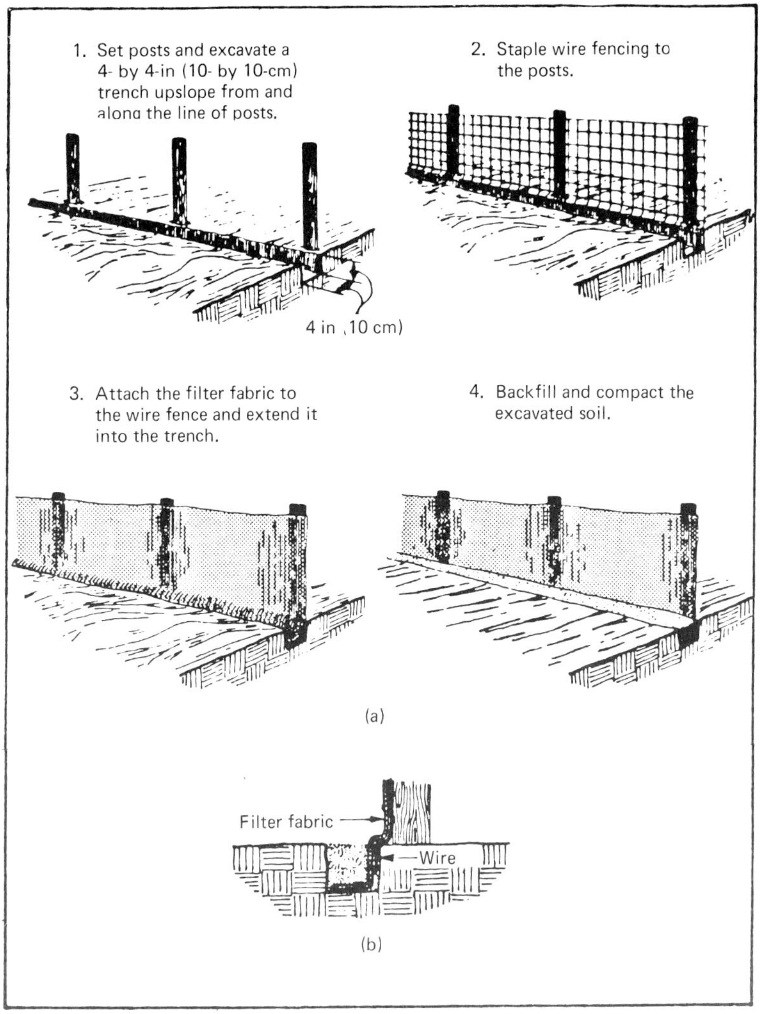

Fig. 8.42 Construction of a silt fence. (*a*) Installation sequence. (*b*) Extension of fabric and wire into the trench.

1. Lay out a suitable fence line and set posts along it. On slopes, align the fence *along the contour* as closely as possible. In small swales, curve the fence line upstream at the sides to direct the flow toward the middle of the fence. The sides should be higher than the center as illustrated in Fig. 8.36.
   Space posts a maximum of 10 ft (3 m) apart and drive them at least 12 in (30 cm) into the ground. [When extra-strength fabric is used without the wire support fence, post spacing must not exceed 6 ft (1.8 m).] Posts for silt fences can be either 4-in- (10-cm-) diameter wood or 1.33 lb/ft (1.97 kg/m) steel with

a minimum length of 5 ft (1.5 m). Steel posts must have projections for fastening wire to them.
Excavate a trench approximately 4 in (10 cm) wide and 4 in (10 cm) deep along the line of posts and upslope from the barrier.

2. Fasten wire mesh securely to the upslope side of the posts. Use heavy-duty wire staples at least 1 in (2.5 cm) long and tie wires or hog rings. Extend the wire 6 in (15 cm) into the trench. Wire fence reinforcement for silt fences must be a minimum of 42 in (107 cm) wide, be a minimum of 14 gauge, and have a maximum mesh spacing of 6 in (15 cm). The 42-in (107-cm) length is needed so that 6 in (15 cm) can be extended into the trench and leave a 36-in (92-cm) support fence above the ground. (*Note:* When extra-strength fabric is used and fence posts are more closely spaced, the wire mesh can be omitted.)
3. Fasten the filter fabric to the uphill side of the fence posts, and extend it 6 to 8 in (15 to 20 cm) into the trench. The height of the fence should not exceed 36 in (0.9 m). Do *not* staple fabric onto trees. Cut the filter fabric from a continuous roll to avoid the use of joints. When joints are necessary, splice the filter cloth at a support post, with a minimum 6-in (15-cm) overlap, and securely fasten both ends to the post.
4. Backfill the trench over the toe of the fabric and compact the soil.

8.5c Straw Bale–Filter Fabric Combinations

Straw bales and filter fabric can be used together to construct a sediment barrier. The combination, although more expensive than either material used separately, compensates for the shortcomings of each. Straw bale dikes are frequently ineffective because they are not firmly staked and are not butted tightly together. When wrapped and secured with fabric, the bales have additional support and the gaps between bales are covered with filter material.

Figure 8.10 shows a straw bale–filter sediment barrier across a swale. Figure 8.43 shows a pair of straw bale–filter fabric barriers placed above and below a storm drain inlet on a paved street. Fabric has been secured on the upstream side of the first row of bales. To avoid damaging the pavement by staking, gravel has been piled behind the bales to hold them in place. Note that the bales extend across the curb. Loose straw has been packed under the bale in the gutter to prevent silt from escaping there.

Installation Procedure

1. Excavate a trench a few inches wider than the bales. Place the bales against the downslope side of the trench and anchor as described in Sec. 8.5a.
2. Place filter fabric or burlap against the upstream face of the bales and extend it into the trench. Staple the fabric to the bales with 6- to 9-in (15- to 23-cm) U-shaped wires.
3. Backfill the trench and compact the soil against the fabric and bales.

Fig. 8.43 Straw bale–filter fabric sediment barrier anchored with gravel.

8.5d Storm Drain Inlet Protection

A storm drain often carries runoff before its drainage area is stabilized, and it can convey large amounts of sediment to a stream or lake. If erosion is extensive, the storm drain itself may clog and lose a major portion of its capacity. To avoid these problems, it is necessary to prevent sediment from entering the storm drain inlets.

The best way to prevent sediment from entering the storm drain system is to stabilize the site with vegetation as quickly as possible, trap sediment near its source with sediment barriers, and pave streets and install curbs and gutters on schedule. That is not always possible, so inlet protection should be provided to reduce the sediment load entering the storm drain system. Common materials used for that purpose include straw bales, filter fabric, gravel, and sand bags. Several types of inlet filters are described below. The choice of filter structure depends upon site conditions and type of inlet.

Sometimes it is convenient and cost-effective to construct the permanent storm drain system at the beginning of a project and use certain inlets as the risers for sediment basins or traps. The area around the inlet is excavated to form the storage area of the trap (Fig. 8.31).

The following inlet protection devices are for drainage areas of *less than 1 acre (0.4 ha)*. They are designed to keep sediment out of the storm drain, and they do *not* have a sediment storage area. Excavating an area around the inlet for deposition of sediment will improve the capture rate, reduce frequency of maintenance, and allow the device to serve an area larger than 1 acre (0.4 ha).

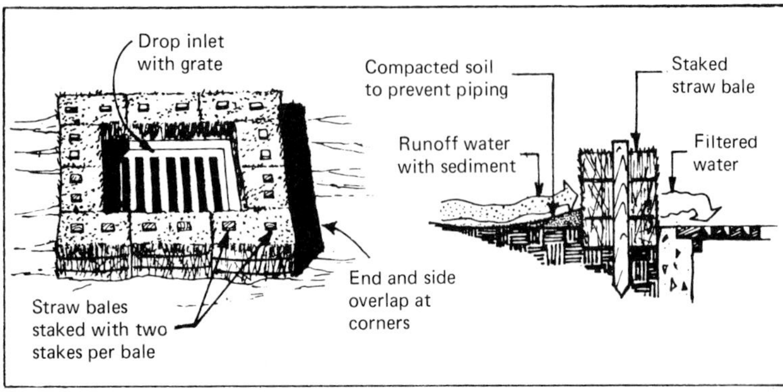

Fig. 8.44 Straw bale drop inlet sediment barrier. (11)

The following sections describe how to construct seven different inlet protection structures. Site planners are encouraged to develop other, innovative techniques for accomplishing the same purpose.

Straw Bale Drop Inlet Sediment Barrier

A straw bale drop inlet sediment barrier can be used where the inlet drains a relatively flat disturbed area (slopes no greater than 5 percent) in which sheet flow [not exceeding 0.5 ft^3/sec (0.014 m^3/sec)] occurs. Barriers of this type should not be placed around inlets receiving concentrated flows such as those along major streets or highways. (11)

Installation Procedure

1. Excavate a 4-in- (10-cm-) deep trench around the inlet. Make the trench as wide as a straw bale (Fig. 8.33).
2. Orient straw bales with the bindings around the sides of the bales rather than over and under the bales (Fig. 8.44).
3. Place bales lengthwise around the inlet and press the ends of adjacent bales together (Fig. 8.44).
4. Drive two 2- by 2-in (5- by 5-cm) stakes through each bale to anchor the bale securely in place.
5. Backfill the excavated soil and compact it against the bales.
6. Wedge loose straw between bales to prevent water from flowing between bales.

Note: Figure 8.45 appeared in a nationally distributed publication as a sample drawing for straw bale inlet protection. As shown in this drawing, the bales are neither overlapped nor abutted. Figure 8.46 is a photograph of a structure

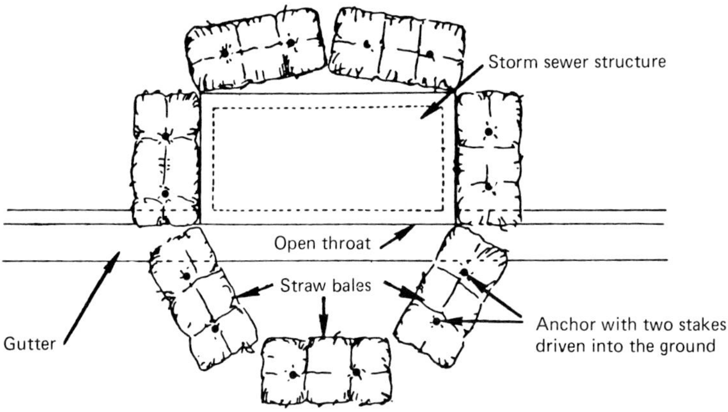

Fig. 8.45 Incorrect method for straw bale placement around inlet.

Fig. 8.46 Improper installation of straw bales; the bales are neither overlapped nor abutted.

based on the drawing. Sediment can freely enter the storm drain. As an alternative to overlapping the bales, as shown in Fig. 8.44, Fig. 8.47 depicts use of filter fabric around the bales.

Burlap or Filter Fabric Drop Inlet Sediment Barrier

Like the straw bale drop inlet sediment barrier, the burlap or filter fabric barrier may be used where the inlet drains a relatively flat disturbed area (slopes no

Fig. 8.47 Straw bale–filter fabric inlet protection.

greater than 5 percent) in which sheet flow [not exceeding 0.5 ft^3/sec (0.014 m^3/sec)] occurs. It should not be placed around an inlet receiving a concentrated flow such as that along a major street or highway.

The filter fabric should meet the specifications in Table 8.4. It must be at least 24 in (0.6 m) wide. If burlap is used, it should weigh at least 10 oz/yd^2 (340 g/m^2). Cut fabric or burlap from a continuous roll to avoid joints.

Installation Procedure

1. Place stakes around the perimeter of the inlet a maximum of 3 ft (0.9 m) apart and drive them at least 8 in (20 cm) into the ground (Fig. 8.48). It is best to use 2- by 2-in (5- by 5-cm) wooden stakes, but metal stakes of equivalent strength also may be used. The stakes must be at least 3 ft (0.9 m) long.
2. Excavate a trench approximately 4 in (10 cm) wide and 4 in (10 cm) deep around the outside perimeter of the stakes.
3. Staple the burlap or filter fabric to the wooden stakes so that 8 in (20 cm) of the fabric extends into the trench. Use heavy-duty wire staples at least ½ in (1.3 cm) long.
4. Backfill the trench and compact the soil over the fabric.

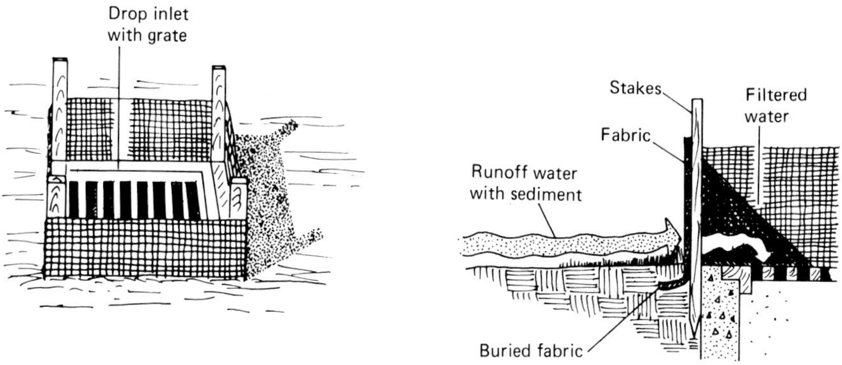

Fig. 8.48 Burlap or filter fabric drop inlet sediment barrier. (11)

Gravel and Wire Mesh Drop Inlet Sediment Barrier

A gravel and wire mesh drop inlet sediment barrier can be used where heavy, concentrated flows are expected, as at drop inlets in unpaved streets. Because this structure has no means for handling overflows, it is likely to cause ponding, especially if sediment is not removed regularly (Fig. 8.49). Therefore, it should not be used where an overflow would endanger an exposed fill slope or where ponding would interfere with traffic movement or construction work or would damage adjacent structures or property. In locations where ponding would be a problem, use a block-and-gravel drop inlet structure.

INSTALLATION PROCEDURE

1. Place wire mesh over the drop inlet so that the wire extends a minimum of 1 ft (0.3 m) beyond each side of the inlet structure. Use hardware cloth or comparable wire mesh with ½-in (1.30-cm) openings. If more than one strip of mesh is necessary, overlap the strips.

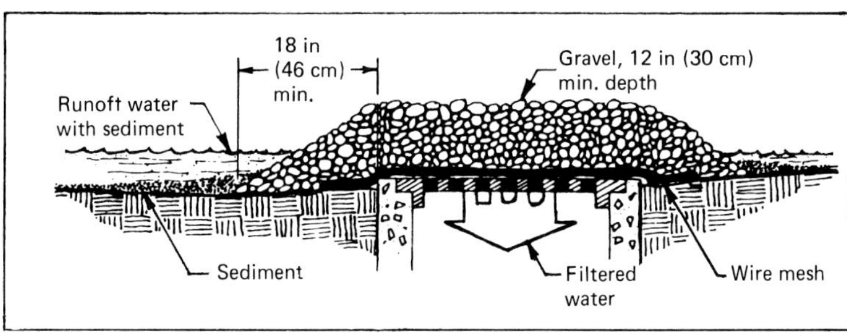

Fig. 8.49 Gravel and wire mesh drop inlet sediment barrier. (11)

2. Place 2- to 3-in (5- to 8-cm) gravel over the wire mesh (Fig. 8.49). The depth of stone should be at least 12 in (30 cm) over the entire inlet opening. Extend the stone beyond the inlet opening at least 18 in (46 cm) on all sides.
3. If the stone filter becomes clogged with sediment, the stones must be pulled away from the inlet and cleaned or replaced. Since cleaning of gravel at a construction site may be difficult, an alternative approach would be to use the clogged stone as fill and put fresh stone around the inlet.

Block and Gravel Drop Inlet Sediment Barrier

A block-and-gravel drop inlet sediment barrier can be used where heavy flows are expected. It has an overflow mechanism to prevent excessive ponding around the structure.

INSTALLATION PROCEDURE

1. Place concrete blocks lengthwise on their sides in a single row around the perimeter of the inlet, so that the open ends face outward, not upward. The ends of adjacent blocks should abut. The height of the barrier can be varied, depending on design needs, by stacking combinations of blocks that are 4 in (10 cm), 8 in (20 cm), and 12 in (30 cm) wide. The row of blocks should be at least 12 in (30 cm) but no greater than 24 in (61 cm) high (Fig. 8.9).
2. Place wire mesh over the outside vertical face (open end) of the concrete blocks to prevent stone from being washed through the blocks. Use hardware cloth or comparable wire mesh with ½-in (1.3-cm) openings.
3. Pile stone against the wire mesh to the top of the blocks. Use 2- to 3-in (5- to 8-cm) gravel.
4. If the stone filter becomes clogged with sediment, the stone must be pulled away from the blocks, cleaned, and replaced.

Gravel Curb Inlet Sediment Barrier

A gravel curb inlet sediment barrier can be used at a curb inlet if ponding in front of the inlet is not likely to cause inconvenience or damage to adjacent structures and unprotected areas (Fig. 8.50). If ponding presents a problem, a block-and-gravel curb inlet sediment barrier should be installed.

INSTALLATION PROCEDURE

1. Place hardware cloth or comparable wire mesh with ½-in (1.3-cm) openings over the curb inlet opening so that at least 12 in (30 cm) of wire extends beyond both the top and bottom of the inlet opening as illustrated in Fig. 8.50.
2. Pile stone against the wire to anchor the wire against the gutter and cover the inlet opening completely. Use 2- to 3-in (5- to 8-cm) gravel.

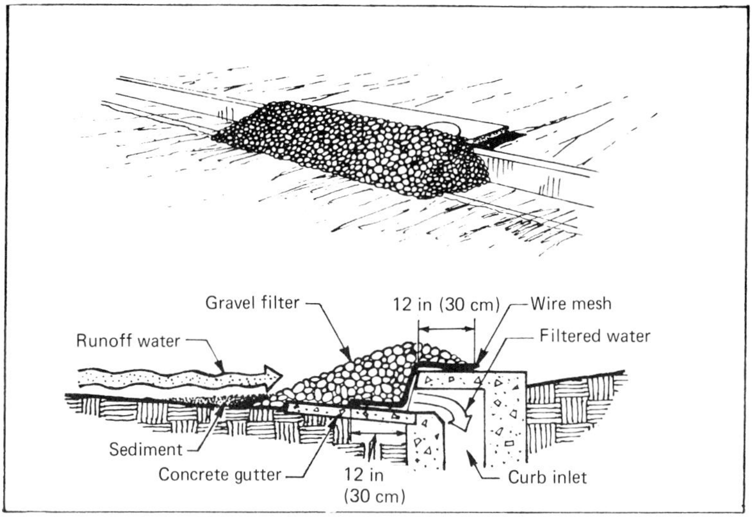

Fig. 8.50 Gravel curb inlet sediment barrier. (11)

3. If the stone filter becomes clogged with sediment, the stone must be pulled away from the block and cleaned or replaced.

Block and Gravel Curb Inlet Sediment Barrier

A block-and-gravel curb inlet sediment barrier can be used at a curb inlet if an overflow mechanism is needed to prevent excessive ponding in front of the inlet (Fig. 8.51).

Installation Procedure

1. Place two concrete blocks on their sides perpendicular to the curb at either end of the inlet opening (Fig. 8.51). These will serve as spacer blocks.
2. Place concrete blocks on their sides across the front of the inlet and abutting the spacer blocks, as illustrated in Fig. 8.51. The openings in the blocks should face outward, not upward.
3. Cut a 2- by 4-in (5- by 10-cm) stud the length of the curb inlet plus the width of the two spacer blocks. Place the stud through the outer hole of each spacer block to help keep the front blocks in place.
4. Place wire mesh over the outside vertical face (open ends) of the concrete blocks to prevent stone from being washed through the blocks. Use chicken wire or hardware cloth with ½-in (1.3-cm) openings.
5. Pile 2- to 3-in (5- to 8-cm) gravel against the wire to the top of the barrier.

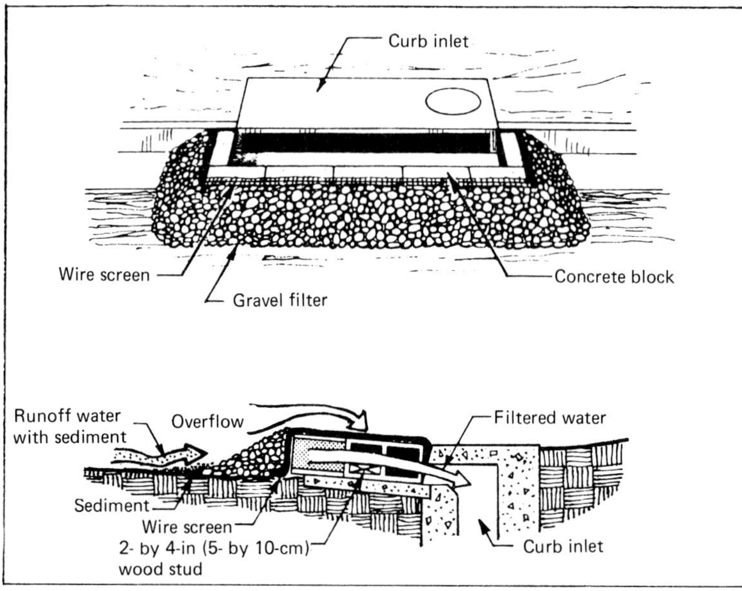

Fig. 8.51 Block and gravel curb inlet sediment barrier. (11)

6. If the stone filter becomes clogged with sediment, the stone must be pulled away from the blocks, cleaned, and replaced.

Sandbag Curb Inlet Sediment Barrier

A sandbag curb inlet sediment barrier can be used at a curb inlet on a paved street which receives relatively small runoff flows [less than 0.5 ft^3/sec (0.014 m^3/sec)]. Although simple to construct, it is not as effective as the inlet protection structures previously described. Once the small catchment area behind the sandbags fills with sediment, future sediment-laden runoff will enter the storm drain without being desilted. Therefore, sediment must be removed from such a structure during or after *every storm*. Additional storage capacity can be obtained by constructing a series of sandbag barriers along a gutter so that each barrier traps small amounts of sediment.

Installation Procedure

1. Place sandbags in a curved row out from the curb and away from the inlet. The row should be at least 6 ft (1.8 m) from the curb inlet and should overlap onto the curb as necessary to divert runoff through the barrier. It should extend into the street a distance sufficient to intercept runoff but at least 3 ft (0.9 m).

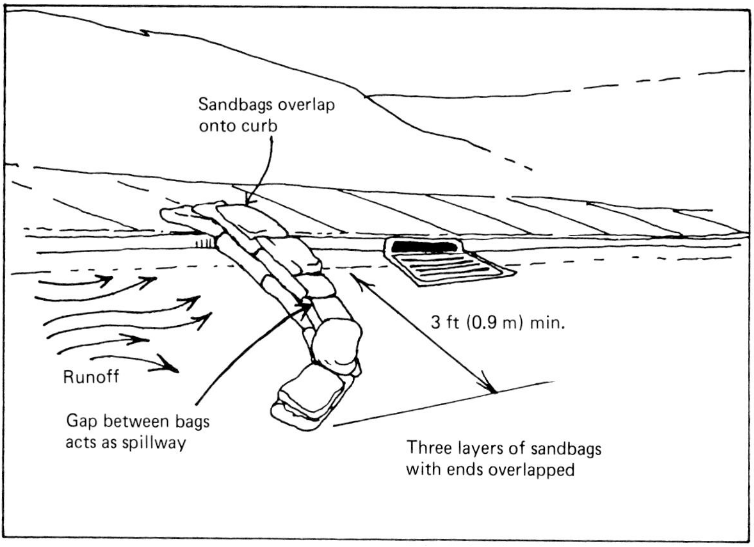

Fig. 8.52 Sand bag curb inlet sediment barrier.

2. Place several layers of sandbags over the first, overlapping bags and pack them tightly together to minimize the space between bags (Fig. 8.52).
3. Leave a gap of one sandbag in the middle of the top row of sandbags to serve as the spillway.
4. Remove sediment when it reaches the top row of sandbags and place it where it will not enter the storm drain.

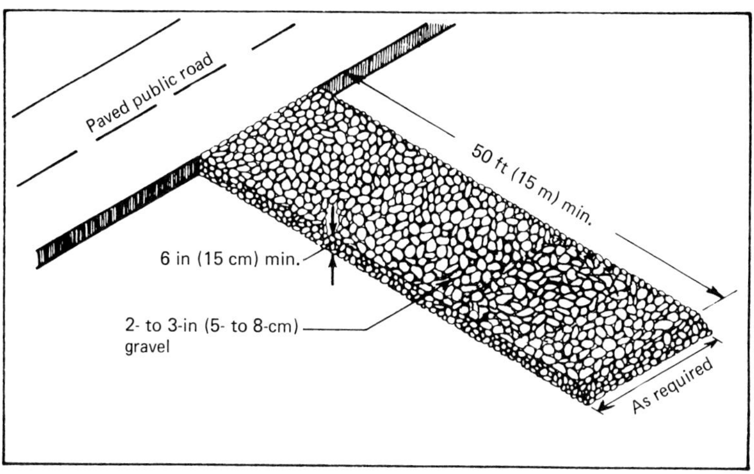

Fig. 8.53 Gravel construction entrance.

8.5e Gravel Construction Entrance

A gravel construction entrance is a pad of crushed stone that reduces the tracking of mud onto a paved street. To construct the pad, place a layer of 2- to 3-in (5- to 7.6-cm) stone across the full width of the vehicle ingress and egress area (Fig. 8.53). The stone pad should be at least 50 ft (15 cm) long and at least 6 in (15 cm) thick. Additional stone may have to be added periodically to maintain the proper functioning of the pad.

If the crushed stone does not adequately remove the mud from vehicle wheels, the wheels should be hosed off before the vehicle enters a public street. The washing should be done on an area covered with crushed stone, and the wash water should drain to a sediment trap or sediment barrier.

REVIEW QUESTIONS

1. What is the basic function of a sediment retention structure?
2. How do sediment basins and sediment traps differ from sediment barriers?
3. What is the difference between a sediment basin and a sediment trap?
4. When should sediment retention structures be built? Where on a site should they be located?
5. What is the surface area formula? How is it used to design sediment basins and traps?
6. Why does the design particle size have such a strong influence on basin area?
7. What is the minimum settling depth of a sediment basin? Why is it important to the functioning of the basin?
8. What is the purpose of baffles in a sediment basin?
9. Why is it important that basin risers not be perforated?
10. What alternatives to riser perforations are available to dewater a basin?
11. What is the theoretical basin efficiency?
12. Does design storm or design particle size have the stronger influence on basin efficiency?
13. Discuss the trade-offs between providing storage volume for one season and scheduling regular cleanings if adequate storage volume is not provided.
14. Determine a standard sediment trap size for your location.
15. What are the limitations to use of straw bale dikes and silt fences? Why are these limitations important?
16. Where on a site should sediment barriers be located?
17. What are the steps in constructing a straw bale dike? A silt fence?
18. What are some simple designs for the protection of individual storm drain inlets? Why are these devices limited to such small drainage areas?

19. ***Given:*** The 15-acre (6-ha) site in Review Question 17 of Chap. 5.
Find: (1) The surface area of a sediment basin designed to capture the 0.02-mm particle in a basin serving the entire site.
(2) The depth necessary to provide storage for one year's predicted soil loss. (Answer provided in Appendix C.)

20. Compare the effectiveness of sediment basins and traps with on-slope measures such as vegetation and sediment barriers.

REFERENCES

1. Association of Bay Area Governments, *Manual of Standards for Erosion and Sediment Control Measures,* Oakland, Calif., 1981.
2. G. M. Fair, J. C. Geyer, and D. A. Okun, *Water and Wastewater Engineering,* vol. 2, John Wiley & Sons, Inc., New York, 1966.
3. A. Hazen, "On Sedimentation," *Transactions ASCE,* vol. 53, 1904.
4. M. McMillan, "Selection of Filter Fabrics for Use in Silt Fences," *Water Quality Technical Memorandum No. 63,* Association of Bay Area Governments, Oakland, Calif., 1981.
5. Metcalf and Eddy, Inc., *Wastewater Engineering,* McGraw-Hill Book Company, New York, 1972.
6. T. R. Mills and M. L. Clar, *Erosion and Sediment Control, Surface Mining in the Eastern U.S.,* U.S. Environmental Protection Agency, Washington, D.C., 1976.
7. E. L. Pemberton and J. M. Lara, *A Procedure to Determine Sediment Deposition in a Settling Basin,* U.S. Department of the Interior, Bureau of Reclamation Sedimentation Investigations Technical Guidance Series, Section E, Part 2, Denver, Colo., 1971.
8. V. L. Streeter, *Fluid Mechanics,* McGraw-Hill Book Company, New York, 1958.
9. U.S. Department of Agriculture, Soil Conservation Service, *Standards and Specifications for Soil Erosion and Sediment Control in Developing Areas,* USDA, SCS, College Park, Md., 1975.
10. U.S. Department of the Interior, Bureau of Reclamation, *Design of Small Dams,* GPO, Washington, D.C., 1973.
11. Virginia Soil and Water Conservation Commission, *Virginia Erosion and Sediment Control Handbook,* Richmond, Va., 1980.

chapter 9

Preparing and Evaluating an Erosion and Sediment Control Plan

Erosion and sediment control plans are commonly prepared *after* a site development plan has been proposed. After street alignments, building pads, cuts, and fills have been drawn on a site map, a junior staff person is asked to draw in some straw bale dikes and sediment basins because the city or county requires erosion control. This is *not* the ideal way to control erosion, and it often results in totally ineffective plans. If grading is proposed in areas of high erosion potential, as on steep slopes and highly erodible soils, it may be impossible to prevent erosion. If soils high in clay and fine silt are allowed to erode, it will be extremely difficult to recapture these soil particles on the site. The control measures described in earlier chapters of this book will not do the job without good planning.

Careful planning is an essential first step in the control of erosion and sediment; it should be closely integrated with site development planning. A thorough site analysis should precede the development of either type of plan. Identify areas most and least suited for development. Try to avoid grading or building in areas of high erosion hazard. What are the alternatives to disturbing those areas? How can erosion be prevented in areas that must be disturbed? When is the best time to begin grading from an erosion control standpoint? Can construction be staged so that only a portion of the site is stripped and made vulnerable to erosion at any one time? How will site drainage be controlled? Where does runoff enter the site, how does it cross the site, and where will it leave the site? Are channel linings and outlet protection devices needed? How will runoff be controlled during the construction period before permanent storm drain systems are in place? Can vegetation and mulches be used to prevent erosion of graded areas? What other control measures are needed? Where should they be placed, and what are their design specifications? How will sediment basins be cleaned out when they fill up with sediment? What will happen if certain control measures malfunction or fail? These are some of the questions that must be answered if erosion and sediment are to be successfully controlled.

The only way the above questions can properly be answered is on paper in the form of a comprehensive, detailed plan. Erosion and sediment control is too complex and too difficult a process to play it by ear. You must work out the details carefully beforehand and then arrange for the necessary materials and services so the control measures can be installed on schedule.

Planning is necessary for small projects as well as large ones. For a small site the plan may be brief and simple—possibly only a sketch and notes on a single sheet of paper. For large sites a comprehensive and detailed plan is imperative.

9.1 PREPARING AN EROSION AND SEDIMENT CONTROL PLAN

Preparing an erosion and sediment control plan is a four-step process:

- **Step 1** Collect data
- **Step 2** Analyze data
- **Step 3** Develop site plan
- **Step 4** Develop erosion and sediment control plan

This process is described step by step in the following section. It is primarily designed for relatively large projects (i.e., more than one building) on several acres or more. For very small sites, such as a single-home site, a more streamlined process may be appropriate. For example, doing a soil particle size analysis for sizing a sediment basin for a very small site would be overkill. It would also be unnecessary to do runoff calculations for sizing drainageways, since runoff from the site would probably be very minimal and a standard drainageway design would do the job in most cases.

A real sample site is used to illustrate the planning process. The site is a 4-acre (1.6-ha) parcel located in a valley in northern California. The property is to be developed with 38 condominium units in 11 separate buildings. The sample project is based on an actual project that was recently constructed. Since this is a real-world example, it is not always possible to apply all the textbook procedures. For example, it is recommended that grading be minimized and natural vegetation be preserved whenever feasible. However, as is typical in real-world developments, a large number of units must be squeezed in on a rather small site. Thus it is not always possible to avoid extensive grading. Nevertheless, with good planning, it is still possible to protect important site features and to minimize erosion and sedimentation.

9.1a Step 1: Collect Data

The purpose of data collection is to gather the information on site conditions that will enable you to develop an effective erosion and sediment control plan. Most of this data describes the natural environment of the site. Drainage information is particularly important (see Step 2).

It is best to collect all data in map form, if possible, and to plot it on one or more site maps at the same scale. Mapping the data at the same scale greatly facilitates the planning process by enabling you to overlay different maps and read through them on a light table.

Topography

A good topographic map should form the basis of any kind of land planning, including site development planning and erosion and sediment control planning. From a topographic base map, you can determine drainage patterns, slope lengths and slope angles, and locations of sensitive features on or adjacent to the site such as water bodies, buildings, and streets. All of these are critical concerns in erosion control.

Prepare a topographic map of the site which shows the existing contours at a suitable interval for determining drainage patterns over small areas. The contour lines must be close enough together to show which way water will flow. On relatively flat sites, a 2-ft (0.6-m) or smaller interval will probably be needed. On a sloping site, a 10-ft (3-m) interval may be acceptable. Figure 9.1 is a topographic base map for our sample site. The 2-ft (0.6-m) coutour lines were derived from an aerial survey of the site.

Drainage

The drainage pattern of a site has two components: overland flow and channel flow. Both are important in erosion control. In the data collection stage, it is helpful to clearly mark all existing streams and major swales on the topographic base map. (Major watercourses are shown as blue lines on U.S. Geological Survey topographic maps, but lesser drainageways will also be important to show.) Delineating drainageways now will make it easier to determine watershed boundaries in the data analysis stage (Step 2). There are no significant drainageways crossing the site except Crow Creek, which is located at the base of the slope along the east side of the property (Fig. 9.1).

Rainfall

In erosion control, rainfall data is primarily used for sizing large drainageways and sediment basins. Rainfall frequency and intensity are the key types of data. Rainfall intensity determines the i value used in runoff calculations. Rainfall intensity is also a component of the R factor in the universal soil loss equation. This equation can be used to estimate the sediment storage requirements of sediment basins. Rainfall frequency data is used for determining “a design storm.”

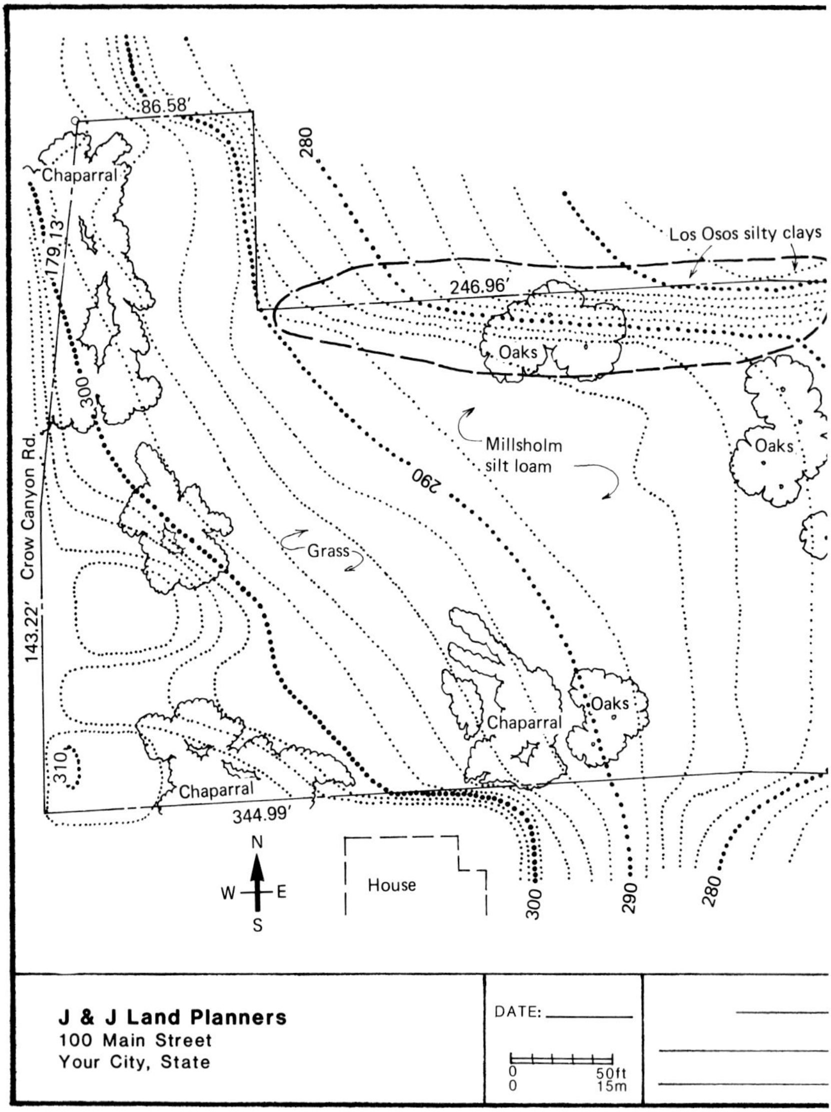

Fig. 9.1 Existing topography, vegetation, and soils.

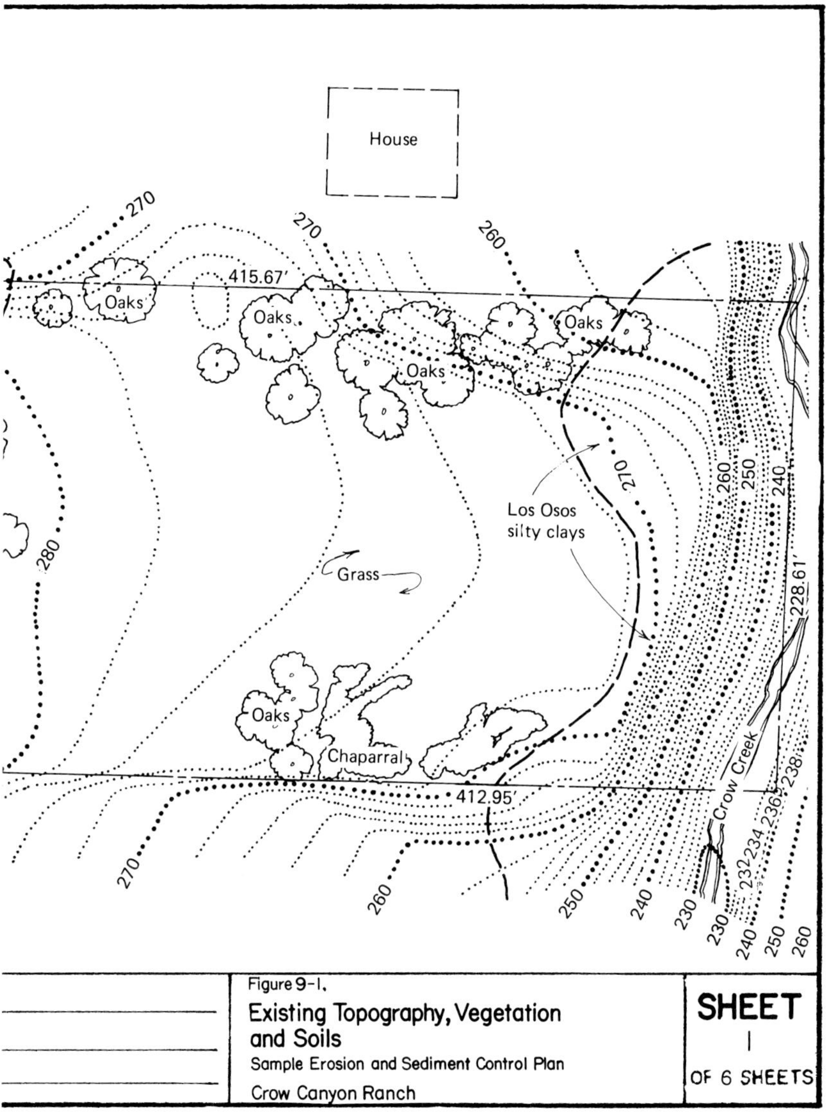

Figure 9-1.

Existing Topography, Vegetation and Soils

Sample Erosion and Sediment Control Plan

Crow Canyon Ranch

SHEET

1

OF 6 SHEETS

The use of rainfall data for designing drainage and erosion and sediment control measures is described in detail in Chap. 4.

On small sites and in small drainage areas it is often unnecessary to size control measures by using rainfall data. Most of the control measures described in this handbook have been designed to handle a major storm in a small drainage area [typically 1 to 5 acres (0.4 to 2 ha)]. Since a fairly large margin of error has been incorporated in the standard designs, these structures, if used properly in small watersheds within the specified size limits, should be able to withstand major storms anywhere in the United States.

Project planners should use their knowledge of the yearly pattern of rainfall to schedule construction during the times of year when erosion potential is lowest. They should pay special attention to control measure installation and maintenance when grading or construction must take place during periods of high erosion hazard (Chap. 2).

Rainfall data can be obtained from the state climatologist (Table 4.2), from the National Climatic Data Center in Asheville, North Carolina, and the National Technical Information Service in Springfield, Virginia (Sec. 4.1f), from local flood control and water supply agencies, and from other sources. If you are not aware of the best source of rainfall data in your area, start by contacting the city or county public works department or the local flood control district.*

Soils

Soils data is used to locate highly erodible areas on a site, where extra erosion control precautions may be needed. It also shows the distribution of particle sizes in the soil, a critical factor in sizing sediment basins and traps. A high content of clay and fine silt in a soil should suggest a strategy of erosion control by using vegetation and mulch rather than a strategy of sediment control by using straw bales and sediment basins.

Soils data for many parts of the country can be obtained from soil surveys published by the U.S. Soil Conservation Service. For many projects, a soils report is specially prepared by a soils engineer. On hillside sites, many jurisdictions routinely require a soils report. If a soils report is to be prepared, it is desirable to include in it a particle size analysis for sediment basin or trap design (see Sec. 8.2c). Such an analysis can be performed for a nominal extra cost.†

---

*To facilitate design of drainage control measures, the authors recommend that each local jurisdiction prepare a brief summary of local rainfall data to be used for drainage system design. Such a summary might look like Table 4.3, which lists precipitation depths for various storm return periods and durations. A copy of the rain data summary could then be handed to each project applicant, which would save developers the trouble of tracking down the data themselves.

†The authors recommend that each local jurisdiction specify a design particle size for sediment traps and basins. This design particle size, which would be based on particle size distribution data for soils in the local area, would provide for construction of cost-effective sediment traps and basins with appropriate trap efficiencies (see Sec. 8.2c). If the local jurisdiction specifies a design particle size, it should not be necessary for developers to obtain soil particle analyses for individual sites for the purpose of sediment control. To further simplify the developer's job, the local authority may wish to provide a standard sizing guideline for sediment *traps* (not basins), in square feet of trap surface area per acre of drainage to the trap (Sec. 8.4a). This guideline should be based on both the desired trap efficiency and the maximum runoff rate expected from a hypothetical 5-acre (2-ha) drainage area

Because our sample site is located in a hilly area, the county required a soils report. The soils report revealed that most of the site is covered with a silt loam, except for the slope on the eastern edge, which is covered with a silty clay. The data from the soils report was transferred to the base map (Fig. 9.1).

Ground Cover

"Ground cover" primarily refers to existing vegetation, which should be preserved to the greatest extent possible because it is the most effective form of erosion control. Many communities also wish to preserve trees and certain vegetation for aesthetic and other reasons. Ground cover, along with other physical characteristics of the watershed, is used to determine the *C* factor in the rational method for calculating runoff (Chap. 4). It is also used to calculate the erosion rate in the universal soil loss equation.

A field survey of the sample site revealed that, except for a few clusters of oak trees, most of the property is covered with grass and brush. The vegetation was mapped on the base map (Fig. 9.1).

Adjacent Areas

Off-site features, such as streams, lakes, buildings, and roads, are particularly sensitive to erosion and sediment damage. Such areas should therefore be noted on the site map. If including the off-site features on the same map would result in an unwieldy document, one of the following options can be chosen:

- Describe on the margin of the map the nature and location of the adjacent feature [e.g., Smith Reservoir 1600 ft (500 m) SSW].
- Draw a smaller-scale map (vicinity map) showing the site and all the pertinent adjacent features (Fig. 9.2).

Our sample base map shows Crow Canyon road on the west edge of the site, Crow Creek on the east edge, and two nearby residences—one to the north and one to the south of the property. Parker Reservoir is located about 1 mi (1.6 km) southwest of the site.

9.1b Step 2: Analyze Data

The purpose of this step is to interpret the data collected in Step 1 for its significance in erosion and sediment control. This interpretation may require stating the data in a different form (e.g., translating a topographic map into a slope map). The result of this step is a map or maps that highlights or highlight areas of importance in erosion and sediment control.

---

located in the local jurisdiction [5 acres (2 ha) is the maximum drainage area for a sediment trap]. Because sediment *basins* have drainage areas ranging from 5 to 150 acres (2 to 60 ha), each sediment basin should be custom-designed for the runoff rate calculated for the actual watershed that drains to it.

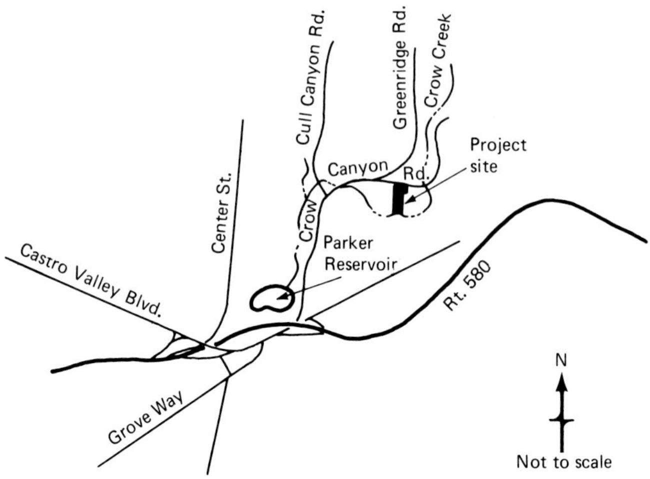

Fig. 9.2 Vicinity map of sample erosion and sediment control plan, Crow Canyon Ranch.

Drainage Areas

The most important part of Step 2 is to understand the site's drainage pattern. You must determine:

- Where concentrated and sheet flows will enter the site
- How runoff, both concentrated and sheet flow, will travel across the site
- Where runoff will leave the site and whether it will be concentrated or sheet flow
- How much water will flow

Map the drainage boundaries of each of the water courses delineated in Step 1, and then estimate the area of each major watershed. If the site is large [more than 5 acres (2 ha)], you may have to subdivide the watersheds into smaller units. Bear in mind that many control measures discussed in this handbook have a 5-acre (2-ha) maximum drainage area (Chaps. 7 and 8) and that straw bale dikes, silt fences, and most inlet protection structures have a 1-acre (0.4-ha) limit. Define watersheds that are appropriate for the control measures to be used. If grading will alter natural watershed boundaries, you will later need to map the drainage boundaries that will exist after grading is completed (see Step 4). If grading will not be completed before the rainy season, you may have to have several interim drainage plans.

The drainage survey of our sample site in Step 1 showed that there were no streams or distinct swales crossing the project area (Fig. 9.1). Drainage from west of the site is intercepted by Crow Canyon Road. A small amount of sheet runoff will enter the site from a ½-acre (0.2-ha) parcel to the southwest of the site. Rain falling on the site will drain to the east and north. The approximate drainage boundaries are shown in Fig. 9.3. Each of the three drainage areas is roughly 1.3 acres (0.5 ha).

Rainfall and Runoff

Examine the rainfall data collected in Step 1 to determine the times of year when erosion potential is at its lowest and highest. Try to schedule grading during times of low erosion potential and take extra precautions during times when heavy, intense rainfalls are likely.

If a project will require permanent waterways and sediment basins draining large areas, rainfall frequency and intensity data will be used to calculate runoff volumes to be expected. Since these calculations must be based on specific watershed areas that drain to each planned facility, the calculations must be done at a later stage of plan development (see Step 4).

We know from experience that our sample site is located in an area with a summer dry season. The data for San Francisco (Fig. 2.4), which is located about 40 mi (24 km) from the site, shows that very little rain falls from May through September. We should therefore plan to begin grading in May, at the beginning of the dry season.

Slope Steepness and Slope Length

Slope steepness and slope length are critical factors in erosion control. The longer and steeper the slope, the greater the erosion potential (see Sec. 5.2d). If an existing long or steep slope is disturbed or a new one is created by grading, carefully designed and installed erosion control measures will be required. These measures may include benches or ditches at regular intervals or a covering of punched straw.

Erosion potential is closely related to slope steepness. The following slope categories can be used as a rough guide for evaluating erosion potential:

0–7 percent slope	Low to moderate potential
7–15 percent slope	Moderate to high potential
Over 15 percent slope	High to very high potential

It is a good idea to outline on the topographic base map the above slope categories. Slopes that are over 15 percent and 7 to 15 percent slopes that are very long [over 100 ft (30 m)] should be highlighted as critical.

The slope of our sample site is gentle except on the eastern, northern, and southern edges, where there is an abrupt change. The slope categories outlined on the site map, Fig. 9.3, were determined by measuring the distances between contour lines. The resulting slope map shows that the central portion of the site has a 0 to 7 percent slope. The most critical slope on the site is the one on the eastern edge, above Crow Creek. Because this slope is greater than 15 percent, it has high erosion potential. The slopes on the north side of the site are also critical

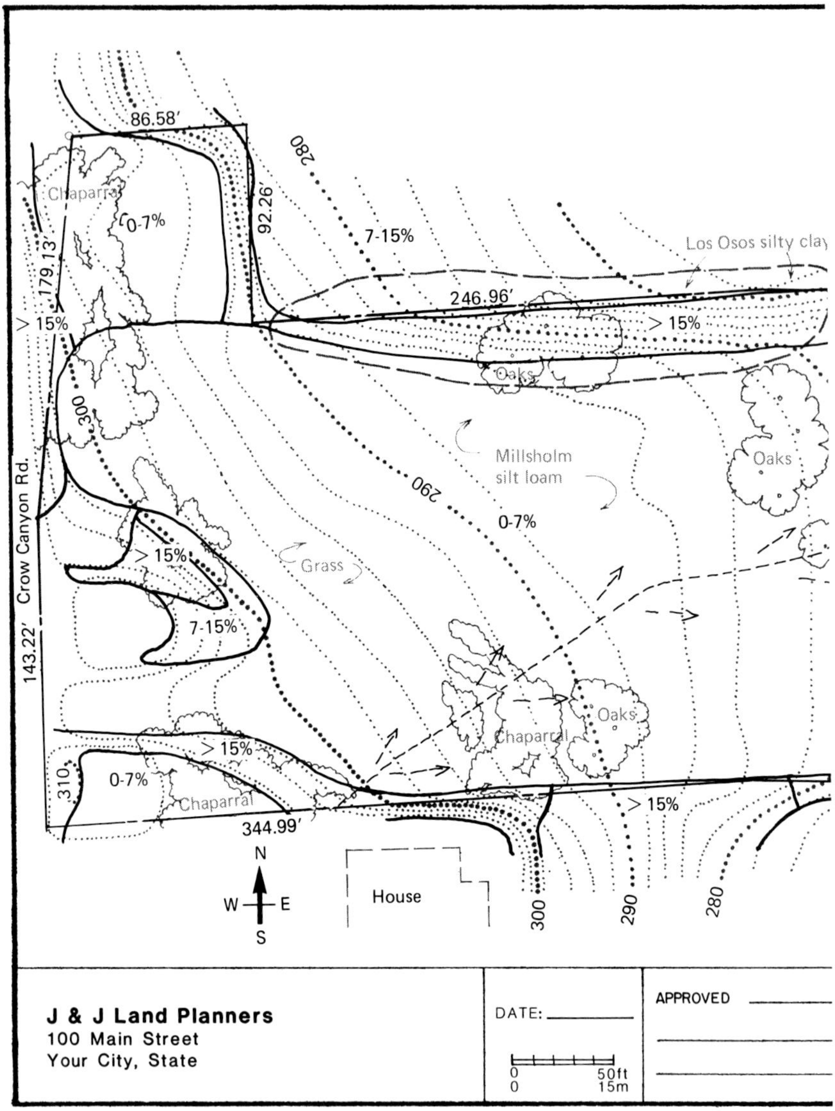

Fig. 9.3 Slope steepness, slope length, and drainage.

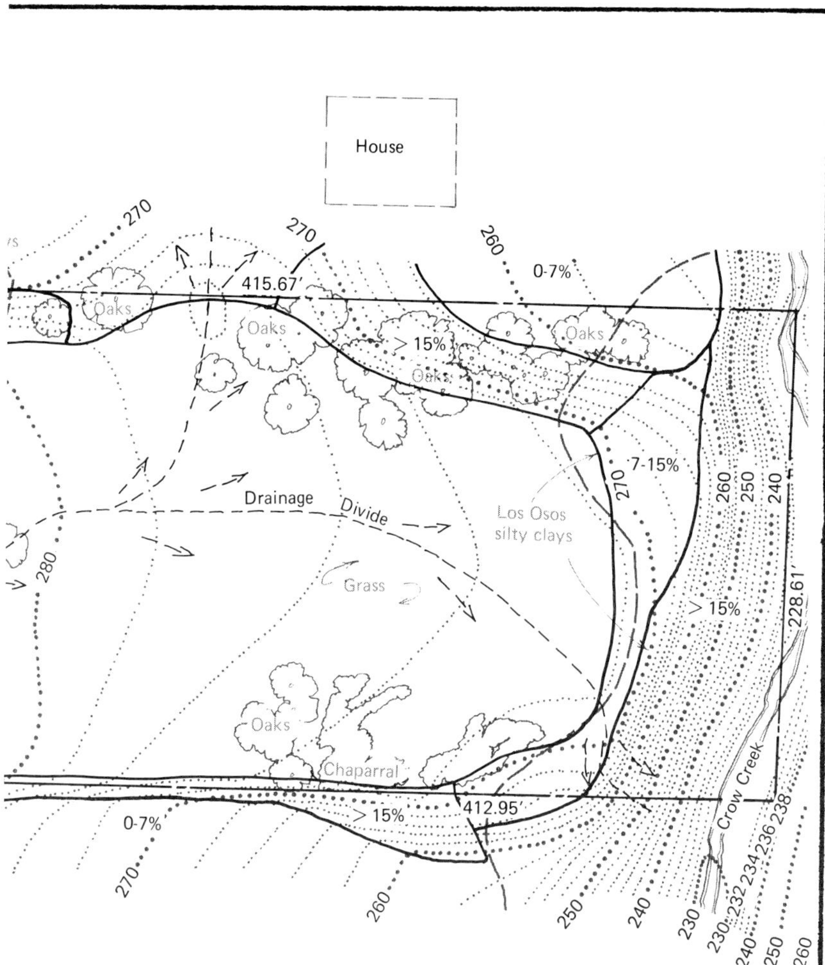

Figure 9-3.

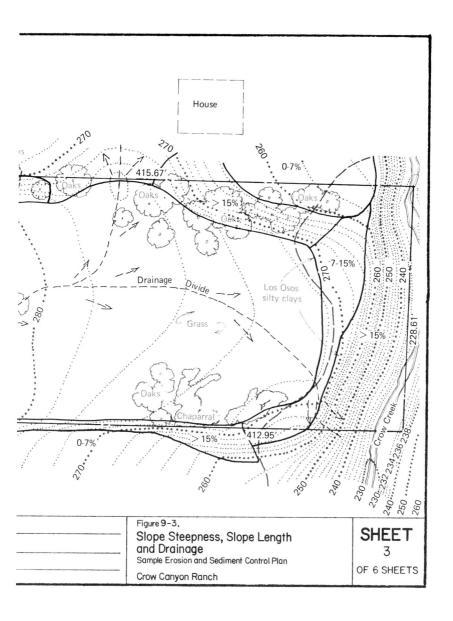

Slope Steepness, Slope Length and Drainage

Sample Erosion and Sediment Control Plan

Crow Canyon Ranch

SHEET
3
OF 6 SHEETS

because they are steeper than 15 percent and because any soil eroded from them will be deposited on a neighbor's property.

Soils

A soils report or soil survey covering the site should indicate soil erodibility. The *K* factor in SCS soil surveys is an estimate of soil erodibility (Chap. 5). Highly erodible soils should be left undisturbed. If they must be disturbed, they should be revegetated and mulched as soon as possible after grading is completed.

If the soils report gives a soil particle size distribution, check what percent of the soil is composed of fine particles (typically 0.02 mm or smaller). If a high percentage of the soil is smaller than 0.02 mm, much of any suspended sediment will escape capture unless a very large sediment basin is constructed. (Chapter 8 describes how to calculate trap efficiency.) Remember that grading will mix topsoils with subsoils and move them around the site. If fill will be imported, this material should be analyzed also.

The soils report for our sample site did not contain a particle size distribution, but it did show that the slope on the eastern side of the property is covered by a silty clay (Fig. 9.1). This slope is particularly critical because it is both steep and long and is above a creek. If the soil on this slope is allowed to erode, it will be nearly impossible to stop it from entering the creek.

Ground Cover

Note any areas of critical vegetation. Vegetation on or above long or steep slopes and on highly erodible soils is particularly important for erosion control.

The site survey for our sample site showed that the property is covered with grass and shrubs and very scattered clusters of oak trees (Fig. 9.1). Our site planning strategy will be to leave the vegetation on the steeper slopes undisturbed (for erosion control) and to try to retain as many of the oak trees as possible (for slope stability and aesthetic value).

Adjacent Areas

Examine areas downslope from the project. Note any watercourses or other sensitive features which receive runoff from the site. Analyze the potential for sediment pollution of these watercourses and the potential for downstream channel erosion due to increased volume, velocity, and peak flow of storm runoff from the site (Chap. 4).

The drainage analysis of our sample site reveals that the house below the northeast corner of the property is sensitive because it receives runoff from the steep slope on the north boundary of the property. The vicinity map (Fig. 9.2) shows that Crow Creek eventually receives the runoff from the entire site. The creek is particularly sensitive because it drains to Parker Reservoir, a regional recreation and water supply facility, which the county is anxious to protect. This fact will justify a high level of erosion and sediment control.

9.1c Step 3: Develop Site Plan

When a site plan is developed, erosion and sediment control should be considered along with such traditional planning criteria as economics, utility access, and traffic patterns. After analyzing the erosion hazards on site in Step 2, develop a site plan with erosion control in mind. Consider the following points when preparing a site plan.

Fit Development to the Terrain

Tailor the locations of building pads and roads to the existing contours of the land as much as possible. Locate them to take advantage of the natural strengths of the site and to minimize disturbance, particularly in hazardous areas.

Figure 9.4 shows several ways in which buildings can be sited to minimize slope disturbance. If a long building is oriented with its major axis parallel to the slope contours, considerably less grading is required (Fig. 9.4*a*, top). Modified rear entries and staggered floor levels also can reduce grading needs (Fig. 9.4*b*).

Confine Construction Activities to the Least Critical Areas

Land disturbance in critically erodible areas, as on steep slopes, will require installation of costly control measures. Keeping construction out of these areas will minimize the costs.

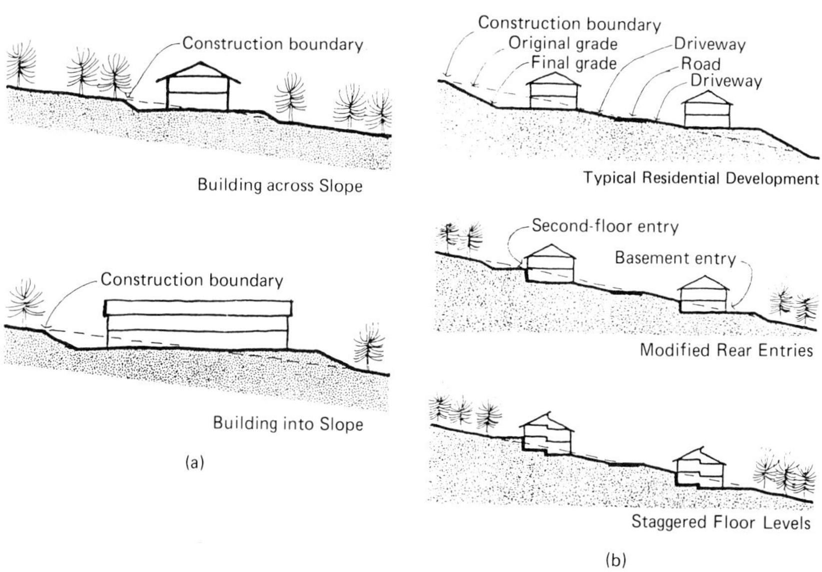

Fig. 9.4 Ways in which buildings can be (*a*) oriented and (*b*) adapted to the slope to minimize slope disturbance. (4)

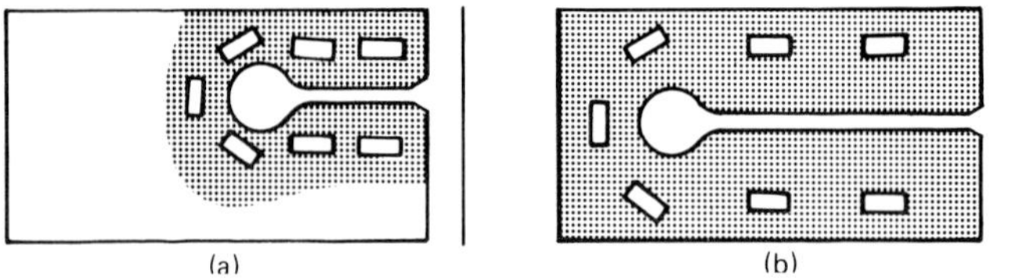

Fig. 9.5 Clustering buildings reduces the area disturbed. (*a*) Clustered and (*b*) conventional single-family homes.

Cluster Buildings Together

Clustering buildings minimizes land disturbance for roads and utilities and reduces erodible area (Fig. 9.5). Other benefits of clustering are reduced runoff, preservation of open space, and reduced development costs.

Minimize Impervious Areas

Make paved areas, such as streets, driveways, and parking lots, as small as possible. Preserve trees, grass, and other natural vegetation. Consider paving driveways with gravel or porous paving stones. French drains (Fig. 9.6), infiltration trenches, and dry wells can be used to percolate runoff from impervious surfaces into the soil. Gravel-filled trenches can be located along drip lines below roof eaves (Fig. 9.7). These measures will keep runoff volumes low and minimize the need for conventional storm drains—drop inlets and underground pipes. In the Lake Tahoe basin, for example, local requirements mandate that all runoff from impervious areas be infiltrated on-site. The advantages of so reducing runoff must be balanced against the risk of destabilizing slopes by water saturation.

In many communities, residential streets are wider than they need to be. Typically, these streets are designed to carry two lanes of traffic and two rows of parked cars, one on each side of the street. An alternative approach, if parking is not a critical problem, is to eliminate the space for the two rows of parked cars and, instead, provide parking bays at regular intervals (Fig. 9.8). This approach will substantially reduce the size of paved areas.

Retain the Natural Drainage System

Use the natural drainage system to convey runoff from the site wherever possible, rather than construct storm drains or concrete channels. If impervious surfaces are kept to a minimum and runoff from these surfaces is percolated into the soils on-site, it may be possible, without installing channel protection measures, to use the natural drainage system to drain a development. The cost of using the natural drainage system can be substantially lower than the cost of constructing a

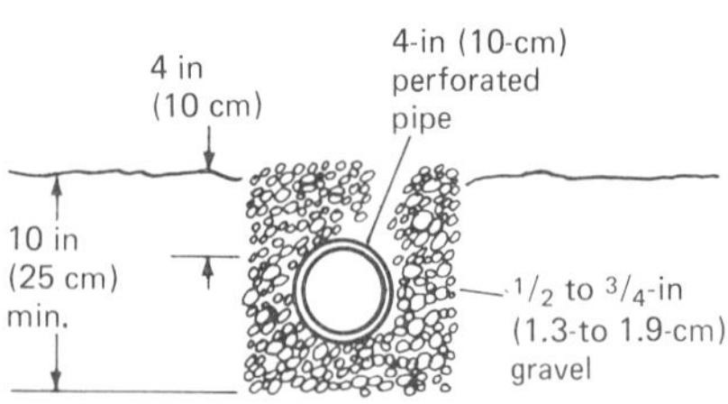

Fig. 9.6 French drain.

conventional storm drain system. Preserving the natural drainage system can also retain a visual amenity that will enhance the value of a development.

If runoff flows will be increased by development, route these augmented flows into a storm drain system and preserve the natural drainage system in its preexisting condition. If the stability of the natural system is upset, it may be very difficult to prevent a long-term erosion process from beginning. A man-made storm drain system can be designed to resemble a natural creek (Fig. 7.8).

Figure 9.9 shows the finished contours and improvements planned for our sample site. Because of the large number of units that must be accommodated, a high percentage of the site must be graded. The buildings have been located on the flatter portions of the site to avoid disturbing the steep slopes on the northern and eastern sides. The long slope running from west to east has been broken up by creating a series of small terraces stair-stepping down the hill. Runoff can thus be easily directed toward the street running down the center of the property.

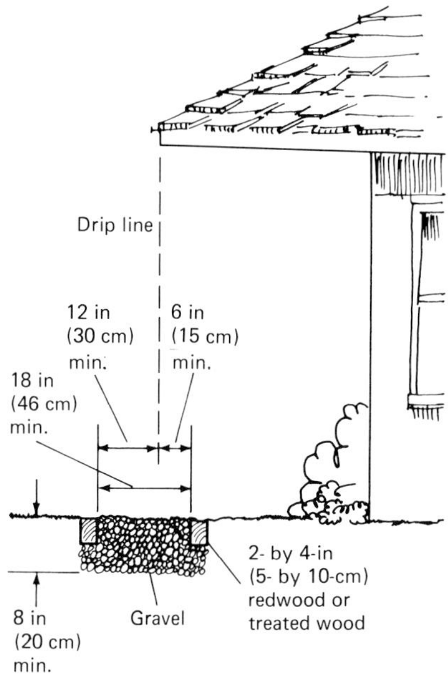

Fig. 9.7 Roof drip line infiltration trench. (2)

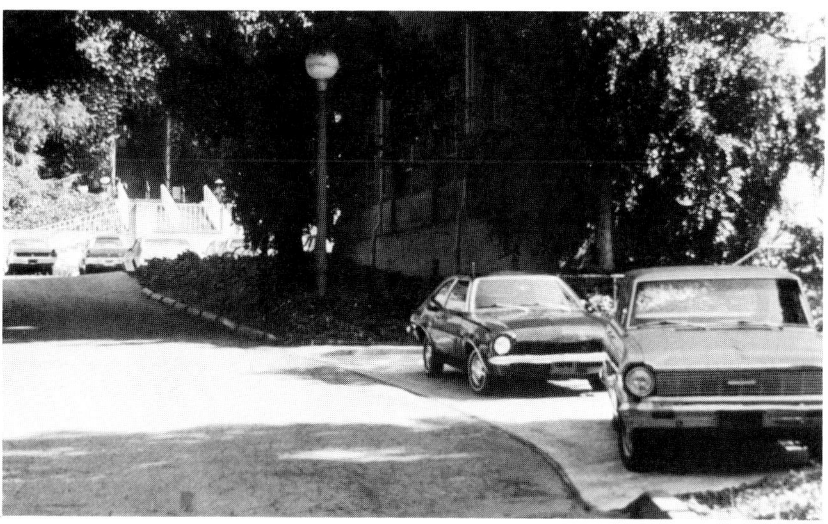

Fig. 9.8 Parking bays allow streets to be narrower.

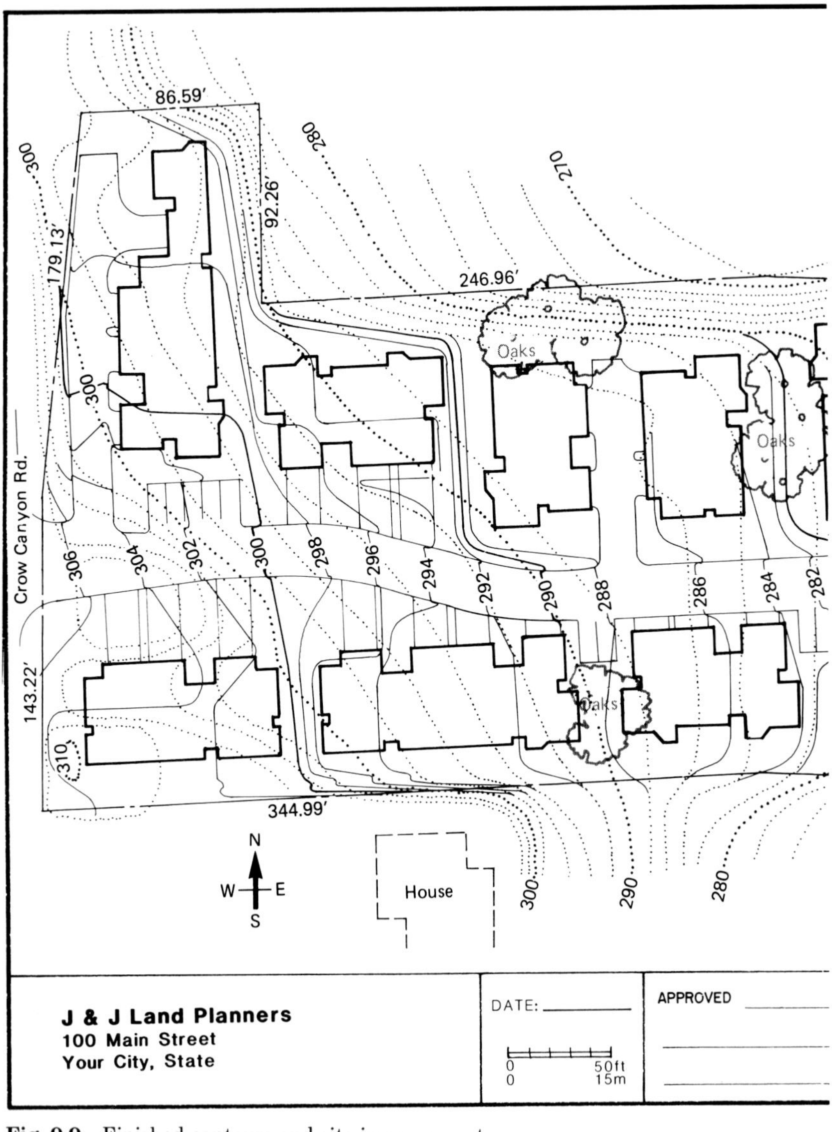

Fig. 9.9 Finished contours and site improvements.

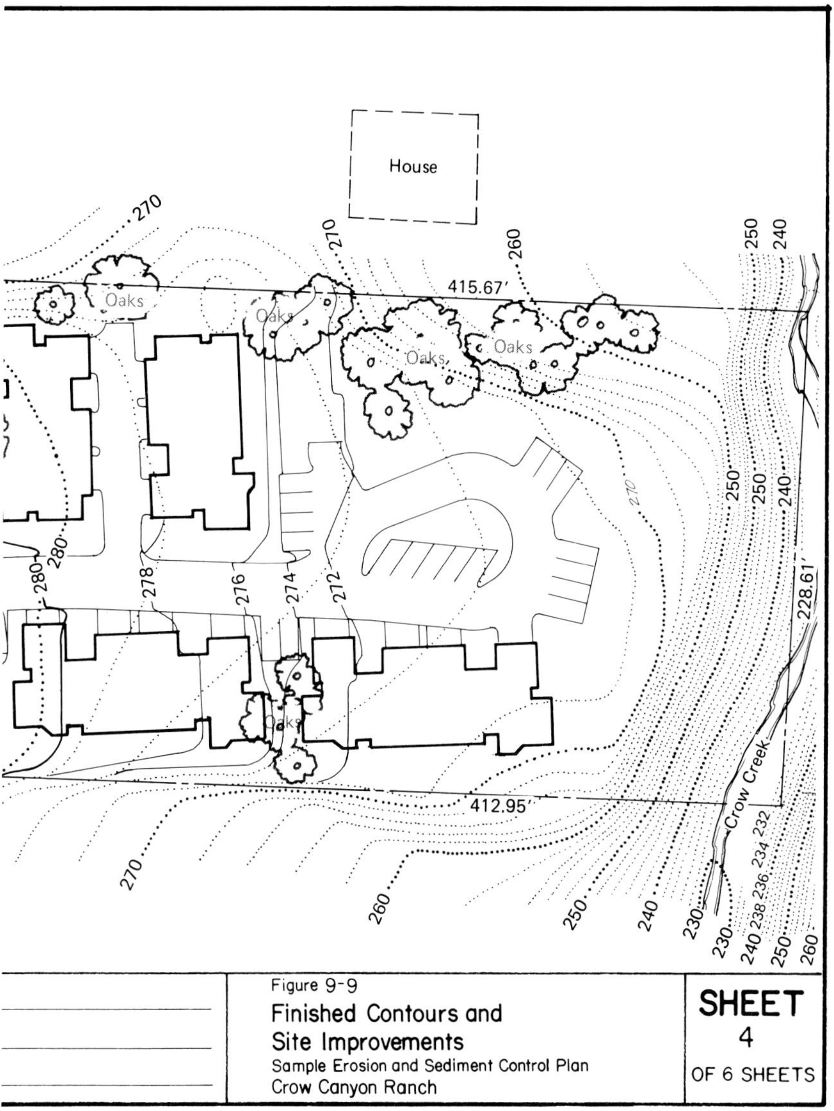

Figure 9-9
Finished Contours and Site Improvements
Sample Erosion and Sediment Control Plan
Crow Canyon Ranch

SHEET 4 OF 6 SHEETS

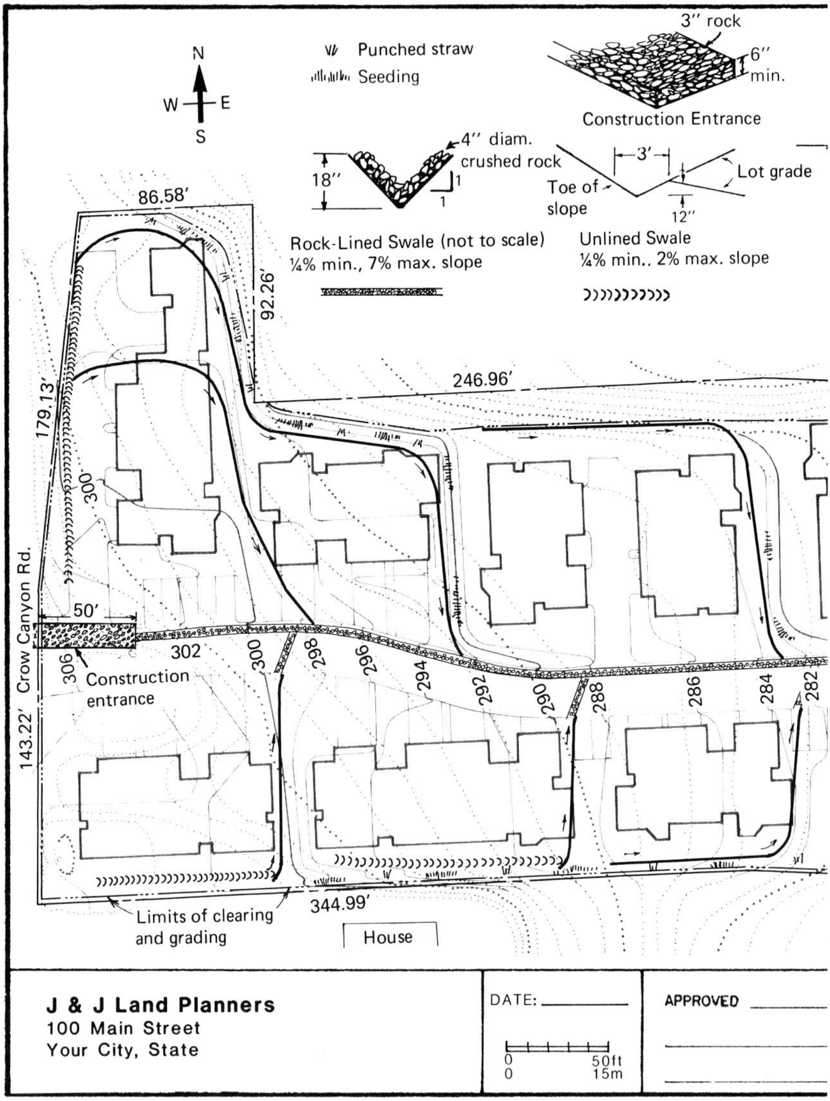

Fig. 9.10 Erosion and sediment control plan map and details.

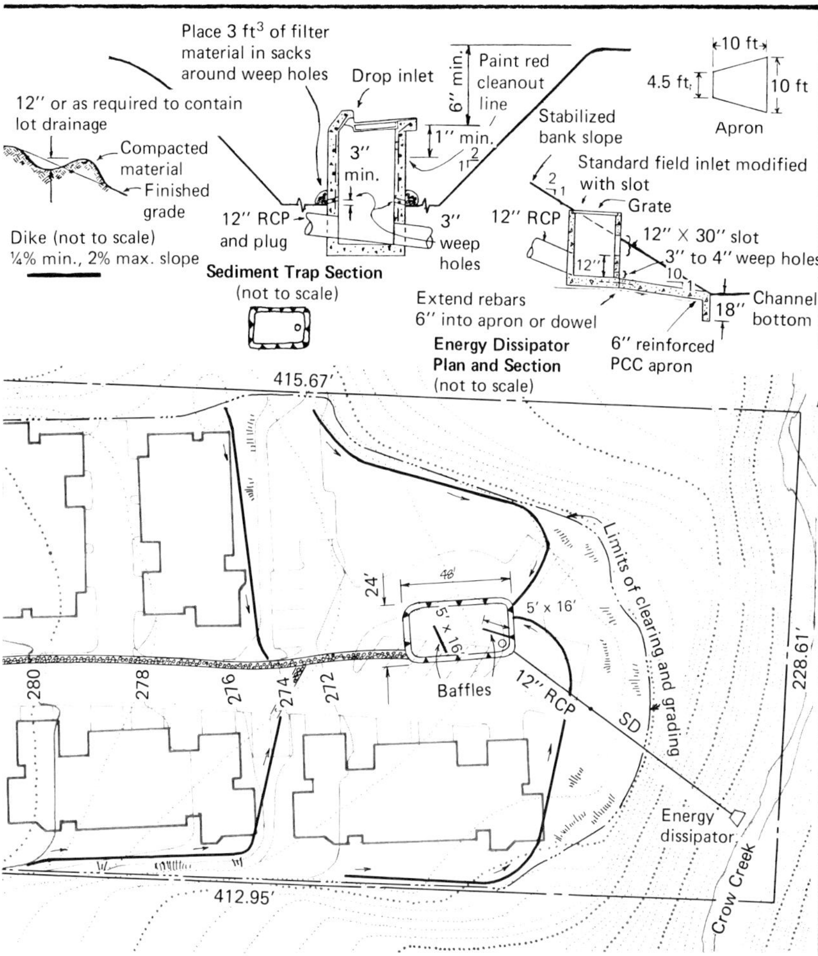

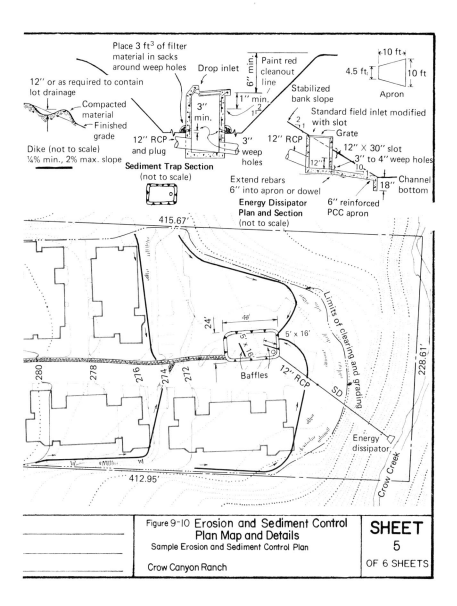

Figure 9-10 Erosion and Sediment Control Plan Map and Details
Sample Erosion and Sediment Control Plan

Crow Canyon Ranch

SHEET
5
OF 6 SHEETS

Seeding Specifications

1. Seed and mulch will be applied by October 1 to all disturbed slopes steeper than 2% and higher than 3 feet, and to all cut and fill slopes within or adjacent to public rights-of-way as directed by the county.

2. Seed and fertilizer will be applied hydraulically or by hand at the rates specified below. On slopes, straw will be applied by blower or by hand and anchored in place by punching.

Item	Pounds per Acre
'Blando' brome	30
Annual ryegrass	20
Fertilizer (16-20-0 and 15% sulfur)	500
Straw mulch	4,000

(This seed mix is presented for sample purposes only.)

Maintenance Notes

1. During the rainy season (October 1 to April 15), all erosion and sediment control measures will be inspected and repaired at the end of each working day and, in addition, after each storm.

2. The sediment trap will be cleaned out to its original dimensions when sediment accumulates up to the red line on the inlet. The spoil material will be deposited so that it does not directly re-enter the basin or cause sedimentation damage on- or off-site. The spoil heap will be seeded when formed and whenever more deposits are added to it.

3. Seeded areas will be repaired, reseeded and mulched as soon as possible after being damaged.

J & J Land Planners 100 Main Street Your City, State	DATE: ______ 0 50ft 0 15m	APPROVED ______

Fig. 9.11 Plan notes.

General Notes

1. By August, a standard drop inlet, energy dissipator and outfall structure will be constructed as shown and will remain as permanent tract improvements.

2. A construction entrance will be installed prior to commencement of grading. Location of the entrance may be adjusted by the contractor to facilitate grading operations. All construction traffic entering the paved road must cross the construction entrance.

3. The erosion and sediment control measures will be operable during the rainy season, October 1 to April 15. By October 1, grading and installation of storm drainage and erosion and sediment control facilities will be completed. No grading will occur between October 1 and April 15 unless authorized by the Director of Public Works.

4. Changes to this erosion and sediment control plan to meet field conditions will be made only with the approval of or at the direction of the Director of Public Works.

5. During the rainy season, all paved areas will be kept clear of earth material and debris. The site will be maintained so that a minimum of sediment-laden runoff enters the storm drainage system. This plan covers only the first winter following grading. Plans shall be resubmitted for approval prior to September 1 of each subsequent year until the tract improvements are accepted by the county.

6. The contractor will inform all construction site workers about the major provisions of the erosion and sediment control plan and seek their cooperation in avoiding the disturbance of these control measures.

Figure 9-11

Plan Notes

Sample Erosion and Sediment Control Plan

Crow Canyon Ranch

SHEET
6
OF 6 SHEETS

9.1d Step 4: Develop Erosion and Sediment Control Plan

Determine Limits of Clearing and Grading

Start with a topographic base map that shows existing and finished contours and proposed improvements (Fig. 9.9). On this base map, delineate the limits of the disturbed area. This line defines the area that must be protected. Figure 9.10 shows the limits of clearing and grading on the sample site.

Reexamine Drainage Areas

Check to see if the drainage boundaries defined in Step 2 have been altered by the development plan. If so, outline the drainage areas that will exist after grading. Since many control measures have a 1-acre (0.4-ha) or a 5-acre (2-ha) drainage area limit, you may want to break large watersheds into smaller units (see the following subsection). Check for and avoid augmented drainage to natural swales or water courses, which may be caused by increased drainage area or increased impervious surface.

As was done in Step 2, determine where concentrated flows will originate on- and off-site, how runoff will cross the site, and where runoff will leave the site. Check for and avoid unnaturally concentrated flows in natural swales created by pipes, ditches, berms, etc.

The improvement plan for our sample site (Fig. 9.9) has substantially changed the natural drainage pattern (Fig. 9.3). Each building pad has essentially become a small subwatershed. We will use this feature to advantage in developing the erosion and sediment control plan.

Apply the Principles of Erosion and Sediment Control

Fit Development to the Terrain

This principle is applied in the site development process (see Step 3).

Time Grading and Construction to Minimize Soil Exposure

Schedule the project so that grading is done during a time of low erosion potential (Chap. 2). On lage projects, stage the construction, if possible, so that one area can be stabilized before another is disturbed. Apply erosion control measures as soon after land disturbance is completed as is practical.

Because our sample site is located in California, which has a summer dry season, grading will take place between April 15 and October 1. All control measures will be in place before the start of the rainy season (Fig. 9.11, general note 3).

Retain Existing Vegetation Wherever Feasible

When laying out site improvements, try to site buildings between existing tree clusters and build roads around trees. Route construction traffic to avoid existing or newly planted vegetation. Avoid unnecessary clearing of vegetation around building pads, where construction will not be taking place (Fig. 2.3). Also avoid disturbing vegetation on steep slopes or in other critical areas.

Physically mark off the limits of land disturbance on the site with rope, fencing, surveyors' flags, or signs so that workers can clearly see areas to be protected (Fig. 9.12). A bulldozer operator will probably not know to protect a clump of trees that is only noted on a set of plans.

Fig. 9.12 Areas to be left undisturbed are roped off.

On our sample site (Fig. 9.9), the oak trees are left standing between the buildings and the vegetation on the steep slopes on the north and east sides of the property is left undisturbed.

Vegetate and Mulch Denuded Areas

Reestablish vegetation on all denuded areas that will not be covered with buildings or pavement. If graded areas are to be paved or built upon at a later date but will be exposed to rain for several months, consider establishing a temporary vegetative cover on those areas. It is often cheaper to establish and remove a temporary cover on such an area than to repair the gullying and sediment damage that is likely to occur. Before seeding an area, make sure necessary drainage controls are installed (see the following subsection). Plant establishment will be more successful if graded slopes are roughened or scarified before seeding Chap. 6).

On the sample site the slopes between the building pads and the slopes on the edges of the pads on the north, south, and east sides of the site will be seeded and mulched with punched straw (Fig. 9.10). The seeding specifications are given in the plan notes (Fig. 9.11).

Divert Runoff Away from Denuded Areas

Determine how runoff should be conveyed from the top to bottom of each drainage area. Is a channel or pipe required? If so, locate it where it can intercept potentially erosive flows and route them to a well-protected outlet such as a storm drain or a lined channel. Do not allow runoff to cross a denuded or newly seeded slope or other critical areas except within a drainage facility. If there is a significant drainage area above a cut or fill slope [Chapter 70 of the Uniform Building Code (5) defines this as a drainage path greater than 40 ft (12 m)], construct a dike or ditch at the top of the slope to convey the water to the bottom without causing erosion (Figs. 2.7 and 2.8). Dikes and ditches also can be used at the base of a disturbed slope to protect downstream areas by diverting sediment-laden runoff to a sediment trap or basin. It is often good strategy to construct a dike or swale all the way around a disturbed area to prevent clean runoff from entering the area and also to prevent silt-laden runoff from escaping before being desilted (Fig. 7.4).

On our sample site, dikes and swales are located around the perimeter of the disturbed area (Fig. 9.10). The dikes and swales along the south property line prevent clean runoff from entering the graded area. The dikes along the north and east edges of the construction site prevent polluted runoff from leaving the property. This runoff will be desilted by the sediment basin at the end of the street (see below). The swales and dikes are located to minimize interference with construction. Where drainage must cross a road, rock-lined swales, rather than dikes, are used because dikes could easily be damaged by construction vehicles.

Minimize Length and Steepness of Slopes

On long or steep disturbed or man-made slopes construct benches or ditches at regular intervals to intercept runoff. Each bench should be tilted at a gentle grade into the hill to channel the flow along the inner edge of the bench (Fig. 9.13). Route the intercepted runoff to a protected outlet.

On our sample site, the long slope running from west to east is broken up by the terraces for building pads and by the series of north-to-south-oriented dikes (Fig. 9.10). The longest west-to-east flow path will be roughly 150 ft (46 m), an acceptable slope length for 0 to 7 percent slopes.

Keep Runoff Velocities Low

Keep runoff velocities low by:

- Minimizing flow path lengths
- Constructing channels with gentle gradients
- Lining channels with rough surfaces

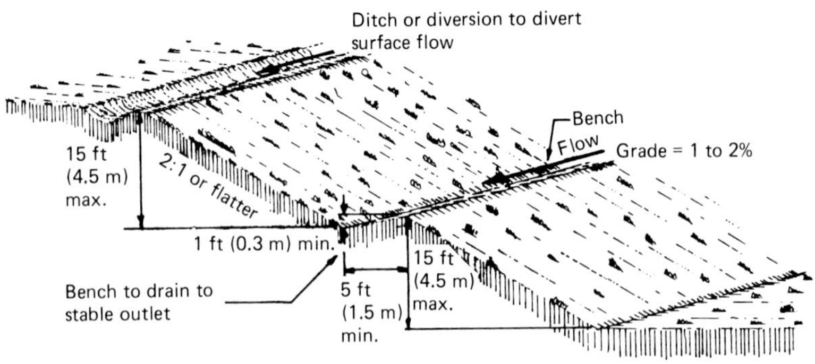

Fig. 9.13 Detail of slope with bench. (1)

On the sample site, flow path lengths were kept short by intercepting runoff at frequent intervals as described above. Except for the main channel down the center of the site, the dikes and swales have a 2 percent maximum grade (Fig. 9.10). The rock lining in the main channel will help keep the velocities there low.

Prepare Drainageways and Outlets to Handle Concentrated or Increased Runoff

Design storm water conveyance channels to withstand the expected flow volume and velocity from a design storm. Compute the expected discharge and velocity for both existing and newly constructed swales and for on- *and off-site* channels which will carry increased flow as a result of the project (see Chap. 4). By using these calculations, determine whether any drainage channels will require protection (Chap. 7). If the computations indicate the runoff flow will be erosive, first determine whether a vegetative lining will be sufficient (Chap. 6). If the expected velocity exceeds the limit for the specified grasses, choose between a rock, asphalt, or concrete lining. Remember, grass and rock linings are desirable because they keep velocities low, allow runoff to percolate into the soil, and are aesthetically pleasing. Because they resemble natural drainageways, they can enhance the appearance of a development.

Determine whether outlet protection will be needed. Pay particular attention to transitions from pipes or paved channels to natural or unlined channels. Locate riprap aprons or other energy-dissipating devices at discharge points where erosion is likely.

The main channel down the center of our sample site has a 7 percent slope. By using the method described in Chap. 4 and assuming a triangular channel with 1:1 side slopes, we calculate the flow in the channel will be about 1.5 ft^3/sec (0.042 m^3/sec), but the velocity will be above 6 ft/sec (1.8 m/sec). Since this velocity is above the maximum allowed for either unlined or grass-lined channels (Tables 7.1 and 6.7), we need to select a more durable lining material. From Fig. 7.31 we determine that 4-in (10-cm) rock will provide adequate erosion protection.

The main channel flows into a sediment trap that is drained by a permanent storm drain inlet in the cul-de-sac at the end of the street. From there it enters a concrete pipe which discharges into Crow Creek (see the following subsection). Because this pipe is on a steep gradient with an outlet onto a natural streambank, an engineered energy dissipator will be constructed at the outfall (Fig. 9.10).

Trap Sediment on Site

Install sediment basins or traps, straw bale dikes, or silt fences below denuded areas so that runoff will be detained long enough for suspended sediment to settle out. Try to locate sediment barriers in relatively level areas or in natural depressions. A flat area at the base of a slope (Fig. 8.34) is a good location for a silt fence or straw bale dike because the runoff can slow down before reaching the barrier and the sediment has a place to settle. Avoid placing sediment barriers where their construction would cause excessive soil disturbance. For example, excavating a sediment trap on a hillside is likely to cause more soil erosion and sedimentation than the device was intended to prevent.

Also, locate sediment barriers above such sensitive areas as streams, lakes, public streets, and adjacent properties. Make sure there will be adequate access in wet weather for maintenance and sediment removal.

Individual lots can be surrounded by dikes to create small sediment traps called lot ponds (Fig. 9.14). Gravel- or fabric-covered driveway aprons can serve as the outlets. If standing water on lots will interfere with construction activities, this type of sediment control should not be used. However, lots are often graded but are not built upon for months or even years. On west coast sites, construction sometimes ceases during the rainy season. In these situations, lot ponding may be a good approach. It should be realized, however, that lot ponding may necessitate recompaction of pads prior to building construction. Consult the soils report for the soil engineer's recommendation on this issue.

The size of the drainage area determines which type of sediment barrier should be used. Straw bale dikes and silt fences have a 1-acre (0.4-ha) drainage area limit. A sediment trap is generally adequate in drainage areas of less than 5 acres (2 ha). A sediment basin is needed if the drainage area exceeds 5 acres (2 ha). Unless the basin is designed as a permanent pond, its maximum drainage area should be less than 150 acres (60 ha). Drainage areas larger than 150 acres

Fig. 9.14 Lot pond sediment trap with fabric-lined spillway.

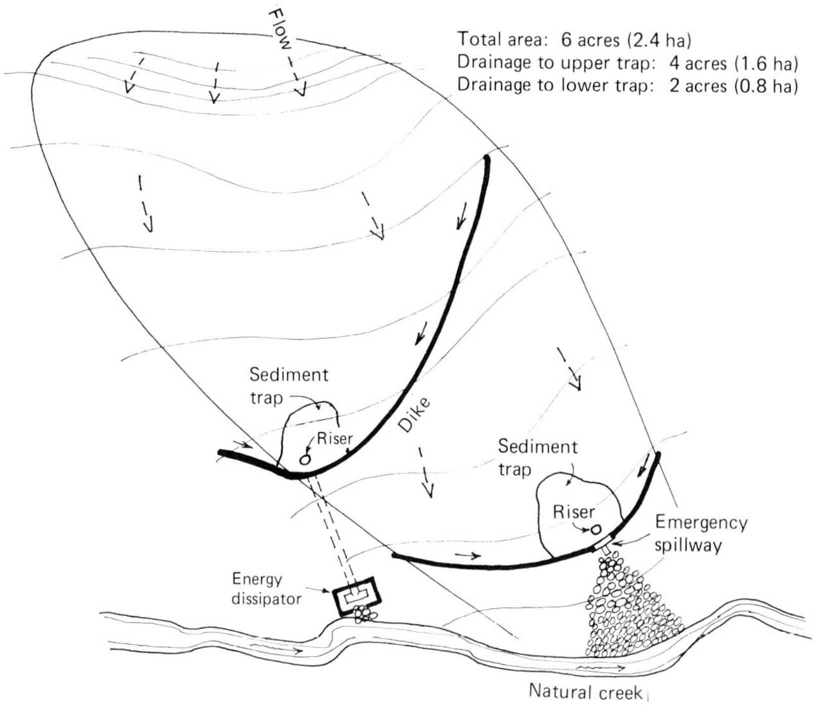

Fig. 9.15 Division of a watershed into smaller subcatchments. *Caution:* The spillway from the upper sediment trap should not discharge to the lower trap. Allowing sediment traps or basins to discharge to other traps or basins in series serves no constructive purpose, since the lower ones must then be designed as if the upper ones did not exist and failure of one basin can cause a domino effect failure of lower basins.

(60 ha) can be subdivided into smaller subcatchments by creating barriers to trap runoff in stages, perhaps in a group of basins. *Basins must drain in parallel, not in series* (Fig. 9.15). When a watershed is subdivided into smaller drainage areas, each with its own sediment basin, the degree of risk is likely to be substantially lower. (That is, the damage which could be caused by the failure of a small basin in a small watershed is minor compared to the damage potential of the failure of a large basin in a large watershed.)

Dividing a watershed into smaller drainage areas also can save money. Sediment basins are more costly than simple sediment traps to construct. In addition, sediment basins require an engineered design, whereas sediment traps are typically based on standard designs. Figure 9.15 shows a 6-acre (2.4-ha) watershed. Since the size of this watershed exceeds the 5-acre (2-ha) limit for sediment traps, a sediment basin would normally be required. However, by dividing the watershed into 4-acre (1.6-ha) and 2-acre (0.8-ha) subunits using a single dike, we are able to construct two simple sediment traps and save the expense of custom designing a sediment basin.

Sediment basins and traps are commnonly located below large disturbed areas, at the lowest point in a watershed, and in swales and small drainageways. Do *not* locate a sediment basin in a major stream, such as one designated with a blue line on a U.S. Geological Survey topographic map. It is unnecessary, costly, and dangerous to impound runoff from large, undisturbed areas. Trap the sediment-laden runoff *before* it enters a stream.

On the sample site, a sediment trap will be constructed in the cul-de-sac at the east end of the site (Fig. 9.10). Runoff from each group of pads is routed with dikes and swales to the main channel down the center of the property. The dikes along the east edge of the graded area will prevent runoff from spilling over this critical slope. Thus the sediment trap will impound runoff from the entire disturbed area. The cul-de-sac is an ideal location for the trap because it will not interfere with construction activities and can remain until the street is paved. It is also easily accessible for maintenance.

Since this site is only 4 acres (1.6 ha) in size, a sediment trap rather than a sediment basin is acceptable. We will size it according to the standard which was developed for this region: 263 ft^2/acre (60 m^2/ha) of drainage area (see Sec. 8.4a). The total surface area needed is thus (263 ft^2/acre)(4 acres) = 1052 ft^2[(60 m^2/ha)(1.6 ha) = 96 m^2]. We therefore specify trap dimensions of 24 by 48 ft (7.3 by 14.6 m) or 1152 ft^2 (107 m^2), which allows a margin for error.

Since the final drainage plan for this project calls for a storm drain inlet in the cul-de-sac, we will install the inlet early so it can serve as the outlet to the sediment trap. This will save the cost of constructing a temporary outlet structure. The inlet will be connected to an underground concrete pipe which discharges through an energy dissipator on the bank of Crow Creek. The sediment trap, dikes, storm drain inlet, outlet pipe, and energy dissipator will be constructed during the initial stages of grading so that the site will be protected from the first rains.

The area around the inlet will be excavated to a depth of 4 ft (1.2 m). Of this depth, 3 ft (0.9 m) is for sediment storage and the remaining 1 ft (0.3 m) serves as a settling zone.

A crushed rock construction entrance will be located at the west end of the site (Fig. 9.10). The crushed rock will prevent construction vehicles from tracking mud onto Crow Canyon Road.

Inspect and Maintain Control Measures

Develop a maintenance schedule and instructions for maintaining control measures. The instructions should specify where sediment dredge spoils should be placed, what spare materials (such as extra filter fabric, straw bales, stakes, and gravel) are needed, and where they should be stockpiled.

It is the responsibility of the contractor to make sure that all workers understand the major provisions of the erosion and sediment control plan. If they understand the plan, they are less likely to disturb drainage patterns and control measures, as by running over a dike with a truck. A routine end-of-day maintenance check is strongly advised. All maintenance procedures and the maintenance schedule should be specified on the plans. The plans should also remind the contractor of his or her responsibility to inform construction site workers about the erosion and sediment control features on the site.

The maintenance notes on Fig. 9.11 outline the maintenance program for our sample site. A red line will be painted on the storm drain inlet 1 ft (0.3 m) below the top of the inlet so that workers will know to clean the sediment trap when the sediment level reaches that line. During the rainy season (October 1 to April 15), all erosion and sediment control measures will be checked, and repaired if necessary, at the end of each working day and after each storm (Fig. 9.11, maintenance note 1). In particular, the site superintendent should look for breaches in dikes and for erosion or sedimentation on the slopes on the northern and eastern sides of the property.

9.1e Writing the Erosion and Sediment Control Plan

The erosion and sediment control plan submitted to the approving agency with the project application should contain all the pertinent information from Steps 1 through 4. The following elements should be present and are required in many communities: (1, 7)

- Narrative
- Map
- Construction details
- Calculations

The Narrative

The narrative is a brief description of the overall strategy for erosion and sediment control. It should summarize for the plan reviewer and the project superintendent the aspects of the project that are important for erosion control and should include:

- A brief description of the proposed land-disturbing activities, existing site conditions, and adjacent areas (such as creeks and buildings) that might be affected by the land disturbance
- A description of critical areas on the site—areas that have potential for serious erosion problems
- A construction schedule that includes the date grading will begin and the expected date of stabilization
- A brief description of the measures that will be used to control erosion and sediment on the site, when they will be installed, and where they will be located
- A maintenance program, including frequency of inspection and provisions for repair of damaged structures

The Map

The map is the key item in an erosion and sediment control plan. It should show:

- Existing and final elevation contours at an interval and scale sufficient for distinguishing runoff patterns before and after disturbance
- Critical areas within or near the project area, such as streams, lakes, wetlands, highly erodible soils, public streets, and residences
- Existing vegetation
- Limits of clearing and grading
- Locations and names of erosion and sediment control measures, with dimensions (Fig. 9.10)

It is strongly recommended that standard symbols be used on the map to denote erosion and sediment control measures. Use of standardized symbols will speed up plan review time and make it easier for site superintendents and

TABLE 9.1 Standard Symbols for Erosion and Sediment Control Measures

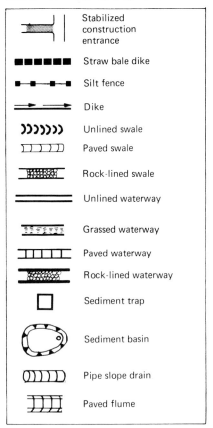

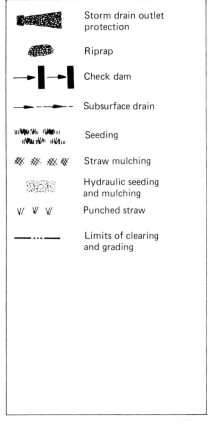

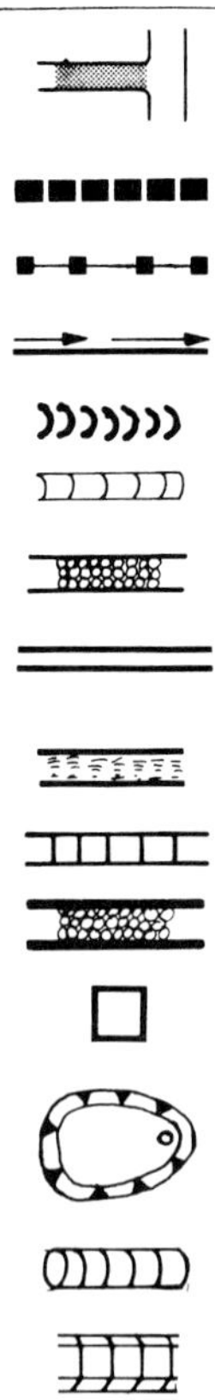

Stabilized construction entrance	Storm drain outlet protection
Straw bale dike	Riprap
Silt fence	Check dam
Dike	Subsurface drain
Unlined swale	Seeding
Paved swale	Straw mulching
Rock-lined swale	Hydraulic seeding and mulching
Unlined waterway	Punched straw
Grassed waterway	Limits of clearing and grading
Paved waterway	
Rock-lined waterway	
Sediment trap	
Sediment basin	
Pipe slope drain	
Paved flume	

inspectors to understand plans quickly. Table 9.1 shows the standard symbols that were developed for use in the San Francisco Bay Area. (1) These symbols were designed to be both pictorially representative of the control measures and easy to draw.

Construction Details, Specifications, and Notes

Construction details, often in large-scale, detailed drawings, provide key dimensions and spatial information that will not fit on the map (see top of Fig. 9.10). Other important information should also be provided; examples are seeding and mulching specifications; equivalent opening size (EOS) and strength requirements for filter fabric; specifications for wire mesh, fence posts, and staples; installation procedures for control measures; and maintenance instructions (Fig. 9.11).

Calculations

Include the calculations used to size the control measures, particularly the data for the design storm (recurrence interval, duration and magnitude, and i_{peak} for the time of concentration) and the design assumptions for sediment basins and traps (design particle size, trap efficiency, discharge rate, and dewatering time). Also include calculations to support the sizing of storm drain systems when an engineered design was necessary.

Table 9.2 is a checklist of items that should be included in an erosion and sediment control plan.

9.2 EVALUATING AN EROSION AND SEDIMENT CONTROL PLAN

The following section is written from the perspective of a public agency plan reviewer; it describes what to look for when evaluating an erosion and sediment control plan.

9.2a General Approach

Responsibility

It is not the responsibility of the plan reviewer to see that the plan is the best possible one. The reviewer can only ensure that the plan meets the minimum standards set by the reviewing agency and its authorizing ordinance. Do not set standards not supported by an ordinance.

Communications

Encourage informal communications between the plan reviewer and the plan preparer. This will enable the reviewer to make informal suggestions that may save the developer money and the preparer time, and it may result in a better, more effective plan. It will also enable the preparer to explain and justify the plan.

Incomplete Plans

Do not review seriously incomplete plans. Send them back with a request for the missing information.

First Review

The first review should be a complete review. In subsequent reviews, deal only with items identified in the first review. It is unfair to the developer to keep injecting new requirements in subsequent reviews.

9.2b Required Information

Make sure all the required information has been submitted. Table 9.2 can be used as a checklist by both plan reviewers and plan preparers. However, checklists can encourage laziness. Having everything checked off does not necessarily mean that everything is in order.

9.2c Plan Concept

The concept should be examined first, starting with the general and moving to the specific. Does the plan make sense?

Schedule

Examine the construction schedule. Will grading be completed before the rainy season (west coast) or before the summer thunderstorm months (east coast)?

TABLE 9.2 Checklist for Erosion and Sediment Control Plans

Narrative
☐ *Project description* A brief description of the nature and purpose of the land-disturbing activity and the amount of grading involved
☐ *Existing site conditions* A description of the existing topography, vegetation, and drainage
☐ *Adjacent areas* A description of neighboring areas, such as streams, lakes, residential areas, and roads that might be affected by the land disturbance
☐ *Soils* A brief description of the soils on the site including erodibility and particle size distribution

- ☐ *Critical areas* A description of areas within the developed site that have potential for serious erosion or sediment problems
- ☐ *Erosion and sediment control measures* A description of the methods that will be used to control erosion and sediment on the site
- ☐ *Permanent stabilization* A brief description of how the site will be stabilized after construction is completed
- ☐ *Maintenance* A schedule of regular inspections and repairs of erosion and sediment control structures

---

Map

The following information should appear on one or more maps:

- ☐ *Existing contours* Existing elevation contours of the site at an interval sufficient to determine drainage patterns
- ☐ *Preliminary and final contours* Proposed changes in the existing elevation contours for each stage of grading
- ☐ *Existing vegetation* Locations of trees, shrubs, grass, and unique vegetation
- ☐ *Soils* Boundaries of the different soil types within the proposed development
- ☐ *North arrow*
- ☐ *Critical areas* Areas within or near the proposed development with potential for serious erosion or sediment problems
- ☐ *Existing and final drainage patterns* A map showing the dividing lines and the direction of flow for the different drainage areas before and after development
- ☐ *Limits of clearing and grading* A line showing the area to be disturbed
- ☐ *Erosion and sediment control measures* Locations, names, and dimensions of the proposed temporary and permanent erosion and sediment control measures
- ☐ *Storm water management system* Location of permanent storm drain inlets, pipes, outlets, and other permanent drainage facilities (swales, waterways, etc.), and sizes of pipes and channels

---

Details

- ☐ *Detailed drawings* Enlarged, dimensioned drawings of such key features as sediment basin risers, energy dissipators, and waterway cross sections
- ☐ *Seeding and mulching specifications* Seeding dates, seeding, fertilizing, and mulching rates in pounds per acre (kilograms per hectare), and application procedures
- ☐ *Maintenance program* Inspection schedule, spare materials needed, stockpile locations, and instructions for sediment removal and disposal and for repair of damaged structures

---

Calculations

- ☐ *Calculations and assumptions* Data for design storm used to size pipes and channels and sediment basins and traps [e.g., 10-year, 6-hr storm = 3.1 in (79 mm); i_{peak} = 2.6 in/hr (66 mm)], design particle size for sediment traps and basins, estimated trap efficiencies, basin discharge rates, size and strength characteristics for filter fabric, wire mesh, fence posts, etc., and other calculations necessary to support drainage, erosion, and sediment control systems

---

When will storm drainage facilities, paving, and utilities be installed in reference to the rainy season? If grading will take place during months when there is a high probability of heavy rains, what extra precautions will be taken to protect against erosion, sedimentation, and changing drainage patterns?

Site Drainage

Make sure you understand where all drainage comes from on and above the site, where it goes, and how it traverses the site. For large sites, require or prepare a drainage area map. If drainage patterns are unclear, ask for clarification.

Sediment Basins and Traps

Locate all sediment basins and traps and define their tributary areas. Erosion control within areas that drain to sediment barriers need not be as intensive as within areas not so protected.

Erosion Control

Check the method used to prevent erosion. Hydraulic seeding and mulching may adequately stabilize some areas, but other areas, because of their proximity to sensitive features such as watercourses or their steepness or lack of backup protection such as sediment basins, may need far more intensive revegetation efforts. On critical slopes, a reliable backup system for hydraulic seeding, such as punched straw, is strongly recommended.

Channels and Outlets

Examine all drainageways where concentrated flows will occur. Be sure adequate erosion protection is provided both along channels and at channel and pipe outlets. Check the sources of runoff to be sure that all the runoff comes from undisturbed or stabilized areas or has been desilted by sediment basins or other sediment retention devices.

Miscellaneous

Look for haul roads, stockpile areas, and borrow areas. They are often overlooked and can have a substantial effect on drainage patterns. Look at all points of vehicle access to the site and be sure mud and dirt will not be tracked onto paved streets and that sediment-laden runoff will not escape from the site at these points. Pay particular attention to watercourses and their protection.

9.2d Plan Details

Once the plan concept has been shown to be sound, check the details to be sure the concept is adequately executed in the plans.

Structural Details

Be sure that sufficiently detailed drawings of each structure (sediment basin, dike, ditch, silt fence, etc.) are included so there is no doubt about locations, dimensions, or method of construction.

Calculations

See if calculations have been submitted to support the capacity and structural integrity of all structures. Were the calculations made correctly?

Vegetation

Look at seed, fertilizer, and mulch specifications. Check quantities and methods of application to be sure they are appropriate and consistent with local guidelines.

Maintenance

Be sure that general maintenance requirements and, where necessary, specific maintenance criteria, such as the frequency of sediment basin cleaning, are included. Are there stockpiles of spare materials (filter fabric, straw bales, stakes, gravel, etc.) to repair damaged control measures? Routine maintenance inspections should be part of the plans.

Contingencies

The plan must provide for unforeseen field conditions, scheduling delays, and other situations that may affect the assumed conditions.

Technical Review

The erosion and sediment control plan should be reviewed by the soils or geotechnical consultant for the project, if there is one.

Signature

The erosion and sediment control plan should be signed by a qualified individual.

REVIEW QUESTIONS

1. In preparing an erosion and sediment control plan, what is the most important part of Step 2. Analyze data?
2. Why is it important to map watershed boundaries?
3. Why is it important to map steep slopes and existing vegetation?

4. List five site-planning techniques for minimizing land disturbance.

5. What are the advantages of using the natural drainage system for permanent site drainage?

6. List six general approaches (not control measures) to the control of erosion.

7. List four measures for controlling sediment.

8. Under what conditions should a watershed be subdivided into smaller drainage areas?

9. Why is it important to physically mark off the limits of land disturbance on the site and to make sure that all workers understand the erosion control plan?

10. What three elements should an erosion and sediment control plan contain?

11. Why should standard symbols be used to delineate control measures on plans?

12. What should a plan reviewer do if an incomplete set of plans is submitted?

13. What should a plan reviewer look for when evaluating site drainage on an erosion and sediment control plan?

REFERENCES

1. Association of Bay Area Governments, *Manual of Standards for Erosion and Sediment Control Measures,* Oakland, Calif., 1981.
2. Fairfax County, Virginia, *Inspector's Check List for Erosion and Sediment Control,* Fairfax County, 1978.
3. Bruce Ferguson, "Erosion and Sedimentation Control in Regional and Site Planning," *Journal of Soil and Water Conservation,* July–August, 1981, pp. 199–204.
4. International Conference of Building Officials, *Uniform Building Code,* Whittier, Calif., 1982.
5. Tahoe Regional Planning Agency and Association of Bay Area Governments, *How to Protect Your Property from Erosion—A Guide for Homebuilders in the Lake Tahoe Basin,* South Lake Tahoe, Calif., 1982.
6. Virginia Soil and Water Conservation Commission, *Field Manual,* Richmond, Va., 1982.
7. Virginia Soil and Water Conservation Commission, *Virginia Erosion and Sediment Control Handbook,* Richmond, Va., 1980.

chapter 10

Maintaining Erosion and Sediment Control Measures

Regular maintenance is vital to the success of an erosion and sediment control system. Control measures must be inspected frequently and repaired as soon as problems arise. Many of the measures described in this handbook are short-term, temporary devices that can break down during a single storm. Often a device fails during the first storm after installation.

If a problem such as a wheel track cut through a dike, is caught early, it can be repaired quickly and simply and an erosion problem can thereby be avoided. If a problem is not discovered until after several storms, repair and cleanup costs can be very high. Figure 10.1 shows gully erosion below a breached dike. The dike was intended to route runoff into the far end of a sediment basin. After breaching the dike, the polluted runoff bypassed the basin completely and entered a storm drain untreated. A few minutes of timely repair work with a shovel would have avoided this damage.

This chapter presents guidelines for maintaining the control measures described in Chaps. 6, 7, and 8 of this handbook. Common problems with each measure are highlighted. Section 10.4 discusses how to plan for emergencies. Table 10.1 is a summary checklist of common maintenance problems and remedies. Various public agency manuals, such as References 1, 2, and 3, also contain checklists and guidelines for field inspectors.

Fig. 10.1 Breach in dike causes gully erosion.

10.1 MAINTAINING VEGETATION

The critical period for vegetative stabilization is the first year. If the ground cover is not complete, rills or gullies, which often grow progressively worse, may develop. Unfortunately, there are no precise performance standards for erosion control effectiveness, such as "at least an 80 percent cover shall be maintained." On some sites, an 80 percent cover may provide adequate protection, whereas on other sites it may be quite inadequate. Achieving close to 100 percent plant cover on any site is extremely difficult. Nevertheless, the following general maintenance guidelines can be followed.

Inspect sites within 30 days after planting or immediately after the first rain. Follow-up inspections after each major storm are advisable. Once the site is well stabilized, no further inspections should be necessary.

One of the most common causes of erosion on a newly seeded slope is runoff from above the slope. If gullies are forming on such a slope, check to see if a diversion dike or swale is located at the top. If a diversion structure is present, check for breaches, vehicle tracks (Fig. 10.2), and low spots (see Secs. 10.2a and 10.2b). If a diversion is not present, install a dike, swale, or pipe slope drain as described in Chap. 7. Figure 10.3 shows a gully beginning to cut through grass seedlings on a slope below a house pad. The pad should have been sloped away from the slope shown. A simple remedy would be to construct an earth dike around the edge of the pad.

TABLE 10.1 Maintenance Checklist

Control measure	Problems to look for	Possible remedies
Vegetation	Rills or gullies forming	Check for top-of-slope diversion and install if needed.
	Bare soil patches	Fill rills and regrade gullied slopes.
	Sediment at toe of slope	Reseed, fertilize, and mulch bare areas.
Dikes	Gully on slope below dike breach; wheel track or low spot in dike	Add soil to breaches or low spots and compact.
	Loose soil	Compact loose soil.
	Erosion of dike face	Seed and mulch dike or line upslope face with crushed rock.
Swales	Gully on slope below swale	Repair breaches.
	Wheel track; low point (water ponded in swale)	Build up low points with compacted soil or sandbags or rebuild swales with positive drainage.
	Sediment or debris in channel	Remove obstructions.
	Erosion of unlined channel surface	Seed and mulch swale and anchor with netting; or line swale with crushed rock; or install check dams; or realign swale on gentler gradient; or divert some or all of swale's drainage to a more stable facility.
	Erosion of channel lining	Install larger riprap; or reseed, mulch, and anchor with netting; or install check dams; or pave channel.
Pipe slope drain or chute	Blocked inlet or outlet	Remove sediment and debris.
	Runoff bypassing inlet	Enlarge headwall. Flare out entrance section.
	Erosion at outlet	Enlarge riprap apron and use larger stones; or convey runoff to more stable outlet.
Grassed waterways	Bare areas	Reseed, mulch, and anchor with netting. Divert flows, if possible, during establishment period.
	Channel capacity reduced by tall growth	Mow grass.
Riprap-lined waterway	Scour beneath stones	Install proper filter fabric or graded bedding. Make sure edges of filter fabric are buried.

TABLE 10.1 Maintenance Checklist (Continued)

Control measure	Problems to look for	Possible remedies
	Dislodged stones	Replace with larger stones.
Outlet protection	Erosion below outlet	Enlarge riprap apron; or line receiving channel below outlet; or convey runoff directly to a more stable outlet, such as a storm drain. Make sure discharge point is on level or nearly level grade.
	Outlet scour	Install proper filter fabric or graded bedding beneath riprap apron.
	Dislodged stones	Replace with larger stones.
Sediment traps and basins	Sediment level near outlet elevation	In traps, remove sediment if less than 1 ft (0.3 m) below outlet elevation. In basins, remove sediment if less than 2 ft (0.6 m) below top of riser.
	Obstructed outlet	Remove debris from trash rack.
	Basin not dewatering between storms	Clear holes. Clean or replace sediment-choked gravel surrounding dewatering hole or subsurface drain.
	Damaged embankments	Rebuild and compact damaged areas.
	Spillway erosion	Line spillway with rock, filter fabric, or pavement.
	Outlet erosion	Make sure outlet is flush with ground and on level grade. Install, extend, or repair riprap apron as required; or convey discharge directly to a more stable outlet such as a storm drain.
	Riser flotation	Anchor riser in concrete footing.
	Excessive discharge to and from basin or trap	Check drainage patterns for consistency with plans. Reroute part of drainage to another outlet; or enlarge basin surface area.
	Sediment storage zone fills too quickly	Increase depth of basin or trap; or stabilize drainage area, as by seeding and mulching bare soils, paving streets, and installing storm drains.
Straw bale dike	Bale displacement	Anchor bales securely with proper stakes. Check drainage area, slope length, and gradient behind barrier.
	Undercutting of bales	Entrench bales to proper depth, backfill, and compact soil.

TABLE 10.1 Maintenance Checklist (Continued)

Control measure	Problems to look for	Possible remedies
	Gaps between bales	Restake bales. Drive first stake in each bale at angle to force the bale toward the adjacent bale.
	Baling wire broken	Retie bale or replace with fresh bale.
	Bale disintegrating	Replace bale.
	Runoff escaping around barrier	Extend barrier or reposition in center of flow path.
	Sediment level near top of bales	Remove sediment when level reaches half of barrier height.
Silt fence	Undercutting of fence	Entrench wire mesh and fabric to proper depth, backfill, and compact.
	Fence collapsing	Check fence post size and spacing, gauge of wire mesh, and fabric strength. Check drainage area, slope length, and gradient behind barrier. Correct any substandard condition.
	Torn fabric	Replace with continuous piece of fabric from post to post. Securely anchor with proper staples.
	Runoff escaping around barrier	Extend fence.
	Sediment level near top of fence	Remove sediment when level reaches half of fence height.
Check dam	Sediment accumulation	Remove sediment after each storm.
	Flow escaping around sides of check dam	Build up ends of dam and provide low center area for spillway.
	Displacement of sandbags, stones, or straw bales	Check drainage area and peak flows. Reinforce dam with larger stones, etc.; or divert portion of flow to another outlet.
Inlet protection	Flooding around or below inlet	Remove accumulated sediment; or convert sediment barrier to an excavated sediment trap; or reroute runoff to a more suitable outlet such as a sediment basin.
	Undercutting of bales or silt fence, bale displacement, torn fabric, etc.	See remedies for straw bale dikes and silt fences.

Fig. 10.2 Tire tracks on newly seeded slope.

Repair, mulch and reseed slopes if there is erosion on bare areas of a slope, rills or gullies are forming (Fig. 10.4), or there is sediment build-up at the toe of a slope. Fill or smooth over rills before reseeding. If rill occurrence is extensive, the entire slope may require regrading before reseeding.

10.2 MAINTAINING WATER CONVEYANCE STRUCTURES

Be sure that water conveyance structures are not carrying more drainage than they were designed to carry. Excess flow can cause erosion, flooding, and other problems. If flow in a drainageway appears to be excessive, look for unplanned diversions upstream and changes in drainage patterns caused by grading. Also, recheck the runoff and flow calculations to see that they were made correctly.

10.2a Dikes

Chief among the common causes of dike failure are breaches made by vehicle wheels. Trucks and other construction equipment can damage a dike merely by driving over it. The wheels compress the soft dike soil and thereby create low points in the dike where water can escape. Runoff pouring through breaches can cause gullies to form below (Fig. 10.1).

Sometimes dikes are not present where they should be (Fig. 10.3). Whenever

Fig. 10.3 Gully forming below house pad.

Fig. 10.4 Slope requiring regrading and reseeding.

a slope is disturbed or created, a diversion should be constructed at the top. Make sure runoff does not escape around the end of a dike.

When dikes are not constructed high enough or wide enough, or the soil material in them is not compacted, they are susceptible to washout. Dike erosion also occurs if a dike is constructed down too steep a grade or if too much flow is carried along its face (Fig. 10.5).

Dike repair is relatively simple. Most repairs can be made with a shovel. The soil can be compacted by tamping with the back side of the shovel blade. If erosion of a dike is severe or persistent, the dike should be lined with a suitable material (see Chap. 7).

It is best to inspect during a storm. A few minutes of timely maintenance during a storm will prevent problems like the one depicted in Fig. 10.1.

10.2b Swales

Maintenance of swales is similar to that of dikes. A common problem is lack of positive drainage at all points. Low spots accumulate water, which eventually

Fig. 10.5 Erosion along dike and swale.

spills over the side and causes erosion below (Fig. 7.22). Erosion of the soil in unlined swales occurs where the gradient is too steep or the flow too great for the swale material (Fig. 10.5). In such cases, a lining is required. If erosion is occurring in a lined swale, the lining may require repair or a stronger lining is needed.

Soil accumulation is common in swales that drain bare slopes (Fig. 10.6). This soil reduces the capacity of the channel. Large quantities of soil or debris can completely obstruct flows and divert them onto exposed slopes. The debris must be removed if the swale is to function.

The following are the more common swale maintenance problems and their solutions:

Low Point

If the low point is only slightly low, one possible temporary solution is to build up the banks at the low spot. Figure 10.7 shows sand bags used for that purpose. If the dip in the swale is pronounced or the gullying below the swale is serious, the only permanent solution is to tear out the low section and reconstruct it with positive drainage.

Erosion of Unlined Swale

Several solutions are possible. One is to line the channel with an erosion-resistant material such as rock, asphalt, or concrete. Plastic sheeting and fabric are other possible lining materials, but they are far less reliable. Before choosing a lining, estimate the expected peak flow (Chap. 4). If rock is used as a lining, the selection of stone size should be based on the flow calculations (Chap. 7).

Fig. 10.6 Soil accumulation in swale.

Fig. 10.7 Building up swale low point.

Two alternatives to lining an eroding earth swale are check dams and diversions. A series of check dams (sand bags, straw bales, logs, etc.) can be installed in a drainageway to reduce velocities to nonerosive rates. However, check dams must be so constructed as not to reduce the capacity of the swale below that necessary to contain the expected peak flow. The other approach is to reduce flows by installing upstream diversions. When a diversion is installed, two structures carry the flow that was formerly carried by one structure. If further flow reduction is needed, several swales can be constructed in parallel (such as a series of cross-road drains).

Erosion of Lined Swale

Erosion may occur on swales lined with vegetation, fabric, or rock. Erosion of a paved swale may occur if the pavement cracks or the swale is undercut.

Small gaps or holes in grass or rock linings should be patched before they increase in size. The bare spots in a grass lining should be reseeded and covered with a mulch and netting until the grass is established. If erosion continues to occur, a more durable lining may be required. Possible options are:

- Larger stones
- Paving
- Plastic sheeting (temporary only)

An alternative to installing a more durable lining might be to install check dams or an upstream diversion.

If a plastic- or fabric-covered swale is eroding, check for proper overlap of fabric layers (Fig. 7.38). Also, check to see that the plastic or fabric is securely anchored. If the lining material is installed correctly and water is still undercutting the channel, replace the plastic or fabric with a more durable material, such as rock or pavement. Check also for unanticipated sources of the water.

Vehicle Damage

If a swale is located across a road or pathway for construction traffic, it is susceptible to damage. Wheel tracks can cause water to escape from the swale and flow down the road. Vehicles can also disturb the swale lining, particularly if it is rock or grass. If the swale has been damaged by vehicles, determine if it was designed to be crossed by construction traffic. Open swales across roadways should be broad and shallow so vehicles can roll gently across them without being jarred. If the traffic across the swale is heavy, possible solutions are the following:

- Installing a culvert underneath the road (Fig. 7.20)
- Installing an open-top wooden box culvert (Fig. 7.21)
- Rerouting the swale away from the traffic area
- Paving the swale
- Enlarging the swale so that vehicles do not create a problem

Sediment in Channel

Remove soil and debris from the channel and place it where it will not reenter the channel, a storm drain, or another water body. Soil placed directly above a channel will just wash back into the channel with the next storm. If a swale drains an exposed slope and continues to fill with sediment, additional soil protection may be required. Seeding and straw mulching the exposed soils will substantially reduce sedimentation in the channel (Chap. 6).

Erosion Along Uphill Edge of Swale

If a paved swale (usually concrete) is so installed that its uphill edge is higher than the ground surface (Fig. 10.8), drainage from the slope above cannot enter the channel. The water will flow along the upper side of the swale and carve a gully. Unfortunately, this problem is difficult to remedy. One option is to fill and compact the upslope hill to the level of the upper edge of the swale, but the fill may wash out in subsequent storms. Another option is to construct small check

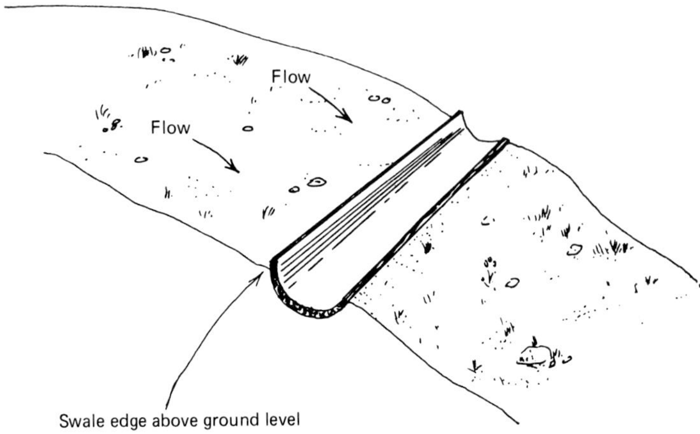

Fig. 10.8 Incorrectly installed concrete swale.

Fig. 10.9 Earth berm diverting gully flow into paved gutter.

Fig. 10.10 Clogged inlet of pipe slope drain (left) causes flow to bypass pipe and erode slope (right).

dams with sandbags or other materials to force the upslope runoff up into the swale (Fig. 10.9). Of course, the best solution is to construct the swale properly in the first place.

EROSION AT OUTLET

If erosion is occurring at the swale outlet, identify the cause. Is the swale lining continuous to the outlet? If it is not, it should be made so. Also see if outlet protection has been installed. If not, install it as described in Chap. 7. If outlet protection is present but not working, see Sec. 10.2e.

10.2c Pipe Slope Drains and Chutes

The most critical points for pipe slope drains are the inlet and the outlet. A clogged or incorrectly constructed inlet will cause run-off to bypass the pipe and erode the slope. If the inlet is installed too high, water will not be able to get into it. Check the inlets and outlets of pipes and chutes for debris or sediment accumulation (Figs. 10.10 and 10.11). Remove obstructions. In the photograph at the left in Fig. 10.10, sediment deposited during a very heavy storm completely obstructed a pipe inlet. Runoff was diverted over the headwall, flowed alongside the pipe, and caused serious erosion on the slope below (Fig. 10.10, right).

Check to see if inflow is undercutting or going around the entry point of the pipe or chute. A cutoff wall at least 6 in (15 cm) deep should be present. If erosion has occurred around the entrance, the headwall should be reinforced with sand-

Fig. 10.11 Clogged pipe outlet.

bags or compacted earth so that it is flush with the pipe entrance. If the pipe entrance protrudes uphill from the headwall, the efficiency of flow into the pipe will be greatly reduced.

Check the outlet of the pipe or chute for erosion. Has outlet protection been installed? If erosion is occurring and no outlet protection is present, install one of the outlet protection measures described in Chap. 7. If outlet protection has been installed but is not working, see Sec. 10.2e.

Check pipes and chutes for cracks or leaks. If a pipe is leaking at a joint, check to see if the connections are secure. If a paved chute is cracked, determine the cause. The chute may be undermined from erosion below. Divert any runoff that is undercutting the structure, refill the base with compacted soil or other suitable materials, and repair the crack.

The concrete chute shown in Fig. 10.12 was the spillway for a sediment basin. An early inspection revealed small cracks near the top of the chute. After further examination, it was discovered that runoff from the hill to the left of the chute was undermining the base of the spillway. The problem was not corrected, and the entire spillway eventually collapsed. Check pipe inlets and outlets for constrictions. Check also if pipes are securely anchored to the slope.

10.2d Permanent Waterways

Maintenance of permanent waterways is similar to maintenance of swales. The primary differences between swales and waterways are that waterways normally

Fig. 10.12 Undercutting causing spillway collapse.

are engineered for a design storm, are designed to function permanently, and are usually lined. Because permanent waterways are more carefully designed and constructed, they are less prone to some of the maintenance problems of swales. However, most of the maintenance guidelines for swales do apply to waterways. The following guidelines highlight the maintenance checks for waterways. They should be used in combination with the maintenance checks for swales.

Grass-Lined Channels

During the initial grass establishment period, inspect the channel frequently. Reseed and mulch bare spots immediately and anchor with netting. After establishment, check the channel during or after major storms to determine if grass is holding. If damage is minor, reseed bare areas. If erosion is severe, a more durable lining is probably needed. An effective alternative is to line the flow area with rock and start the grass at the high-water mark. This will provide erosion protection while maintaining the appearance of a natural waterway.

If the grass lining requires it, mow at regular intervals so that the grass does not impede channel flow. Be careful not to damage the grass during mowing operations.

Riprap-Lined Channels

A properly installed riprap lining should require very little maintenance. Check channels after major storms to see if stones have been dislodged or if scour has occurred beneath the riprap. If stones have been dislodged, larger stones are required. If scour has occurred, see if the proper type of filter cloth or a graded layer of sand and gravel was placed beneath the riprap. If filter cloth was used, check to see if the ends of the fabric are buried so water cannot flow underneath the fabric.

Concrete-Lined Channels

Check concrete-lined channels periodically to ensure that there is no undermining of the channel (Fig. 10.13). If undermining is occurring, first determine the cause and then divert or otherwise control the flow that is undercutting the channel. Backfill, seed, and mulch the eroded area.

Check channel outlets for scour. If scour is occurring, install appropriate outlet protection measures (Chap. 7). If outlet protection is already present and erosion is occurring, see Sec. 10.2e.

Natural Channels

Streambanks are always vulnerable to damage, particularly in natural channels below developed areas. Repairs may be needed periodically . Check banks after every high-water event. Fill gaps in the vegetative cover at once by adding plants and mulching. Fresh cuttings from other plants on the banks can be used, or

Fig. 10.13 Gully undermining paved channel.

cuttings can be taken from mother-stock plantings. If the vegetation will not hold, riprap or other more durable linings may be necessary to prevent bank erosion.

10.2e Outlet Protection (Energy Dissipators)

Erosion at outlets is common, even when energy dissipators or other outlet protection measures have been installed. Check the outlet points of dikes, swales, waterways, pipe slope drains, chutes, sediment basins and traps, storm drains, and channels to make sure they are not eroding. These inspections should be made both before and after storms. Sometimes a light storm will not cause a problem but a later storm will. Reinspect all measures during or after every major storm.

Erosion may occur either at or below an outlet. A common mistake is to install an energy dissipator on too steep a slope (Figs. 7.41 and 7.42). After the water leaves the dissipator, it rapidly picks up momentum again until it begins to erode the channel. If erosion is occurring on the slope below an energy dissipator, relocate the dissipator or construct a new one at the bottom of the slope (where the slope is under 5 percent). Line the eroding portion of the channel with a suitable material or install a pipe.

If a pipe or swale with outlet protection discharges to a natural channel, check if erosion is occurring downstream. Development in the watershed may have increased runoff volumes so that the creek at the bottom is no longer stable, even though energy dissipators have been installed on the tributaries. If downstream erosion is occurring, verify the drainage area and runoff calculations. The receiving channel may have to be lined for some distance downstream of the tributary drainage (Chap. 7).

Riprap Aprons

See if stones have been dislodged or if scour has occurred beneath the riprap layer. If stones have been displaced by flowing water, replace them with larger stones. If scour has occurred, see if filter cloth or a properly graded layer of sand and gravel has been placed beneath the riprap apron.

See if erosion is occurring at the edges of the apron. If it is, enlarge the apron or raise its edges. In addition, the receiving channel may need to be lined for some distance downstream of the tributary drainage.

10.3 MAINTENANCE OF SEDIMENT RETENTION STRUCTURES

10.3a Sediment Basins and Traps

Sediment basins and traps must be cleaned out when the storage zone is full. Remove the sediment from a trap when it is within 1 ft (0.3 m) of the outlet elevation. Remove the sediment from a basin when it reaches the top of the storage zone. This point should be specified in the erosion control plan for the project as a specific distance below the top of the riser. The best way to denote the proper clean-out level in the field is to paint a line on the riser. *In no case should sediment accumulation in a basin be closer than 2 ft (0.6 m) from the top of the riser.*

Fig. 10.14 Poorly placed dredge spoils washing into gutter.

Inspect sediment basins and traps after each significant rainfall. Check both the principal and emergency spillways to see that they are unclogged and not eroding. If erosion is occurring at the outlet, see Sec. 10.3e.

Inspect the embankments to ensure that they are structurally sound and not damaged by erosion or construction equipment. Rebuild low areas in earth embankments and compact the soil. The elevation of the outlet or principal spillway should be at least 1 ft (0.3 m) below the top of the embankment.

When removing sediment from a basin or trap, place the sediment where it will not enter a storm drain or watercourse and where it will not immediately reenter the basin. Figure 10.14 shows a poorly placed pile of soil washing into a gutter just above a storm drain.

See if sediment basins are draining between storms. The settling zone should be completely drained. If it is not, the dewatering system or the riser outlet may be clogged. It is not necessary to dewater the stored sediment, but doing so will facilitate sediment removal. The only practical way to dewater the sediment is with a subsurface drain.

Remember that the drainage area of a sediment trap should not exceed 5 acres (2 ha). Wheel tracks, earth moving, or erosion can make changes in the drainage pattern that will cause runoff to bypass a sediment trap (Fig. 10.1) or too much runoff to enter a trap. Be sure that drainage boundaries are the same as those shown on the erosion and sediment control plan.

10.3b Straw Bale Dikes

Straw bale dikes have very short useful lives and are prone to breakdown. Common problems with this control measure are undercutting of bales (Fig. 10.15), end runs (Fig. 8.37), gaps between bales, bale disintegration, and bale dislodgement (Fig. 8.38).

Inspect straw bale dikes during or immediately after each rainfall. A routine end-of-day check is also advisable, particularly during prolonged rainfall. See if bales are properly trenched, staked, and abutted as described in Chap. 8. The first stake driven into each bale should be angled toward the adjacent bale to force the bales together (Fig. 8.35). If an end run is occurring, the barrier should be relocated or extra bales should be added on the end. See if bales have been moved from their proper position by construction vehicles or by flowing water. Check the condition of individual bales. The baling wire should be intact. Repair or replace damaged bales promptly.

Remove sediment deposits when the level of deposition reaches approximately one-half the height of the barrier. Place the sediment where it will not enter a watercourse nor immediately refill the area behind the barrier. After the straw bale dike is removed, dress the sediment deposits to conform to the surrounding grade and then seed them.

Remember that the drainage area to a straw bale dike should be less than 1 acre (0.4 ha), the slope length behind the barrier should be less than 100 ft (30 m), and the slope gradient should be less than 2:1.

Fig. 10.15 Straw bales undercut by runoff in swale.

10.3c Silt Fences

Silt fences have performance characteristics similar to those of straw bale dikes. Common silt fence problems are undercutting (Fig. 10.16), end runs, holes or tears in filter fabric, and fence collapse (Fig. 8.41).

Undercutting of a silt fence is often due to failure to entrench and backfill the filter fabric. The filter fabric should extend into a 4- by 4-in (10- by 10-cm) trench at the base of the fence (Fig. 8.42). Soil should be compacted on top of the entrenched fabric. Another problem is improper installation of the wire mesh support. Check to see if the wire backing goes completely down into the ground and is securely fastened to the fence posts. If an end run has occurred, reposition or extend the fence.

Fabric bursting or tearing is relatively rare. Holes or tears in fabric may be due to vandalism. However, some early versions of filter fabric were subject to disintegration in the presence of sunlight. When repairing torn fabric, it is better to replace the entire length of fence with a continuous piece of fabric to avoid having joints. If replacement is not feasible, splice new pieces of fabric only between support posts, with a minimum of 6 in (15 cm) overlap, and fasten both ends securely to the posts.

Fence collapse may be due to any of the following:

Fig. 10.16 Undercut silt fence.

- Excessive sediment accumulation (Fig. 8.41)
- Lack of wire mesh support
- Wire mesh too light or improperly secured
- Fence posts too small
- Fence posts too far apart
- Fence posts not driven deep enough into ground
- Fence too high
- Fence below slope that is too long and steep
- Fence below drainage area that is too large

Inspect silt fences during or immediately after prolonged rainfall. Remove sediment deposits before they reach a depth of 1.5 ft (0.5 m). See if wire mesh support is at least 14 gauge and has a maximum mesh spacing of 6 in (15 cm). Chicken wire is not acceptable. If no mesh is present, the fabric must be extra strength and the posts no more than 6 ft (1.8 m) apart.

Check to see if the fence posts are the proper size [4-in-(10-cm) diameter wood or 1.33 lb/ft (1.98 kg/m) steel] and are driven at least 12 in (30 cm) into the ground. Silt fence failure is often due to fence posts collapsing in rain-softened ground. The posts should be no more than 10 ft (3 m) apart, and the *fence should be no more than 3 ft (0.9 m) high.* If a fence is built too high, excessive weights of water and sediment will build up behind it and cause its collapse.

When a silt fence is removed, dress any sediment deposits to conform to the existing grade and then seed them.

Remember that the drainage area to a silt fence should be less than 1 acre (0.4 ha), the slope length behind the fence should be less than 100 ft (30 m), and the slope gradient should be less than 2:1.

10.3d Check Dams

Because check dams slow the velocity of flowing water, sediment accumulates behind them. Inspect check dams after each rainfall and at least daily during prolonged rainfall. Remove sediment when its level is one-half the dam height. Deposit the sediment where it will not enter a storm drain or watercourse.

See if the dam center is at least 6 in (15 cm) lower than its edges (Fig. 7.26). If erosion is occurring around the edges of the check dam, extend the dam length.

Look for dislodgement of check dam materials—sandbags, stones, or straw bales. Straw bale check dams should be trenched, abutted, and staked as described in Chap. 8. Reanchor or reposition displaced materials. If displacement recurs, a more stable control measure, such as a riprap channel lining, may be required. Other alternatives may be heavier stones (for stone check dams) or diversion of flow upstream. Remember that the drainage area for a straw bale check dam should not exceed 2 acres (0.8 ha) and that no check dam should be placed across a perennial stream.

10.3e Inlet Protection

Commonly used forms of inlet protection are straw bales, silt fences, sandbag check dams, and gravel filters. With the exception of gravel filters, each of these structures should be maintained according to the guidelines given in their respective sections of this chapter.

See if straw bales placed around inlets are entrenched and tightly abutted end-to-end, not corner-to-corner (Figs. 8.36, 8.43, and 8.45). Also check to see if straw bales are staked down, even on asphalt.

When a gravel filter is used around an inlet (Figs. 8.9 and 8.49 to 8.51), check during storms to see if runoff can enter the inlet. If the gravel becomes clogged with sediment, remove it from the inlet, wash it off in a protected area, and then reposition it around the inlet. Make sure the wash water does not enter a stream, lake, or storm drain. If washing is impractical, replace the gravel with fresh rock and use the older material as fill.

Sometimes a sediment barrier placed around an inlet causes runoff to bypass the inlet. Check all protected inlets during storms to see if water is entering the storm drains. If water is not able to enter the inlet, it may cause erosion below or cause ponding that interferes with construction. One solution to this problem is to replace the inlet protection structure with a sediment trap excavated around the inlet so the inlet serves as the trap outlet (Figs. 8.5 and 8.31).

10.4 PLANNING FOR EMERGENCIES

10.4a Communications

An important aspect of erosion and sediment control is maintaining good communications. The developer or contractor can be spared considerable expense if he can be reached immediately if a potential control measure failure is observed by an inspector or an interested citizen. Since many construction sites are not manned during wet weather, good communications may mean providing the local authority with the home phone number of a site foreman or company official. It may be far cheaper for the developer to be called out some Sunday morning to repair an overtopping sediment basin than to come to the site on Monday and be presented with a lawsuit by flooded adjacent homeowners.

10.4b Stockpiled Supplies

To prepare for unexpected conditions in the planned performance of the erosion and sediment control measures, materials should be stockpiled at various locations on the construction site. At least one piece of equipment such as a backhoe, loader, or bulldozer should be left on the site to transport materials for repairing control measures during muddy conditions. It is important to have the supplies and equipment at the site before the first heavy rains. The first major storm will test the effectiveness of the control measures and identify potential problems. Weakened or damaged structural measures should be repaired as soon as feasible after the storm, since once the ground is wet, a larger proportion of additional rain will form runoff. The following are a few of the types of supplies that could be of value in correcting erosion problems in an emergency.

Sand and Bags

Place piles of sand at strategic locations on the site, especially where there are a number of water conveyance structures. Since site access roads are usually unpaved, it is important to have the sand delivered before rain softens the ground. Leave burlap or other bags at the sand pile or keep them in a materials trailer.

Filter Fabric and Fence Posts

If the erosion control plan specifies a number of silt fences, stockpile the spare filter fabric, wire mesh, staples, and posts that would be necessary to repair any damage. The need for additional fences may become apparent after the first storms.

Straw Bales and Stakes

If the erosion control plan calls for straw bale dikes, stockpile extra straw bales and stakes. Keep the spare bales covered with plastic sheeting to prolong their life while in storage.

Channel-Lining Materials

A number of channel-lining materials, such as jute netting, burlap, filter fabric, and crushed rock, can be kept on the project site to remedy excessive spillway or channel erosion. All can be used quickly to protect an earthen spillway which is handling more than its design flow or to reinforce a dike or swale that is eroding. With the fabric lining materials, stock a supply of 6- to 12-in (15- to 30-cm) U-shaped staples.

REVIEW QUESTIONS

1. What are the most common failure points of straw bale dikes?
2. What is the first thing to look for if you see a gully forming on a newly seeded slope?
3. List three ways to correct erosion of an unlined swale.
4. When should a sediment basin be cleaned out? When should a sediment trap be cleaned out?
5. What are the possible causes of erosion at an outlet protected by an energy dissipator or riprap? Suggest ways to correct these problems.
6. How should a tear in a silt fence be repaired?
7. How often should control measures be inspected? Which measures should be inspected most frequently? Why?
8. List five common problems with dikes and swales.
9. What are the most likely causes of undercutting of a silt fence or straw bale dike? How can this problem be corrected?
10. Discuss the reasons for stockpiling erosion control supplies at a construction site.

REFERENCES

1. Association of Bay Area Governments, *Manual of Standards for Erosion and Sediment Control Measures,* Oakland, Calif., 1981.
2. Fairfax County, Virginia, *Construction Superintendent's Check List for Erosion and Sediment Control,* Fairfax County, 1978.
3. Virginia Soil and Water Conservation Commission, *Field Manual,* Richmond, Va., 1982.

appendix A

A Model Grading and Erosion and Sediment Control Ordinance

The model ordinance printed in this appendix integrates comprehensive erosion and sediment control provisions into a commonly used grading ordinance: Chapter 70 of the Uniform Building Code (UBC). Portions of this model ordinance were reproduced from Chapter 70 of the 1979 edition of the UBC with permission of the publisher, the International Conference of Building Officials.* Some modification of Chapter 70 was necessary to fit the format of this model.

Article I

Title, Purpose and General Provisions

1 Title. This ordinance shall be known as the "[City/Town/County of ____________] Grading and Erosion and Sediment Control Ordinance" and may be so cited.

*In the text of the model, passages or blanks enclosed in brackets are intended to be filled in by local jurisdictions. Provisions noted as "optional" may be deleted. Substantive changes to other parts of the model may undermine the effectiveness of the erosion and sediment control program.

2 Purpose. The purpose of this ordinance is to provide for safe grading operations, to safeguard life, limb and property, and to preserve and enhance the natural environment, including but not limited to water quality, by regulating clearing and grading on private property.

3 Scope. This chapter sets forth rules and regulations to control land disturbances, land fill, soil storage, and erosion and sedimentation resulting from such activities. This chapter establishes procedures for issuance, administration and enforcement of a permit.

4 Definitions. When used in this chapter, the following words shall have the meanings ascribed to them in this Section:

(a) Applicant: any person, corporation, partnership, association of any type, public agency or any other legal entity who submits an application to the Director for a permit pursuant to this Chapter.

(b) As-Graded: the surface conditions extant on completion of grading.

(c) Bedrock: in-place solid rock.

(d) Bench: a relatively level step excavated into earth material on which fill is to be placed.

(e) Best Management Practices: a technique or series of techniques which, when used in an erosion control plan, is proven to be effective in controlling construction-related runoff, erosion and sedimentation.

(f) Borrow: earth material acquired from an off-site location for use in grading on a site.

(g) Building Official: [the Public Works Director of the City/Town/County of ________, or the person responsible for the administrative and operational control of grading activities in the City/Town/County of ________] and his/her duly authorized designees.

(h) Chapter: this ordinance in its entirety.

(i) Civil Engineer: a professional engineer registered in the State of California to practice in the field of civil works.

(j) Civil Engineering: the application of the knowledge of the forces of nature, principles of mechanics and the properties of materials to the evaluation, design and construction of civil works for the beneficial uses of mankind.

(k) Compaction: the densification of a fill by mechanical means.

(l) Drainageway: a natural or manmade channel which collects and intermittently or continuously conveys stormwater runoff.

(m) Earth Material: any rock, natural soil or fill and/or combination thereof.

(n) Engineering Geologist: a geologist experienced and knowledgeable in engineering geology and certified by the State of California to practice engineering geology.

(o) Engineering Geology: the application of geologic knowledge and principles in the investigation and evaluation of naturally occurring rock and soil for use in the design of civil works.

(p) Erosion: the wearing away of the ground surface as a result of the movement of wind, water and/or ice.

(q) Final Erosion and Sediment Control Plan (Final Plan): a set of best management practices or equivalent measures designed to control surface runoff and erosion and to retain sediment on a particular site after all other planned final structures and permanent improvements have been erected or installed.

(r) Grade: the vertical location of the ground surface.
Existing Grade - the grade prior to grading.
Rough Grade - the stage at which the grade approximately conforms to the approved plan.
Finish Grade - the final grade of the site which conforms to the approved plan.

(s) Grading: any land disturbance or land fill, or combination thereof.

(t) Interim Erosion and Sediment Control Plan (Interim Plan): a set of best management practices or equivalent measures designed to control surface runoff and erosion and to retain sediment on a particular site during the period in which pre-construction and construction-related land disturbances, fills and soil storage occur, and before final improvements are completed.

(u) Key: a designed compacted fill placed in a trench excavated in earth material beneath the toe of a proposed fill slope.

(v) Land Disturbance/Land-Disturbing Activities: any moving or removing by manual or mechanical means of the soil mantle or top 6 inches of soil whichever is shallower, including but not limited to excavations.

(w) Land Fill: any human activity depositing soil or other earth materials.

(x) Manual of Standards: a compilation of technical standards and design specifications adopted by the building official as being proven methods of controlling construction-related surface runoff, erosion and sedimentation.

(y) Permittee: the applicant in whose name a valid permit is duly issued pursuant to this chapter and his/her agents, employees and others acting under his/her direction.

(z) Sediment: earth material deposited by water or wind.

(aa) Site: a parcel or parcels of real property owned by one or more than one person which is being or is capable of being developed as a single project.

(bb) Slope: an inclined ground surface the inclination of which is expressed as a ratio of horizontal distance to vertical distance.

(cc) Soil: naturally occurring superficial deposits overlying bed rock.

(dd) Soil Engineer: a civil engineer experienced and knowledgeable in the practice of soil engineering.

(ee) Soil Engineering: the application of the principles of soil mechanics in the investigation, evaluation and design of civil works involving the use of earth materials and the inspection and testing of the construction thereof.

(ff) Wet Season: the period from October 15 to April 15.

5 Hazards. Whenever the building official determines that any existing excavation or embankment or fill on private property has become a hazard to life and limb, or endangers property, or adversely affects the safety, use or stability of a public way or drainage channel, the owner of the property upon which the excavation or fill is located, or other person or agency in control of said property, upon receipt of notice in writing from the building official, shall within the period specified therein repair or eliminate such excavation or embankment so as to eliminate the hazard and be in conformance with the requirements of this code.

6 Other Laws. Neither this ordinance nor any administrative decision made under it:

(a) Exempts the Permittee from procurring other required permits or complying with the requirements and conditions of such a permit; or

(b) Limits the right of any person to maintain, at any time, any appropriate action, at law or in equity, for relief or damages against the Permittee arising from the permitted activity.

7 Severability and Validity. If any part of this ordinance is found not valid, the remainder of this ordinance shall remain in effect.

8-9 Reserved.

Article II

Permit Application Procedures

10 Scope. No person may grade, fill, excavate, store or dispose of soil and earth materials or perform any other land-disturbing or land-filling activity without first obtaining a Permit as set forth in this chapter.

11 General Exemptions. All land-disturbing or land-filling activities or soil storage shall be undertaken in a manner designed to minimize surface runoff, erosion and sedimentation and to safeguard life, limb, property, and the public welfare. A person performing such activities need not apply for a Permit pursuant to this chapter, if all the following criteria are met:

(a) The site upon which land area is disturbed or filled is 10,000 square feet or less.

(b) Natural and finished slopes are less than 10%.

(c) Volume of soil or earth materials stored is 50 cubic yards or less.

(d) Rainwater runoff is diverted, either during or after construction, from an area smaller than 5,000 square feet.

(e) An impervious surface, if any, of less than 5,000 square feet is created.

(f) No drainageway is blocked or has its stormwater carrying capacities or characteristics modified.

(g) The activity does not take place within 100 feet by horizontal measurement from the top of the bank of a watercourse, the mean high watermark (line of vegetation) of a body of water or within the wetlands associated with a watercourse or water body, whichever distance is greater.

12 Categorical Exemptions. Sections 10 and 11(a)-(f) notwithstanding, the following activities are exempt from the permit requirements:

(a) An excavation below finished grade for basements and footings of a building, retaining wall or other structure authorized by a valid building permit. This shall not exempt any fill made with the material from such excavation nor exempt any excavation having an unsupported height greater than 5 feet after the completion of the structure.

(b) Cemetery graves.

(c) Refuse disposal sites controlled by other regulations.

(d) Excavations for wells or tunnels.

(e) Mining, quarrying, excavating, processing, stockpiling of rock, sand, gravel, aggregate or clay where established and provided for by law, provided such operations do not affect the lateral support or increase the stresses in or pressure upon any adjacent or contiguous property.

(f) Exploratory excavations under the direction of soil engineers or engineering geologists.

(g) Routine agricultural crop management practices.

(h) Emergencies posing an immediate danger to life or property, or substantial flood or fire hazards.

(i) Any activity where total volume of material disturbed, stored, disposed of or used as fill does not exceed 50 cubic yards and which does not obstruct a drainage course.

(j) Sections 10 and 11(a)-(g) notwithstanding, any activity where total volume of material disturbed, stored, disposed of or used as fill does not exceed 5 cubic yards is always exempt from the permit requirements.

13 Reserved.

14 Application. The application for a Permit must include all of the following items:

(a) Application form.

(b) Site Map and Grading Plan.

(c) Interim Erosion and Sediment Control Plan.

(d) Final Erosion and Sediment Control Plan, where required.

(e) Soil Engineering Report, where required.

(f) Engineering Geology Report, where required.

(g) Work schedule.

(h) Application fees.

(i) Performance bond or other acceptable security (see § 22).

(j) Any supplementary material required by the building official.

15 Application Form. The following information is required on the application form:

(a) Name, address and telephone number of the Applicant.

(b) Names, addresses and telephone numbers of any and all contractors, subcontractors or persons actually doing the land-disturbing and land-filling activities and their respective tasks.

(c) Name(s), address(es) and telephone number(s) of the person(s) responsible for the preparation of the Site Map and Grading Plan.

(d) Name(s), address(es) and telephone number(s) of the person(s) responsible for the preparation of the Interim and/or Final Erosion and Sediment Control Plan.

(e) Name(s), address(es) and telephone number(s) of the registered engineer(s) responsible for the preparation of the soil engineering and engineering geology reports, where required.

(f) A vicinity map showing the location of the site in relationship to the surrounding area's watercourses, water bodies and other significant geographic features, and roads and other significant structures.

(g) Date of the application.

(h) Signature(s) of the owner(s) of the site or of an authorized representative.

16 Site Map and Grading Plan (Grading Plan). The Site Map and Grading Plan shall contain all the following information:

(a) Existing and proposed topography of the site taken at a contour interval sufficiently detailed to define the topography over the entire site. Ninety percent (90%) of the contours shall be plotted within one contour interval of the true location.

(b) Two contour intervals that extend a minimum of [100 feet off-site, or sufficient to show on- and off-site drainage].

(c) Site's property lines shown in true location with respect to the plan's topographic information.

(d) Location and graphic representation of all existing and proposed natural and manmade drainage facilities.

(e) Detailed plans of all surface and subsurface drainage devices, walls, cribbing, dams and other protective devices to be constructed with or as a part of the proposed work, together with a map showing the drainage area and the estimated runoff of the area served by any drain.

(f) Location and graphic representation of proposed excavations and fills, of on-site storage of soil and other earth material, and of on-site disposal.

(optional)(g) [Location of existing vegetation types and the] location and type of vegetation to be left undisturbed.

(h) Location of proposed final surface runoff, erosion and sediment control measures.

(optional)(i) Quantity of soil or earth material in tons and cubic yards to be excavated, filled, stored or otherwise utilized on-site.

(optional)(j) Outline of the methods to be used in clearing vegetation, and in storing and disposing of the cleared vegetative matter.

(k) Proposed sequence and schedule of excavation, filling and other land-disturbing and filling activities, and soil or earth material storage and disposal.

(l) Location of any buildings or structures on the property where the work is to be performed and the location of any buildings or structures on land of adjacent owners which are within 15 feet of the property or which may be affected by the proposed grading operations.

Specifications shall contain information covering construction and material requirements.

17 Interim Erosion and Sediment Control Plan (Interim Plan). All the following information shall be provided with respect to conditions existing on the site during land-disturbing or filling activities or soil storage:

(a) Maximum surface runoff from the site shall be calculated using the method approved by the building official and maintained in the Manual of Standards, or any other method proven to the building official to be as or more accurate.

(b) The Interim Plan shall also contain the following information:

(1) a delineation and brief description of the measures to be undertaken to retain sediment on the site, including, but not limited to, the designs and specifications for sediment detention basins and traps, and a schedule for their maintenance and upkeep;

(2) a delineation and brief description of the surface runoff and erosion control measures to be implemented, including, but not limited to, types and method of applying mulches, and designs and specifications for diverters, dikes and drains, and a schedule for their maintenance and upkeep;

(3) a delineation and brief description of the vegetative measures to be used, including, but not limited to, types of seeds and fertilizer and their application rates, the type, location and extent of pre-existing and undisturbed vegetation types, and a schedule for maintenance and upkeep.

(c) The location of all the measures listed by the Applicant under Subsection (b) above, shall be depicted on the Grading Plan, or on a separate plan at the discretion of the building official.

(d) An estimate of the cost of implementing and maintaining all interim erosion and sediment control measures must be submitted in a form acceptable to the building official.

(e) The Applicant may propose the use of any erosion and sediment control techniques in the Interim Plan provided such techniques are proven to be as or more effective than the equivalent best management practices contained in the Manual of Standards.

18 Final Erosion and Sediment Control Plan (Final Plan). All the following information shall be provided with respect to conditions existing on the site after final structures and improvements (except those required under this Section) have been completed and where these final structures have not been covered by an Interim Plan (see § 29):

(a) Maximum runoff from the site shall be calculated using the method approved by the building official and maintained in the Manual of Standards, or any other method proven to the building official to be as or more accurate.

(b) The Final Plan shall also contain the following information:

(1) a description of and specifications for sediment retention devices;

(2) a description of and specifications for surface runoff and erosion control devices;

(3) a description of vegetative measures;

(4) a graphic representation of the location of all items in Subsections (1)-(3) above (see § 16(h));

(c) An estimate of the costs of implementing all final erosion and sediment control measures must be submitted in a form acceptable to the building official.

(d) The Applicant may propose the use of any erosion and sediment control techniques in the Final Plan provided such techniques are proven to be as or more effective than the equivalent best management practices contained in the Manual of Standards.

19 Soil Engineering Report. A soil engineering report, when required by the building official, shall be based on adequate and necessary test borings, and shall contain all the following information:

(a) Data regarding the nature, distribution, strength, and erodibility of existing soils.

(b) Data regarding the nature, distribution, strength and erodibility of soil to be placed on the site, if any.

(c) Conclusions and recommendations for grading procedures.

(d) Conclusions and recommended designs for interim soil stabilization devices and measures and for permanent soil stabilization after construction is completed.

(e) Design criteria for corrective measures when necessary.

(f) Opinions and recommendations covering adequacy of sites to be developed by the proposed grading.

Recommendations included in the report and approved by the building official shall be incorporated in the grading plans or specifications.

20 Engineering Geology Report. An engineering geology report, when required by the building official, shall be based on adequate and necessary test borings and shall contain the following information:

(a) An adequate description of the geology of the site.

(b) Conclusions and recommendations regarding the effect of geologic conditions on the proposed development.

(c) Opinions and recommendations covering the adequacy of sites to be developed by the proposed grading.

Recommendations included in the report and approved by the building official shall be incorporated in the grading plans or specifications.

21 Work Schedule. The Applicant must submit a master work schedule showing the following information:

(a) Proposed grading schedule.

(b) Proposed conditions of the site on each July 15, August 15, September 15, October 1 and October 15 during which the Permit is in effect.

(c) Proposed schedule for installation of all interim erosion and sediment control measures including, but not limited to, the stage of completion of erosion and sediment control devices and vegetative measures on each of the dates set forth in Subsection (b).

(d) Schedule for construction of final improvements, if any.

(e) Schedule for installation of permanent erosion and sediment control devices where required.

22 Security.

(a) The Applicant shall provide security for the performance of the work described and delineated on the approved Grading Plan in an amount to be set by the building official. The form of security shall be one or a combination of the following to be determined by the building official.

(1) bond or bonds issued by one or more duly authorized corporate sureties. The form of the bond or bonds shall be subject to the approval of the City Attorney;

(2) deposit, either with the City or a responsible escrow agent or trust company at the option of the City, of money, negotiable bonds of the kind approved for securing deposits of public monies, or other instrument of credit from one or more financial institutions subject to regulation by the State or Federal government wherein said financial institution pledges funds are on deposit and guaranteed for payment;

(3) cash in U.S. currency.

(b) The Applicant shall provide security for the performance of the work described and delineated in the Interim Plan in an amount to be determined by the building official but not less than 100% of the approved estimated cost of performing said work. The form of the security shall be as set forth in Subsections (a)(2) and (3).

(c) The Applicant shall provide security for the performance of the work described and delineated in the Final Plan in an amount to be determined by the building official but not less than 100% of the approved estimated cost of performing said work. The form of the security shall be as set forth in Subsections (a)(2) and (3).

23 Fees. Fees are to be paid pursuant to a schedule of fees adopted, and amended from time to time by separate resolution of the [Council/Supervisors].

24 Decision on a Permit. The building official shall review all documents submitted pursuant to this chapter and, if necessary, request additional data, clarification of submitted data or correction of defective submissions within 10 working days after the date of submission. The building official shall notify Applicant of his/her decision on the Permit within 20 working days of the initial submission or of the corrected submissions, whichever is later.

25 Notice. Applicant shall be notified of building official's decision on the application within 3 working days of the decision.

26 Permit Duration. Permits issued under this chapter shall be valid for the period during which the proposed land-disturbing or filling activities and soil storage takes place or is scheduled to take place, whichever is shorter. Permittee shall commence permitted activities within 60 days of the scheduled commencement date for grading or the Permittee shall resubmit all required application forms, maps, plans, schedules and security to the building official except where an item to be resubmitted is waived by the building official. The building official may require additional fees.

27 Permit Denial. The Applicant may request a hearing before the City Council within 5 working days of notification of a permit denial. The hearing shall be held at the next regularly scheduled City Council meeting following the date of the request for a hearing.

28 Assignment of Permit. A Permit issued pursuant to this chapter may be assigned, provided:

(a) The Permittee notifies the building official of the proposed assignment.

(b) The proposed assignee:

(1) submits an application form pursuant to Section 15; and

(2) agrees in writing to all the conditions and duties imposed by the Permit; and

(3) agrees in writing to assume responsibility for all work performed prior to the assignment; and

(4) provides security pursuant to Section 22; and

(5) agrees to pay all applicable fees.

(c) The building official approves the assignment.

The building official shall set forth in writing the reasons for his/her approval or disapproval of an assignment.

29 No Improvements Planned. Where an Applicant does not plan to construct permanent improvements on the site, or plans to leave portions of the site graded but unimproved, Applicant must:

(a) Meet all the requirements of this chapter except that an Interim Plan designed to control runoff and erosion on the site for the period of time during which the site, or portions thereof, remain unimproved must be submitted in lieu of a Final Plan; and

(b) Submit executed contract(s) as defined in Section 33(a) after completion of grading.

Article III

Implementation and Enforcement

30 Issuance of Permits. Building official shall issue a Permit upon approval of a Grading Plan, Interim Plan, and where required, Final Plan, soil engineering report, and engineering geology report, deposit of appropriate security and payment of fees. Permit shall be issued subject to the following conditions:

(a) The Permittee shall maintain a copy of the Permit, approved plans and reports required under Section 31 on the work site and available for public inspection during all working hours.

(b) The Permittee shall, at all times, be in conformity with approved Grading Plan, Interim and Final Plans.

31 Implementation of Permits--Permittee's Duties. In addition to performing as required under Section 30:

(a) Unless this requirement is waived by the building official, Permittee shall notify the building official within 72 hours of:

(1) the beginning of the permitted activity;

(2) the completion of rough grading;

(3) the completion of finished grading;

(4) the installation of all erosion control devices and the completion of planting requirements;

(5) readiness of the site for final inspection, including, but not limited to, finished grading, installation of drainage devices and final erosion control measures.

(b) Permittee shall submit to the building official, reports if:

(1) there are delays in obtaining materials, machinery, services or manpower necessary to the implementation of the Grading, Interim or Final Plans as scheduled;

(2) there are any delays in land-disturbing or filling activities or soil storage;

(3) the work is not being done in conformance with the approved Grading, Interim or Final Plans;

(4) there are any departures from the approved Grading Plan which may affect implementation of the Interim or Final Plans as scheduled;

(5) there are any delays in the implementation of the Interim or Final Plans;

(6) there are any other departures from implementation of the Interim or Final Plans;

(c) Unless this requirement is waived by the building official, Permittee shall submit recommendations for corrective measures, if necessary and appropriate, with the reports made under Subsection (b).

32 Implementation of Permits.

(a) The building official shall review all reports submitted by Permittee. The building official may require Permittee to modify the Grading Plan, Interim or Final Plans, and maintenance methods and schedules. The building official shall notify the Permittee in writing of the requirement and specify a reasonable period of time within which Permittee must comply. All modifications are subject to building official's approval.

(b) The building official may inspect the site:

(1) upon receipt of a report by Permittee under provisions of Section 31(a) and (b);

(2) to verify completion of modifications required under Section 32(a);

(3) during and following any rainfall;

(4) at any other time, at the building official's discretion.

(c) Upon completion of the rough grading work and at the final completion of the work, the building official may require the following reports and drawings and supplements thereto:

(1) an as-graded grading plan prepared by the civil engineer including original ground surface elevations, as-graded ground surface elevations, lot drainage patterns and locations and elevations of all surface and subsurface drainage facilities. He/she shall provide approval that the work was done in accordance with the final approved Grading Plan.

(2) a soil grading report prepared by the soil engineer including locations and elevations of field density tests, summaries of field and laboratory tests and other substantiating data and comments on any changes made during grading and their effect on the recommendations made in the soil engineering investigation report. He/she shall provide approval as to the adequacy of the site for the intended use.

(3) a geologic grading report prepared by the engineering geologist including a final description of the geology of the site including any new information disclosed during the grading and the effect of same on recommendations incorporated in the approved Grading Plan. He/she shall provide approval as to the adequacy of the site for the intended use as affected by geologic factors.

33 Post-Grading Procedures: Upon completion of final grading and permanent improvements, where such permanent improvements are planned at the time grading is performed, Permittee shall submit:

(a) Executed contract(s) for maintenance and upkeep of Final Plan runoff and erosion control measures for an [x] year(s) period.

[Less desirable alternatives: deed restrictions requiring maintenance; instructions on maintenance provided subsequent owners.]

34 <u>Suspension or Revocation of Permit</u>. The building official shall first have resort to the procedures set forth in this Section before any other enforcement procedure set forth in this Article.

(a) The building official shall suspend the Permit and issue a stop work order, and Permittee shall cease all work on the work site, except work necessary to remedy the cause of the suspension, upon notification of such suspension when:

(1) the building official determines that the permit was issued in error or on the basis of incorrect information supplied, or in violation of any ordinance or regulation or the provisions of this ordinance;

(2) Permittee fails to submit reports when required under Sections 31 and 32(c);

(3) inspection by the building official under Section 32(b) reveals that the work or the work site:

(i) is not in compliance with the conditions set forth in Section 30, or

(ii) is not in conformity with the Grading Plan, Interim or Final Plan as approved or as modified under Section 32(a), or

(iii) is not in compliance with an order to modify under Section 32(a);

(4) Permittee fails to comply with an order to modify within the time limits imposed by the building official (see § 32(a));

(5) Permittee fails to obtain permission for wet season activity under Section 41.

(b) The building official shall revoke the Permit and issue a stop work order, and Permittee shall cease work if Permittee fails or refuses to cease work, as required under Section 34(a) above, after suspension of the Permit and receipt of a stop work order and notification thereof.

(c) The building official shall reinstate a suspended Permit upon Permittee's correction of the cause of the suspension.

(d) The building official shall not reinstate a revoked Permit.

35 <u>Fines and Penalties</u>. Any person, firm, corporation or agency acting as principal agent, employee or otherwise, who fails to

comply with the provisions of this ordinance shall be guilty of a misdemeanor and upon conviction thereof shall be punishable by a fine of not less than One Hundred Dollars ($100.00) and not more than Five Hundred Dollars ($500.00), or by imprisonment in the county jail for not more than 30 days, or by both, for each separate offense. Each day any violation of this chapter shall continue shall constitute a separate offense.

36 Action against the Security. The building official may act against the appropriate security if any of the conditions listed in Subsections (a)-(d) below exists. The building official shall use funds from the appropriate security to finance remedial work undertaken by the city or a private contractor under contract to the city, and to reimburse the city for all direct costs incurred in the process of the remedial work.

(a) The Permittee ceases land-disturbing activities and/or filling and abandons the work site prior to completion of the Grading Plan.

(b) The Permittee fails to conform to the Interim Plan or Final Plan as approved or as modified under Section 32(a) and has had his/her permit revoked under Section 34.

(c) The techniques utilized under the Interim or Final Plan fail within 1 year of installation, or before a Final Plan is implemented for the site or portions of the site, whichever is later.

(d) The building official determines that action by the city is necessary to prevent excessive erosion from occurring on the site.

37 Release of Security. Security deposited with the city for faithful performance of the grading and erosion control work and to finance necessary remedial work shall be released according to the following schedule:

(a) Securities held against the successful completion of the Grading Plan and the Interim Plan, except for Interim Plans described in Section 29, shall be released to the Permittee at the termination of the Permit, provided no action against such security is filed prior to that date.

(b) Securities held against the successful completion of the Final Plan and an Interim Plan described in Section 29 shall be released to the Permittee either 1 year after termination of the Permit or when a Final Plan is submitted for the unimproved site, whichever is later, provided no action against such security has been filed prior to that date.

38 Cumulative Enforcement Procedures. The procedures for enforcement of a Permit, as set forth in this Article, are cumulative and not exclusive.

39 Reserved.

Article IV

Special Circumstances

40 Grading Designation. All grading in excess of 5,000 cubic yards shall be performed in accordance with the approved Grading Plan prepared by a civil engineer, and shall be designated as "engineered grading." Grading involving less than 5,000 cubic yards shall be designated "regular grading" unless the Permittee, with the approval of the building official, chooses to have the grading performed as "engineered grading."

(a) Engineered Grading Requirements. For engineered grading, it shall be the responsibility of the civil engineer who prepares the approved Grading Plan to incorporate all recommendations from the soil engineering and engineering geology reports into the Grading Plan. He/she also shall be responsible for the professional inspection and approval of the grading within his/her area of technical specialty. This responsibility shall include, but need not be limited to, inspection and approval as to the establishment of line, grade and drainage of the development area. The civil engineer shall act as the coordinating agent in the event the need arises for liaison between the other professionals, the contractor and the building official. The civil engineer also shall be responsible for the preparation of revised plans and the submission of as-graded grading plans upon completion of the work. The grading contractor shall submit in a form prescribed by the building official a statement of compliance to said as-built plan.

Soil engineering and engineering geology reports shall be required at the discretion of the building official. During grading all necessary reports, compaction data and soil engineering and engineering geology recommendations shall be submitted to the civil engineer and the building official by the soil engineer and the engineering geologist.

The soil engineer's area of responsibility shall include, but need not be limited to, the professional inspection and approval concerning the preparation of ground to receive fills, testing for required compaction, stability of all finish slopes and the design of buttress fills, where required, incorporating data supplied by the engineering geologist.

The engineering geologist's area of responsibility shall include, but need not be limited to, professional inspection and approval of the adequacy of natural ground for receiving fills and the stability of cut slopes with respect to geological matters and the need for subdrains or other ground water drainage devices. He/she shall report his/her findings to the soil engineer and the civil engineer for engineering analysis.

The building official shall inspect the project as required under Section 32 and at any more frequent intervals necessary to determine that adequate control is being exercised by the professional consultants.

(b) Regular Grading Requirements. The building official may require inspection and testing by an approved testing agency.

The testing agency's responsibility shall include, but need not be limited to, approval concerning the inspection of cleared areas and benches to receive fill, and the compaction of fills.

When the building official has cause to believe that geologic factors may be involved, the grading operation will be required to conform to "engineered grading" requirements.

(c) If, in the course of fulfilling their responsibility under this chapter, the civil engineer, the soil engineer, the engineering geologist or the testing agency finds that the work is not being done in conformance with this chapter or the approved Grading Plans, the discrepancies shall be reported immediately in writing to the person in charge of the grading work and to the building official (see § 31).

(d) If the civil engineer, the soil engineer, the engineering geologist or the testing agency of record is changed during the course of the work, the work shall be stopped until the replacement has agreed to accept the responsibility within the area of their technical competence for approval upon completion of the work.

41 <u>Wet Season Work</u>.

(a) For commencement of land-disturbing or filling activity during the wet season, Applicant shall demonstrate that land disturbance is relatively minor and that erosion and sedimentation can be controlled.

(b) For continuation of land-disturbing or filling activities, other than installation, maintenance or repair of measures in the Interim or Final Plans, during the wet season, Permittee must apply for and receive, every 5 working days, special permission to proceed.

(c) The building official shall grant permission under this section on the basis of weather forecasts, experience and other pertinent factors which indicate the activity may commence or continue without excessive erosion occurring.

(d) Applicant/Permittee's failure to obtain permission for wet season activity shall result in the imposition of suspension/revocation, and action against the security or criminal penalties as described in Sections 34-36.

42-49 Reserved.

Article V

Additional Requirements

50 Cuts.

(a) General. Unless otherwise recommended in the approved soil engineering and/or engineering geology report, cuts shall conform to the provisions of this Section.

(b) Slope. The slope of cut surfaces shall be no steeper than is safe for the intended use. Cut slopes shall be no steeper than two horizontal to one vertical.

(c) Drainage and Terracing. Drainage and terracing shall be provided as required by Section 53.

51 Fills.

(a) General. Unless otherwise recommended in the approved soil engineering report, fills shall conform to the provisions of this Section.

In the absence of an approved soil engineering report, these provisions may be waived for minor fills not intended to support structures.

(b) Fill Location. Fill slopes shall not be constructed on natural slopes steeper than two to one.

(c) Preparation of Ground. The ground surface shall be prepared to receive fill by removing vegetation, noncomplying fill, top-soil and other unsuitable materials, scarifying to provide a bond with the new fill, and, where slopes are steeper than five to one, and the height is greater than 5 feet, by benching into sound bedrock or other competent material as determined by the soil engineer. The bench under the toe of a fill on a slope steeper than five to one shall be at least 10 feet wide. The area beyond the toe of fill shall be sloped for sheet overflow or a paved drain shall be provided. Where fill is to be placed over a cut, the bench under the toe of fill shall be at least 10 feet wide but the cut must be made before placing fill and approved by the soil engineer and engineering geologist as suitable foundation for fill. Unsuitable soil is soil which, in the opinion of the building official or the civil engineer or the soil engineer or the geologist,is not competent to support other soil or fill, to support structures or to satisfactorily perform the other functions for which the soil is intended.

(d) Fill Material. Detrimental amounts of organic material shall not be permitted in fills. Except as permitted by the building official, no rock or similar irreducible material with a maximum dimension greater than 12 inches shall be buried or placed in fills.

EXCEPTION: The building official may permit placement of larger rock when the soil engineer properly devises a method of placement, continuously inspects its placement and approves the fill stability. The following conditions shall also apply:

(i) prior to issuance of the grading permit, potential rock disposal areas shall be delineated on the Grading Plan,

(ii) rock sizes greater than 12 inches in maximum dimension shall be 10 feet or more below grade, measured vertically,

(iii) rocks shall be placed so as to assure filling of all voids with fines.

(e) Compaction. All fills shall be compacted to a minimum of 90% of maximum density as determined by UBC Standard No. 70-1. Field density shall be determined in accordance with UBC Standard No. 70-2 or equivalent as approved by the building official.

(f) Slope. The slope of fill surfaces shall be no steeper than is safe for the intended use. Fill slopes shall be no steeper than two horizontal to one vertical.

(g) Drainage and Terracing. Drainage and terracing shall be provided and the area above fill slopes and the surfaces of terraces shall be graded and paved as required by Section 53.

52 Setbacks.

(a) General. The setbacks and other restrictions specified by this Section are minimum and may be increased by the building official or by the recommendations of a civil engineer, soil engineer or engineering geologist, if necessary for safety and stability, or to prevent damage of adjacent properties from deposition or erosion, or to provide access for slope maintenance and drainage. Retaining walls may be used to reduce the required setbacks when approved by the building official.

(b) Setbacks from Property Lines. The tops of cuts and toes of fill slopes shall be set back from the outer boundaries of the permit area, including slope-right areas and easements, in accordance with Figure B-1 and Table B-1.

FIGURE B-1

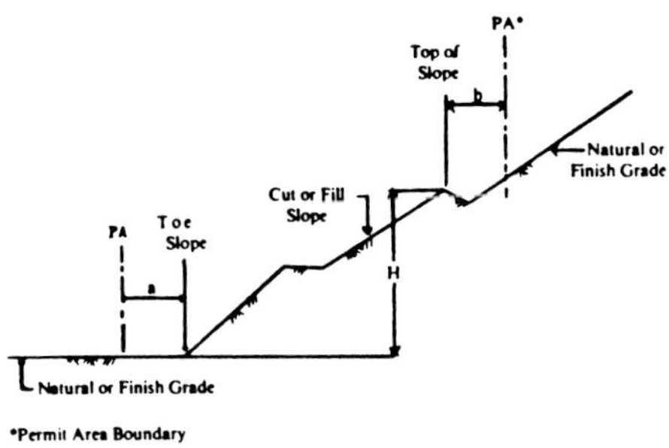

TABLE B-1. REQUIRED SETBACKS FROM PERMIT AREA BOUNDARY (FEET)

	SETBACKS	
H	a	b[1]
Under 5	0	1
5 - 30	H/2	H/5
Over 30	15	6

[1]Additional width may be required for interceptor drain.

(c) Design Standards for Setbacks. Setbacks between graded slopes (cut or fill) and structures shall be provided in accordance with Figure B-2.

FIGURE B-2

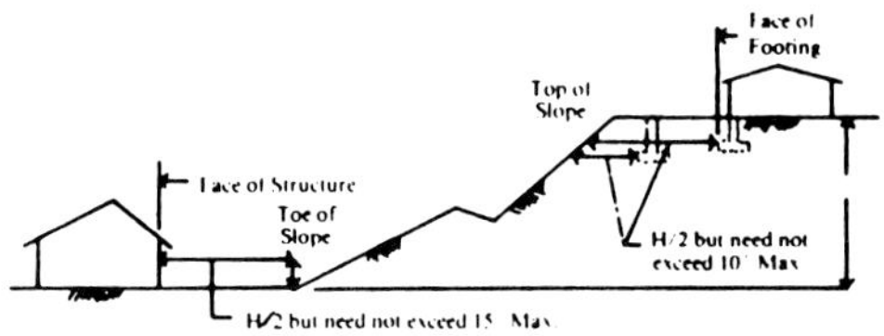

53 Drainage and Terracing.

(a) General. Unless otherwise indicated on the approved Grading Plan, drainage facilities and terracing shall conform to the provision of this Section.

(b) Terrace. Terraces at least 6 feet in width shall be established at not more than 30-foot vertical intervals on all cut or fill slopes to control surface drainage and debris except that where only one terrace is required, it shall be at mid-height. For cut or fill slopes greater than 60 feet and up to 120 feet in vertical height, one terrace at approximately mid-height shall be 12 feet in width. Terrace widths and spacing for cut and fill slopes greater than 120 feet in height shall be designed by the civil engineer and approved by the building official. Suitable access shall be provided to permit proper cleaning and maintenance.

Swales or ditches on terraces shall have a minimum gradient of 5% and must be paved with reinforced concrete not less than 3 inches in thickness or an approved equal paving. They shall have a minimum depth at the deepest point of 1 foot and a minimum paved width of 5 feet.

A single run of swale or ditch shall not collect runoff from a tributary area exceeding 13,500 square feet (projected) without discharging into a downdrain.

(c) Subsurface Drainage. Cut and fill slopes shall be provided with subsurface drainage as necessary for stability.

(d) Disposal. All drainage facilities shall be designed to carry waters to the nearest practicable drainageway approved by the building official and/or other appropriate jurisdiction as a safe place to deposit such waters. Erosion of ground in the area of discharge shall be prevented by installation of nonerosive downdrains or other devices.

Building pads shall have a drainage gradient of 2% toward approved drainage facilities, unless waived by the building official.

EXCEPTION: The gradient from the building pad may be 1% if all the following conditions exist throughout the permit area:

(i) no proposed fills are greater than 10 feet in maximum depth,

(ii) no proposed finish cut or fill slope faces have a vertical height in excess of 10 feet,

(iii) no existing slope faces, which have a slope face steeper than 10 horizontally to 1 vertically, have a vertical height in excess of 10 feet.

(e) Interceptor Drains. Paved interceptor drains shall be installed along the top of all cut slopes where the tributary drainage area above slopes towards the cut and has a drainage path greater than 40 feet measured horizontally. Interceptor drains shall be paved with a minimum of 3 inches of concrete or gunite and reinforced. They shall have a minimum depth of 12 inches and a minimum paved width of 30 inches measured horizontally across the drain. The slope of drain shall be approved by the building official.

appendix B

Grassed Waterway Design Tables

The tables printed in this appendix are used in the design of grass-lined waterways. The design of such waterways is complicated by the fact that Manning's roughness coefficient n changes with the height of the grass. The tables include the effect of grass retardance on flow. The use of these tables is illustrated in Sec. 7.5c.

When you design trapezoidal channels with potential for a B or A retardance (i.e., taller grasses, see Table B.1), or channels with bottom or side slopes other than those covered in these design charts, you will have to apply Manning's equation in a trial-and-error process as follows:

1. By using the continuity equation, determine a cross-sectional area for V_{max}, choose a base width, and solve for depth and hydraulic radius.
2. Use Fig. B.1 to estimate a value for Manning's n at the lowest retardance class and size the channel for V_{max}.
3. Repeat the sizing process at a higher retardance class to calculate channel dimensions that will carry the peak flow when the grass grows tall.

Except for Tables B.1 and B.5, the tables and figures in this appendix are reprinted from U.S. Department of Agriculture, Soil Conservation Service, *Engineering Field Manual for Conservation Practices,* Washington, D.C., 1979.

TABLE B.1 Grass Retardance Classes
Grassed Waterway and Diversion Design Table

Retardance	Cover*	Stand	Condition and height
A	Reed canarygrass	Excellent	Tall; avg. 36 in (91cm)
	Kentucky 31 tall fescue	Excellent	Tall; avg. 36 in (91 cm)
B	Tufcote, midland, and coastal Bermuda grass	Good	Tall; avg. 12 in (30 cm)
	Reed canarygrass	Good	Mowed; avg. 12–15 in (30–38 cm)
	Kentucky 31 tall fescue	Good	Unmowed; avg. 18 in (46 cm)
	Red fescue	Good	Unmowed; avg. 16 in (41 cm)
	Kentucky bluegrass	Good	Unmowed; avg. 16 in (41 cm)
	Redtop	Good	Average; 22 in (56 cm)
C	Kentucky bluegrass	Good	Headed; 6–12 in (15–30 cm)
	Red fescue	Good	Headed; 6–12 in (15–30 cm)
	Tufcote, midland, and coastal Bermuda grass	Good	Mowed; avg. 6 in (15 cm)
	Redtop	Good	Headed; 15–20 in (38–51 cm)
D	Tufcote, midland, and coastal Bermuda grass	Good	Mowed; 2½ in (6 cm)
	Red fescue	Good	Mowed; 2½ in (6 cm)
	Kentucky bluegrass	Good	Mowed; 2–5 in (5–13 cm)

*Classification of vegetal cover in waterways and diversions is based on degree of flow retardance.

Note: Grasses not listed above can be classified according to the height of growth: retardance D, less than 6 in (15 cm); retardance C, 6–10 in (15–25 cm); retardance B, 10–24 in (25–61 cm); retardance A, above 24 in (61 cm).

Source: U.S. Department of Agriculture, Soil Conservation Service, *Standards and Specifications for Erosion and Sediment Control in Developing Areas,* USDA, SCS, College Park, Md., 1975.

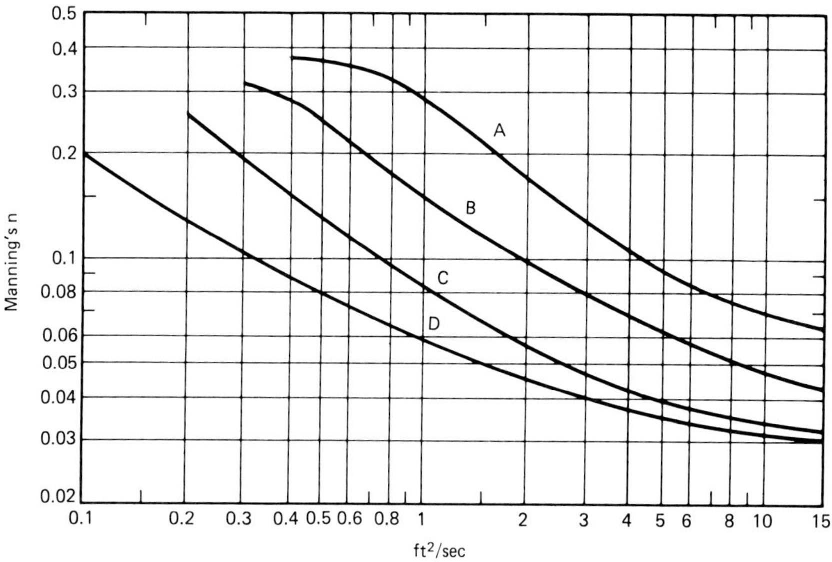

Fig. B.1 Manning's roughness coefficient n as a function of grass retardance class, velocity, and hydraulic radius.

TABLE B.2 Elements of Channel Cross Sections

Section	Area a	Wetted Perimeter p	Hydraulic Radius r	Top Width T
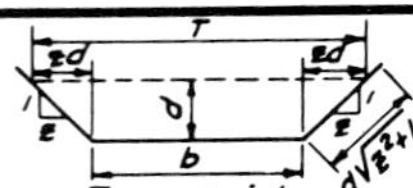 Trapezoid	$bd+zd^2$	$b+2d\sqrt{z^2+1}$	$\dfrac{bd+zd^2}{b+2d\sqrt{z^2+1}}$	$b+2zd$
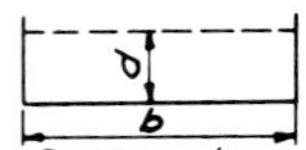 Rectangle	bd	$b+2d$	$\dfrac{bd}{b+2d}$	b
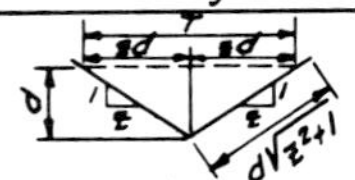 Triangle	zd^2	$2d\sqrt{z^2+1}$	$\dfrac{zd}{2\sqrt{z^2+1}}$	$2zd$
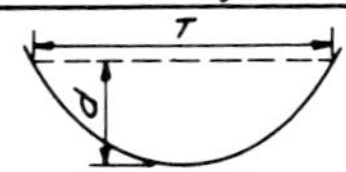 Parabola	$\frac{2}{3}dT$	$T+\dfrac{8d^2}{3T}$ [1]	$\dfrac{2dT^2}{3T^2+8d^2}$ [1]	$\dfrac{3a}{2d}$
 Circle - <½ full [2]	$\dfrac{D^2}{8}\left(\dfrac{\pi\theta}{180}-\sin\theta\right)$	$\dfrac{\pi D\theta}{360}$	$\dfrac{45D}{\pi\theta}\left(\dfrac{\pi\theta}{180}-\sin\theta\right)$	$D\sin\dfrac{\theta}{2}$ or $2\sqrt{d(D-d)}$
 Circle - >½ full [3]	$\dfrac{D^2}{8}\left(2\pi-\dfrac{\pi\theta}{180}+\sin\theta\right)$	$\dfrac{\pi D(360-\theta)}{360}$	$\dfrac{45D}{\pi(360-\theta)}\left(2\pi-\dfrac{\pi\theta}{180}+\sin\theta\right)$	$D\sin\dfrac{\theta}{2}$ or $2\sqrt{d(D-d)}$

[1] Satisfactory approximation for the interval $0 < \frac{d}{T} \leq 0.25$
When $d/T > 0.25$, use $p = \frac{1}{2}\sqrt{16d^2+T^2} + \frac{T^2}{8d}\sinh^{-1}\frac{4d}{T}$

[2] $\theta = 4\sin^{-1}\sqrt{d/D}$
[3] $\theta = 4\cos^{-1}\sqrt{d/D}$
Insert θ in degrees in above equations

TABLE B.3 Parabolic Waterway Design
V_1 for *RETARDANCE "D"*. Top Width (T), Depth (D) and V_2 for *RETARDANCE "B"*.
Grade 0.25 Percent

Q cfs	V_1 = 2.0			V_1 = 2.5			V_1 = 3.0			V_1 = 3.5			V_1 = 4.0			V_1 = 4.5			V_1 = 5.0			V_1 = 5.5			V_1 = 6.0		
	T	D	V_2	T	D	V_2	T	D	V_2	T	D	V_2	T	D	V_2	T	D	V_2	T	D	V_2	T	D	V_2	T	D	V_2
15																											
20																											
25	11.3	3.27	1.00																								
30	13.2	3.09	1.09																								
35	15.2	3.01	1.13																								
40	17.3	2.99	1.15	12.1	3.61	1.36																					
45	19.3	2.94	1.18	13.4	3.49	1.42																					
50	21.4	2.93	1.18	14.7	3.41	1.48																					
55	23.5	2.92	1.19	16.1	3.38	1.50																					
60	25.5	2.89	1.21	17.5	3.35	1.52																					
65	27.6	2.89	1.21	18.8	3.30	1.56																					
70	29.7	2.89	1.21	20.2	3.28	1.57	14.4	3.98	1.81																		
75	31.7	2.87	1.23	21.6	3.27	1.58	15.3	3.91	1.86																		
80	33.8	2.87	1.23	23.0	3.25	1.58	16.3	3.90	1.87																		
90	38.0	2.87	1.23	25.8	3.25	1.60	18.1	3.80	1.94																		
100	42.1	2.85	1.24	28.6	3.23	1.61	20.0	3.76	1.98																		
110	46.3	2.85	1.24	31.4	3.22	1.62	21.9	3.73	2.01																		
120	50.4	2.84	1.25	34.1	3.20	1.64	23.9	3.73	2.00	17.0	4.47	2.34															
130	54.6	2.85	1.24	36.9	3.19	1.64	25.8	3.70	2.02	18.3	4.42	2.39															
140	58.7	2.84	1.25	39.7	3.19	1.65	27.7	3.68	2.04	19.6	4.37	2.43															
150	62.9	2.85	1.25	42.5	3.19	1.65	29.6	3.67	2.06	20.9	4.33	2.47															
160	67.0	2.84	1.25	45.3	3.18	1.65	31.6	3.68	2.05	22.2	4.30	2.50															
170	71.1	2.84	1.26	48.1	3.18	1.65	33.5	3.66	2.07	23.5	4.27	2.53	18.5	4.95	2.76												
180	75.3	2.84	1.25	50.9	3.18	1.66	35.4	3.65	2.08	24.8	4.24	2.55	19.5	4.90	2.80												
190	79.4	2.84	1.26	53.7	3.18	1.66	37.4	3.66	2.07	26.1	4.22	2.57	20.5	4.87	2.84												
200	83.5	2.84	1.26	56.5	3.18	1.66	39.3	3.65	2.08	27.5	4.23	2.56	21.5	4.83	2.87												
220	91.8	2.84	1.26	62.1	3.18	1.66	43.2	3.65	2.08	30.1	4.19	2.60	23.5	4.77	2.92												
240	100.0	2.83	1.26	67.6	3.17	1.67	47.0	3.63	2.10	32.7	4.15	2.64	25.5	4.72	2.97												
260	108.3	2.84	1.26	73.2	3.17	1.67	50.9	3.63	2.10	35.4	4.15	2.64	27.5	4.68	3.01	21.4	5.50	3.29									
280	116.6	2.84	1.26	78.8	3.17	1.67	54.8	3.63	2.10	38.1	4.14	2.64	29.5	4.64	3.05	22.9	5.44	3.35									
300	124.8	2.84	1.26	84.4	3.17	1.67	58.6	3.62	2.11	40.8	4.14	2.65	31.5	4.61	3.08	24.5	5.42	3.37									

TABLE B.3 *(Continued)* Parabolic Waterway Design
V_1 for *RETARDANCE "D"*. Top Width (T), Depth (D) and V_2 for *RETARDANCE "B"*.
Grade 0.50 Percent

Q	V_1 = 2.0			V_1 = 2.5			V_1 = 3.0			V_1 = 3.5			V_1 = 4.0			V_1 = 4.5			V_1 = 5.0			V_1 = 5.5			V_1 = 6.0		
cfs	T	D	V_2	T	D	V_2	T	D	V_2	T	D	V_2	T	D	V_2	T	D	V_2	T	D	V_2	T	D	V_2	T	D	V_2
15	10.2	2.28	0.95																								
20	13.3	2.18	1.02																								
25	16.5	2.15	1.05	10.5	2.60	1.35																					
30	19.7	2.12	1.06	12.4	2.51	1.42	9.5	2.91	1.60																		
35	22.8	2.09	1.09	14.3	2.45	1.48	10.9	2.81	1.69																		
40	26.0	2.08	1.09	16.3	2.44	1.49	12.3	2.74	1.76																		
45	29.2	2.08	1.10	18.2	2.41	1.52	13.7	2.69	1.81																		
50	32.4	2.08	1.10	20.2	2.40	1.53	15.1	2.64	1.86	11.1	3.22	2.07															
55	35.6	2.08	1.11	22.1	2.38	1.55	16.6	2.64	1.86	12.0	3.11	2.18															
60	38.8	2.08	1.11	24.1	2.38	1.55	18.0	2.61	1.90	13.0	3.07	2.23															
65	42.0	2.08	1.11	26.0	2.36	1.57	19.5	2.61	1.89	14.0	3.03	2.27															
70	45.2	2.08	1.11	28.0	2.36	1.57	20.9	2.59	1.92	15.0	3.01	2.31															
75	48.4	2.08	1.11	29.9	2.35	1.59	22.4	2.59	1.92	16.0	2.98	2.34	12.7	3.48	2.52												
80	51.6	2.08	1.11	31.9	2.36	1.58	23.8	2.58	1.94	17.0	2.96	2.36	13.4	3.41	2.60												
90	57.9	2.07	1.12	35.8	2.35	1.59	26.7	2.56	1.96	19.1	2.95	2.37	15.0	3.37	2.64												
100	64.3	2.07	1.11	39.7	2.34	1.60	29.6	2.55	1.97	21.1	2.92	2.42	16.5	3.31	2.72	13.3	3.77	2.96									
110	70.7	2.08	1.11	43.6	2.34	1.61	32.6	2.56	1.96	23.2	2.92	2.42	18.1	3.29	2.75	14.5	3.70	3.05									
120	77.0	2.07	1.12	47.5	2.33	1.61	35.5	2.56	1.97	25.2	2.89	2.45	19.6	3.24	2.81	15.7	3.64	3.12									
130	83.4	2.08	1.12	51.5	2.34	1.61	38.4	2.55	1.98	27.3	2.90	2.45	21.2	3.23	2.82	16.9	3.60	3.18									
140	89.7	2.08	1.12	55.4	2.34	1.61	41.3	2.55	1.98	29.3	2.88	2.47	22.8	3.22	2.84	18.1	3.56	3.23									
150	96.0	2.08	1.12	59.3	2.34	1.61	44.2	2.55	1.99	31.4	2.88	2.47	24.3	3.19	2.88	19.4	3.55	3.23	15.5	4.09	3.52						
160	102.3	2.08	1.12	63.2	2.34	1.61	47.1	2.54	1.99	33.5	2.89	2.47	25.9	3.19	2.88	20.6	3.53	3.27	16.4	4.02	3.60						
170	108.6	2.08	1.12	67.1	2.34	1.62	50.0	2.54	1.99	35.5	2.87	2.48	27.5	3.19	2.89	21.9	3.54	3.27	17.4	4.01	3.62						
180	114.9	2.08	1.12	70.9	2.33	1.62	52.9	2.54	1.99	37.6	2.88	2.48	29.1	3.19	2.89	23.1	3.51	3.30	18.3	3.96	3.69						
190	121.2	2.08	1.12	74.8	2.33	1.62	55.8	2.54	2.00	39.6	2.87	2.49	30.6	3.16	2.92	24.3	3.49	3.34	19.3	3.96	3.70						
200	127.4	2.08	1.13	78.7	2.33	1.62	58.7	2.54	2.00	41.7	2.87	2.49	32.2	3.16	2.93	25.6	3.50	3.33	20.2	3.91	3.77	16.8	4.48	3.96			
220	140.0	2.08	1.13	86.5	2.33	1.62	64.5	2.54	2.00	45.8	2.87	2.50	35.4	3.16	2.93	28.1	3.49	3.35	22.1	3.87	3.83	18.3	4.39	4.08			
240	152.6	2.08	1.13	94.3	2.33	1.63	70.3	2.54	2.01	49.9	2.86	2.51	38.6	3.16	2.93	30.6	3.48	3.36	24.1	3.87	3.84	19.9	4.36	4.12			
260	165.2	2.08	1.13	102.1	2.33	1.63	76.1	2.54	2.01	54.0	2.86	2.51	41.7	3.15	2.95	33.1	3.47	3.38	26.0	3.84	3.88	21.5	4.34	4.15			
280	177.7	2.08	1.13	109.8	2.33	1.63	81.9	2.54	2.01	58.2	2.87	2.50	44.9	3.15	2.95	35.6	3.46	3.39	28.0	3.84	3.88	23.0	4.28	4.24	19.0	4.88	4.50
300	190.3	2.08	1.13	117.6	2.33	1.63	87.7	2.54	2.01	67.3	2.86	2.51	48.0	3.14	2.97	38.1	3.45	3.40	29.9	3.82	3.92	24.6	4.27	4.26	20.2	4.80	4.61

TABLE B.3 *(Continued)* Parabolic Waterway Design

V_1 for *RETARDANCE "D"*. Top Width (T), Depth (D) and V_2 for *RETARDANCE "B"*.

Grade 0.75 Percent

Q cfs	V_1 = 2.0			V_1 = 2.5			V_1 = 3.0			V_1 = 3.5			V_1 = 4.0			V_1 = 4.5			V_1 = 5.0			V_1 = 5.5			V_1 = 6.0		
	T	D	V_2	T	D	V_2	T	D	V_2	T	D	V_2	T	D	V_2	T	D	V_2	T	D	V_2	T	D	V_2	T	D	V_2
15	13.7	1.76	0.92	8.0	2.22	1.24																					
20	18.2	1.75	0.93	10.4	2.10	1.35																					
25	22.6	1.73	0.95	12.8	2.03	1.42	9.5	2.33	1.66																		
30	27.1	1.73	0.95	15.3	2.02	1.44	11.2	2.25	1.76	8.9	2.56	1.94															
35	31.5	1.72	0.96	17.8	2.01	1.45	13.0	2.23	1.79	10.2	2.47	2.05															
40	36.0	1.72	0.96	20.2	1.98	1.48	14.8	2.21	1.81	11.6	2.44	2.09															
45	40.4	1.71	0.96	22.7	1.98	1.49	16.5	2.17	1.86	12.9	2.39	2.16															
50	44.9	1.72	0.96	25.2	1.98	1.49	18.3	2.16	1.87	14.3	2.38	2.18	10.6	2.82	2.47												
55	49.3	1.72	0.96	27.6	1.96	1.51	20.1	2.16	1.88	15.7	2.37	2.19	11.6	2.80	2.51												
60	53.7	1.72	0.97	30.1	1.96	1.51	21.9	2.16	1.89	17.1	2.36	2.20	12.5	2.73	2.61												
65	58.1	1.72	0.97	32.5	1.95	1.52	23.6	2.14	1.92	18.4	2.33	2.25	13.5	2.71	2.63	11.2	3.04	2.83									
70	62.5	1.72	0.97	35.0	1.96	1.52	25.4	2.14	1.92	19.8	2.33	2.25	14.4	2.66	2.71	12.0	3.01	2.87									
75	66.9	1.72	0.97	37.4	1.95	1.53	27.2	2.14	1.92	21.2	2.33	2.26	15.4	2.65	2.73	12.8	2.99	2.91									
80	71.2	1.71	0.97	39.9	1.95	1.53	29.0	2.14	1.92	22.5	2.31	2.29	16.4	2.65	2,74	13.5	2.92	3.01									
90	80.0	1.71	0.97	44.8	1.95	1.53	32.5	2.12	1.94	25.3	2.31	2.29	18.4	2.63	2.76	15.1	2.89	3.07	12.4	3.26	3.30						
100	88.8	1.72	0.98	49.7	1.95	1.54	36.1	2.13	1.94	28.1	2.31	2.29	20.4	2.62	2.78	16.7	2.86	3.11	13.7	3.22	3.36						
110	97.6	1.72	0.98	54.7	1.95	1.53	39.7	2.13	1.94	30.8	2.30	2.31	22.4	2.61	2.79	18.3	2.84	3.15	14.9	3.15	3.48						
120	106.3	1.72	0.98	59.6	1.95	1.54	43.2	2.12	1.95	33.6	2.30	2.31	24.4	2.61	2.81	19.9	2.82	3.18	16.2	3.13	3.52	13.4	3.49	3.81			
130	115.0	1.72	0.98	64.5	1.95	1.54	46.8	2.12	1.95	36.4	2.30	2.31	26.3	2.58	2.85	21.5	2.81	3.20	17.5	3.11	3.55	14.5	3.49	3.82			
140	123.7	1.72	0.98	69.4	1.95	1.54	50.3	2.12	1.96	39.1	2.29	2.33	28.3	2.58	2.85	23.1	2.80	3.22	18.7	3.07	3.63	15.5	3.44	3.91			
150	132.4	1.72	0.98	74.2	1.95	1.55	53.9	2.12	1.95	41.9	2.30	2.32	30.3	2.58	2.86	24.7	2.79	3.24	20.0	3.06	3.65	16.5	3.39	3.98			
160	141.1	1.72	0.98	79.1	1.95	1.55	57.4	2.12	1.96	44.6	2.29	2.33	32.3	2.58	2.86	26.3	2.78	3.26	21.3	3.05	3.67	17.6	3.39	3.98	14.4	3.77	4.37
170	149.7	1.72	0.98	84.0	1.95	1.55	60.9	2.12	1.97	47.4	2.30	2.33	34.3	2.58	2.86	28.0	2.80	3.24	22.6	3.04	3.68	18.6	3.36	4.05	15.3	3.77	4.38
180	158.3	1.72	0.98	88.8	1.95	1.55	64.5	2.12	1.96	50.1	2.29	2.34	36.3	2.58	2.86	29.6	2.79	3.25	23.9	3.04	3.69	19.6	3.33	4.11	16.1	3.73	4.46
190	166.9	1.72	0.98	93.7	1.95	1.55	68.0	2.12	1.97	52.8	2.29	2.34	38.3	2.58	2.86	31.2	2.78	3.26	25.2	3.03	3.70	20.7	3.33	4.10	17.0	3.73	4.45
200	175.5	1.72	0.99	98.5	1.95	1.55	71.5	2.12	1.97	55.6	2.29	2.34	40.2	2.57	2.89	32.8	2.78	3.27	26.5	3.03	3.71	21.7	3.31	4.15	17.8	3.70	4.52
220	192.8	1.72	0.99	108.3	1.95	1.55	78.6	2.12	1.97	61.1	2.29	2.34	44.2	2.57	2.89	36.0	2.77	3.29	29.1	3.02	3.73	23.8	3.29	4.18	19.5	3.67	4.58
240	210.1	1.72	0.99	118.0	1.95	1.56	85.6	2.12	1.98	66.6	2.29	2.35	48.2	2.57	2.89	39.2	2.76	3.31	31.7	3.01	3.75	25.9	3.28	4.22	21.2	3.64	4.63
260	227.3	1.72	0.99	127.7	1.95	1.56	92.7	2.12	1.98	72.1	2.29	2.35	52.2	2.57	2.89	42.5	2.77	3.30	34.3	3.01	3.76	28.0	3.26	4.24	22.9	3.62	4.67
280	244.5	1.72	0.99	137.4	1.95	1.56	99.7	2.12	1.98	77.6	2.29	2.35	56.2	2.57	2.89	45.7	2.76	3.31	36.9	3.00	3.77	30.1	3.25	4.26	24.7	3.63	4.65
300	261.7	1.72	0.99	147.1	1.95	1.56	106.8	2.12	1.98	83.0	2.29	2.36	60.1	2.57	2.90	48.9	2.76	3.32	39.5	3.00	3.78	32.2	3.25	4.28	26.4	3.62	4.68

TABLE B.3 *(Continued)* Parabolic Waterway Design

V_1 for *RETARDANCE "D"*. Top Width (T), Depth (D) and V_2 for *RETARDANCE "B"*.

Grade 1.0 Percent

Q cfs	V_1 = 2.0			V_1 = 2.5			V_1 = 3.0			V_1 = 3.5			V_1 = 4.0			V_1 = 4.5			V_1 = 5.0			V_1 = 5.5			V_1 = 6.0		
	T	D	V_2	T	D	V_2	T	D	V_2	T	D	V_2	T	D	V_2	T	D	V_2	T	D	V_2	T	D	V_2	T	D	V_2
15	15.7	1.55	0.91	9.9	1.80	1.24																					
20	20.9	1.54	0.92	13.0	1.74	1.31	8.8	2.04	1.65																		
25	26.0	1.53	0.93	16.2	1.73	1.32	10.9	1.99	1.70	8.9	2.21	1.87															
30	31.1	1.52	0.94	19.3	1.70	1.35	12.9	1.94	1.78	10.5	2.14	1.98															
35	36.2	1.52	0.94	22.5	1.70	1.36	15.0	1.93	1.80	12.1	2.08	2.06	9.2	2.37	2.37												
40	41.3	1.52	0.95	25.7	1.70	1.36	17.1	1.92	1.81	13.7	2.04	2.12	10.4	2.32	2.45												
45	46.4	1.52	0.95	28.8	1.69	1.37	19.2	1.91	1.82	15.4	2.04	2.12	11.7	2.32	2.46												
50	51.5	1.52	0.95	32.0	1.70	1.37	21.2	1.89	1.85	17.0	2.02	2.16	12.9	2.28	2.52	9.9	2.51	2.86									
55	56.5	1.51	0.95	35.1	1.69	1.38	23.3	1.89	1.86	18.7	2.02	2.16	14.1	2.25	2.57	10.8	2.57	2.93									
60	61.6	1.52	0.95	38.3	1.69	1.37	25.4	1.89	1.86	20.3	2.00	2.19	15.3	2.23	2.61	11.7	2.54	3.00									
65	66.6	1.52	0.96	41.4	1.69	1.38	27.5	1.89	1.86	22.0	2.01	2.19	16.6	2.24	2.60	12.7	2.55	2.98	10.6	2.88	3.16						
70	71.6	1.51	0.96	44.6	1.70	1.38	29.5	1.88	1.88	23.6	1.99	2.21	17.8	2.22	2.63	13.6	2.52	3.03	11.3	2.82	3.26						
75	76.6	1.51	0.96	47.7	1.69	1.38	31.6	1.88	1.88	25.3	2.00	2.21	19.0	2.21	2.66	14.5	2.50	3.08	12.0	2.77	3.35						
80	81.6	1.52	0.96	50.8	1.69	1.38	33.7	1.88	1.88	26.9	1.99	2.22	20.3	2.22	2.64	15.4	2.48	3.11	12.7	2.73	3.43						
90	91.7	1.52	0.96	57.1	1.69	1.39	37.8	1.88	1.89	30.2	1.98	2.24	22.8	2.21	2.65	17.3	2.47	3.13	14.2	2.69	3.49	11.8	2.99	3.78			
100	101.7	1.52	0.96	63.4	1.69	1.39	42.0	1.88	1.89	33.5	1.98	2.24	25.2	2.19	2.69	19.2	2.47	3.14	15.7	2.67	3.55	13.0	2.94	3.88			
110	111.7	1.52	0.97	69.6	1.69	1.39	46.1	1.87	1.90	36.8	1.98	2.25	27.7	2.19	2.70	21.0	2.44	3.20	17.3	2.68	3.53	14.3	2.94	3.88	12.4	3.23	4.07
120	121.7	1.52	0.97	75.8	1.69	1.39	50.2	1.87	1.90	40.1	1.98	2.26	30.2	2.19	2.70	22.9	2.44	3.20	18.8	2.66	3.57	15.5	2.91	3.96	13.4	3.17	4.19
130	131.6	1.51	0.97	82.1	1.69	1.39	54.4	1.87	1.90	43.4	1.97	2.26	32.7	2.19	2.70	24.8	2.44	3.20	20.3	2.64	3.61	16.7	2.88	4.03	14.4	3.12	4.30
140	141.5	1.51	0.97	88.3	1.69	1.39	58.5	1.87	1.90	46.7	1.97	2.26	35.2	2.19	2.70	26.6	2.42	3.24	21.8	2.63	3.64	17.9	2.85	4.08	15.5	3.12	4.31
150	151.4	1.52	0.97	94.5	1.69	1.40	62.6	1.87	1.91	50.0	1.98	2.26	37.6	2.18	2.72	28.5	2.42	3.23	23.3	2.62	3.66	19.2	2.86	4.07	16.5	3.08	4.39
160	161.3	1.52	0.97	100.7	1.69	1.40	66.7	1.87	1.91	53.3	1.98	2.26	40.1	2.18	2.72	30.4	2.43	3.23	24.8	2.61	3.68	20.4	2.84	4.11	17.6	3.08	4.39
170	171.1	1.52	0.97	106.8	1.69	1.40	70.8	1.87	1.91	56.5	1.97	2.28	42.6	2.19	2.72	32.2	2.41	3.26	26.3	2.60	3.70	21.7	2.85	4.10	18.6	3.05	4.46
180	180.9	1.52	0.98	113.0	1.69	1.40	74.9	1.87	1.91	59.8	1.97	2.27	45.0	2.18	2.73	34.1	2.42	3.25	27.9	2.61	3.68	22.9	2.83	4.14	19.6	3.02	4.52
190	190.6	1.52	0.98	119.1	1.69	1.40	79.0	1.87	1.91	63.1	1.98	2.27	47.5	2.18	2.73	35.9	2.41	3.27	29.4	2.61	3.69	24.1	2.81	4.17	20.7	3.03	4.51
200	200.4	1.52	0.98	125.3	1.69	1.40	83.0	1.87	1.92	66.3	1.97	2.28	49.9	2.18	2.74	37.8	2.41	3.27	30.9	2.60	3.71	25.4	2.82	4.15	21.7	3.00	4.57
220	220.1	1.52	0.98	137.6	1.69	1.41	91.2	1.87	1.92	72.9	1.97	2.28	54.9	2.18	2.74	41.5	2.41	3.28	33.9	2.59	3.73	27.9	2.82	4.17	23.8	2.99	4.61
240	239.8	1.52	0.98	150.0	1.69	1.41	99.4	1.87	1.93	79.5	1.98	2.28	59.8	2.18	2.75	45.3	2.41	3.28	37.0	2.60	3.72	30.4	2.81	4.18	25.9	2.97	4.64
260	259.4	1.52	0.98	162.3	1.69	1.41	107.6	1.87	1.93	86.0	1.97	2.28	64.8	2.18	2.74	49.0	2.41	3.29	40.0	2.59	3.75	32.9	2.81	4.20	28.0	2.96	4.67
280	279.0	1.52	0.98	174.6	1.70	1.41	115.8	1.87	1.93	92.5	1.97	2.29	69.7	2.18	2.75	52.7	2.40	3.30	43.0	2.58	3.76	35.3	2.79	4.24	30.2	2.97	4.65
300	298.5	1.52	0.98	186.9	1.70	1.41	123.9	1.87	1.93	99.1	1.98	2.29	74.6	2.18	2.75	56.4	2.40	3.30	46.1	2.59	3.75	37.8	2.79	4.24	32.3	2.97	4.67

TABLE B.3 *(Continued)* Parabolic Waterway Design

V_1 for *RETARDANCE "D"*. Top Width (T), Depth (D) and V_2 for *RETARDANCE "B"*.

Grade 1.25 Percent

Q	V_1 = 2.0			V_1 = 2.5			V_1 = 3.0			V_1 = 3.5			V_1 = 4.0			V_1 = 4.5			V_1 = 5.0			V_1 = 5.5			V_1 = 6.0		
cfs	T	D	V_2	T	D	V_2	T	D	V_2	T	D	V_2	T	D	V_2	T	D	V_2	T	D	V_2	T	D	V_2	T	D	V_2
15	18.1	1.40	0.88	11.5	1.59	1.21	7.7	1.85	1.55																		
20	24.0	1.38	0.89	15.2	1.56	1.25	10.1	1.79	1.64	7.7	2.03	1.89															
25	30.0	1.38	0.89	19.0	1.56	1.25	12.5	1.75	1.69	9.5	1.97	1.97	7.8	2.19	2.16												
30	35.9	1.38	0.90	22.7	1.55	1.27	14.9	1.73	1.73	11.2	1.90	2.09	9.2	2.11	2.28												
35	41.8	1.38	0.90	26.4	1.54	1.28	17.3	1.71	1.75	13.0	1.88	2.12	10.5	2.02	2.44	8.3	2.36	2.64									
40	47.7	1.38	0.90	30.1	1.54	1.28	19.7	1.70	1.77	14.8	1.87	2.15	12.0	2.02	2.44	9.4	2.32	2.72									
45	53.6	1.38	0.90	33.9	1.54	1.28	22.1	1.69	1.78	16.6	1.86	2.17	13.4	1.99	2.50	10.5	2.28	2.78									
50	59.4	1.38	0.91	37.6	1.54	1.28	24.5	1.69	1.79	18.4	1.85	2.18	14.9	1.99	2.50	11.5	2.22	2.90	9.4	2.49	3.16						
55	65.3	1.38	0.91	41.3	1.54	1.28	26.9	1.69	1.80	20.2	1.84	2.19	16.3	1.97	2.54	12.6	2.20	2.94	10.3	2.47	3.20						
60	71.1	1.38	0.91	44.9	1.53	1.29	29.3	1.68	1.81	22.0	1.84	2.20	17.7	1.96	2.57	13.7	2.19	2.97	11.2	2.45	3.23						
65	76.9	1.38	0.91	48.6	1.54	1.29	31.7	1.68	1.81	23.7	1.82	2.24	19.2	1.97	2.56	14.8	2.18	2.99	12.0	2.40	3.34	9.8	2.76	3.55			
70	82.7	1.38	0.91	52.3	1.54	1.29	34.1	1.68	1.81	25.5	1.82	2.24	20.6	1.95	2.58	15.9	2.17	3.01	12.9	2.39	3.36	10.5	2.74	3.61			
75	88.4	1.38	0.91	55.9	1.53	1.30	36.5	1.68	1.82	27.3	1.82	2.24	22.1	1.96	2.57	17.0	2.16	3.03	13.8	2.39	3.38	11.1	2.67	3.76			
80	94.2	1.38	0.91	59.6	1.54	1.30	38.9	1.68	1.82	29.1	1.82	2.24	23.5	1.95	2.59	18.1	2.16	3.05	14.7	2.38	3.39	11.8	2.65	3.80	10.3	2.92	3.94
90	105.8	1.38	0.91	66.9	1.53	1.30	43.7	1.68	1.82	32.7	1.82	2.25	26.4	1.95	2.60	20.3	2.14	3.07	16.4	2.34	3.48	13.2	2.62	3.87	11.4	2.83	4.14
100	117.3	1.38	0.92	74.3	1.54	1.30	48.5	1.68	1.83	36.3	1.82	2.25	29.3	1.95	2.61	22.5	2.14	3.10	18.2	2.34	3.49	14.6	2.59	3.92	12.6	2.80	4.21
110	128.9	1.38	0.92	81.6	1.54	1.31	53.3	1.68	1.83	39.8	1.81	2.27	32.2	1.95	2.61	24.7	2.13	3.11	20.0	2.34	3.50	16.0	2.58	3.97	13.8	2.77	4.27
120	140.4	1.38	0.92	88.9	1.54	1.31	58.1	1.68	1.83	43.4	1.81	2.27	35.0	1.94	2.64	26.9	2.12	3.13	21.7	2.31	3.56	17.4	2.56	4.00	15.0	2.75	4.32
130	151.8	1.38	0.92	96.1	1.53	1.31	62.9	1.68	1.83	47.0	1.82	2.27	37.9	1.94	2.64	29.1	2.12	3.14	23.5	2.31	3.56	18.8	2.55	4.04	16.2	2.74	4.36
140	163.2	1.38	0.92	103.4	1.54	1.31	67.6	1.68	1.84	50.5	1.81	2.28	40.8	1.94	2.64	31.3	2.12	3.14	25.3	2.31	3.56	20.2	2.54	4.06	17.4	2.72	4.39
150	174.6	1.38	0.92	110.6	1.53	1.31	72.4	1.68	1.83	54.1	1.81	2.28	43.7	1.94	2.64	33.5	2.12	3.15	27.0	2.30	3.60	21.6	2.53	4.08	18.6	2.71	4.42
160	186.0	1.38	0.92	117.8	1.53	1.32	77.1	1.68	1.84	57.6	1.81	2.29	46.5	1.93	2.65	35.7	2.12	3.16	28.8	2.30	3.60	22.9	2.50	4.16	19.8	2.70	4.45
170	197.3	1.39	0.92	125.0	1.54	1.32	81.9	1.68	1.84	61.2	1.81	2.28	49.4	1.94	2.65	37.9	2.12	3.16	30.6	2.30	3.59	24.3	2.50	4.17	21.0	2.69	4.47
180	208.5	1.38	0.93	132.2	1.54	1.32	86.6	1.68	1.84	64.7	1.81	2.29	52.2	1.93	2.66	40.1	2.12	3.16	32.3	2.29	3.62	25.7	2.49	4.18	22.2	2.69	4.49
190	219.8	1.39	0.93	139.3	1.54	1.32	91.3	1.68	1.84	68.2	1.81	2.29	55.1	1.94	2.65	42.3	2.12	3.16	34.1	2.30	3.62	27.1	2.49	4.19	23.4	2.68	4.51
200	231.0	1.39	0.93	146.5	1.54	1.32	96.0	1.68	1.84	71.8	1.82	2.29	57.9	1.93	2.66	44.5	2.12	3.17	35.8	2.29	3.64	28.5	2.49	4.20	24.6	2.68	4.52
220	253.7	1.39	0.93	160.9	1.54	1.32	105.5	1.68	1.84	78.9	1.82	2.29	63.7	1.94	2.66	48.9	2.12	3.17	39.4	2.29	3.63	31.3	2.48	4.22	27.0	2.67	4.55
240	276.3	1.39	0.93	175.3	1.54	1.33	115.0	1.69	1.84	85.9	1.81	2.30	69.4	1.94	2.66	53.3	2.11	3.18	42.9	2.29	3.65	34.1	2.48	4.23	29.4	2.66	4.57
260	298.9	1.39	0.93	189.7	1.54	1.33	124.4	1.68	1.85	93.0	1.81	2.30	75.1	1.93	2.67	57.6	2.11	3.19	46.5	2.29	3.64	36.9	2.48	4.24	31.8	2.65	4.59
280	321.3	1.39	0.93	204.0	1.54	1.33	133.9	1.69	1.85	100.1	1.82	2.30	80.8	1.93	2.67	62.0	2.11	3.19	50.0	2.29	3.65	39.7	2.47	4.25	34.2	2.65	4.61
300	343.7	1.39	0.93	218.2	1.54	1.33	143.3	1.69	1.85	107.1	1.81	2.30	86.5	1.94	2.67	66.4	2.11	3.19	53.5	2.28	3.66	42.5	2.47	4.26	36.6	2.65	4.62

TABLE B.3 *(Continued)* Parabolic Waterway Design
V_1 for *RETARDANCE "D"*. Top Width (T), Depth (D) and V_2 for *RETARDANCE "B"*.
Grade 1.50 Percent

Q	V_1 = 2.0			V_1 = 2.5			V_1 = 3.0			V_1 = 3.5			V_1 = 4.0			V_1 = 4.5			V_1 = 5.0			V_1 = 5.5			V_1 = 6.0		
cfs	T	D	V_2	T	D	V_2	T	D	V_2	T	D	V_2	T	D	V_2	T	D	V_2	T	D	V_2	T	D	V_2	T	D	V_2
15	20.1	1.29	0.86	13.2	1.43	1.17	8.9	1.65	1.51																		
20	26.8	1.29	0.86	17.5	1.42	1.19	11.7	1.60	1.58	8.1	1.86	1.95	6.7	2.09	2.10												
25	33.4	1.28	0.86	21.8	1.41	1.21	14.5	1.57	1.63	9.9	1.78	2.10	8.1	1.95	2.34												
30	40.0	1.28	0.87	26.2	1.42	1.20	17.4	1.57	1.63	11.8	1.75	2.15	9.6	1.90	2.43	8.0	2.13	2.60									
35	46.5	1.28	0.87	30.5	1.41	1.21	20.2	1.56	1.65	13.7	1.74	2.18	11.1	1.86	2.50	9.2	2.06	2.72									
40	53.1	1.28	0.87	34.8	1.41	1.21	23.0	1.55	1.67	15.6	1.73	2.20	12.6	1.84	2.56	10.4	2.02	2.82	8.2	2.38	3.03						
45	59.6	1.28	0.88	39.0	1.40	1.22	25.9	1.55	1.66	17.5	1.72	2.22	14.2	1.85	2.55	11.7	2.02	2.82	9.1	2.31	3.16						
50	66.1	1.28	0.88	43.3	1.41	1.22	28.7	1.55	1.67	19.4	1.71	2.24	15.7	1.83	2.58	12.9	1.99	2.89	10.0	2.26	3.27						
55	72.6	1.28	0.88	47.6	1.41	1.22	31.5	1.55	1.68	21.3	1.71	2.25	17.2	1.82	2.61	14.2	1.99	2.89	10.9	2.22	3.36	9.3	2.47	3.54			
60	79.0	1.28	0.88	51.8	1.41	1.22	34.3	1.54	1.69	23.2	1.70	2.25	18.8	1.82	2.60	15.4	1.97	2.94	11.9	2.22	3.36	10.1	2.45	3.59			
65	85.5	1.28	0.88	56.0	1.40	1.23	37.1	1.54	1.69	25.1	1.70	2.26	20.3	1.82	2.62	16.6	1.95	2.98	12.8	2.19	3.43	10.8	2.39	3.73			
70	91.9	1.28	0.88	60.3	1.41	1.22	39.9	1.54	1.69	27.0	1.70	2.26	21.8	1.81	2.64	17.9	1.96	2.96	13.8	2.20	3.42	11.6	2.38	3.76	9.8	2.62	4.03
75	98.2	1.28	0.89	64.5	1.41	1.23	42.7	1.54	1.69	28.9	1.70	2.27	23.3	1.80	2.66	19.1	1.95	3.00	14.7	2.18	3.48	12.4	2.37	3.79	10.5	2.62	4.04
80	104.6	1.28	0.89	68.7	1.41	1.23	45.5	1.54	1.70	30.8	1.70	2.27	24.9	1.81	2.64	20.4	1.95	2.98	15.7	2.18	3.46	13.2	2.36	3.81	11.1	2.58	4.15
90	117.5	1.28	0.89	77.1	1.41	1.23	51.1	1.54	1.70	34.6	1.70	2.28	27.9	1.80	2.67	22.9	1.95	3.00	17.6	2.17	3.50	14.7	2.31	3.93	12.4	2.54	4.24
100	130.3	1.28	0.89	85.6	1.41	1.23	56.7	1.54	1.70	38.3	1.69	2.30	31.0	1.80	2.66	25.4	1.94	3.02	19.5	2.16	3.53	16.3	2.30	3.96	13.7	2.51	4.32
110	143.0	1.28	0.89	94.0	1.41	1.24	62.3	1.54	1.71	42.1	1.69	2.30	34.0	1.79	2.68	27.9	1.94	3.03	21.4	2.15	3.56	17.9	2.30	3.98	15.0	2.49	4.38
120	155.8	1.28	0.89	102.4	1.41	1.24	67.9	1.54	1.71	45.9	1.69	2.30	37.1	1.80	2.68	30.4	1.94	3.03	23.3	2.14	3.57	19.5	2.29	3.99	16.3	2.47	4.43
130	168.4	1.28	0.90	110.7	1.41	1.24	73.4	1.54	1.71	49.7	1.69	2.30	40.1	1.79	2.69	32.9	1.94	3.04	25.2	2.14	3.59	21.1	2.29	4.01	17.6	2.45	4.48
140	181.0	1.28	0.90	119.1	1.41	1.24	79.0	1.54	1.71	53.4	1.69	2.31	43.2	1.80	2.68	35.3	1.93	3.07	27.1	2.14	3.60	22.7	2.29	4.02	18.9	2.44	4.51
150	193.6	1.28	0.90	127.4	1.41	1.24	84.5	1.54	1.72	57.2	1.69	2.31	46.2	1.79	2.70	37.8	1.93	3.07	29.0	2.13	3.61	24.2	2.27	4.07	20.2	2.43	4.54
160	206.2	1.28	0.90	135.7	1.41	1.24	90.0	1.54	1.72	60.9	1.69	2.32	49.3	1.80	2.69	40.3	1.93	3.07	30.9	2.13	3.62	25.8	2.27	4.08	21.6	2.45	4.51
170	218.6	1.28	0.90	144.0	1.41	1.24	95.6	1.54	1.72	64.7	1.69	2.31	52.3	1.80	2.69	42.8	1.93	3.07	32.8	2.13	3.62	27.4	2.27	4.08	22.9	2.44	4.53
180	231.1	1.28	0.90	152.2	1.41	1.25	101.0	1.54	1.72	68.4	1.69	2.32	55.3	1.80	2.70	45.3	1.93	3.07	34.7	2.13	3.63	29.0	2.27	4.08	24.2	2.43	4.56
190	243.5	1.28	0.90	160.4	1.41	1.25	106.5	1.54	1.72	72.1	1.69	2.32	58.3	1.80	2.71	47.7	1.92	3.08	36.5	2.12	3.66	30.6	2.27	4.08	25.5	2.42	4.58
200	255.8	1.28	0.91	168.6	1.41	1.25	112.0	1.54	1.72	75.9	1.69	2.32	61.3	1.79	2.71	50.2	1.93	3.08	38.4	2.12	3.66	32.1	2.26	4.12	26.8	2.42	4.60
220	280.9	1.28	0.91	185.2	1.41	1.25	123.0	1.54	1.73	83.4	1.69	2.32	67.4	1.80	2.71	55.2	1.93	3.08	42.2	2.12	3.67	35.3	2.26	4.12	29.5	2.42	4.58
240	305.8	1.28	0.91	201.7	1.41	1.25	134.1	1.54	1.73	90.9	1.69	2.32	73.5	1.80	2.71	60.1	1.92	3.09	46.0	2.12	3.67	38.5	2.26	4.12	32.1	2.41	4.62
260	330.7	1.28	0.91	218.2	1.41	1.26	145.1	1.54	1.73	98.3	1.69	2.33	79.5	1.80	2.72	65.1	1.93	3.09	49.8	2.12	3.68	41.7	2.26	4.12	34.7	2.41	4.64
280	355.5	1.28	0.91	234.7	1.41	1.26	156.0	1.54	1.73	105.8	1.69	2.33	85.5	1.79	2.72	70.0	1.93	3.10	53.6	2.12	3.68	44.8	2.25	4.14	37.4	2.41	4.63
300	380.2	1.28	0.91	251.1	1.41	1.26	167.0	1.54	1.73	113.2	1.69	2.33	91.6	1.80	2.72	74.9	1.92	3.11	57.4	2.12	3.68	48.0	2.25	4.14	40.0	2.41	4.65

TABLE B.3 *(Continued)* Parabolic Waterway Design
V_1 for *RETARDANCE "D"*. Top Width (T), Depth (D) and V_2 for *RETARDANCE "B"*.
Grade 1.75 Percent

Q cfs	V_1 = 2.0 T	D	V_2	V_1 = 2.5 T	D	V_2	V_1 = 3.0 T	D	V_2
15	21.8	1.21	0.84	14.2	1.33	1.17	10.0	1.50	1.48
20	29.0	1.20	0.85	18.9	1.33	1.18	13.2	1.46	1.53
25	36.2	1.20	0.85	23.5	1.32	1.20	16.4	1.45	1.56
30	43.4	1.21	0.85	28.2	1.32	1.20	19.6	1.44	1.58
35	50.5	1.20	0.85	32.8	1.31	1.21	22.8	1.43	1.59
40	57.6	1.20	0.85	37.4	1.31	1.21	26.0	1.43	1.60
45	64.6	1.20	0.86	42.0	1.31	1.21	29.2	1.43	1.61
50	71.6	1.20	0.86	46.6	1.31	1.21	32.4	1.43	1.61
55	78.6	1.20	0.86	51.2	1.31	1.22	35.6	1.43	1.61
60	85.6	1.20	0.87	55.8	1.31	1.22	38.8	1.43	1.61
65	92.6	1.20	0.87	60.3	1.31	1.22	41.9	1.42	1.62
70	99.5	1.20	0.87	64.9	1.32	1.22	45.1	1.42	1.62
75	106.4	1.20	0.87	69.4	1.31	1.22	48.2	1.42	1.63
80	113.3	1.20	0.87	73.9	1.31	1.22	51.4	1.43	1.62
90	127.2	1.20	0.87	83.0	1.31	1.23	57.7	1.42	1.63
100	141.0	1.20	0.87	92.1	1.32	1.23	64.0	1.42	1.64
110	154.9	1.21	0.87	101.2	1.32	1.23	70.4	1.43	1.63
120	168.6	1.20	0.88	110.2	1.32	1.23	76.6	1.42	1.64
130	182.3	1.21	0.88	119.2	1.32	1.23	82.9	1.42	1.64
140	196.0	1.21	0.88	128.1	1.32	1.23	89.2	1.43	1.64
150	209.6	1.21	0.88	137.1	1.32	1.23	95.4	1.42	1.64
160	223.1	1.21	0.88	146.0	1.32	1.24	101.6	1.42	1.65
170	236.6	1.21	0.88	154.9	1.32	1.24	107.9	1.43	1.64
180	250.0	1.21	0.89	163.7	1.32	1.24	114.0	1.43	1.65
190	263.4	1.21	0.89	172.6	1.32	1.24	120.2	1.43	1.65
200	276.7	1.21	0.89	181.4	1.32	1.24	126.4	1.43	1.65
220	303.8	1.21	0.89	199.2	1.32	1.24	138.8	1.43	1.65
240	330.8	1.21	0.89	217.0	1.32	1.24	151.2	1.43	1.66
260	357.7	1.21	0.89	234.7	1.32	1.25	163.6	1.43	1.66
280	384.5	1.21	0.89	252.3	1.32	1.25	176.0	1.43	1.66
300	411.2	1.21	0.90	269.9	1.32	1.25	188.3	1.43	1.66

Q cfs	V_1 = 3.5 T	D	V_2	V_1 = 4.0 T	D	V_2	V_1 = 4.5 T	D	V_2
15	6.9	1.74	1.84						
20	9.0	1.66	1.98	7.6	1.83	2.12			
25	11.2	1.64	2.01	9.3	1.75	2.27	7.6	1.93	2.51
30	13.3	1.60	2.08	11.1	1.72	2.32	9.0	1.88	2.62
35	15.5	1.60	2.09	12.8	1.68	2.40	10.4	1.84	2.70
40	17.6	1.58	2.13	14.6	1.68	2.42	11.8	1.81	2.77
45	19.8	1.58	2.13	16.4	1.68	2.43	13.2	1.79	2.82
50	21.9	1.57	2.16	18.2	1.67	2.44	14.7	1.80	2.80
55	24.1	1.58	2.15	19.9	1.66	2.48	16.1	1.79	2.83
60	26.2	1.57	2.17	21.7	1.66	2.48	17.5	1.78	2.86
65	28.4	1.57	2.16	23.5	1.66	2.48	18.9	1.77	2.88
70	30.5	1.57	2.18	25.2	1.65	2.50	20.3	1.76	2.90
75	32.7	1.57	2.17	27.0	1.65	2.50	21.8	1.78	2.88
80	34.8	1.57	2.18	28.8	1.66	2.50	23.2	1.77	2.90
90	39.1	1.57	2.18	32.3	1.65	2.52	26.0	1.76	2.93
100	43.4	1.57	2.18	35.8	1.64	2.53	28.9	1.76	2.92
110	47.7	1.57	2.19	39.4	1.65	2.52	31.7	1.76	2.94
120	51.9	1.57	2.20	42.9	1.64	2.53	34.6	1.76	2.93
130	56.2	1.57	2.20	46.4	1.64	2.54	37.4	1.76	2.95
140	60.4	1.57	2.21	49.9	1.64	2.55	40.2	1.75	2.96
150	64.7	1.57	2.20	53.4	1.64	2.55	43.1	1.76	2.95
160	68.9	1.57	2.21	56.9	1.64	2.55	45.9	1.76	2.96
170	73.1	1.57	2.21	60.4	1.64	2.55	48.7	1.76	2.96
180	77.4	1.57	2.21	63.9	1.64	2.55	51.5	1.75	2.97
190	81.6	1.57	2.21	67.4	1.64	2.55	54.3	1.75	2.97
200	85.7	1.57	2.22	70.8	1.64	2.56	57.1	1.75	2.98
220	94.2	1.57	2.22	77.8	1.64	2.57	62.8	1.76	2.97
240	102.7	1.57	2.22	84.8	1.64	2.57	68.4	1.75	2.98
260	111.1	1.57	2.22	91.8	1.64	2.57	74.1	1.76	2.98
280	119.5	1.57	2.23	98.7	1.64	2.58	79.7	1.76	2.98
300	127.9	1.57	2.23	105.7	1.64	2.57	85.3	1.76	2.99

Q cfs	V_1 = 5.0 T	D	V_2	V_1 = 5.5 T	D	V_2	V_1 = 6.0 T	D	V_2
15									
20									
25									
30									
35	8.2	2.11	2.98						
40	9.3	2.08	3.06						
45	10.3	2.01	3.21	8.6	2.26	3.41			
50	11.4	2.00	3.25	9.5	2.24	3.48			
55	12.5	1.99	3.28	10.3	2.17	3.64	8.6	2.51	3.77
60	13.6	1.98	3.31	11.2	2.16	3.67	9.2	2.41	4.00
65	14.7	1.97	3.33	12.1	2.15	3.70	9.9	2.38	4.08
70	15.8	1.96	3.35	13.0	2.14	3.73	10.6	2.35	4.15
75	16.9	1.96	3.36	13.9	2.13	3.75	11.4	2.37	4.11
80	17.9	1.94	3.43	14.8	2.13	3.77	12.1	2.35	4.17
90	20.1	1.93	3.45	16.6	2.12	3.81	13.5	2.31	4.28
100	22.3	1.93	3.46	18.4	2.11	3.83	14.9	2.28	4.37
110	24.5	1.92	3.47	20.1	2.08	3.91	16.4	2.29	4.36
120	26.7	1.92	3.48	21.9	2.08	3.93	17.8	2.26	4.43
130	28.9	1.92	3.49	23.7	2.07	3.93	19.2	2.25	4.48
140	31.1	1.92	3.49	25.5	2.07	3.94	20.7	2.25	4.47
150	33.3	1.92	3.49	27.3	2.07	3.95	22.1	2.24	4.51
160	35.4	1.91	3.52	29.1	2.07	3.95	23.6	2.25	4.49
170	37.6	1.91	3.52	30.9	2.07	3.95	25.0	2.24	4.53
180	39.8	1.92	3.52	32.6	2.06	3.99	26.4	2.23	4.56
190	42.0	1.92	3.52	34.4	2.06	3.99	27.9	2.23	4.54
200	44.1	1.91	3.54	36.2	2.06	3.99	29.3	2.23	4.57
220	48.5	1.91	3.53	39.8	2.07	3.99	32.2	2.22	4.58
240	52.9	1.92	3.53	43.3	2.06	4.02	35.1	2.22	4.58
260	57.2	1.91	3.54	46.9	2.06	4.01	38.0	2.22	4.59
280	61.6	1.92	3.54	50.5	2.06	4.01	40.9	2.22	4.59
300	65.9	1.91	3.55	54.0	2.06	4.03	43.7	2.21	4.62

TABLE B.3 *(Continued)* Parabolic Waterway Design
V_1 for *RETARDANCE "D"*. Top Width (T), Depth (D) and V_2 for *RETARDANCE "B"*.
Grade 2.0 Percent

Q	V_1 = 2.0			V_1 = 2.5			V_1 = 3.0			V_1 = 3.5			V_1 = 4.0			V_1 = 4.5			V_1 = 5.0			V_1 = 5.5			V_1 = 6.0		
cfs	T	D	V_2	T	D	V_2	T	D	V_2	T	D	V_2	T	D	V_2	T	D	V_2	T	D	V_2	T	D	V_2	T	D	V_2
15	24.7	1.14	0.79	15.0	1.25	1.18	11.0	1.40	1.43	7.8	1.56	1.81															
20	32.8	1.14	0.79	20.0	1.25	1.18	14.5	1.37	1.49	10.3	1.53	1.88	7.5	1.71	2.29	6.3	1.91	2.44									
25	41.0	1.14	0.79	24.9	1.24	1.20	18.1	1.37	1.50	12.7	1.48	1.96	9.3	1.68	2.36	7.7	1.82	2.62									
30	49.0	1.14	0.80	29.8	1.24	1.20	21.6	1.35	1.52	15.2	1.48	1.98	11.0	1.63	2.47	9.1	1.77	2.76	7.6	1.97	2.95						
35	57.1	1.14	0.80	34.7	1.24	1.21	25.1	1.35	1.54	17.7	1.47	1.99	12.8	1.62	2.50	10.5	1.73	2.86	8.8	1.94	3.03						
40	65.1	1.14	0.80	39.6	1.24	1.21	28.7	1.35	1.53	20.2	1.47	2.00	14.6	1.62	2.51	12.0	1.73	2.86	9.9	1.88	3.18	8.2	2.13	3.38			
45	73.1	1.14	0.80	44.5	1.24	1.21	32.2	1.35	1.54	22.6	1.46	2.03	16.4	1.61	2.52	13.5	1.73	2.86	11.1	1.87	3.21	9.1	2.08	3.52			
50	81.0	1.14	0.81	49.3	1.24	1.22	35.7	1.35	1.55	25.1	1.46	2.03	18.1	1.59	2.57	14.9	1.71	2.92	12.3	1.86	3.24	10.0	2.03	3.64	8.6	2.29	3.76
55	88.9	1.14	0.81	54.2	1.24	1.21	39.2	1.34	1.55	27.6	1.46	2.03	19.9	1.60	2.57	16.4	1.71	2.91	13.5	1.85	3.26	11.0	2.03	3.65	9.3	2.21	3.96
60	96.8	1.14	0.81	59.0	1.24	1.22	42.7	1.34	1.55	30.0	1.45	2.05	21.7	1.60	2.57	17.8	1.69	2.95	14.7	1.85	3.28	11.9	2.00	3.74	10.1	2.19	4.01
65	104.6	1.14	0.81	63.8	1.24	1.22	46.2	1.35	1.55	32.5	1.46	2.04	23.4	1.58	2.60	19.3	1.70	2.94	15.8	1.82	3.36	12.9	2.00	3.73	10.9	2.18	4.06
70	112.4	1.14	0.81	68.6	1.24	1.22	49.7	1.35	1.56	34.9	1.45	2.05	25.2	1.59	2.60	20.7	1.69	2.98	17.0	1.82	3.36	13.8	1.98	3.81	11.7	2.16	4.10
75	120.2	1.14	0.81	73.4	1.24	1.22	53.1	1.34	1.56	37.3	1.45	2.06	27.0	1.59	2.60	22.2	1.69	2.96	18.2	1.82	3.37	14.8	1.98	3.79	12.5	2.15	4.14
80	127.9	1.14	0.82	78.1	1.24	1.23	56.6	1.35	1.56	39.8	1.45	2.06	28.7	1.58	2.62	23.6	1.69	2.99	19.4	1.82	3.37	15.8	1.99	3.78	13.3	2.14	4.17
90	143.6	1.14	0.82	87.8	1.24	1.23	63.5	1.34	1.57	44.7	1.45	2.06	32.3	1.59	2.61	26.5	1.68	3.00	21.8	1.82	3.38	17.7	1.97	3.83	14.9	2.12	4.22
100	159.2	1.14	0.82	97.4	1.24	1.23	70.5	1.34	1.57	49.6	1.45	2.07	35.8	1.58	2.63	29.4	1.68	3.01	24.1	1.80	3.43	19.6	1.96	3.87	16.5	2.11	4.26
110	174.8	1.14	0.82	106.9	1.24	1.23	77.4	1.34	1.57	54.5	1.45	2.07	39.3	1.58	2.64	32.3	1.68	3.02	26.5	1.80	3.43	21.5	1.95	3.90	18.1	2.10	4.30
120	190.3	1.14	0.82	116.5	1.24	1.23	84.3	1.34	1.58	59.4	1.45	2.07	42.8	1.58	2.65	35.2	1.68	3.02	28.9	1.80	3.43	23.4	1.94	3.92	19.7	2.09	4.33
130	205.7	1.14	0.82	126.0	1.24	1.23	91.2	1.34	1.58	64.2	1.45	2.08	46.4	1.58	2.64	38.1	1.68	3.03	31.3	1.80	3.43	25.4	1.95	3.90	21.3	2.09	4.35
140	221.0	1.14	0.82	135.4	1.24	1.24	98.0	1.34	1.58	69.1	1.45	2.08	49.9	1.58	2.64	41.0	1.68	3.03	33.6	1.80	3.46	27.3	1.95	3.92	22.9	2.08	4.37
150	236.3	1.14	0.83	144.9	1.25	1.24	104.9	1.34	1.58	73.9	1.45	2.08	53.4	1.58	2.65	43.9	1.68	3.03	36.0	1.80	3.45	29.2	1.95	3.93	24.5	2.08	4.38
160	251.5	1.14	0.83	154.3	1.25	1.24	111.7	1.34	1.59	78.7	1.45	2.09	56.9	1.58	2.65	46.7	1.67	3.05	38.3	1.79	3.47	31.1	1.94	3.94	26.1	2.08	4.40
170	266.6	1.14	0.83	163.7	1.25	1.24	118.5	1.34	1.59	83.6	1.45	2.08	60.3	1.58	2.66	49.6	1.68	3.05	40.7	1.80	3.46	33.0	1.94	3.95	27.7	2.07	4.41
180	281.7	1.14	0.83	173.0	1.25	1.24	125.2	1.34	1.59	88.4	1.45	2.09	63.8	1.58	2.66	52.5	1.68	3.04	43.0	1.79	3.48	34.9	1.94	3.96	29.3	2.07	4.41
190	296.7	1.14	0.83	182.3	1.25	1.24	132.0	1.34	1.59	93.2	1.45	2.09	67.3	1.58	2.66	55.3	1.68	3.06	45.4	1.80	3.47	36.8	1.94	3.97	30.8	2.06	4.46
200	311.7	1.14	0.83	191.6	1.25	1.24	138.7	1.34	1.60	97.9	1.45	2.10	70.7	1.58	2.67	58.2	1.68	3.05	47.7	1.79	3.49	38.7	1.94	3.97	32.4	2.06	4.47
220	342.1	1.14	0.83	210.4	1.25	1.25	152.4	1.35	1.60	107.6	1.45	2.10	77.7	1.58	2.67	63.9	1.68	3.06	52.5	1.80	3.47	42.5	1.93	3.99	35.6	2.06	4.48
240	372.4	1.14	0.84	229.2	1.25	1.25	165.9	1.34	1.60	117.2	1.45	2.10	84.7	1.58	2.67	69.7	1.68	3.06	57.2	1.80	3.48	46.4	1.94	3.97	38.8	2.06	4.49
260	402.5	1.14	0.84	247.9	1.25	1.25	179.5	1.34	1.60	126.8	1.45	2.10	91.7	1.58	2.67	75.4	1.68	3.07	61.9	1.80	3.49	50.2	1.94	3.99	42.0	2.06	4.49
280	432.6	1.14	0.84	266.5	1.25	1.25	193.0	1.34	1.61	136.4	1.45	2.10	98.6	1.58	2.68	81.1	1.68	3.07	66.6	1.80	3.49	54.0	1.94	3.99	45.2	2.06	4.49
300	462.5	1.14	0.84	285.1	1.25	1.25	206.5	1.35	1.61	146.0	1.46	2.10	105.5	1.58	2.69	86.8	1.68	3.07	71.2	1.79	3.51	57.8	1.94	4.00	48.4	2.06	4.50

TABLE B.3 *(Continued)* Parabolic Waterway Design
V_1 for *RETARDANCE "D"*. Top Width (T), Depth (D) and V_2 for *RETARDANCE "B"*.
Grade 3.0 Percent

Q	V_1 = 2.0			V_1 = 2.5			V_1 = 3.0			V_1 = 3.5			V_1 = 4.0		
cfs	T	D	V_2	T	D	V_2	T	D	V_2	T	D	V_2	T	D	V_2
15	28.1	0.98	0.80	19.2	1.05	1.10	13.4	1.15	1.44	10.3	1.24	1.74	7.6	1.37	2.12
20	37.4	0.98	0.80	25.6	1.05	1.10	17.8	1.14	1.46	13.7	1.23	1.76	10.0	1.33	2.22
25	46.7	0.99	0.80	31.9	1.05	1.11	22.2	1.14	1.47	17.0	1.21	1.80	12.5	1.33	2.23
30	55.9	0.99	0.81	38.2	1.05	1.11	26.6	1.14	1.47	20.3	1.20	1.82	14.9	1.31	2.27
35	65.0	0.99	0.81	44.4	1.04	1.12	31.0	1.14	1.47	23.7	1.21	1.81	17.3	1.30	2.30
40	74.1	0.99	0.81	50.6	1.04	1.12	35.3	1.14	1.48	27.0	1.20	1.83	19.7	1.29	2.33
45	83.2	0.99	0.81	56.8	1.04	1.13	39.7	1.14	1.48	30.3	1.20	1.84	22.1	1.29	2.34
50	92.2	0.99	0.81	63.0	1.04	1.13	44.0	1.14	1.48	33.6	1.20	1.84	24.6	1.30	2.32
55	101.1	0.99	0.82	69.1	1.04	1.13	48.3	1.14	1.49	36.9	1.20	1.84	27.0	1.30	2.34
60	110.1	0.99	0.82	75.3	1.05	1.13	52.6	1.14	1.49	40.2	1.20	1.85	29.4	1.29	2.34
65	118.9	0.99	0.82	81.4	1.05	1.13	56.9	1.14	1.49	43.4	1.20	1.86	31.8	1.29	2.35
70	127.8	0.99	0.82	87.4	1.04	1.14	61.1	1.14	1.50	46.7	1.20	1.86	34.2	1.29	2.35
75	136.6	0.99	0.82	93.5	1.05	1.14	65.4	1.14	1.50	49.9	1.20	1.87	36.6	1.29	2.35
80	145.3	0.99	0.83	99.5	1.05	1.14	69.6	1.14	1.50	53.1	1.20	1.87	38.9	1.29	2.37
90	163.1	0.99	0.83	111.7	1.05	1.14	78.2	1.14	1.50	59.7	1.20	1.87	43.7	1.29	2.38
100	180.8	0.99	0.83	123.8	1.05	1.15	86.7	1.14	1.51	66.2	1.20	1.88	48.5	1.29	2.38
110	198.3	0.99	0.83	135.9	1.05	1.15	95.2	1.14	1.51	72.7	1.20	1.88	53.3	1.29	2.38
120	215.8	0.99	0.83	148.0	1.05	1.15	103.7	1.14	1.51	79.2	1.20	1.88	58.1	1.29	2.38
130	233.3	0.99	0.83	160.0	1.05	1.15	112.1	1.14	1.51	85.7	1.20	1.88	62.8	1.29	2.39
140	250.6	0.99	0.84	171.9	1.05	1.15	120.5	1.14	1.52	92.1	1.20	1.89	67.6	1.29	2.38
150	267.8	0.99	0.84	183.8	1.05	1.16	128.9	1.14	1.52	98.5	1.20	1.89	72.3	1.29	2.39
160	285.0	0.99	0.84	195.7	1.05	1.16	137.2	1.14	1.52	104.9	1.20	1.89	77.0	1.29	2.39
170	302.0	0.99	0.84	207.5	1.05	1.16	145.5	1.14	1.52	111.3	1.20	1.89	81.7	1.29	2.40
180	319.0	0.99	0.84	219.2	1.05	1.16	153.8	1.14	1.53	117.7	1.20	1.89	86.4	1.29	2.40
190	335.9	0.99	0.85	231.0	1.05	1.16	162.1	1.14	1.53	124.0	1.20	1.90	91.1	1.30	2.40
200	352.7	0.99	0.85	242.6	1.05	1.16	170.3	1.14	1.53	130.3	1.20	1.90	95.7	1.29	2.40
220	387.1	0.99	0.85	266.3	1.05	1.17	187.0	1.14	1.53	143.1	1.20	1.90	105.1	1.29	2.41
240	421.2	0.99	0.85	289.9	1.05	1.17	203.6	1.14	1.53	155.9	1.20	1.90	114.5	1.29	2.41
260	455.2	0.99	0.85	313.4	1.05	1.17	220.2	1.14	1.54	168.6	1.20	1.91	123.9	1.30	2.41
280	489.0	0.99	0.85	336.9	1.05	1.17	236.7	1.14	1.54	181.3	1.20	1.91	133.3	1.30	2.41
300	522.6	0.99	0.86	360.2	1.05	1.17	253.2	1.15	1.54	193.9	1.20	1.91	142.6	1.30	2.42

Q	V_1 = 4.5			V_1 = 5.0			V_1 = 5.5			V_1 = 6.0		
cfs	T	D	V_2	T	D	V_2	T	D	V_2	T	D	V_2
15	5.8	1.56	2.44									
20	7.6	1.50	2.59	6.8	1.58	2.74						
25	9.4	1.46	2.69	8.4	1.54	2.85	6.9	1.68	3.18			
30	11.2	1.44	2.75	9.9	1.48	3.02	8.2	1.64	3.28	6.7	1.88	3.50
35	13.0	1.43	2.80	11.5	1.47	3.07	9.5	1.62	3.36	7.7	1.82	3.68
40	14.8	1.42	2.83	13.1	1.46	3.10	10.8	1.60	3.42	8.7	1.78	3.82
45	16.6	1.41	2.85	14.7	1.45	3.12	12.1	1.59	3.46	9.7	1.75	3.92
50	18.4	1.40	2.87	16.3	1.45	3.14	13.4	1.58	3.50	10.7	1.72	4.01
55	20.2	1.40	2.88	17.9	1.45	3.15	14.7	1.57	3.53	11.7	1.70	4.09
60	22.0	1.40	2.89	19.5	1.45	3.16	16.0	1.57	3.55	12.7	1.69	4.15
65	23.8	1.40	2.90	21.1	1.45	3.16	17.3	1.57	3.56	13.8	1.70	4.11
70	25.6	1.40	2.90	22.7	1.45	3.17	18.5	1.55	3.63	14.8	1.69	4.16
75	27.4	1.40	2.91	24.2	1.44	3.21	19.8	1.55	3.64	15.8	1.68	4.20
80	29.2	1.40	2.91	25.8	1.44	3.21	21.1	1.55	3.64	16.9	1.69	4.16
90	32.7	1.39	2.94	29.0	1.44	3.21	23.7	1.54	3.66	18.9	1.67	4.23
100	36.3	1.39	2.95	32.2	1.44	3.21	26.3	1.54	3.67	21.0	1.67	4.23
110	39.9	1.39	2.95	35.4	1.44	3.21	28.9	1.54	3.67	23.1	1.68	4.22
120	43.5	1.39	2.95	38.5	1.43	3.23	31.5	1.54	3.68	25.1	1.67	4.27
130	47.0	1.39	2.96	41.7	1.44	3.23	34.1	1.54	3.68	27.2	1.67	4.26
140	50.6	1.39	2.96	44.8	1.43	3.24	36.7	1.54	3.68	29.3	1.67	4.25
150	54.1	1.39	2.97	48.0	1.44	3.24	39.3	1.55	3.68	31.3	1.67	4.28
160	57.7	1.39	2.96	51.1	1.44	3.25	41.8	1.54	3.70	33.4	1.67	4.27
170	61.2	1.39	2.97	54.3	1.44	3.24	44.4	1.54	3.70	35.4	1.67	4.29
180	64.7	1.39	2.98	57.4	1.44	3.25	47.0	1.55	3.69	37.5	1.67	4.28
190	68.2	1.39	2.98	60.5	1.44	3.25	49.5	1.54	3.71	39.5	1.67	4.30
200	71.7	1.39	2.99	63.6	1.44	3.26	52.1	1.55	3.70	41.6	1.67	4.28
220	78.8	1.39	2.99	69.9	1.44	3.26	57.2	1.54	3.71	45.7	1.67	4.29
240	85.9	1.39	2.99	76.2	1.44	3.26	62.3	1.54	3.72	49.8	1.67	4.30
260	92.9	1.39	2.99	82.4	1.44	3.27	67.5	1.55	3.72	53.9	1.67	4.31
280	99.9	1.39	3.00	88.6	1.44	3.27	72.6	1.55	3.72	58.0	1.67	4.31
300	106.9	1.39	3.00	94.9	1.44	3.27	77.7	1.55	3.73	62.0	1.67	4.33

TABLE B.3 *(Continued)* Parabolic Waterway Design

V_1 for *RETARDANCE "D"*. Top Width (T), Depth (D) and V_2 for *RETARDANCE "B"*.

Grade 4.0 Percent

Q cfs	V_1 = 2.0 T	D	V_2	V_1 = 2.5 T	D	V_2	V_1 = 3.0 T	D	V_2
15	33.1	0.87	0.77	23.6	0.93	1.01	16.3	1.00	1.36
20	44.0	0.87	0.77	31.4	0.93	1.01	21.7	1.00	1.37
25	54.9	0.87	0.77	39.1	0.93	1.02	27.1	1.00	1.37
30	65.7	0.87	0.78	46.8	0.93	1.02	32.4	0.99	1.38
35	76.4	0.87	0.78	54.5	0.93	1.02	37.7	0.99	1.39
40	87.1	0.87	0.78	62.1	0.93	1.03	43.0	0.99	1.39
45	97.7	0.87	0.78	69.7	0.93	1.03	48.3	0.99	1.39
50	108.3	0.87	0.78	77.3	0.93	1.03	53.5	0.99	1.40
55	118.8	0.87	0.78	84.7	0.93	1.04	58.7	0.99	1.40
60	129.2	0.87	0.79	92.2	0.93	1.04	63.9	0.99	1.40
65	139.6	0.87	0.79	99.6	0.93	1.04	69.1	0.99	1.41
70	149.9	0.88	0.79	107.0	0.93	1.04	74.2	0.99	1.41
75	160.2	0.88	0.79	114.3	0.93	1.05	79.3	0.99	1.42
80	170.4	0.88	0.79	121.6	0.93	1.05	84.4	0.99	1.42
90	191.1	0.88	0.80	136.5	0.93	1.05	94.8	0.99	1.42
100	211.8	0.88	0.80	151.2	0.93	1.05	105.1	0.99	1.42
110	232.3	0.88	0.80	165.9	0.93	1.06	115.3	0.99	1.43
120	252.7	0.88	0.80	180.5	0.93	1.06	125.5	0.99	1.43
130	273.0	0.88	0.80	195.1	0.93	1.06	135.6	0.99	1.43
140	293.2	0.88	0.80	209.5	0.93	1.06	145.7	0.99	1.44
150	313.3	0.88	0.81	223.9	0.93	1.07	155.8	0.99	1.44
160	333.2	0.88	0.81	238.2	0.93	1.07	165.8	1.00	1.44
170	353.1	0.88	0.81	252.4	0.93	1.07	175.8	1.00	1.44
180	372.8	0.88	0.81	266.5	0.93	1.07	185.7	1.00	1.45
190	392.4	0.88	0.81	280.6	0.93	1.08	195.6	1.00	1.45
200	411.9	0.88	0.82	294.6	0.93	1.08	205.4	1.00	1.45
220	451.9	0.88	0.82	323.2	0.93	1.08	225.4	1.00	1.46
240	491.6	0.88	0.82	351.6	0.93	1.09	245.3	1.00	1.46
260	531.0	0.88	0.82	379.9	0.93	1.09	265.2	1.00	1.46
280	570.3	0.88	0.82	408.0	0.93	1.09	284.9	1.00	1.47
300	609.3	0.88	0.83	436.0	0.93	1.09	304.6	1.00	1.47

Q cfs	V_1 = 3.5 T	D	V_2	V_1 = 4.0 T	D	V_2	V_1 = 4.5 T	D	V_2
15	12.1	1.08	1.69	9.2	1.15	2.08	7.4	1.27	2.34
20	16.1	1.08	1.70	12.3	1.16	2.08	9.7	1.23	2.48
25	20.1	1.08	1.71	15.3	1.15	2.11	12.1	1.22	2.50
30	24.0	1.07	1.73	18.3	1.14	2.13	14.4	1.20	2.57
35	27.9	1.07	1.74	21.3	1.14	2.14	16.8	1.20	2.57
40	31.9	1.07	1.74	24.3	1.14	2.15	19.2	1.21	2.56
45	35.8	1.07	1.74	27.2	1.13	2.17	21.5	1.20	2.59
50	39.7	1.07	1.75	30.2	1.13	2.17	23.9	1.20	2.58
55	43.6	1.07	1.75	33.2	1.13	2.17	26.2	1.20	2.60
60	47.4	1.07	1.76	36.1	1.13	2.18	28.5	1.19	2.62
65	51.3	1.07	1.76	39.0	1.13	2.19	30.9	1.20	2.61
70	55.1	1.07	1.77	42.0	1.13	2.19	33.2	1.20	2.62
75	58.9	1.07	1.77	44.9	1.13	2.19	35.5	1.20	2.62
80	62.7	1.07	1.77	47.8	1.13	2.20	37.8	1.20	2.63
90	70.4	1.07	1.78	53.7	1.13	2.20	42.4	1.19	2.65
100	78.1	1.07	1.78	59.6	1.13	2.20	47.1	1.20	2.64
110	85.8	1.07	1.78	65.4	1.13	2.21	51.7	1.19	2.65
120	93.4	1.07	1.78	71.2	1.13	2.22	56.3	1.19	2.66
130	101.0	1.07	1.79	77.0	1.13	2.22	60.9	1.19	2.66
140	108.5	1.07	1.79	82.8	1.13	2.22	65.5	1.20	2.66
150	116.0	1.07	1.80	88.6	1.13	2.22	70.1	1.20	2.66
160	123.5	1.07	1.80	94.3	1.13	2.23	74.6	1.20	2.67
170	131.0	1.07	1.80	100.0	1.13	2.23	79.2	1.20	2.67
180	138.4	1.07	1.80	105.7	1.13	2.24	83.7	1.20	2.68
190	145.8	1.07	1.81	111.4	1.13	2.24	88.2	1.20	2.68
200	153.2	1.07	1.81	117.1	1.14	2.24	92.7	1.20	2.68
220	168.2	1.07	1.81	128.6	1.14	2.24	101.9	1.20	2.68
240	183.2	1.07	1.81	140.0	1.14	2.25	110.9	1.20	2.69
260	198.0	1.07	1.82	151.5	1.14	2.25	120.0	1.20	2.69
280	212.9	1.08	1.82	162.8	1.14	2.25	129.0	1.20	2.70
300	227.6	1.08	1.83	174.2	1.14	2.25	138.1	1.20	2.69

Q cfs	V_1 = 5.0 T	D	V_2	V_1 = 5.5 T	D	V_2	V_1 = 6.0 T	D	V_2
15	5.7	1.43	2.69						
20	7.4	1.35	2.94	6.4	1.47	3.12			
25	9.2	1.33	3.01	7.9	1.43	3.26	6.5	1.56	3.62
30	11.0	1.32	3.05	9.3	1.38	3.46	7.7	1.52	3.78
35	12.7	1.30	3.14	10.9	1.39	3.42	8.9	1.49	3.90
40	14.5	1.30	3.15	12.4	1.38	3.47	10.2	1.50	3.87
45	16.3	1.30	3.16	13.9	1.37	3.51	11.4	1.48	3.95
50	18.1	1.30	3.16	15.4	1.36	3.54	12.6	1.46	4.01
55	19.8	1.29	3.21	16.9	1.36	3.56	13.8	1.45	4.06
60	21.6	1.29	3.20	18.4	1.36	3.57	15.1	1.46	4.02
65	23.4	1.29	3.19	19.9	1.35	3.58	16.3	1.46	4.06
70	25.1	1.28	3.23	21.4	1.35	3.59	17.5	1.45	4.09
75	26.9	1.29	3.22	22.9	1.35	3.60	18.7	1.45	4.12
80	28.6	1.28	3.24	24.4	1.35	3.60	20.0	1.46	4.08
90	32.1	1.28	3.26	27.4	1.35	3.62	22.4	1.45	4.13
100	35.7	1.29	3.24	30.4	1.35	3.63	24.9	1.45	4.12
110	39.2	1.29	3.25	33.4	1.35	3.63	27.3	1.44	4.16
120	42.7	1.28	3.26	36.3	1.34	3.66	29.8	1.45	4.14
130	46.2	1.29	3.26	39.3	1.34	3.66	32.2	1.44	4.16
140	49.6	1.28	3.28	42.3	1.35	3.66	34.6	1.44	4.18
150	53.1	1.28	3.28	45.3	1.35	3.66	37.1	1.45	4.17
160	56.6	1.28	3.28	48.2	1.35	3.67	39.5	1.44	4.18
170	60.0	1.28	3.29	51.2	1.35	3.67	41.9	1.44	4.19
180	63.5	1.28	3.29	54.1	1.35	3.68	44.3	1.44	4.20
190	66.9	1.28	3.30	57.0	1.34	3.69	46.7	1.44	4.21
200	70.3	1.28	3.30	60.0	1.35	3.68	49.1	1.44	4.21
220	77.2	1.28	3.31	65.9	1.35	3.69	54.0	1.44	4.21
240	84.1	1.28	3.32	71.8	1.35	3.70	58.8	1.44	4.22
260	91.0	1.28	3.32	77.7	1.35	3.70	63.7	1.44	4.21
280	97.9	1.28	3.32	83.5	1.35	3.71	68.5	1.44	4.22
300	104.8	1.29	3.32	89.4	1.35	3.71	73.3	1.44	4.23

TABLE B.3 *(Continued)* Parabolic Waterway Design
V_1 for *RETARDANCE "D"*. Top Width (T), Depth (D) and V_2 for *RETARDANCE "B"*.
Grade 5.0 Percent

Q cfs	V_1 = 2.0			V_1 = 2.5			V_1 = 3.0			V_1 = 3.5			V_1 = 4.0			V_1 = 4.5			V_1 = 5.0			V_1 = 5.5			V_1 = 6.0		
	T	D	V_2	T	D	V_2	T	D	V_2	T	D	V_2	T	D	V_2	T	D	V_2	T	D	V_2	T	D	V_2	T	D	V_2
15	34.8	0.80	0.79	25.2	0.86	1.02	17.7	0.92	1.36	14.3	0.96	1.61	10.6	1.04	2.00	8.4	1.12	2.34	6.7	1.23	2.67	5.4	1.36	3.00			
20	46.3	0.80	0.79	33.5	0.86	1.02	23.5	0.92	1.37	19.0	0.96	1.62	14.1	1.04	2.02	11.1	1.10	2.41	8.8	1.19	2.81	7.0	1.28	3.29	6.1	1.40	3.44
25	57.7	0.80	0.80	41.7	0.86	1.03	29.3	0.92	1.38	23.7	0.96	1.63	17.5	1.03	2.06	13.8	1.09	2.46	10.9	1.17	2.90	8.7	1.26	3.37	7.5	1.35	3.64
30	69.0	0.80	0.80	49.9	0.86	1.04	35.0	0.91	1.39	28.3	0.95	1.65	21.0	1.03	2.06	16.6	1.10	2.44	13.0	1.15	2.96	10.4	1.25	3.41	8.9	1.32	3.78
35	80.3	0.80	0.80	58.1	0.86	1.04	40.8	0.91	1.39	33.0	0.95	1.65	24.4	1.02	2.08	19.3	1.09	2.46	15.2	1.16	2.94	12.1	1.24	3.45	10.3	1.30	3.87
40	91.5	0.81	0.80	66.2	0.86	1.04	46.5	0.91	1.40	37.6	0.95	1.66	27.8	1.02	2.09	22.0	1.09	2.47	17.3	1.15	2.97	13.8	1.24	3.47	11.8	1.30	3.85
45	102.6	0.81	0.81	74.2	0.86	1.05	52.2	0.91	1.40	42.2	0.95	1.66	31.3	1.03	2.08	24.7	1.09	2.48	19.4	1.15	2.99	15.4	1.22	3.55	13.2	1.29	3.91
50	113.7	0.81	0.81	82.3	0.86	1.05	57.9	0.92	1.40	46.8	0.95	1.66	34.7	1.03	2.08	27.3	1.08	2.51	21.5	1.15	3.01	17.1	1.22	3.55	14.7	1.30	3.88
55	124.7	0.81	0.81	90.2	0.86	1.05	63.5	0.91	1.41	51.4	0.96	1.66	38.0	1.02	2.11	30.0	1.08	2.51	23.7	1.15	2.98	18.8	1.22	3.55	16.1	1.29	3.93
60	135.7	0.81	0.81	98.2	0.86	1.05	69.1	0.91	1.41	56.0	0.96	1.66	41.4	1.02	2.11	32.7	1.08	2.51	25.8	1.15	3.00	20.5	1.22	3.55	17.5	1.28	3.97
65	146.5	0.81	0.81	106.1	0.86	1.05	74.7	0.92	1.41	60.5	0.96	1.67	44.8	1.02	2.10	35.3	1.08	2.53	27.9	1.15	3.00	22.1	1.22	3.59	18.9	1.28	4.00
70	157.3	0.81	0.82	113.9	0.86	1.06	80.2	0.91	1.42	65.0	0.96	1.67	48.1	1.02	2.11	38.0	1.08	2.53	30.0	1.15	3.01	23.8	1.22	3.58	20.4	1.29	3.96
75	168.1	0.81	0.82	121.7	0.86	1.06	85.8	0.92	1.42	69.5	0.96	1.67	51.4	1.02	2.12	40.6	1.08	2.54	32.0	1.15	3.04	25.4	1.21	3.62	21.8	1.28	3.98
80	178.8	0.81	0.82	129.4	0.86	1.06	91.3	0.92	1.42	74.0	0.96	1.67	54.8	1.02	2.12	43.3	1.09	2.53	34.1	1.15	3.04	27.1	1.22	3.61	23.2	1.28	4.00
90	200.5	0.81	0.82	145.2	0.86	1.07	102.4	0.92	1.43	83.0	0.96	1.68	61.5	1.02	2.12	48.6	1.08	2.54	38.3	1.15	3.05	30.4	1.21	3.63	26.1	1.28	4.00
100	222.2	0.81	0.82	160.9	0.86	1.07	113.5	0.92	1.43	92.1	0.96	1.68	68.2	1.02	2.13	53.9	1.08	2.54	42.5	1.15	3.05	33.8	1.22	3.61	28.9	1.28	4.03
110	243.7	0.81	0.82	176.5	0.86	1.07	124.6	0.92	1.43	101.0	0.96	1.69	74.8	1.02	2.14	59.2	1.09	2.55	46.7	1.15	3.05	37.1	1.22	3.63	31.8	1.28	4.02
120	265.0	0.81	0.83	192.0	0.86	1.07	135.6	0.92	1.43	110.0	0.96	1.69	81.4	1.02	2.15	64.4	1.08	2.56	50.8	1.15	3.07	40.4	1.21	3.64	34.6	1.28	4.04
130	286.3	0.81	0.83	207.5	0.86	1.08	146.6	0.92	1.43	118.9	0.96	1.69	88.1	1.02	2.14	69.7	1.09	2.56	55.0	1.15	3.06	43.7	1.21	3.65	37.5	1.28	4.03
140	307.4	0.81	0.83	222.8	0.86	1.08	157.5	0.92	1.44	127.8	0.96	1.69	94.6	1.02	2.15	74.9	1.08	2.57	59.1	1.15	3.07	47.0	1.21	3.65	40.3	1.28	4.04
150	328.4	0.81	0.83	238.1	0.87	1.08	168.3	0.92	1.44	136.6	0.96	1.70	101.2	1.02	2.15	80.1	1.08	2.57	63.2	1.15	3.08	50.2	1.21	3.67	43.1	1.28	4.05
160	349.3	0.81	0.83	253.2	0.86	1.08	179.1	0.92	1.44	145.4	0.96	1.70	107.7	1.02	2.16	85.3	1.09	2.57	67.3	1.15	3.09	53.5	1.21	3.67	45.9	1.28	4.06
170	370.0	0.81	0.84	268.3	0.87	1.09	189.9	0.92	1.45	154.1	0.96	1.70	114.2	1.02	2.16	90.4	1.08	2.58	71.4	1.15	3.09	56.8	1.21	3.67	48.7	1.28	4.07
180	390.7	0.82	0.84	283.3	0.87	1.09	200.6	0.92	1.45	162.9	0.96	1.70	120.7	1.02	2.17	95.6	1.09	2.58	75.5	1.15	3.09	60.0	1.21	3.69	51.5	1.28	4.07
190	411.2	0.82	0.84	298.2	0.87	1.09	211.2	0.92	1.45	171.5	0.96	1.71	127.1	1.02	2.17	100.7	1.08	2.59	79.6	1.15	3.09	63.3	1.21	3.68	54.3	1.28	4.07
200	431.6	0.82	0.84	313.0	0.87	1.09	221.8	0.92	1.46	180.2	0.96	1.71	133.5	1.02	2.18	105.8	1.09	2.59	83.6	1.15	3.10	66.5	1.21	3.69	57.1	1.28	4.08
220	473.3	0.82	0.84	343.4	0.87	1.10	243.4	0.92	1.46	197.8	0.97	1.71	146.6	1.02	2.18	116.2	1.09	2.60	91.9	1.15	3.10	73.1	1.22	3.69	62.7	1.28	4.09
240	514.8	0.82	0.84	373.6	0.87	1.10	264.9	0.92	1.46	215.3	0.97	1.72	159.6	1.02	2.18	126.5	1.09	2.60	100.1	1.15	3.10	79.6	1.21	3.70	68.3	1.28	4.09
260	556.1	0.82	0.85	403.6	0.87	1.10	286.3	0.92	1.46	232.7	0.97	1.72	172.6	1.03	2.19	136.8	1.09	2.61	108.2	1.15	3.11	86.1	1.21	3.71	73.9	1.28	4.10
280	597.1	0.82	0.85	433.5	0.87	1.10	307.6	0.92	1.47	250.1	0.97	1.72	185.5	1.03	2.19	147.1	1.09	2.61	116.4	1.15	3.11	92.6	1.21	3.71	79.5	1.28	4.10
300	638.0	0.82	0.85	463.1	0.87	1.11	328.7	0.92	1.47	267.4	0.97	1.73	198.4	1.03	2.19	157.3	1.09	2.61	124.5	1.15	3.12	99.1	1.22	3.71	85.1	1.28	4.10

TABLE B.3 *(Continued)* Parabolic Waterway Design
V_1 for *RETARDANCE "D"*. Top Width (T), Depth (D) and V_2 for *RETARDANCE "B"*.
Grade 6.0 Percent

Q	V_1 = 2.0			V_1 = 2.5			V_1 = 3.0		
cfs	T	D	V_2	T	D	V_2	T	D	V_2
15	40.5	0.72	0.76	26.9	0.81	1.02	19.6	0.85	1.34
20	53.9	0.72	0.76	35.7	0.81	1.03	26.1	0.85	1.34
25	67.1	0.72	0.76	44.5	0.81	1.03	32.5	0.85	1.35
30	80.3	0.72	0.76	53.2	0.80	1.04	38.9	0.85	1.35
35	93.4	0.72	0.76	61.9	0.81	1.04	45.3	0.85	1.35
40	106.4	0.73	0.77	70.6	0.81	1.04	51.6	0.85	1.36
45	119.3	0.73	0.77	79.2	0.81	1.04	57.9	0.85	1.36
50	132.1	0.73	0.77	87.7	0.81	1.05	64.1	0.84	1.37
55	144.9	0.73	0.77	96.2	0.81	1.05	70.4	0.85	1.37
60	157.5	0.73	0.77	104.6	0.81	1.05	76.5	0.84	1.38
65	170.1	0.73	0.78	113.0	0.81	1.06	82.7	0.85	1.38
70	182.6	0.73	0.78	121.3	0.81	1.06	88.8	0.85	1.38
75	195.1	0.73	0.78	129.6	0.81	1.06	94.9	0.85	1.39
80	207.4	0.73	0.78	137.8	0.81	1.06	100.9	0.85	1.39
90	232.6	0.73	0.78	154.5	1.81	1.07	113.2	0.85	1.40
100	257.7	0.73	0.78	171.2	0.81	1.07	125.5	0.85	1.40
110	282.6	0.74	0.78	187.8	0.81	1.07	137.7	0.85	1.40
120	307.3	0.74	0.79	204.2	0.81	1.08	149.8	0.85	1.40
130	331.9	0.74	0.79	220.6	0.81	1.08	161.9	0.85	1.41
140	356.3	0.74	0.79	236.8	0.81	1.08	173.9	0.85	1.41
150	380.5	0.74	0.79	253.0	0.81	1.09	185.8	0.85	1.41
160	404.6	0.74	0.79	269.0	0.81	1.09	197.7	0.85	1.42
170	428.6	0.74	0.79	285.0	0.81	1.09	209.5	0.85	1.42
180	452.4	0.74	0.79	300.8	0.81	1.09	221.3	0.85	1.42
190	476.1	0.74	0.80	316.6	0.81	1.10	233.0	0.85	1.42
200	499.6	0.74	0.80	332.3	0.81	1.10	244.6	0.85	1.43
220	547.7	0.74	0.80	364.5	0.81	1.10	268.4	0.85	1.43
240	595.6	0.74	0.80	396.5	0.81	1.11	292.1	0.85	1.43
260	643.2	0.74	0.80	428.3	0.81	1.11	315.7	0.85	1.43
280	690.5	0.75	0.81	459.9	0.81	1.11	339.1	0.85	1.44
300	737.5	0.75	0.81	491.4	0.81	1.11	362.4	0.85	1.44

Q	V_1 = 3.5			V_1 = 4.0			V_1 = 4.5		
cfs	T	D	V_2	T	D	V_2	T	D	V_2
15	14.8	0.90	1.66	11.7	0.96	1.97	9.4	1.02	2.31
20	19.6	0.89	1.69	15.6	0.96	1.97	12.5	1.01	2.33
25	24.5	0.90	1.69	19.4	0.95	2.00	15.5	1.00	2.39
30	29.3	0.89	1.70	23.3	0.96	1.99	18.6	1.00	2.39
35	34.1	0.89	1.70	27.1	0.96	2.00	21.6	0.99	2.42
40	38.9	0.89	1.71	30.9	0.95	2.01	24.7	1.00	2.41
45	43.7	0.90	1.71	34.6	0.95	2.03	27.7	1.00	2.42
50	48.4	0.89	1.71	38.4	0.95	2.03	30.7	0.99	2.43
55	53.1	0.89	1.72	42.2	0.95	2.03	33.7	0.99	2.44
60	57.8	0.89	1.72	45.9	0.95	2.04	36.7	1.00	2.44
65	62.5	0.90	1.72	49.6	0.95	2.04	39.7	1.00	2.44
70	67.2	0.90	1.72	53.3	0.95	2.05	42.7	1.00	2.44
75	71.8	0.90	1.73	57.0	0.95	2.05	45.6	1.00	2.46
80	76.4	0.90	1.73	60.6	0.95	2.06	48.6	1.00	2.45
90	85.8	0.90	1.73	68.1	0.95	2.06	54.6	1.00	2.45
100	95.1	0.90	1.74	75.5	0.95	2.06	60.5	1.00	2.46
110	104.4	0.90	1.74	82.8	0.95	2.07	66.4	1.00	2.47
120	113.6	0.90	1.74	90.2	0.95	2.07	72.3	1.00	2.47
130	122.8	0.90	1.75	97.5	0.95	2.08	78.2	1.00	2.48
140	131.9	0.90	1.75	104.7	0.95	2.09	84.0	1.00	2.48
150	141.0	0.90	1.75	112.0	0.96	2.09	89.9	1.00	2.48
160	150.1	0.90	1.76	119.2	0.96	2.09	95.7	1.00	2.49
170	159.1	0.90	1.76	126.4	0.96	2.09	101.5	1.00	2.49
180	168.1	0.90	1.76	133.5	0.96	2.10	107.2	1.00	2.50
190	177.0	0.90	1.77	140.6	0.96	2.10	113.0	1.00	2.50
200	185.9	0.90	1.77	147.7	0.96	2.11	118.7	1.00	2.50
220	204.1	0.91	1.77	162.1	0.96	2.11	130.3	1.00	2.51
240	222.1	0.91	1.78	176.5	0.96	2.11	141.9	1.00	2.51
260	240.1	0.91	1.78	190.7	0.96	2.12	153.4	1.00	2.52
280	258.0	0.91	1.78	205.0	0.96	2.12	164.9	1.00	2.52
300	275.8	0.91	1.78	219.1	0.96	2.13	176.3	1.00	2.52

Q	V_1 = 5.0			V_1 = 5.5			V_1 = 6.0		
cfs	T	D	V_2	T	D	V_2	T	D	V_2
15	7.6	1.09	2.67	6.1	1.18	3.05	5.0	1.32	3.33
20	10.1	1.08	2.71	8.0	1.14	3.23	6.5	1.24	3.63
25	12.5	1.06	2.80	10.0	1.14	3.24	8.1	1.23	3.69
30	15.0	1.06	2.80	11.9	1.12	3.32	9.6	1.20	3.84
35	17.4	1.05	2.84	13.9	1.13	3.31	11.2	1.20	3.84
40	19.9	1.05	2.83	15.8	1.12	3.36	12.7	1.19	3.93
45	22.3	1.05	2.86	17.7	1.11	3.40	14.3	1.19	3.92
50	24.7	1.05	2.87	19.7	1.12	3.38	15.8	1.18	3.98
55	27.1	1.04	2.89	21.6	1.11	3.40	17.4	1.18	3.96
60	29.6	1.05	2.87	23.5	1.11	3.42	18.9	1.18	4.01
65	32.0	1.05	2.88	25.4	1.11	3.43	20.5	1.18	3.98
70	34.4	1.05	2.88	27.4	1.11	3.41	22.0	1.18	4.01
75	36.7	1.04	2.91	29.3	1.11	3.42	23.6	1.18	3.99
80	39.1	1.05	2.91	31.2	1.11	3.42	25.1	1.18	4.02
90	43.9	1.05	2.92	35.0	1.11	3.44	28.2	1.18	4.02
100	48.7	1.05	2.92	38.8	1.11	3.46	31.3	1.18	4.03
110	53.5	1.05	2.92	42.6	1.11	3.47	34.4	1.18	4.03
120	58.3	1.05	2.92	46.4	1.11	3.47	37.4	1.18	4.06
130	63.0	1.05	2.93	50.2	1.11	3.48	40.5	1.18	4.05
140	67.7	1.05	2.94	54.0	1.11	3.48	43.5	1.18	4.07
150	72.5	1.05	2.93	57.8	1.11	3.48	46.6	1.18	4.06
160	77.2	1.05	2.94	61.5	1.11	3.49	49.6	1.18	4.08
170	81.8	1.05	2.95	65.3	1.11	3.49	52.7	1.18	4.07
180	86.5	1.05	2.95	69.0	1.11	3.49	55.7	1.18	4.08
190	91.1	1.05	2.96	72.7	1.11	3.50	58.7	1.18	4.08
200	95.8	1.05	2.96	76.4	1.11	3.51	61.7	1.18	4.09
220	105.2	1.05	2.96	83.9	1.11	3.51	67.8	1.18	4.09
240	114.5	1.05	2.97	91.4	1.11	3.52	73.8	1.18	4.11
260	123.9	1.05	2.97	98.9	1.11	3.52	79.9	1.18	4.10
280	133.2	1.05	2.97	106.3	1.11	3.53	85.9	1.18	4.11
300	142.4	1.05	2.98	113.7	1.11	3.53	91.9	1.18	4.12

TABLE B.3 *(Continued)* Parabolic Waterway Design
V_1 for *RETARDANCE "D"*. Top Width (T), Depth (D) and V_2 for *RETARDANCE "B"*.
Grade 8.0 Percent

Q	V_1 = 2.0			V_1 = 2.5			V_1 = 3.0			V_1 = 3.5			V_1 = 4.0			V_1 = 4.5			V_1 = 5.0			V_1 = 5.5			V_1 = 6.0		
cfs	T	D	V_2	T	D	V_2	T	D	V_2	T	D	V_2	T	D	V_2	T	D	V_2	T	D	V_2	T	D	V_2	T	D	V_2
15	43.1	0.64	0.80	31.6	0.71	0.98	22.2	0.76	1.31	18.0	0.79	1.56	13.7	0.84	1.92	11.0	0.89	2.26	9.2	0.95	2.54	7.4	0.99	3.00	6.2	1.09	3.26
20	57.3	0.64	0.80	42.0	0.71	0.99	29.5	0.76	1.32	24.0	0.79	1.56	18.2	0.84	1.94	14.7	0.89	2.25	12.2	0.93	2.59	9.8	0.98	3.08	8.1	1.04	3.49
25	71.4	0.65	0.80	52.3	0.71	0.99	36.7	0.76	1.33	29.9	0.79	1.57	22.7	0.84	1.95	18.3	0.89	2.28	15.2	0.93	2.62	12.2	0.97	3.12	10.1	1.04	3.53
30	85.4	0.65	0.80	62.6	0.71	0.99	43.9	0.76	1.33	35.8	0.79	1.57	27.2	0.84	1.95	21.9	0.89	2.29	18.1	0.92	2.67	14.6	0.97	3.15	12.1	1.03	3.55
35	99.3	0.65	0.80	72.8	0.72	0.99	51.1	0.76	1.34	41.6	0.79	1.58	31.7	0.84	1.95	25.5	0.89	2.30	21.1	0.92	2.67	17.0	0.96	3.16	14.0	1.02	3.64
40	113.1	0.65	0.81	82.9	0.72	1.00	58.2	0.76	1.34	47.4	0.79	1.59	36.1	0.84	1.96	29.0	0.88	2.32	24.1	0.92	2.67	19.4	0.96	3.17	16.0	1.02	3.63
45	126.8	0.65	0.81	93.0	0.72	1.00	65.3	0.76	1.34	53.2	0.79	1.59	40.5	0.84	1.97	32.6	0.88	2.32	27.0	0.92	2.69	21.8	0.97	3.17	18.0	1.02	3.62
50	140.4	0.65	0.81	102.9	0.72	1.01	72.3	0.76	1.35	59.0	0.79	1.59	44.9	0.84	1.97	36.1	0.88	2.33	30.0	0.92	2.68	24.2	0.97	3.17	19.9	1.02	3.67
55	153.9	0.65	0.81	112.8	0.72	1.01	79.3	0.76	1.35	64.7	0.79	1.60	49.3	0.84	1.97	39.6	0.88	2.34	32.9	0.92	2.70	26.5	0.96	3.20	21.9	1.02	3.65
60	167.3	0.65	0.81	122.7	0.72	1.01	86.2	0.76	1.36	70.4	0.79	1.60	53.7	0.84	1.97	43.1	0.88	2.34	35.8	0.92	2.70	28.9	0.97	3.19	23.8	1.02	3.69
65	180.6	0.65	0.82	132.4	0.72	1.01	93.1	0.76	1.36	76.0	0.79	1.61	58.0	0.84	1.98	46.6	0.88	2.35	38.7	0.92	2.71	31.3	0.97	3.18	25.7	1.01	3.71
70	193.9	0.65	0.82	142.2	0.72	1.01	100.0	0.76	1.37	81.7	0.79	1.61	62.3	0.84	1.98	50.1	0.88	2.35	41.6	0.92	2.71	33.6	0.97	3.20	27.7	1.02	3.69
75	207.0	0.65	0.82	151.8	0.72	1.02	106.8	0.76	1.37	87.2	0.79	1.62	66.6	0.84	1.99	53.5	0.88	2.36	44.5	0.92	2.72	35.9	0.96	3.22	29.6	1.02	3.71
80	220.1	0.66	0.82	161.4	0.72	1.02	113.6	0.76	1.37	92.8	0.79	1.62	70.9	0.84	1.99	56.9	0.88	2.37	47.3	0.92	2.73	38.2	0.96	3.23	31.5	1.01	3.72
90	246.8	0.66	0.82	180.9	0.72	1.02	127.4	0.76	1.38	104.1	0.79	1.62	79.5	0.84	2.00	63.9	0.88	2.37	53.1	0.92	2.74	42.9	0.96	3.23	35.4	1.02	3.72
100	273.3	0.66	0.82	200.3	0.72	1.03	141.1	0.76	1.38	115.4	0.79	1.62	88.1	0.84	2.00	70.8	0.88	2.38	58.9	0.92	2.74	47.6	0.97	3.24	39.2	1.01	3.75
110	299.6	0.66	0.83	219.6	0.72	1.03	154.8	0.76	1.38	126.5	0.79	1.63	96.7	0.84	2.01	77.7	0.88	2.38	64.7	0.92	2.74	52.3	0.97	3.23	43.1	1.02	3.74
120	325.7	0.66	0.83	238.8	0.72	1.03	168.4	0.76	1.39	137.7	0.79	1.63	105.3	0.84	2.01	84.6	0.88	2.39	70.4	0.92	2.75	56.9	0.97	3.25	46.9	1.01	3.75
130	351.7	0.66	0.83	257.9	0.72	1.04	181.9	0.76	1.39	148.7	0.79	1.64	113.8	0.84	2.01	91.4	0.88	2.40	76.1	0.92	2.75	61.6	0.97	3.24	50.7	1.01	3.76
140	377.5	0.66	0.83	276.8	0.72	1.04	195.3	0.76	1.39	159.7	0.79	1.64	122.2	0.84	2.02	98.2	0.88	2.40	81.8	0.92	2.76	66.2	0.97	3.25	54.5	1.01	3.77
150	403.1	0.66	0.83	295.6	0.72	1.04	208.6	0.76	1.40	170.7	0.80	1.64	130.6	0.84	2.02	105.0	0.88	2.40	87.4	0.92	2.77	70.8	0.97	3.25	58.3	1.01	3.78
160	428.5	0.66	0.83	314.2	0.72	1.04	221.9	0.76	1.40	181.5	0.79	1.65	139.0	0.85	2.02	111.7	0.88	2.41	93.1	0.92	2.77	75.3	0.97	3.27	62.1	1.01	3.78
170	453.8	0.66	0.84	332.8	0.72	1.05	235.1	0.76	1.40	192.4	0.80	1.65	147.3	0.85	2.03	118.4	0.88	2.42	98.7	0.92	2.77	79.9	0.97	3.27	65.9	1.02	3.78
180	478.9	0.66	0.84	351.2	0.72	1.05	248.2	0.76	1.41	203.1	0.80	1.65	155.6	0.85	2.03	125.1	0.88	2.42	104.3	0.93	2.78	84.4	0.97	3.28	69.6	1.02	3.79
190	503.9	0.66	0.84	369.5	0.72	1.05	261.2	0.77	1.41	213.8	0.80	1.66	163.9	0.85	2.03	131.8	0.89	2.42	109.8	0.92	2.79	89.0	0.97	3.27	73.4	1.02	3.79
200	528.5	0.67	0.84	387.7	0.73	1.05	274.1	0.77	1.42	224.4	0.80	1.66	172.1	0.85	2.04	138.4	0.89	2.43	115.3	0.92	2.79	93.5	0.97	3.28	77.1	1.02	3.80
220	579.5	0.67	0.84	425.1	0.73	1.06	300.7	0.77	1.42	246.2	0.80	1.66	188.8	0.85	2.04	151.8	0.88	2.44	126.6	0.93	2.80	102.6	0.97	3.29	84.7	1.02	3.80
240	630.1	0.67	0.84	462.2	0.73	1.06	327.0	0.77	1.42	267.9	0.80	1.67	205.5	0.85	2.05	165.2	0.89	2.44	137.8	0.93	2.80	111.8	0.97	3.29	92.2	1.02	3.81
260	680.4	0.67	0.85	499.1	0.73	1.06	353.3	0.77	1.42	289.4	0.80	1.67	222.1	0.85	2.05	178.6	0.89	2.45	149.0	0.93	2.80	120.8	0.97	3.30	99.7	1.02	3.82
280	730.4	0.67	0.85	535.6	0.73	1.07	379.3	0.77	1.43	310.8	0.80	1.68	238.6	0.85	2.05	191.9	0.89	2.45	160.1	0.93	2.81	129.9	0.97	3.30	107.2	1.02	3.82
300	780.0	0.67	0.85	572.1	0.73	1.07	405.2	0.77	1.43	332.1	0.80	1.68	255.0	0.85	2.06	205.1	0.89	2.46	171.2	0.93	2.81	138.9	0.97	3.31	114.7	1.02	3.82

TABLE B.3 *(Continued)* Parabolic Waterway Design
V_1 for *RETARDANCE "D"*. Top Width (T), Depth (D) and V_2 for *RETARDANCE "B"*.
Grade 10.0 Percent

Q	V_1 = 2.0			V_1 = 2.5			V_1 = 3.0			V_1 = 3.5			V_1 = 4.0			V_1 = 4.5			V_1 = 5.0			V_1 = 5.5			V_1 = 6.0		
cfs	T	D	V_2	T	D	V_2	T	D	V_2	T	D	V_2	T	D	V_2	T	D	V_2	T	D	V_2	T	D	V_2	T	D	V_2
15	51.4	0.56	0.77	38.3	0.63	0.92	27.2	0.69	1.18	19.6	0.73	1.56	15.7	0.76	1.86	12.6	0.81	2.18	10.6	0.85	2.45	8.7	0.90	2.83	7.2	0.94	3.26
20	68.2	0.55	0.78	50.9	0.63	0.92	36.2	0.69	1.18	26.1	0.73	1.56	20.9	0.76	1.87	16.8	0.81	2.18	14.0	0.84	2.53	11.5	0.88	2.91	9.5	0.92	3.39
25	85.0	0.56	0.78	63.4	0.63	0.92	45.1	0.69	1.19	32.5	0.73	1.57	26.1	0.76	1.87	20.9	0.80	2.20	17.5	0.84	2.52	14.3	0.87	2.96	11.9	0.92	3.37
30	101.6	0.56	0.78	75.8	0.63	0.92	53.9	0.69	1.20	38.8	0.72	1.58	31.2	0.76	1.88	25.0	0.80	2.22	20.9	0.83	2.55	17.1	0.87	2.98	14.2	0.91	3.42
35	118.1	0.56	0.78	88.1	0.63	0.93	62.7	0.69	1.20	45.2	0.73	1.58	36.3	0.76	1.88	29.1	0.80	2.22	24.3	0.83	2.57	19.9	0.87	3.00	16.5	0.91	3.46
40	134.4	0.56	0.79	100.3	0.64	0.93	71.3	0.69	1.21	51.5	0.73	1.59	41.3	0.76	1.90	33.2	0.80	2.22	27.7	0.83	2.58	22.7	0.87	3.00	18.9	0.91	3.43
45	150.7	0.56	0.79	112.5	0.64	0.93	80.0	0.69	1.21	57.7	0.72	1.60	46.4	0.76	1.90	37.2	0.80	2.24	31.1	0.83	2.58	25.5	0.87	3.01	21.2	0.91	3.45
50	166.8	0.56	0.79	124.5	0.64	0.93	88.5	0.69	1.21	63.9	0.72	1.60	51.3	0.76	1.91	41.3	0.80	2.23	34.5	0.83	2.58	28.3	0.87	3.01	23.5	0.91	3.46
55	182.8	0.56	0.79	136.4	0.64	0.93	97.1	0.69	1.21	70.1	0.73	1.60	56.3	0.76	1.92	45.3	0.80	2.24	37.8	0.83	2.60	31.1	0.87	3.00	25.8	0.91	3.47
60	198.6	0.56	0.79	148.2	0.64	0.94	105.5	0.69	1.22	76.2	0.72	1.61	61.2	0.76	1.92	49.3	0.80	2.25	41.1	0.83	2.61	33.8	0.87	3.02	28.1	0.91	3.47
65	214.3	0.57	0.79	160.0	0.64	0.94	113.9	0.69	1.22	82.3	0.73	1.62	66.2	0.76	1.92	53.2	0.80	2.26	44.4	0.83	2.62	36.5	0.87	3.04	30.3	0.91	3.51
70	229.9	0.57	0.79	171.6	0.64	0.94	122.2	0.69	1.23	88.4	0.73	1.62	71.0	0.76	1.93	57.2	0.80	2.26	47.7	0.83	2.63	39.3	0.87	3.03	32.6	0.91	3.50
75	245.4	0.57	0.80	183.2	0.64	0.94	130.5	0.69	1.23	94.4	0.73	1.62	75.9	0.76	1.94	61.1	0.80	2.27	51.0	0.83	2.63	42.0	0.87	3.04	34.9	0.91	3.50
80	260.7	0.57	0.80	194.6	0.64	0.95	138.6	0.69	1.24	100.3	0.73	1.63	80.7	0.76	1.94	65.0	0.80	2.27	54.3	0.83	2.63	44.7	0.87	3.04	37.1	0.91	3.52
90	292.1	0.57	0.80	218.1	0.64	0.95	155.4	0.69	1.24	112.5	0.73	1.64	90.5	0.76	1.95	73.0	0.81	2.27	60.9	0.83	2.64	50.1	0.87	3.06	41.7	0.91	3.52
100	323.3	0.57	0.80	241.5	0.64	0.95	172.1	0.69	1.24	124.6	0.73	1.64	100.3	0.76	1.95	80.9	0.81	2.28	67.5	0.83	2.65	55.6	0.87	3.06	46.2	0.91	3.53
110	354.3	0.57	0.80	264.6	0.64	0.96	188.6	0.69	1.25	136.7	0.73	1.64	110.0	0.76	1.96	88.7	0.81	2.29	74.1	0.83	2.65	61.0	0.87	3.07	50.7	0.91	3.54
120	385.0	0.57	0.80	287.6	0.64	0.96	205.0	0.69	1.25	148.7	0.73	1.65	119.6	0.76	1.97	96.5	0.81	2.29	80.6	0.83	2.66	66.4	0.87	3.08	55.2	0.91	3.55
130	415.5	0.57	0.81	310.3	0.65	0.96	221.3	0.69	1.26	160.6	0.73	1.65	129.3	0.76	1.97	104.3	0.81	2.30	87.1	0.83	2.67	71.8	0.87	3.08	59.7	0.91	3.55
140	445.7	0.57	0.81	333.0	0.65	0.96	237.5	0.69	1.26	172.5	0.73	1.65	138.8	0.76	1.97	112.1	0.81	2.30	93.5	0.83	2.68	77.1	0.87	3.09	64.2	0.91	3.55
150	475.7	0.58	0.81	355.4	0.65	0.97	253.6	0.69	1.26	184.2	0.73	1.66	148.4	0.76	1.97	119.8	0.81	2.30	100.0	0.83	2.68	82.5	0.88	3.09	68.6	0.91	3.57
160	505.5	0.58	0.81	377.7	0.65	0.97	269.5	0.69	1.27	195.9	0.73	1.66	157.8	0.76	1.98	127.4	0.81	2.31	106.4	0.83	2.69	87.8	0.88	3.10	73.0	0.91	3.58
170	535.0	0.58	0.81	399.7	0.65	0.97	285.3	0.69	1.27	207.6	0.73	1.66	167.2	0.76	1.98	135.0	0.81	2.32	112.8	0.83	2.69	93.0	0.87	3.11	77.5	0.91	3.57
180	564.3	0.58	0.81	421.7	0.65	0.97	301.0	0.70	1.28	219.1	0.73	1.67	176.6	0.76	1.99	142.6	0.81	2.32	119.1	0.83	2.70	98.3	0.88	3.11	81.8	0.91	3.59
190	593.4	0.58	0.82	443.4	0.65	0.98	316.6	0.70	1.28	230.6	0.73	1.67	185.9	0.76	1.99	150.1	0.81	2.33	125.4	0.83	2.71	103.5	0.88	3.12	86.2	0.91	3.59
200	622.2	0.58	0.82	465.0	0.65	0.98	332.1	0.70	1.28	242.1	0.73	1.68	195.2	0.76	1.99	157.6	0.81	2.33	131.7	0.83	2.71	108.7	0.88	3.13	90.6	0.92	3.59
220	681.9	0.58	0.82	509.6	0.65	0.98	364.1	0.70	1.29	265.5	0.73	1.68	214.1	0.76	2.00	173.0	0.81	2.33	144.6	0.83	2.71	119.3	0.88	3.13	99.4	0.91	3.60
240	741.1	0.58	0.82	553.8	0.65	0.99	395.9	0.70	1.29	288.8	0.73	1.68	233.0	0.76	2.00	188.2	0.81	2.34	157.4	0.84	2.72	129.9	0.88	3.13	108.3	0.92	3.60
260	799.9	0.58	0.82	597.7	0.65	0.99	427.4	0.70	1.29	312.0	0.73	1.69	251.7	0.77	2.01	203.4	0.81	2.34	170.1	0.84	2.72	140.4	0.88	3.14	117.0	0.92	3.61
280	858.2	0.58	0.83	641.3	0.65	0.99	458.7	0.70	1.30	335.0	0.73	1.69	270.4	0.77	2.01	218.5	0.81	2.35	182.7	0.84	2.73	150.9	0.88	3.15	125.8	0.92	3.62
300	916.0	0.59	0.83	684.6	0.65	0.99	489.8	0.70	1.30	357.9	0.73	1.69	289.0	0.77	2.01	233.5	0.81	2.35	195.4	0.84	2.73	161.3	0.88	3.15	134.5	0.92	3.62

TABLE B.4 Parabolic Waterway Design
V_1 for *RETARDANCE "D"*. Top Width (T), Depth (D) and V_2 for *RETARDANCE "C"*.
Grade 0.25 Percent

Q cfs	V_1 = 2.0			V_1 = 2.5			V_1 = 3.0			V_1 = 3.5			V_1 = 4.0			V_1 = 4.5			V_1 = 5.0			V_1 = 5.5			V_1 = 6.0		
	T	D	V_2	T	D	V_2	T	D	V_2	T	D	V_2	T	D	V_2	T	D	V_2	T	D	V_2	T	D	V_2	T	D	V_2
15																											
20																											
25	9.6	2.36	1.63																								
30	11.4	2.31	1.68																								
35	13.2	2.27	1.73																								
40	15.0	2.25	1.76	10.4	2.67	2.13																					
45	16.8	2.23	1.78	11.6	2.62	2.19																					
50	18.6	2.21	1.80	12.8	2.59	2.24																					
55	20.4	2.20	1.82	14.0	2.56	2.28																					
60	22.2	2.19	1.83	15.2	2.53	2.31																					
65	24.0	2.18	1.84	16.5	2.54	2.30																					
70	25.8	2.18	1.85	17.7	2.52	2.33	12.6	3.05	2.70																		
75	27.6	2.17	1.86	18.9	2.51	2.35	13.4	3.00	2.76																		
80	29.4	2.17	1.87	20.1	2.50	2.37	14.3	3.01	2.76																		
90	33.1	2.17	1.86	22.6	2.49	2.38	16.0	2.97	2.81																		
100	36.7	2.17	1.87	25.1	2.49	2.38	17.7	2.95	2.85																		
110	40.3	2.16	1.88	27.5	2.47	2.41	19.4	2.93	2.88																		
120	43.9	2.16	1.89	30.0	2.47	2.41	21.1	2.91	2.91	15.2	3.58	3.28															
130	47.6	2.16	1.88	32.5	2.48	2.41	22.8	2.89	2.93	16.4	3.55	3.32															
140	51.2	2.16	1.88	34.9	2.46	2.43	24.6	2.91	2.91	17.6	3.53	3.35															
150	54.8	2.16	1.89	37.4	2.47	2.42	26.3	2.90	2.93	18.8	3.51	3.39															
160	58.4	2.16	1.89	39.9	2.47	2.42	28.0	2.89	2.95	20.0	3.49	3.41															
170	62.0	2.16	1.89	42.3	2.46	2.43	29.7	2.88	2.96	21.2	3.47	3.44	16.7	4.03	3.75												
180	65.6	2.16	1.90	44.8	2.47	2.43	31.4	2.87	2.97	22.4	3.46	3.46	17.6	4.00	3.81												
190	69.2	2.16	1.90	47.2	2.46	2.44	33.1	2.87	2.98	23.6	3.45	3.48	18.5	3.97	3.85												
200	72.8	2.16	1.90	49.7	2.46	2.44	34.9	2.88	2.97	24.8	3.44	3.49	19.4	3.94	3.90												
220	80.0	2.16	1.90	54.6	2.46	2.44	38.3	2.87	2.99	27.2	3.42	3.53	21.3	3.92	3.92												
240	87.3	2.16	1.90	59.5	2.46	2.45	41.7	2.86	3.00	29.6	3.40	3.55	23.1	3.88	3.99												
260	94.5	2.16	1.90	64.5	2.46	2.44	45.2	2.86	3.00	32.1	3.41	3.54	25.0	3.87	4.01	19.5	4.57	4.34									
280	101.7	2.16	1.90	69.4	2.46	2.45	48.6	2.85	3.01	34.5	3.40	3.56	26.9	3.86	4.02	21.0	4.57	4.34									
300	108.9	2.16	1.90	74.3	2.46	2.45	52.1	2.86	3.00	36.9	3.39	3.58	28.7	3.83	4.07	22.4	4.53	4.40									

TABLE B.4 *(Continued)* Parabolic Waterway Design
V_1 for *RETARDANCE "D"*. Top Width (T), Depth (D) and V_2 for *RETARDANCE "C"*.
Grade 0.50 Percent

Q cfs	V_1 = 2.0			V_1 = 2.5			V_1 = 3.0			V_1 = 3.5			V_1 = 4.0			V_1 = 4.5			V_1 = 5.0			V_1 = 5.5			V_1 = 6.0		
	T	D	V_2	T	D	V_2	T	D	V_2	T	D	V_2	T	D	V_2	T	D	V_2	T	D	V_2	T	D	V_2	T	D	V_2
15	8.6	1.63	1.58																								
20	11.3	1.58	1.66																								
25	14.1	1.57	1.67	9.0	1.91	2.14																					
30	16.9	1.56	1.68	10.7	1.87	2.21	8.2	2.18	2.48																		
35	19.6	1.55	1.71	12.4	1.85	2.26	9.4	2.10	2.62																		
40	22.4	1.55	1.71	14.1	1.83	2.30	10.7	2.08	2.66																		
45	25.1	1.54	1.73	15.8	1.82	2.33	11.9	2.03	2.76																		
50	27.9	1.54	1.73	17.5	1.80	2.35	13.2	2.02	2.78	9.6	2.42	3.19															
55	30.7	1.54	1.72	19.2	1.80	2.37	14.5	2.02	2.79	10.5	2.39	3.25															
60	33.4	1.54	1.74	20.9	1.79	2.38	15.8	2.01	2.80	11.4	2.37	3.30															
65	36.1	1.53	1.75	22.7	1.80	2.36	17.0	1.99	2.86	12.3	2.35	3.34															
70	38.9	1.54	1.74	24.4	1.80	2.37	18.3	1.99	2.86	13.2	2.33	3.38															
75	41.6	1.54	1.75	26.1	[illegible]	2.38	19.6	1.99	2.86	14.1	2.32	3.41	11.2	2.71	3.66												
80	44.3	1.53	1.75	27.8	[illegible]	2.39	20.9	1.99	2.86	15.0	2.31	3.43	11.8	2.65	3.80												
90	49.8	1.53	1.75	31.2	1.78	2.41	23.5	1.99	2.87	16.9	2.31	3.42	13.3	2.65	3.78												
100	55.3	1.53	1.75	34.6	1.78	2.42	26.0	1.97	2.90	18.7	2.29	3.47	14.7	2.63	3.85	11.9	3.02	4.13									
110	60.8	1.54	1.75	38.1	1.78	2.41	28.6	1.97	2.90	20.5	2.28	3.50	16.1	2.60	3.90	13.0	2.98	4.22									
120	66.3	1.54	1.75	41.5	1.78	2.42	31.2	1.98	2.90	22.4	2.29	3.49	17.5	2.58	3.94	14.1	2.94	4.30									
130	71.7	1.53	1.76	44.9	1.78	2.42	33.7	1.97	2.92	24.2	2.28	3.51	18.9	2.57	3.98	15.2	2.91	4.36									
140	77.2	1.54	1.76	48.3	1.78	2.43	36.3	1.97	2.92	26.0	2.27	3.54	20.4	2.58	3.95	16.4	2.93	4.34									
150	82.6	1.54	1.76	51.7	1.78	2.43	38.9	1.97	2.91	27.9	2.28	3.52	21.8	2.57	3.98	17.5	2.90	4.39	14.0	3.34	4.77						
160	88.0	1.53	1.76	55.1	1.78	2.44	41.4	1.97	2.93	29.7	2.27	3.54	23.2	2.56	4.01	18.6	2.88	4.44	14.9	3.33	4.80						
170	93.4	1.53	1.77	58.5	1.78	2.44	44.0	1.97	2.92	31.5	2.26	3.55	24.6	2.55	4.03	19.8	2.89	4.41	15.7	3.27	4.92						
180	98.8	1.53	1.77	61.9	1.78	2.44	46.5	1.96	2.94	33.3	2.26	3.57	26.1	2.56	4.01	20.9	2.88	4.45	16.6	3.26	4.94						
190	104.2	1.54	1.77	65.3	1.78	2.44	49.1	1.97	2.93	35.2	2.27	3.55	27.5	2.56	4.03	22.0	2.86	4.49	17.5	3.26	4.96						
200	109.6	1.54	1.77	68.7	1.78	2.44	51.6	1.96	2.94	37.0	2.26	3.56	28.9	2.55	4.04	23.1	2.85	4.52	18.4	3.25	4.98	15.3	3.72	5.23			
220	120.5	1.54	1.77	75.5	1.78	2.44	56.8	1.97	2.93	40.7	2.26	3.56	31.8	2.55	4.04	25.4	2.85	4.53	20.2	3.24	5.01	16.7	3.66	5.36			
240	131.3	1.54	1.77	82.3	1.78	2.45	61.9	1.97	2.94	44.3	2.26	3.58	34.6	2.54	4.07	27.7	2.85	4.53	22.0	3.23	5.04	18.2	3.65	5.38			
260	142.1	1.54	1.77	89.1	1.78	2.45	67.0	1.97	2.94	48.0	2.26	3.58	37.5	2.55	4.06	30.0	2.85	4.54	23.8	3.22	5.06	19.7	3.64	5.39			
280	152.9	1.54	1.78	95.9	1.78	2.45	72.1	1.97	2.95	51.6	2.25	3.59	40.3	2.54	4.08	32.2	2.83	4.58	25.6	3.21	5.08	21.1	3.61	5.48	17.5	4.14	5.75
300	163.7	1.54	1.78	102.6	1.78	2.46	77.2	1.97	2.95	55.3	2.26	3.59	43.2	2.54	4.08	34.5	2.83	4.58	27.3	3.18	5.15	22.6	3.60	5.49	18.7	4.12	5.80

TABLE B.4 *(Continued)* Parabolic Waterway Design
V_1 for *RETARDANCE "D"*. Top Width (T), Depth (D) and V_2 for *RETARDANCE "C"*.
Grade 0.75 Percent

Q cfs	V_1 = 2.0			V_1 = 2.5			V_1 = 3.0			V_1 = 3.5			V_1 = 4.0			V_1 = 4.5			V_1 = 5.0			V_1 = 5.5			V_1 = 6.0		
	T	D	V_2	T	D	V_2	T	D	V_2	T	D	V_2	T	D	V_2	T	D	V_2	T	D	V_2	T	D	V_2	T	D	V_2
15	11.7	1.29	1.47	6.8	1.61	2.02																					
20	15.5	1.27	1.51	8.9	1.54	2.15																					
25	19.3	1.26	1.52	11.0	1.50	2.23	8.2	1.74	2.58																		
30	23.1	1.25	1.54	13.2	1.50	2.24	9.7	1.69	2.70	7.7	1.92	2.99															
35	27.0	1.26	1.53	15.3	1.48	2.29	11.3	1.68	2.72	8.9	1.88	3.08															
40	30.8	1.26	1.53	17.5	1.49	2.28	12.8	1.65	2.80	10.1	1.86	3.16															
45	34.5	1.25	1.55	19.6	1.47	2.31	14.4	1.65	2.80	11.3	1.84	3.21															
50	38.3	1.25	1.55	21.8	1.48	2.30	15.9	1.63	2.85	12.5	1.82	3.26	9.3	2.18	3.65												
55	42.1	1.25	1.55	24.0	1.48	2.29	17.5	1.64	2.85	13.7	1.81	3.30	10.1	2.12	3.80												
60	45.9	1.25	1.55	26.1	1.48	2.31	19.1	1.64	2.84	15.0	1.82	3.26	11.0	2.12	3.82												
65	49.6	1.25	1.56	28.2	1.47	2.33	20.6	1.63	2.88	16.2	1.81	3.29	11.9	2.11	3.84	9.9	2.38	4.08									
70	53.4	1.25	1.56	30.4	1.48	2.32	22.2	1.63	2.87	17.4	1.80	3.31	12.8	2.11	3.85	10.6	2.35	4.15									
75	57.1	1.25	1.56	32.5	1.47	2.33	23.7	1.62	2.90	18.6	1.80	3.34	13.6	2.07	3.95	11.3	2.33	4.22									
80	60.9	1.25	1.56	34.7	1.48	2.32	25.3	1.63	2.89	19.8	1.79	3.35	14.5	2.07	3.95	12.0	2.31	4.28									
90	68.4	1.25	1.56	38.9	1.47	2.34	28.4	1.62	2.91	22.2	1.78	3.39	16.3	2.07	3.97	13.4	2.28	4.38	11.1	2.62	4.59						
100	75.9	1.25	1.56	43.2	1.47	2.34	31.5	1.62	2.92	24.7	1.79	3.37	18.1	2.07	3.98	14.9	2.28	4.37	12.2	2.56	4.75						
110	83.4	1.25	1.57	47.5	1.47	2.34	34.7	1.63	2.90	27.1	1.78	3.40	19.8	2.04	4.04	16.3	2.26	4.45	13.4	2.55	4.78						
120	90.8	1.25	1.57	51.8	1.47	2.34	37.8	1.62	2.91	29.6	1.79	3.38	21.6	2.05	4.04	17.8	2.26	4.43	14.6	2.54	4.80	12.1	2.85	5.16			
130	98.3	1.25	1.57	56.0	1.47	2.35	40.9	1.62	2.92	32.0	1.78	3.40	23.4	2.05	4.04	19.2	2.24	4.49	15.7	2.51	4.91	13.0	2.81	5.29			
140	105.7	1.25	1.57	60.3	1.47	2.35	44.0	1.62	2.92	34.4	1.78	3.41	25.1	2.03	4.09	20.7	2.25	4.48	16.9	2.51	4.92	14.0	2.81	5.29			
150	113.1	1.25	1.58	64.5	1.47	2.36	47.1	1.62	2.93	36.8	1.77	3.43	26.9	2.04	4.08	22.1	2.23	4.52	18.1	2.50	4.92	15.0	2.81	5.30			
160	120.5	1.25	1.58	68.8	1.47	2.35	50.2	1.62	2.93	39.3	1.78	3.41	28.7	2.04	4.07	23.6	2.24	4.50	19.3	2.50	4.93	15.9	2.77	5.40	13.1	3.13	5.80
170	127.9	1.25	1.58	73.0	1.47	2.36	53.3	1.62	2.93	41.7	1.78	3.42	30.4	2.03	4.11	25.0	2.23	4.54	20.4	2.48	5.00	16.9	2.78	5.39	13.9	3.12	5.83
180	135.2	1.25	1.58	77.2	1.47	2.36	56.4	1.62	2.93	44.1	1.78	3.43	32.2	2.03	4.10	26.5	2.24	4.52	21.6	2.48	5.00	17.8	2.75	5.48	14.7	3.11	5.85
190	142.6	1.25	1.58	81.5	1.47	2.36	59.5	1.62	2.93	46.5	1.77	3.43	34.0	2.04	4.09	27.9	2.23	4.56	22.8	2.48	5.00	18.8	2.75	5.46	15.5	3.11	5.87
200	149.9	1.25	1.58	85.7	1.47	2.36	62.5	1.62	2.95	48.9	1.77	3.44	35.7	2.03	4.12	29.4	2.23	4.54	24.0	2.49	4.99	19.7	2.73	5.54	16.3	3.10	5.88
220	164.7	1.25	1.58	94.2	1.47	2.36	68.7	1.62	2.95	53.8	1.78	3.43	39.3	2.03	4.11	32.3	2.23	4.55	26.3	2.47	5.05	21.7	2.74	5.52	17.9	3.09	5.92
240	179.4	1.25	1.59	102.6	1.47	2.37	74.9	1.62	2.95	58.6	1.77	3.44	42.8	2.03	4.12	35.2	2.23	4.56	28.7	2.47	5.04	23.6	2.72	5.57	19.4	3.05	6.04
260	194.1	1.25	1.59	111.1	1.47	2.37	81.1	1.62	2.95	63.5	1.78	3.44	46.3	2.02	4.14	38.1	2.23	4.57	31.0	2.46	5.09	25.5	2.71	5.61	21.0	3.05	6.05
280	208.8	1.25	1.59	119.5	1.47	2.37	87.3	1.62	2.95	68.3	1.77	3.44	49.9	2.03	4.12	41.0	2.22	4.58	33.4	2.46	5.08	27.4	2.70	5.65	22.6	3.04	6.07
300	223.5	1.26	1.59	127.9	1.47	2.37	93.4	1.62	2.96	73.1	1.77	3.45	53.4	2.03	4.13	43.9	2.22	4.58	35.8	2.47	5.07	29.4	2.71	5.62	24.2	3.04	6.08

TABLE B.4 *(Continued)* Parabolic Waterway Design
V_1 for *RETARDANCE "D"*. Top Width (T), Depth (D) and V_2 for *RETARDANCE "C"*.
Grade 1.0 Percent

Q cfs	V_1 = 2.0			V_1 = 2.5			V_1 = 3.0			V_1 = 3.5			V_1 = 4.0			V_1 = 4.5			V_1 = 5.0			V_1 = 5.5			V_1 = 6.0		
	T	D	V_2	T	D	V_2	T	D	V_2	T	D	V_2	T	D	V_2	T	D	V_2	T	D	V_2	T	D	V_2	T	D	V_2
15	13.4	1.13	1.47	8.4	1.30	2.03																					
20	17.8	1.12	1.49	11.1	1.27	2.10	7.6	1.52	2.55																		
25	22.2	1.11	1.50	13.9	1.27	2.09	9.4	1.49	2.64	7.6	1.62	2.99															
30	26.6	1.11	1.50	16.6	1.26	2.13	11.2	1.46	2.71	9.1	1.61	3.03															
35	30.9	1.11	1.52	19.3	1.25	2.15	13.0	1.45	2.75	10.5	1.57	3.14	8.0	1.80	3.59												
40	35.3	1.11	1.52	22.1	1.26	2.13	14.8	1.44	2.79	12.0	1.57	3.14	9.1	1.78	3.65												
45	39.7	1.11	1.52	24.8	1.25	2.15	16.7	1.45	2.76	13.4	1.55	3.21	10.2	1.76	3.70												
50	44.0	1.11	1.52	27.5	1.25	2.16	18.5	1.44	2.79	14.9	1.55	3.21	11.3	1.75	3.74	8.7	2.C2	4.20									
55	48.3	1.11	1.53	30.2	1.25	2.16	20.3	1.43	2.80	16.3	1.54	3.26	12.4	1.75	3.76	9.5	1.99	4.30									
60	52.7	1.11	1.52	32.9	1.25	2.17	22.1	1.43	2.82	17.8	1.54	3.25	13.5	1.74	3.79	10.4	2.01	4.26									
65	57.0	1.11	1.53	35.6	1.25	2.17	23.9	1.43	2.83	19.2	1.53	3.29	14.6	1.73	3.81	11.2	1.98	4.33	9.3	2.22	4.66						
70	61.3	1.11	1.53	38.3	1.25	2.17	25.7	1.43	2.84	20.7	1.53	3.27	15.6	1.71	3.90	12.0	1.96	4.40	10.0	2.21	4.69						
75	65.6	1.11	1.53	41.0	1.25	2.18	27.5	1.42	2.85	22.1	1.53	3.31	16.7	1.71	3.90	12.8	1.95	4.46	10.7	2.21	4.71						
80	69.8	1.11	1.54	43.7	1.25	2.18	29.3	1.42	2.85	23.6	1.53	3.29	17.8	1.71	3.91	13.7	1.95	4.42	11.3	2.16	4.85						
90	78.5	1.11	1.54	49.1	1.25	2.18	32.9	1.42	2.87	26.5	1.53	3.31	20.0	1.70	3.93	15.3	1.93	4.52	12.7	2.16	4.87	10.6	2.42	5.20			
100	87.1	1.11	1.54	54.5	1.25	2.18	36.6	1.43	2.85	29.4	1.52	3.32	22.2	1.70	3.94	17.0	1.93	4.52	14.1	2.15	4.89	11.7	2.39	5.31			
110	95.6	1.11	1.54	59.9	1.25	2.18	40.2	1.42	2.86	32.3	1.52	3.33	24.4	1.70	3.94	18.7	1.93	4.52	15.4	2.12	5.00	12.9	2.40	5.28	11.1	2.59	5.67
120	104.2	1.11	1.54	65.2	1.25	2.19	43.8	1.42	2.87	35.2	1.52	3.33	26.6	1.70	3.95	20.3	1.92	4.59	16.8	2.12	5.00	14.0	2.37	5.36	12.1	2.59	5.69
130	112.7	1.11	1.55	70.6	1.25	2.19	47.4	1.42	2.87	38.1	1.52	3.34	28.8	1.70	3.95	22.0	1.92	4.58	18.2	2.13	5.00	15.1	2.35	5.44	13.0	2.55	5.83
140	121.2	1.11	1.55	76.0	1.25	2.19	51.0	1.42	2.87	41.0	1.52	3.34	30.9	1.69	3.99	23.7	1.92	4.57	19.6	2.13	5.00	16.2	2.34	5.50	14.0	2.55	5.83
150	129.7	1.11	1.55	81.3	1.25	2.19	54.6	1.42	2.87	43.9	1.52	3.34	33.1	1.69	3.99	25.3	1.91	4.62	20.9	2.11	5.07	17.4	2.35	5.46	15.0	2.55	5.84
160	138.1	1.11	1.55	86.6	1.25	2.20	58.2	1.42	2.88	46.8	1.52	3.34	35.3	1.69	3.99	27.0	1.91	4.61	22.3	2.11	5.06	18.5	2.33	5.51	15.9	2.52	5.95
170	146.6	1.11	1.55	91.9	1.25	2.20	61.7	1.42	2.89	49.7	1.52	3.34	37.5	1.69	3.99	28.7	1.92	4.60	23.7	2.11	5.05	19.6	2.32	5.56	16.9	2.52	5.94
180	155.0	1.11	1.55	97.2	1.25	2.20	65.3	1.42	2.89	52.5	1.52	3.36	39.6	1.69	4.01	30.3	1.91	4.63	25.0	2.10	5.10	20.7	2.31	5.60	17.9	2.52	5.93
190	163.4	1.11	1.55	102.5	1.25	2.20	68.9	1.42	2.89	55.4	1.52	3.36	41.8	1.69	4.01	32.0	1.91	4.62	26.4	2.10	5.09	21.9	2.32	5.56	18.8	2.50	6.02
200	171.7	1.11	1.56	107.8	1.25	2.20	72.4	1.42	2.90	58.3	1.52	3.35	44.0	1.69	4.00	33.6	1.91	4.65	27.8	2.11	5.08	23.0	2.32	5.59	19.8	2.50	6.01
220	188.7	1.11	1.56	118.4	1.25	2.21	79.6	1.42	2.89	64.0	1.52	3.37	48.4	1.70	4.00	37.0	1.91	4.63	30.5	2.10	5.12	25.3	2.32	5.59	21.7	2.48	6.08
240	205.5	1.11	1.56	129.0	1.25	2.21	86.7	1.42	2.90	69.8	1.52	3.37	52.7	1.69	4.01	40.3	1.91	4.65	33.3	2.10	5.11	27.5	2.30	5.65	23.6	2.47	6.13
260	222.4	1.11	1.56	139.6	1.25	2.21	93.9	1.42	2.90	75.5	1.52	3.38	57.1	1.69	4.01	43.6	1.91	4.66	36.0	2.10	5.14	29.8	2.30	5.64	25.6	2.48	6.11
280	239.1	1.11	1.56	150.2	1.25	2.22	101.0	1.42	2.91	81.3	1.52	3.37	61.4	1.69	4.02	46.9	1.90	4.68	38.8	2.10	5.12	32.1	2.31	5.63	27.5	2.47	6.15
300	255.9	1.11	1.56	160.8	1.25	2.22	108.1	1.42	2.91	87.0	1.52	3.38	65.7	1.69	4.03	50.3	1.91	4.66	41.5	2.10	5.14	34.3	2.30	5.68	29.5	2.48	6.12

TABLE B.4 *(Continued)* Parabolic Waterway Design
V_1 for *RETARDANCE "D"*. Top Width (T), Depth (D) and V_2 for *RETARDANCE "C"*.
Grade 1.25 Percent

Q cfs	V_1 = 2.0			V_1 = 2.5			V_1 = 3.0			V_1 = 3.5			V_1 = 4.0			V_1 = 4.5			V_1 = 5.0			V_1 = 5.5			V_1 = 6.0		
	T	D	V_2	T	D	V_2	T	D	V_2	T	D	V_2	T	D	V_2	T	D	V_2	T	D	V_2	T	D	V_2	T	D	V_2
15	15.3	1.00	1.45	9.8	1.15	1.95	6.6	1.37	2.44																		
20	20.4	1.00	1.45	12.9	1.12	2.04	8.7	1.33	2.55	6.6	1.49	2.98															
25	25.4	0.99	1.47	16.1	1.12	2.05	10.8	1.31	2.61	8.2	1.47	3.06	6.7	1.62	3.38												
30	30.5	1.00	1.46	19.3	1.12	2.06	12.9	1.30	2.66	9.7	1.43	3.20	7.9	1.56	3.58												
35	35.5	0.99	1.47	22.5	1.12	2.06	15.0	1.29	2.68	11.3	1.42	3.22	9.2	1.55	3.62	7.2	1.78	4.02									
40	40.5	0.99	1.47	25.7	1.12	2.06	17.1	1.28	2.70	12.9	1.42	3.24	10.5	1.55	3.64	8.2	1.77	4.07									
45	45.5	0.99	1.48	28.8	1.11	2.08	19.2	1.28	2.72	14.4	1.40	3.31	11.7	1.52	3.75	9.1	1.72	4.25									
50	50.4	0.99	1.49	32.0	1.12	2.08	21.3	1.28	2.73	16.0	1.40	3.31	13.0	1.52	3.75	10.1	1.71	4.27	8.3	1.95	4.57						
55	55.4	0.99	1.48	35.1	1.11	2.09	23.4	1.28	2.74	17.6	1.40	3.31	14.3	1.52	3.75	11.1	1.71	4.29	9.1	1.93	4.63						
60	60.3	0.99	1.49	38.3	1.12	2.09	25.5	1.28	2.74	19.1	1.39	3.36	15.6	1.52	3.75	12.1	1.71	4.30	9.9	1.92	4.67						
65	65.2	0.99	1.49	41.4	1.11	2.10	27.6	1.28	2.74	20.7	1.39	3.35	16.8	1.51	3.81	13.0	1.68	4.41	10.6	1.88	4.84	8.7	2.18	5.06			
70	70.1	0.99	1.50	44.6	1.12	2.09	29.7	1.28	2.75	22.3	1.39	3.35	18.1	1.51	3.80	14.0	1.68	4.41	11.4	1.87	4.86	9.3	2.15	5.18			
75	75.0	0.99	1.50	47.7	1.12	2.09	31.8	1.28	2.75	23.8	1.39	3.38	19.4	1.51	3.80	15.0	1.68	4.40	12.2	1.87	4.88	9.9	2.12	5.28			
80	79.9	0.99	1.50	50.8	1.12	2.10	33.8	1.27	2.77	25.4	1.39	3.37	20.6	1.50	3.84	16.0	1.69	4.40	13.0	1.87	4.89	10.5	2.10	5.37	9.2	2.33	5.51
90	89.7	0.99	1.50	5[illegible].1	1.12	2.10	38.0	1.27	2.77	28.5	1.38	3.39	23.2	1.51	3.83	17.9	1.67	4.48	14.6	1.86	4.92	11.8	2.09	5.40	10.3	2.31	5.60
100	99.6	0.99	1.50	63.3	1.11	2.11	42.2	1.27	2.77	31.7	1.39	3.38	25.7	1.50	3.86	19.9	1.67	4.47	16.2	1.85	4.94	13.1	2.09	5.42	11.4	2.29	5.68
110	109.4	0.99	1.50	69.6	1.12	2.11	46.4	1.27	2.77	34.8	1.39	3.39	28.3	1.50	3.85	21.9	1.67	4.46	17.8	1.85	4.96	14.4	2.09	5.43	12.4	2.24	5.88
120	119.1	0.99	1.51	75.8	1.12	2.11	50.5	1.27	2.78	37.9	1.38	3.41	30.8	1.50	3.87	23.8	1.66	4.51	19.3	1.83	5.05	15.6	2.06	5.55	13.5	2.23	5.91
130	128.9	1.00	1.51	82.0	1.12	2.11	54.7	1.27	2.78	41.1	1.39	3.39	33.3	1.49	3.89	25.8	1.67	4.50	20.9	1.83	5.05	16.9	2.06	5.55	14.6	2.22	5.95
140	138.6	1.00	1.51	88.2	1.12	2.12	58.8	1.27	2.79	44.2	1.39	3.40	35.9	1.50	3.87	27.8	1.67	4.48	22.5	1.83	5.06	18.2	2.06	5.55	15.7	2.22	5.97
150	148.2	1.00	1.51	94.4	1.12	2.12	63.0	1.27	2.78	47.3	1.39	3.41	38.4	1.50	3.88	29.7	1.66	4.52	24.1	1.83	5.06	19.4	2.04	5.63	16.8	2.21	6.00
160	157.9	1.00	1.51	100.6	1.12	2.12	67.1	1.27	2.79	50.4	1.39	3.41	40.9	1.50	3.89	31.7	1.67	4.51	25.7	1.83	5.06	20.7	2.05	5.62	17.9	2.21	6.01
170	167.5	1.00	1.51	106.7	1.12	2.12	71.2	1.27	2.79	53.5	1.39	3.41	43.4	1.49	3.90	33.6	1.66	4.53	27.3	1.83	5.06	22.0	2.05	5.62	19.0	2.21	6.03
180	177.1	1.00	1.51	112.9	1.12	2.12	75.3	1.27	2.80	56.6	1.39	3.42	46.0	1.50	3.88	35.6	1.67	4.52	28.9	1.84	5.05	23.3	2.05	5.61	20.1	2.21	6.04
190	186.6	1.00	1.52	119.0	1.12	2.12	79.4	1.27	2.80	59.7	1.39	3.42	48.5	1.50	3.89	37.5	1.66	4.54	30.4	1.83	5.10	24.5	2.04	5.67	21.2	2.20	6.05
200	196.1	1.00	1.52	125.1	1.12	2.13	83.5	1.27	2.80	62.8	1.39	3.42	51.0	1.50	3.90	39.5	1.67	4.52	32.0	1.83	5.09	25.8	2.04	5.66	22.3	2.20	6.06
220	215.4	1.00	1.52	137.4	1.12	2.13	91.8	1.27	2.80	69.0	1.39	3.42	56.0	1.50	3.91	43.4	1.67	4.53	35.2	1.83	5.09	28.3	2.03	5.70	24.5	2.20	6.08
240	234.7	1.00	1.52	149.8	1.12	2.13	100.0	1.27	2.81	75.2	1.39	3.43	61.1	1.50	3.90	47.3	1.67	4.54	38.4	1.83	5.09	30.9	2.04	5.68	26.7	2.20	6.10
260	253.8	1.00	1.52	162.0	1.12	2.13	108.2	1.27	2.81	81.4	1.39	3.43	66.1	1.50	3.91	51.2	1.66	4.55	41.5	1.83	5.12	33.4	2.03	5.72	28.8	2.18	6.18
280	273.0	1.00	1.52	174.3	1.12	2.13	116.4	1.27	2.81	87.6	1.39	3.43	71.1	1.50	3.92	55.1	1.67	4.55	44.7	1.83	5.11	36.0	2.03	5.70	31.0	2.18	6.18
300	292.0	1.00	1.53	186.5	1.12	2.14	124.6	1.27	2.82	93.7	1.39	3.44	76.2	1.50	3.91	59.0	1.67	4.55	47.8	1.82	5.13	38.5	2.03	5.73	33.2	2.18	6.19

TABLE B.4 *(Continued)* Parabolic Waterway Design
V_1 for *RETARDANCE "D"*. Top Width (T), Depth (D) and V_2 for *RETARDANCE "C"*.
Grade 1.50 Percent

Q cfs	V_1 = 2.0			V_1 = 2.5			V_1 = 3.0			V_1 = 3.5			V_1 = 4.0			V_1 = 4.5			V_1 = 5.0			V_1 = 5.5			V_1 = 6.0		
	T	D	V_2	T	D	V_2	T	D	V_2	T	D	V_2	T	D	V_2	T	D	V_2	T	D	V_2	T	D	V_2	T	D	V_2
15	17.0	0.92	1.42	11.3	1.05	1.86	7.6	1.20	2.41																		
20	22.7	0.92	1.41	14.9	1.03	1.93	10.0	1.17	2.53	7.0	1.40	3.01	5.7	1.52	3.39												
25	28.3	0.92	1.42	18.6	1.03	1.94	12.4	1.15	2.59	8.6	1.35	3.19	7.0	1.46	3.60												
30	33.9	0.92	1.43	22.3	1.03	1.94	14.9	1.15	2.59	10.3	1.34	3.22	8.4	1.46	3.62	6.9	1.59	4.03									
35	39.5	0.92	1.43	26.0	1.03	1.95	17.3	1.14	2.62	11.9	1.31	3.32	9.7	1.43	3.74	8.0	1.56	4.13									
40	45.0	0.92	1.44	29.7	1.03	1.94	19.8	1.15	2.61	13.6	1.31	3.32	11.1	1.43	3.73	9.1	1.55	4.20	7.1	1.79	4.64						
45	50.5	0.92	1.44	33.3	1.02	1.96	22.2	1.14	2.63	15.2	1.30	3.39	12.4	1.41	3.81	10.2	1.54	4.25	8.0	1.79	4.63						
50	56.1	0.92	1.44	37.0	1.03	1.95	24.6	1.14	2.65	16.9	1.30	3.38	13.7	1.39	3.88	11.3	1.53	4.29	8.8	1.75	4.78						
55	61.5	0.92	1.45	40.6	1.02	1.96	27.1	1.14	2.64	18.6	1.30	3.37	15.1	1.40	3.86	12.4	1.52	4.32	9.6	1.73	4.91	8.2	1.93	5.14			
60	67.0	0.92	1.45	44.2	1.02	1.97	29.5	1.14	2.65	20.2	1.29	3.41	16.4	1.39	3.91	13.5	1.52	4.35	10.5	1.73	4.88	8.9	1.91	5.23			
65	72.5	0.92	1.45	47.8	1.02	1.98	31.9	1.14	2.66	21.9	1.30	3.40	17.8	1.40	3.88	14.6	1.51	4.37	11.3	1.71	4.97	9.6	1.89	5.30			
70	77.9	0.92	1.45	51.4	1.02	1.98	34.3	1.14	2.66	23.5	1.29	3.43	19.1	1.39	3.92	15.7	1.51	4.39	12.2	1.72	4.94	10.3	1.88	5.36	8.8	2.12	5.55
75	83.3	0.92	1.46	55.0	1.02	1.98	36.7	1.14	2.67	25.2	1.29	3.42	20.5	1.40	3.89	16.8	1.51	4.40	13.0	1.70	5.02	11.0	1.87	5.41	9.4	2.10	5.61
80	88.7	0.92	1.46	58.6	1.02	1.98	39.1	1.14	2.67	26.8	1.29	3.44	21.8	1.39	3.93	17.9	1.50	4.41	13.9	1.71	4.98	11.7	1.86	5.46	9.9	2.05	5.83
90	99.6	0.92	1.46	65.8	1.02	1.99	44.0	1.14	2.66	30.1	1.29	3.46	24.5	1.39	3.93	20.1	1.50	4.43	15.6	1.71	5.02	13.2	1.87	5.42	11.1	2.04	5.90
100	110.5	0.92	1.46	73.0	1.02	1.99	48.8	1.14	2.67	33.4	1.29	3.47	27.2	1.39	3.94	22.3	1.50	4.45	17.3	1.70	5.05	14.6	1.85	5.50	12.3	2.03	5.95
110	121.4	0.92	1.46	80.2	1.02	1.99	53.6	1.14	2.68	36.7	1.28	3.47	29.9	1.39	3.94	24.5	1.50	4.46	19.0	1.70	5.07	16.0	1.84	5.56	13.5	2.02	6.00
120	132.2	0.92	1.46	87.3	1.02	2.00	58.4	1.14	2.68	40.0	1.28	3.48	32.6	1.39	3.95	26.7	1.49	4.47	20.7	1.69	5.09	17.5	1.85	5.52	14.7	2.01	6.03
130	142.9	0.92	1.47	94.5	1.02	2.00	63.2	1.14	2.68	43.3	1.29	3.48	35.3	1.39	3.94	28.9	1.49	4.48	22.4	1.69	5.10	18.9	1.84	5.57	15.9	2.00	6.06
140	153.6	0.92	1.47	101.6	1.02	2.00	68.0	1.14	2.69	46.6	1.29	3.48	37.9	1.38	3.97	31.1	1.49	4.48	24.0	1.68	5.18	20.3	1.83	5.61	17.1	2.00	6.08
150	164.3	0.92	1.47	108.7	1.03	2.00	72.8	1.14	2.68	49.9	1.29	3.48	40.6	1.39	3.97	33.3	1.49	4.49	25.7	1.68	5.18	21.7	1.82	5.64	18.3	2.00	6.10
160	175.0	0.92	1.47	115.7	1.02	2.01	77.6	1.14	2.68	53.2	1.29	3.48	43.3	1.39	3.97	35.5	1.50	4.49	27.4	1.68	5.19	23.2	1.83	5.60	19.5	1.99	6.12
170	185.6	0.92	1.48	122.8	1.03	2.01	82.3	1.14	2.69	56.4	1.29	3.49	45.9	1.38	3.99	37.7	1.50	4.49	29.1	1.68	5.19	24.6	1.83	5.63	20.7	1.99	6.13
180	196.2	0.92	1.48	129.8	1.02	2.01	87.1	1.14	2.69	59.7	1.29	3.49	48.6	1.39	3.98	39.8	1.49	4.52	30.8	1.68	5.19	26.0	1.82	5.66	21.9	1.99	6.14
190	206.7	0.92	1.48	136.8	1.03	2.02	91.8	1.14	2.69	62.9	1.29	3.50	51.3	1.39	3.97	42.0	1.49	4.52	32.5	1.68	5.18	27.4	1.82	5.68	23.1	1.99	6.15
200	217.2	0.92	1.48	143.8	1.03	2.02	96.5	1.14	2.70	66.2	1.29	3.49	53.9	1.39	3.99	44.2	1.49	4.51	34.2	1.68	5.18	28.9	1.83	5.64	24.3	1.99	6.16
220	238.5	0.92	1.48	157.9	1.03	2.02	106.0	1.14	2.70	72.7	1.29	3.50	59.3	1.39	3.98	48.6	1.50	4.51	37.6	1.68	5.19	31.7	1.82	5.68	26.7	1.99	6.18
240	259.7	0.92	1.49	172.0	1.03	2.02	115.6	1.15	2.70	79.3	1.29	3.50	64.6	1.39	3.99	53.0	1.50	4.51	40.9	1.67	5.23	34.6	1.82	5.67	29.1	1.98	6.19
260	280.9	0.92	1.49	186.1	1.03	2.03	125.0	1.15	2.71	85.8	1.29	3.51	69.9	1.39	3.99	57.3	1.49	4.53	44.3	1.68	5.22	37.4	1.82	5.70	31.5	1.98	6.20
280	302.0	0.92	1.49	200.1	1.03	2.03	134.5	1.15	2.71	92.3	1.29	3.51	75.2	1.39	4.00	61.7	1.50	4.53	47.7	1.68	5.22	40.3	1.82	5.69	33.9	1.98	6.21
300	323.0	0.93	1.49	214.1	1.03	2.03	143.9	1.15	2.71	98.8	1.29	3.51	80.5	1.39	4.00	66.0	1.49	4.54	51.1	1.68	5.21	43.1	1.82	5.71	36.3	1.98	6.21

TABLE B.4 *(Continued)* Parabolic Waterway Design
V_1 for *RETARDANCE "D"*. Top Width (T), Depth (D) and V_2 for *RETARDANCE "C"*.
Grade 1.75 Percent

Q cfs	V_1 = 2.0			V_1 = 2.5			V_1 = 3.0			V_1 = 3.5			V_1 = 4.0			V_1 = 4.5			V_1 = 5.0			V_1 = 5.5			V_1 = 6.0		
	T	D	V_2	T	D	V_2	T	D	V_2	T	D	V_2	T	D	V_2	T	D	V_2	T	D	V_2	T	D	V_2	T	D	V_2
15	18.5	0.87	1.38	12.1	0.97	1.89	8.5	1.08	2.40	5.9	1.28	2.92															
20	24.5	0.86	1.41	16.1	0.96	1.91	11.3	1.07	2.43	7.8	1.25	3.03	6.5	1.34	3.38												
25	30.6	0.86	1.41	20.1	0.96	1.91	14.1	1.07	2.45	9.7	1.23	3.09	8.1	1.33	3.43	6.6	1.46	3.81									
30	36.7	0.86	1.40	24.1	0.96	1.92	16.8	1.06	2.50	11.6	1.22	3.13	9.6	1.29	3.57	7.8	1.41	4.01									
35	42.7	0.86	1.41	28.1	0.96	1.92	19.6	1.06	2.50	13.4	1.20	3.23	11.2	1.29	3.58	9.1	1.41	4.02	7.1	1.59	4.57						
40	48.7	0.86	1.42	32.0	0.96	1.93	22.4	1.06	2.50	15.3	1.20	3.24	12.7	1.27	3.67	10.3	1.38	4.15	8.1	1.58	4.61						
45	54.7	0.86	1.42	36.0	0.96	1.93	25.1	1.05	2.53	17.2	1.20	3.24	14.3	1.28	3.66	11.6	1.39	4.14	9.1	1.57	4.64	7.5	1.73	5.12			
50	60.7	0.86	1.42	39.9	0.96	1.94	27.9	1.06	2.52	19.1	1.20	3.25	15.8	1.26	3.72	12.8	1.37	4.22	10.0	1.54	4.80	8.3	1.71	5.20			
55	66.6	0.86	1.42	43.8	0.96	1.94	30.6	1.05	2.53	21.0	1.20	3.25	17.4	1.27	3.70	14.1	1.37	4.21	11.0	1.54	4.80	9.1	1.70	5.25	7.5	1.91	5.65
60	72.5	0.86	1.43	47.7	0.96	1.95	33.3	1.05	2.55	22.8	1.19	3.29	18.9	1.26	3.75	15.4	1.38	4.19	12.0	1.54	4.81	9.9	1.69	5.30	8.2	1.92	5.63
65	78.4	0.86	1.43	51.6	0.96	1.95	36.1	1.06	2.54	24.7	1.19	3.28	20.5	1.26	3.73	16.6	1.37	4.25	12.9	1.52	4.92	10.7	1.68	5.34	8.8	1.88	5.79
70	84.3	0.86	1.43	55.5	0.96	1.95	38.8	1.05	2.54	26.6	1.19	3.28	22.0	1.26	3.76	17.9	1.37	4.23	13.9	1.52	4.91	11.5	1.68	5.38	9.5	1.89	5.76
75	90.2	0.86	1.43	59.4	0.96	1.95	41.5	1.05	2.55	28.5	1.20	3.27	23.6	1.26	3.74	19.1	1.36	4.28	14.9	1.52	4.90	12.3	1.67	5.40	10.1	1.87	5.89
80	96.0	0.86	1.43	63.2	0.96	1.96	44.2	1.05	2.56	30.3	1.19	3.30	25.1	1.26	3.77	20.4	1.37	4.25	15.9	1.53	4.89	13.1	1.67	5.43	10.8	1.88	5.85
90	107.8	0.86	1.44	71.0	0.96	1.96	49.7	1.05	2.55	34.1	1.19	3.29	28.2	1.26	3.78	22.9	1.37	4.28	17.8	1.51	4.96	14.7	1.66	5.47	12.1	1.86	5.93
100	119.5	0.86	1.44	78.8	0.96	1.96	55.1	1.05	2.56	37.8	1.19	3.31	31.3	1.26	3.79	25.4	1.36	4.30	19.8	1.52	4.94	16.3	1.65	5.51	13.4	1.85	5.99
110	131.2	0.86	1.44	86.5	0.96	1.97	60.5	1.05	2.57	41.5	1.19	3.32	34.4	1.26	3.79	27.9	1.36	4.31	21.7	1.51	4.99	17.9	1.65	5.53	14.7	1.84	6.05
120	142.9	0.86	1.44	94.2	0.96	1.97	65.9	1.05	2.58	45.3	1.19	3.31	37.5	1.26	3.79	30.4	1.36	4.32	23.7	1.51	4.97	19.5	1.65	5.55	16.0	1.83	6.09
130	154.5	0.86	1.45	101.9	0.96	1.97	71.3	1.05	2.58	49.0	1.19	3.31	40.6	1.26	3.79	32.9	1.36	4.33	25.6	1.51	5.01	21.1	1.65	5.57	17.3	1.82	6.12
140	166.1	0.87	1.45	109.6	0.96	1.97	76.7	1.05	2.58	52.7	1.19	3.32	43.7	1.26	3.79	35.4	1.36	4.33	27.6	1.51	4.99	22.7	1.64	5.58	18.6	1.82	6.15
150	177.6	0.87	1.45	117.2	0.96	1.98	82.1	1.05	2.58	56.4	1.19	3.32	46.8	1.26	3.79	37.9	1.36	4.33	29.5	1.51	5.02	24.3	1.64	5.59	19.9	1.82	6.17
160	189.1	0.87	1.45	124.8	0.96	1.98	87.5	1.06	2.58	60.1	1.19	3.33	49.8	1.26	3.81	40.4	1.36	4.33	31.4	1.50	5.04	25.9	1.64	5.60	21.3	1.83	6.10
170	200.5	0.87	1.45	132.4	0.96	1.98	92.8	1.05	2.58	63.7	1.19	3.34	52.9	1.26	3.80	42.9	1.36	4.33	33.4	1.51	5.02	27.5	1.64	5.60	22.6	1.83	6.13
180	211.9	0.87	1.46	140.0	0.96	1.98	98.1	1.05	2.59	67.4	1.19	3.34	55.9	1.26	3.82	45.4	1.36	4.33	35.3	1.51	5.04	29.1	1.64	5.61	23.9	1.83	6.14
190	223.3	0.87	1.46	147.5	0.96	1.99	103.4	1.05	2.59	71.1	1.19	3.34	59.0	1.26	3.81	47.8	1.36	4.36	37.2	1.50	5.06	30.7	1.64	5.61	25.2	1.82	6.16
200	234.6	0.87	1.46	155.1	0.97	1.99	108.7	1.06	2.60	74.7	1.19	3.35	62.0	1.26	3.82	50.3	1.36	4.35	39.2	1.51	5.03	32.3	1.64	5.61	26.5	1.82	6.17
220	257.6	0.87	1.46	170.3	0.96	1.99	119.4	1.06	2.60	82.1	1.19	3.35	68.1	1.26	3.83	55.3	1.36	4.35	43.1	1.51	5.03	35.5	1.64	5.62	29.1	1.82	6.20
240	280.4	0.87	1.47	185.5	0.97	1.99	130.1	1.06	2.60	89.5	1.19	3.35	74.3	1.26	3.82	60.3	1.36	4.35	46.9	1.51	5.06	38.7	1.64	5.62	31.7	1.81	6.22
260	303.2	0.87	1.47	200.6	0.97	2.00	140.8	1.06	2.60	96.9	1.19	3.35	80.4	1.26	3.82	65.2	1.36	4.37	50.8	1.51	5.06	41.9	1.64	5.63	34.3	1.81	6.24
280	326.0	0.87	1.47	215.7	0.97	2.00	151.4	1.06	2.61	104.2	1.19	3.36	86.5	1.26	3.83	70.2	1.36	4.36	54.7	1.51	5.05	45.1	1.64	5.63	36.9	1.81	6.25
300	348.6	0.87	1.47	230.8	0.97	2.00	162.0	1.06	2.61	111.5	1.19	3.36	92.5	1.26	3.84	75.1	1.36	4.37	58.5	1.51	5.07	48.3	1.65	5.63	39.5	1.81	6.26

TABLE B.4 *(Continued)* Parabolic Waterway Design
V_1 for *RETARDANCE "D"*. Top Width (T), Depth (D) and V_2 for *RETARDANCE "C"*.
Grade 2.0 Percent

Q cfs	V_1 = 2.0			V_1 = 2.5			V_1 = 3.0			V_1 = 3.5			V_1 = 4.0			V_1 = 4.5			V_1 = 5.0			V_1 = 5.5			V_1 = 6.0		
	T	D	V_2	T	D	V_2	T	D	V_2	T	D	V_2	T	D	V_2	T	D	V_2	T	D	V_2	T	D	V_2	T	D	V_2
15	20.8	0.81	1.32	12.8	0.91	1.90	9.3	1.00	2.37	6.7	1.15	2.85															
20	27.6	0.80	1.33	17.1	0.91	1.89	12.3	0.99	2.43	8.8	1.12	3.00	6.5	1.29	3.51	5.4	1.41	3.84									
25	34.5	0.81	1.33	21.3	0.91	1.91	15.4	0.99	2.43	11.0	1.11	3.01	8.0	1.25	3.69	6.7	1.38	3.96									
30	41.3	0.81	1.34	25.5	0.91	1.92	18.4	0.98	2.46	13.2	1.11	3.02	9.6	1.24	3.71	7.9	1.33	4.20	6.6	1.49	4.48						
35	48.0	0.80	1.35	29.7	0.91	1.93	21.5	0.99	2.44	15.3	1.10	3.08	11.1	1.22	3.82	9.2	1.33	4.23	7.6	1.45	4.68						
40	54.8	0.80	1.34	33.9	0.91	1.93	24.5	0.98	2.46	17.5	1.10	3.07	12.7	1.22	3.81	10.5	1.32	4.26	8.7	1.45	4.67	7.2	1.65	4.96			
45	61.5	0.80	1.35	38.1	0.91	1.93	27.5	0.98	2.47	19.6	1.10	3.11	14.3	1.23	3.80	11.8	1.32	4.27	9.7	1.43	4.80	8.0	1.61	5.16			
50	68.2	0.80	1.35	42.3	0.91	1.93	30.5	0.98	2.48	21.8	1.10	3.09	15.8	1.22	3.86	13.1	1.32	4.28	10.8	1.43	4.78	8.8	1.57	5.33	7.5	1.74	5.64
55	74.9	0.81	1.35	46.4	0.91	1.94	33.5	0.98	2.48	23.9	1.09	3.12	17.4	1.22	3.84	14.4	1.32	4.29	11.8	1.42	4.87	9.7	1.58	5.30	8.2	1.72	5.75
60	81.5	0.81	1.36	50.6	0.91	1.93	36.5	0.98	2.49	26.1	1.10	3.10	18.9	1.21	3.89	15.6	1.30	4.38	12.9	1.42	4.84	10.6	1.59	5.28	9.0	1.74	5.65
65	88.1	0.81	1.36	54.7	0.91	1.94	39.5	0.98	2.49	28.2	1.10	3.12	20.5	1.22	3.87	16.9	1.30	4.38	13.9	1.41	4.92	11.4	1.56	5.40	9.7	1.73	5.74
70	94.7	0.81	1.36	58.8	0.91	1.94	42.5	0.98	2.49	30.3	1.09	3.14	22.0	1.21	3.90	18.2	1.31	4.37	15.0	1.42	4.89	12.3	1.57	5.37	10.4	1.71	5.82
75	101.2	0.81	1.36	62.9	0.91	1.94	45.5	0.99	2.49	32.4	1.09	3.15	23.6	1.22	3.88	19.5	1.31	4.37	16.0	1.41	4.95	13.1	1.55	5.46	11.1	1.70	5.89
80	107.8	0.81	1.36	67.0	0.91	1.95	48.4	0.98	2.50	34.6	1.10	3.13	25.1	1.21	3.91	20.7	1.30	4.42	17.1	1.41	4.91	14.0	1.56	5.43	11.8	1.69	5.95
90	121.0	0.81	1.37	75.2	0.91	1.95	54.4	0.98	2.50	38.8	1.09	3.15	28.2	1.21	3.92	23.3	1.33	4.41	19.2	1.41	4.94	15.7	1.55	5.48	13.3	1.69	5.92
100	134.2	0.81	1.37	83.4	0.91	1.96	60.4	0.99	2.50	43.1	1.10	3.15	31.3	1.21	3.93	25.9	1.30	4.40	21.3	1.41	4.96	17.4	1.55	5.52	14.7	1.68	6.02
110	147.3	0.81	1.37	91.6	0.91	1.96	66.3	0.98	2.51	47.4	1.10	3.15	34.4	1.21	3.93	28.4	1.30	4.44	23.4	1.40	4.98	19.1	1.54	5.55	16.2	1.68	5.99
120	160.3	0.81	1.38	99.8	0.91	1.96	72.2	0.98	2.51	51.6	1.10	3.16	37.5	1.21	3.93	31.0	1.30	4.42	25.5	1.40	4.99	20.8	1.54	5.58	17.6	1.67	6.06
130	173.3	0.81	1.38	107.9	0.91	1.96	78.1	0.98	2.51	55.8	1.09	3.17	40.6	1.21	3.93	33.5	1.30	4.45	27.6	1.40	5.00	22.5	1.53	5.60	19.1	1.68	6.03
140	186.3	0.81	1.38	116.0	0.91	1.97	84.0	0.99	2.52	60.1	1.10	3.16	43.6	1.21	3.96	36.0	1.29	4.47	29.7	1.40	5.00	24.2	1.53	5.62	20.5	1.67	6.08
150	199.2	0.81	1.38	124.1	0.91	1.97	89.9	0.99	2.52	64.3	1.10	3.16	46.7	1.21	3.96	38.6	1.30	4.45	31.8	1.40	5.00	25.9	1.53	5.63	21.9	1.66	6.13
160	212.0	0.81	1.38	132.1	0.91	1.97	95.7	0.99	2.52	68.5	1.10	3.17	49.8	1.21	3.95	41.1	1.30	4.47	33.8	1.40	5.05	27.6	1.53	5.64	23.4	1.67	6.09
170	224.8	0.81	1.39	140.2	0.91	1.97	101.6	0.99	2.52	72.7	1.10	3.17	52.8	1.21	3.97	43.6	1.30	4.48	35.9	1.40	5.05	29.3	1.53	5.65	24.8	1.66	6.13
180	237.5	0.81	1.39	148.2	0.91	1.98	107.4	0.99	2.53	76.8	1.10	3.18	55.9	1.21	3.96	46.2	1.30	4.46	38.0	1.40	5.04	31.0	1.53	5.65	26.3	1.67	6.10
190	250.2	0.81	1.39	156.1	0.91	1.98	113.2	0.99	2.53	81.0	1.10	3.18	58.9	1.21	3.97	48.7	1.30	4.47	40.1	1.40	5.04	32.7	1.53	5.65	27.7	1.67	6.13
200	262.8	0.81	1.39	164.1	0.91	1.98	119.0	0.99	2.53	85.2	1.10	3.18	61.9	1.21	3.98	51.2	1.30	4.48	42.2	1.40	5.03	34.4	1.53	5.66	29.1	1.66	6.16
220	288.5	0.81	1.40	180.2	0.91	1.99	130.7	0.99	2.54	93.6	1.10	3.18	68.1	1.21	3.97	56.3	1.30	4.48	46.3	1.40	5.06	37.8	1.53	5.67	32.0	1.66	6.16
240	314.1	0.81	1.40	196.2	0.91	1.99	142.4	0.99	2.54	102.0	1.10	3.19	74.2	1.21	3.98	61.3	1.30	4.50	50.5	1.40	5.06	41.2	1.53	5.68	34.9	1.66	6.16
260	339.5	0.81	1.40	212.2	0.91	1.99	154.0	0.99	2.54	110.3	1.10	3.20	80.3	1.21	3.98	66.4	1.30	4.49	54.7	1.40	5.05	44.6	1.53	5.68	37.8	1.66	6.16
280	364.9	0.81	1.40	228.2	0.92	1.99	165.6	0.99	2.55	118.7	1.10	3.19	86.3	1.21	4.00	71.4	1.30	4.50	58.8	1.40	5.07	48.0	1.53	5.68	40.6	1.66	6.20
300	390.2	0.81	1.40	244.1	0.92	2.00	177.2	0.99	2.55	127.0	1.10	3.20	92.4	1.21	4.00	76.4	1.30	4.51	63.0	1.40	5.06	51.4	1.53	5.68	43.5	1.66	6.19

TABLE B.4 *(Continued)* Parabolic Waterway Design
V_1 for *RETARDANCE "D"*. Top Width (T), Depth (D) and V_2 for *RETARDANCE "C"*.
Grade 3.0 Percent

Q cfs	V_1 = 2.0			V_1 = 2.5			V_1 = 3.0			V_1 = 3.5			V_1 = 4.0			V_1 = 4.5			V_1 = 5.0			V_1 = 5.5			V_1 = 6.0		
	T	D	V_2	T	D	V_2	T	D	V_2	T	D	V_2	T	D	V_2	T	D	V_2	T	D	V_2	T	D	V_2	T	D	V_2
15	23.6	0.69	1.35	16.3	0.76	1.80	11.4	0.83	2.33	8.8	0.90	2.77	6.5	1.01	3.37	5.0	1.16	3.78									
20	31.4	0.69	1.36	21.7	0.76	1.81	15.2	0.83	2.34	11.7	0.90	2.81	8.6	0.99	3.48	6.6	1.13	3.94	5.9	1.19	4.17						
25	39.2	0.69	1.36	27.0	0.75	1.83	19.0	0.83	2.33	14.6	0.90	2.83	10.8	0.99	3.44	8.1	1.09	4.18	7.3	1.16	4.33	6.0	1.27	4.80			
30	46.9	0.69	1.37	32.4	0.75	1.82	22.7	0.83	2.36	17.4	0.88	2.89	12.9	0.98	3.4[illegible]	9.7	1.08	4.22	8.7	1.15	4.44	7.1	1.24	5.03	5.8	1.41	5.37
35	54.6	0.69	1.37	37.7	0.75	1.83	26.4	0.83	2.38	20.3	0.89	2.88	15.0	0.98	3.53	11.3	1.08	4.25	10.1	1.13	4.51	8.3	1.24	5.02	6.7	1.38	5.55
40	62.2	0.69	1.37	43.0	0.75	1.83	30.2	0.83	2.37	23.2	0.89	2.88	17.1	0.98	3.55	12.9	1.08	4.26	11.5	1.13	4.57	9.4	1.22	5.17	7.6	1.36	5.70
45	69.9	0.70	1.37	48.3	0.75	1.83	33.9	0.83	2.37	26.0	0.88	2.90	19.2	0.97	3.57	14.5	1.08	4.27	12.9	1.12	4.61	10.6	1.22	5.14	8.5	1.34	5.81
50	77.4	0.69	1.38	53.5	0.75	1.84	37.6	0.83	2.38	28.9	0.89	2.89	21.3	0.97	3.58	16.0	1.06	4.36	14.3	1.12	4.63	11.7	1.21	5.24	9.4	1.33	5.90
55	85.0	0.70	1.38	58.7	0.75	1.85	41.2	0.83	2.40	31.7	0.89	2.91	23.4	0.97	3.58	17.6	1.06	4.35	15.7	1.11	4.66	12.9	1.21	5.20	10.4	1.35	5.80
60	92.5	0.70	1.38	64.0	0.75	1.84	44.9	0.83	2.40	34.5	0.88	2.92	25.5	0.97	3.59	19.2	1.07	4.35	17.1	1.11	4.67	14.0	1.20	5.28	11.3	1.34	5.87
65	99.9	0.69	1.39	69.1	0.75	1.85	48.6	0.83	2.39	37.3	0.88	2.93	27.6	0.97	3.59	20.8	1.07	4.34	18.5	1.11	4.69	15.2	1.21	5.24	12.2	1.33	5.93
70	107.3	0.69	1.39	74.3	0.75	1.86	52.2	0.83	2.40	40.1	0.88	2.93	29.7	0.98	3.59	22.3	1.06	4.39	19.9	1.11	4.69	16.3	1.20	5.30	13.1	1.32	5.98
75	114.7	0.70	1.39	79.4	0.75	1.86	55.8	0.83	2.41	42.9	0.88	2.94	31.8	0.98	3.59	23.9	1.06	4.38	21.3	1.11	4.70	17.5	1.21	5.26	14.0	1.32	6.02
80	122.1	0.70	1.40	84.5	0.75	1.87	59.4	0.83	2.42	45.7	0.88	2.94	33.9	0.98	3.58	25.5	1.07	4.36	22.7	1.11	4.70	18.6	1.20	5.31	15.0	1.33	5.94
90	137.0	0.70	1.40	94.9	0.75	1.87	66.7	0.83	2.42	51.4	0.89	2.93	38.0	0.97	3.61	28.6	1.06	4.40	25.5	1.11	4.72	20.9	1.20	5.33	16.8	1.32	6.01
100	151.8	0.70	1.40	105.2	0.75	1.87	74.0	0.83	2.42	57.0	0.89	2.94	42.2	0.98	3.61	31.7	1.06	4.42	28.3	1.11	4.73	23.2	1.20	5.34	18.6	1.31	6.08
110	166.6	0.70	1.41	115.5	0.75	1.87	81.3	0.83	2.42	62.6	0.89	2.95	46.4	0.98	3.61	34.9	1.06	4.40	31.0	1.10	4.78	25.5	1.20	5.34	20.5	1.32	6.04
120	181.3	0.70	1.41	125.7	0.75	1.88	88.5	0.83	2.43	68.2	0.89	2.95	50.5	0.98	3.62	38.0	1.06	4.42	33.8	1.11	4.78	27.7	1.19	5.40	22.3	1.32	6.08
130	195.9	0.70	1.41	135.9	0.76	1.88	95.7	0.83	2.43	73.7	0.89	2.96	54.6	0.98	3.63	41.1	1.06	4.43	36.6	1.11	4.77	30.0	1.19	5.40	24.2	1.32	6.04
140	210.5	0.70	1.41	146.1	0.76	1.88	102.8	0.83	2.44	79.3	0.89	2.96	58.8	0.98	3.62	44.2	1.06	4.44	39.4	1.11	4.77	32.3	1.20	5.39	26.0	1.32	6.08
150	225.0	0.70	1.42	156.2	0.76	1.89	110.0	0.83	2.44	84.8	0.89	2.96	62.9	0.98	3.63	47.3	1.06	4.44	42.1	1.11	4.80	34.6	1.20	5.38	27.8	1.31	6.11
160	239.4	0.70	1.42	166.2	0.76	1.89	117.1	0.83	2.45	90.3	0.89	2.97	67.0	0.98	3.63	50.4	1.06	4.45	44.9	1.11	4.79	36.8	1.19	5.42	29.6	1.31	6.13
170	253.7	0.70	1.42	176.2	0.76	1.90	124.2	0.83	2.45	95.8	0.89	2.97	71.1	0.98	3.64	53.5	1.06	4.45	47.7	1.11	4.78	39.1	1.20	5.41	31.5	1.32	6.09
180	268.0	0.70	1.43	186.2	0.76	1.90	131.2	0.83	2.46	101.3	0.89	2.97	75.2	0.98	3.64	56.6	1.06	4.45	50.4	1.11	4.80	41.3	1.19	5.44	33.3	1.32	6.11
190	282.2	0.70	1.43	196.1	0.76	1.90	138.3	0.83	2.46	106.7	0.89	2.98	79.2	0.98	3.65	59.7	1.07	4.45	53.1	1.11	4.81	43.6	1.20	5.42	35.1	1.32	6.12
200	296.3	0.70	1.43	206.0	0.76	1.90	145.3	0.83	2.46	112.2	0.89	2.98	83.3	0.98	3.65	62.7	1.06	4.47	55.9	1.11	4.80	45.8	1.19	5.45	36.9	1.32	6.14
220	325.1	0.70	1.44	226.1	0.76	1.91	159.5	0.83	2.47	123.2	0.89	2.98	91.5	0.98	3.65	68.9	1.06	4.47	61.4	1.11	4.81	50.4	1.20	5.43	40.6	1.32	6.12
240	353.8	0.70	1.44	246.2	0.76	1.91	173.7	0.83	2.47	134.2	0.89	2.99	99.7	0.98	3.65	75.1	1.07	4.47	66.9	1.11	4.81	54.9	1.20	5.44	44.2	1.32	6.15
260	382.4	0.70	1.44	266.1	0.76	1.92	187.8	0.83	2.48	145.1	0.89	2.99	107.8	0.98	3.67	81.3	1.07	4.47	72.4	1.11	4.82	59.4	1.20	5.45	47.8	1.31	6.17
280	410.8	0.70	1.45	286.0	0.76	1.92	201.9	0.83	2.48	156.0	0.89	3.00	116.0	0.98	3.66	87.4	1.07	4.48	77.9	1.11	4.82	63.9	1.20	5.46	51.5	1.32	6.15
300	439.0	0.70	1.45	305.8	0.76	1.92	215.9	0.83	2.49	166.9	0.89	3.00	124.1	0.98	3.67	93.6	1.07	4.47	83.3	1.11	4.83	68.4	1.20	5.46	55.1	1.32	6.17

TABLE B.4 *(Continued)* Parabolic Waterway Design
V_1 for *RETARDANCE "D"*. Top Width (T), Depth (D) and V_2 for *RETARDANCE "C"*.
Grade 4.0 Percent

Q	V_1 = 2.0			V_1 = 2.5			V_1 = 3.0		
cfs	T	D	V_2	T	D	V_2	T	D	V_2
15	27.9	0.62	1.29	19.9	0.66	1.68	13.9	0.73	2.20
20	37.1	0.62	1.29	26.5	0.66	1.69	18.5	0.72	2.21
25	46.2	0.62	1.30	33.0	0.66	1.70	23.0	0.72	2.24
30	55.3	0.62	1.30	39.5	0.66	1.70	27.6	0.72	2.23
35	64.3	0.62	1.31	46.0	0.66	1.71	32.1	0.72	2.25
40	73.3	0.62	1.31	52.4	0.66	1.71	36.6	0.72	2.25
45	82.2	0.62	1.32	58.8	0.66	1.72	41.1	0.72	2.26
50	91.1	0.62	1.32	65.2	0.66	1.72	45.6	0.72	2.26
55	99.9	0.62	1.32	71.5	0.66	1.73	50.1	0.72	2.26
60	108.7	0.62	1.32	77.8	0.66	1.73	54.5	0.72	2.26
65	117.4	0.62	1.33	84.1	0.66	1.73	58.9	0.72	2.27
70	126.1	0.62	1.33	90.3	0.66	1.74	63.3	0.72	2.27
75	134.7	0.62	1.33	96.5	0.66	1.74	67.7	0.72	2.28
80	143.3	0.62	1.34	102.7	0.66	1.74	72.1	0.72	2.28
90	160.8	0.62	1.34	115.2	0.66	1.75	80.9	0.72	2.28
100	178.2	0.62	1.34	127.7	0.66	1.75	89.7	0.72	2.29
110	195.4	0.62	1.35	140.1	0.66	1.76	98.5	0.72	2.29
120	212.6	0.62	1.35	152.5	0.66	1.76	107.2	0.72	2.30
130	229.6	0.62	1.35	164.8	0.66	1.76	115.9	0.72	2.30
140	246.6	0.62	1.36	177.0	0.66	1.77	124.5	0.72	2.30
150	263.5	0.62	1.36	189.1	0.66	1.77	133.2	0.73	2.30
160	280.3	0.62	1.36	201.2	0.66	1.78	141.7	0.73	2.31
170	296.9	0.62	1.37	213.3	0.67	1.78	150.3	0.73	2.31
180	313.5	0.62	1.37	225.3	0.67	1.78	158.8	0.73	2.32
190	330.0	0.62	1.37	237.2	0.67	1.79	167.3	0.73	2.32
200	346.4	0.62	1.37	249.1	0.67	1.79	175.7	0.73	2.32
220	380.0	0.62	1.38	273.3	0.67	1.79	192.9	0.73	2.33
240	413.3	0.62	1.38	297.4	0.67	1.80	209.9	0.73	2.33
260	446.5	0.62	1.39	321.4	0.67	1.80	227.0	0.73	2.33
280	479.5	0.62	1.39	345.3	0.67	1.80	243.9	0.73	2.34
300	512.3	0.62	1.39	369.0	0.67	1.81	260.8	0.73	2.34

Q	V_1 = 3.5			V_1 = 4.0			V_1 = 4.5		
cfs	T	D	V_2	T	D	V_2	T	D	V_2
15	10.3	0.79	2.73	7.9	0.85	3.28	6.3	0.92	3.78
20	13.7	0.78	2.76	10.5	0.84	3.33	8.4	0.92	3.81
25	17.1	0.78	2.77	13.1	0.84	3.35	10.5	0.92	3.82
30	20.4	0.77	2.82	15.7	0.84	3.36	12.5	0.91	3.92
35	23.8	0.78	2.81	18.3	0.84	3.37	14.6	0.91	3.90
40	27.1	0.77	2.83	20.8	0.83	3.42	16.6	0.90	3.96
45	30.4	0.77	2.85	23.4	0.84	3.41	18.7	0.91	3.94
50	33.7	0.77	2.86	26.0	0.84	3.40	20.7	0.90	3.97
55	37.0	0.77	2.86	28.5	0.84	3.43	22.8	0.91	3.95
60	40.3	0.77	2.87	31.0	0.83	3.45	24.8	0.90	3.97
65	43.6	0.77	2.87	33.6	0.84	3.43	26.8	0.90	3.99
70	46.9	0.77	2.86	36.1	0.84	3.44	28.8	0.90	4.01
75	50.1	0.77	2.88	38.6	0.84	3.45	30.9	0.91	3.98
80	53.3	0.77	2.89	41.1	0.84	3.46	32.9	0.91	3.99
90	59.9	0.77	2.89	46.2	0.84	3.46	36.9	0.90	4.01
100	66.4	0.77	2.90	51.2	0.84	3.47	41.0	0.91	4.00
110	72.9	0.77	2.90	56.2	0.84	3.48	45.0	0.90	4.02
120	79.4	0.77	2.90	61.2	0.84	3.49	49.0	0.90	4.03
130	85.9	0.77	2.91	66.2	0.84	3.49	53.0	0.90	4.03
140	92.3	0.77	2.91	71.2	0.84	3.49	57.0	0.90	4.04
150	98.7	0.77	2.92	76.2	0.84	3.49	61.0	0.91	4.04
160	105.1	0.77	2.92	81.1	0.84	3.50	65.0	0.91	4.04
170	111.5	0.78	2.92	86.0	0.84	3.51	68.9	0.91	4.05
180	117.8	0.78	2.93	90.9	0.84	3.52	72.9	0.91	4.05
190	124.2	0.78	2.93	95.8	0.84	3.52	76.8	0.91	4.06
200	130.5	0.78	2.93	100.7	0.84	3.52	80.7	0.91	4.07
220	143.3	0.78	2.93	110.6	0.84	3.53	88.7	0.91	4.07
240	156.0	0.78	2.94	120.4	0.84	3.53	96.6	0.91	4.07
260	168.7	0.78	2.94	130.2	0.84	3.54	104.5	0.91	4.08
280	181.3	0.78	2.95	140.0	0.84	3.54	112.3	0.91	4.09
300	193.9	0.78	2.95	149.8	0.84	3.55	120.2	0.91	4.09

Q	V_1 = 5.0			V_1 = 5.5			V_1 = 6.0		
cfs	T	D	V_2	T	D	V_2	T	D	V_2
15	4.9	1.06	4.21						
20	6.4	1.01	4.52	5.5	1.09	4.88			
25	8.0	1.01	4.55	6.8	1.06	5.09	5.7	1.20	5.34
30	9.5	0.99	4.71	8.2	1.07	5.03	6.7	1.15	5.71
35	11.1	0.99	4.70	9.5	1.06	5.15	7.8	1.15	5.77
40	12.7	1.00	4.68	10.8	1.04	5.24	8.9	1.14	5.81
45	14.2	0.98	4.77	12.2	1.05	5.17	10.0	1.14	5.83
50	15.8	0.99	4.74	13.5	1.05	5.24	11.1	1.14	5.85
55	17.3	0.98	4.80	14.8	1.04	5.28	12.2	1.14	5.87
60	18.9	0.99	4.77	16.1	1.04	5.32	13.3	1.14	5.88
65	20.4	0.98	4.81	17.5	1.05	5.26	14.3	1.12	6.00
70	21.9	0.98	4.85	18.8	1.04	5.29	15.4	1.12	6.00
75	23.5	0.98	4.82	20.1	1.04	5.31	16.5	1.13	5.99
80	25.0	0.98	4.84	21.4	1.04	5.33	17.6	1.13	5.98
90	28.1	0.98	4.85	24.0	1.04	5.38	19.7	1.12	6.07
100	31.2	0.98	4.85	26.7	1.04	5.35	21.9	1.12	6.05
110	34.3	0.98	4.85	29.3	1.04	5.37	24.1	1.12	6.03
120	37.3	0.98	4.88	31.9	1.04	5.40	26.2	1.12	6.09
130	40.4	0.98	4.87	34.6	1.04	5.36	28.4	1.12	6.07
140	43.4	0.98	4.90	37.2	1.04	5.38	30.5	1.12	6.10
150	46.5	0.98	4.88	39.8	1.04	5.39	32.7	1.12	6.08
160	49.5	0.98	4.90	42.4	1.04	5.40	34.8	1.12	6.11
170	52.5	0.98	4.91	45.0	1.04	5.40	36.9	1.12	6.13
180	55.6	0.98	4.90	47.6	1.04	5.40	39.1	1.12	6.11
190	58.6	0.98	4.90	50.2	1.04	5.40	41.2	1.12	6.12
200	61.6	0.98	4.91	52.7	1.04	5.43	43.3	1.12	6.14
220	67.6	0.98	4.93	57.9	1.04	5.44	47.6	1.12	6.14
240	73.7	0.98	4.93	63.1	1.04	5.44	51.9	1.12	6.14
260	79.7	0.98	4.94	68.3	1.04	5.44	56.2	1.12	6.13
280	85.8	0.99	4.93	73.5	1.04	5.44	60.4	1.12	6.15
300	91.8	0.99	4.94	78.6	1.04	5.46	64.7	1.12	6.14

TABLE B.4 *(Continued)* Parabolic Waterway Design
V_1 for *RETARDANCE "D"*. Top Width (T), Depth (D) and V_2 for *RETARDANCE "C"*.
Grade 5.0 Percent

Q cfs	V_1 = 2.0			V_1 = 2.5			V_1 = 3.0			V_1 = 3.5			V_1 = 4.0		
	T	D	V_2	T	D	V_2	T	D	V_2	T	D	V_2	T	D	V_2
15	29.3	0.57	1.33	21.1	0.60	1.74	15.0	0.66	2.23	12.2	0.70	2.58	9.0	0.75	3.25
20	39.0	0.57	1.33	28.1	0.61	1.74	19.9	0.66	2.26	16.2	0.70	2.62	12.0	0.75	3.26
25	48.6	0.57	1.34	35.1	0.61	1.73	24.8	0.66	2.28	20.3	0.70	2.59	15.0	0.75	3.27
30	58.1	0.57	1.34	42.0	0.61	1.74	29.7	0.66	2.28	24.3	0.70	2.61	18.0	0.76	3.26
35	67.6	0.57	1.35	48.8	0.61	1.75	34.6	0.66	2.28	28.2	0.70	2.64	20.9	0.75	3.30
40	77.0	0.57	1.35	55.7	0.61	1.75	39.5	0.66	2.28	32.2	0.70	2.64	23.9	0.75	3.29
45	86.4	0.57	1.35	62.5	0.61	1.75	44.3	0.66	2.29	36.1	0.70	2.65	26.8	0.75	3.31
50	95.7	0.57	1.36	69.2	0.61	1.76	49.1	0.66	2.30	40.1	0.70	2.64	29.7	0.75	3.32
55	105.0	0.57	1.36	75.9	0.61	1.77	53.9	0.66	2.30	44.0	0.70	2.65	32.6	0.75	3.33
60	114.2	0.57	1.36	82.6	0.61	1.77	58.7	0.66	2.30	47.9	0.70	2.66	35.5	0.75	3.34
65	123.4	0.57	1.36	89.3	0.61	1.77	63.4	0.66	2.31	51.8	0.70	2.66	38.4	0.75	3.34
70	132.4	0.57	1.37	95.9	0.61	1.77	68.2	0.66	2.31	55.6	0.70	2.67	41.3	0.75	3.34
75	141.5	0.57	1.37	102.4	0.61	1.78	72.9	0.66	2.31	59.4	0.70	2.68	44.1	0.75	3.36
80	150.5	0.57	1.37	109.0	0.61	1.78	77.5	0.66	2.32	63.3	0.70	2.68	47.0	0.75	3.36
90	168.8	0.57	1.38	122.3	0.61	1.79	87.0	0.66	2.33	71.0	0.70	2.69	52.8	0.75	3.36
100	187.0	0.57	1.38	135.5	0.61	1.79	96.5	0.66	2.33	78.7	0.70	2.70	58.5	0.75	3.37
110	205.1	0.57	1.38	148.7	0.61	1.79	105.9	0.66	2.33	86.4	0.70	2.70	64.3	0.75	3.37
120	223.1	0.57	1.39	161.8	0.61	1.80	115.3	0.66	2.33	94.1	0.70	2.70	70.0	0.75	3.38
130	240.9	0.57	1.39	174.8	0.61	1.80	124.6	0.66	2.34	101.7	0.70	2.71	75.7	0.76	3.38
140	258.7	0.57	1.40	187.7	0.61	1.81	133.9	0.66	2.34	109.3	0.70	2.71	81.3	0.75	3.39
150	276.4	0.58	1.40	200.6	0.61	1.81	143.1	0.66	2.35	116.8	0.70	2.72	87.0	0.76	3.39
160	293.9	0.58	1.40	213.4	0.61	1.81	152.3	0.66	2.35	124.3	0.70	2.72	92.6	0.76	3.40
170	311.4	0.58	1.40	226.1	0.61	1.82	161.5	0.66	2.35	131.8	0.70	2.73	98.2	0.76	3.41
180	328.7	0.58	1.41	238.8	0.61	1.82	170.6	0.66	2.36	139.2	0.70	2.73	103.8	0.76	3.41
190	346.0	0.58	1.41	251.4	0.61	1.83	179.7	0.67	2.36	146.6	0.70	2.74	109.4	0.76	3.41
200	363.1	0.58	1.42	263.9	0.61	1.83	188.7	0.67	2.37	154.0	0.70	2.74	114.9	0.76	3.42
220	398.3	0.58	1.42	289.6	0.62	1.83	207.1	0.67	2.37	169.0	0.70	2.75	126.1	0.76	3.43
240	433.2	0.58	1.42	315.0	0.62	1.84	225.4	0.67	2.37	184.0	0.70	2.75	137.4	0.76	3.43
260	467.9	0.58	1.43	340.4	0.62	1.84	243.7	0.67	2.38	198.9	0.70	2.76	148.5	0.76	3.44
280	502.5	0.58	1.43	365.6	0.62	1.84	261.8	0.67	2.38	213.7	0.70	2.76	159.7	0.76	3.44
300	536.7	0.58	1.43	390.7	0.62	1.85	279.9	0.67	2.38	228.5	0.71	2.77	170.7	0.76	3.45

Q cfs	V_1 = 4.5			V_1 = 5.0			V_1 = 5.5			V_1 = 6.0		
	T	D	V_2	T	D	V_2	T	D	V_2	T	D	V_2
15	7.2	0.83	3.70	5.8	0.93	4.09	4.6	0.99	4.81			
20	9.5	0.81	3.84	7.6	0.89	4.35	6.1	0.97	4.95	5.3	1.06	5.21
25	11.9	0.81	3.82	9.5	0.89	4.37	7.6	0.96	5.03	6.5	1.02	5.56
30	14.2	0.80	3.89	11.3	0.87	4.49	9.1	0.96	5.08	7.8	1.01	5.59
35	16.6	0.81	3.86	13.2	0.88	4.47	10.5	0.94	5.26	9.1	1.01	5.60
40	18.9	0.80	3.90	15.1	0.88	4.46	12.0	0.94	5.26	10.3	0.99	5.77
45	21.3	0.81	3.87	16.9	0.87	4.52	13.5	0.94	5.25	11.6	1.00	5.75
50	23.6	0.81	3.89	18.8	0.88	4.50	15.0	0.94	5.25	12.9	1.00	5.73
55	25.9	0.81	3.90	20.6	0.87	4.54	16.5	0.94	5.24	14.1	0.99	5.84
60	28.2	0.81	3.92	22.4	0.87	4.57	17.9	0.93	5.32	15.4	0.99	5.81
65	30.5	0.81	3.92	24.3	0.87	4.54	19.4	0.94	5.30	16.7	1.00	5.78
70	32.8	0.81	3.93	26.1	0.87	4.56	20.8	0.93	5.36	17.9	0.99	5.85
75	35.1	0.81	3.93	27.9	0.87	4.58	22.3	0.93	5.34	19.2	1.00	5.82
80	37.4	0.81	3.92	29.7	0.87	4.60	23.8	0.94	5.32	20.4	0.99	5.88
90	42.0	0.81	3.93	33.4	0.87	4.59	26.7	0.94	5.35	22.9	0.99	5.91
100	46.5	0.81	3.96	37.0	0.87	4.62	29.6	0.93	5.38	25.5	0.99	5.86
110	51.1	0.81	3.96	40.7	0.87	4.61	32.5	0.93	5.39	28.0	0.99	5.88
120	55.7	0.81	3.96	44.3	0.87	4.62	35.4	0.93	5.41	30.5	0.99	5.89
130	60.2	0.81	3.97	47.9	0.87	4.64	38.3	0.93	5.41	33.0	0.99	5.90
140	64.7	0.81	3.98	51.5	0.87	4.64	41.2	0.93	5.42	35.5	0.99	5.91
150	69.3	0.81	3.97	55.1	0.87	4.65	44.1	0.93	5.42	37.9	0.99	5.96
160	73.7	0.81	3.99	58.7	0.87	4.65	47.0	0.94	5.42	40.4	0.99	5.95
170	78.2	0.81	3.99	62.3	0.87	4.65	49.9	0.94	5.41	42.9	0.99	5.95
180	82.7	0.81	3.99	65.9	0.87	4.65	52.7	0.93	5.44	45.4	0.99	5.94
190	87.1	0.81	4.00	69.4	0.87	4.67	55.6	0.94	5.43	47.8	0.99	5.97
200	91.6	0.81	4.00	73.0	0.87	4.66	58.4	0.94	5.45	50.3	0.99	5.96
220	100.6	0.81	4.00	80.1	0.87	4.68	64.2	0.94	5.45	55.2	0.99	5.99
240	109.5	0.81	4.01	87.3	0.87	4.68	69.9	0.94	5.46	60.2	0.99	5.98
260	118.5	0.81	4.01	94.4	0.87	4.69	75.6	0.94	5.47	65.1	0.99	5.99
280	127.4	0.81	4.02	101.5	0.87	4.70	81.4	0.94	5.46	70.0	0.99	6.01
300	136.2	0.81	4.03	108.6	0.87	4.70	87.0	0.94	5.48	74.9	0.99	6.01

TABLE B.4 *(Continued)* Parabolic Waterway Design
V_1 for *RETARDANCE "D"*. Top Width (T), Depth (D) and V_2 for *RETARDANCE "C"*.
Grade 6.0 Percent

Q cfs	V_1 = 2.0			V_1 = 2.5			V_1 = 3.0			V_1 = 3.5			V_1 = 4.0			V_1 = 4.5			V_1 = 5.0			V_1 = 5.5			V_1 = 6.0		
	T	D	V_2	T	D	V_2	T	D	V_2	T	D	V_2	T	D	V_2	T	D	V_2	T	D	V_2	T	D	V_2	T	D	V_2
15	34.6	0.53	1.22	22.6	0.57	1.72	16.6	0.61	2.20	12.6	0.65	2.68	10.0	0.70	3.15	8.1	0.76	3.59	6.6	0.82	4.05	5.3	0.90	4.61	4.3	0.98	5.19
20	46.0	0.53	1.22	30.0	0.57	1.73	22.1	0.61	2.20	16.8	0.66	2.68	13.2	0.69	3.25	10.7	0.74	3.71	8.7	0.80	4.22	7.0	0.88	4.79	5.7	0.96	5.36
25	57.2	0.52	1.23	37.4	0.57	1.74	27.6	0.61	2.20	21.0	0.66	2.68	16.5	0.69	3.25	13.3	0.73	3.78	10.8	0.79	4.32	8.7	0.86	4.90	7.0	0.92	5.69
30	68.5	0.53	1.23	44.7	0.57	1.75	33.0	0.61	2.22	25.1	0.66	2.70	19.8	0.69	3.24	16.0	0.74	3.75	13.0	0.80	4.29	10.4	0.86	4.97	8.4	0.92	5.71
35	79.6	0.53	1.24	52.0	0.57	1.76	38.4	0.61	2.22	29.2	0.65	2.71	23.0	0.69	3.28	18.6	0.74	3.78	15.1	0.79	4.34	12.1	0.85	5.01	9.8	0.92	5.72
40	90.6	0.52	1.24	59.3	0.57	1.76	43.8	0.61	2.22	33.3	0.65	2.72	26.3	0.69	3.26	21.2	0.74	3.80	17.2	0.79	4.37	13.8	0.85	5.04	11.2	0.92	5.72
45	101.6	0.53	1.25	66.5	0.57	1.76	49.1	0.61	2.24	37.4	0.66	2.72	29.5	0.69	3.28	23.8	0.73	3.81	19.4	0.79	4.33	15.5	0.85	5.05	12.5	0.91	5.85
50	112.5	0.53	1.25	73.6	0.57	1.77	54.4	0.61	2.24	41.5	0.66	2.72	32.7	0.69	3.29	26.4	0.74	3.82	21.5	0.79	4.35	17.2	0.85	5.06	13.9	0.91	5.83
55	123.3	0.53	1.25	80.8	0.57	1.77	59.7	0.61	2.25	45.5	0.66	2.73	35.9	0.69	3.29	29.0	0.74	3.82	23.6	0.79	4.37	18.8	0.84	5.15	15.3	0.92	5.82
60	134.1	0.53	1.26	87.8	0.57	1.78	65.0	0.61	2.25	49.5	0.66	2.74	39.1	0.69	3.30	31.6	0.74	3.82	25.7	0.79	4.38	20.5	0.84	5.14	16.6	0.91	5.90
65	144.7	0.53	1.26	94.9	0.57	1.78	70.2	0.61	2.26	53.5	0.66	2.75	42.2	0.69	3.32	34.1	0.73	3.85	27.8	0.79	4.38	22.2	0.85	5.14	18.0	0.91	5.87
70	155.3	0.53	1.27	101.9	0.57	1.79	75.4	0.61	2.26	57.5	0.66	2.75	45.4	0.69	3.31	36.7	0.74	3.84	29.9	0.79	4.39	23.9	0.85	5.13	19.3	0.91	5.94
75	165.8	0.53	1.27	108.8	0.57	1.80	80.6	0.61	2.26	61.5	0.66	2.75	48.5	0.69	3.33	39.2	0.73	3.86	31.9	0.79	4.43	25.5	0.84	5.18	20.7	0.91	5.91
80	176.3	0.53	1.27	115.7	0.57	1.80	85.8	0.61	2.27	65.4	0.66	2.76	51.7	0.69	3.32	41.8	0.74	3.85	34.0	0.79	4.42	27.2	0.85	5.16	22.0	0.91	5.96
90	197.6	0.53	1.28	129.8	0.57	1.80	96.2	0.61	2.28	73.4	0.66	2.77	58.0	0.69	3.33	46.9	0.74	3.87	38.2	0.79	4.43	30.5	0.84	5.20	24.8	0.91	5.91
100	218.8	0.53	1.28	143.8	0.57	1.81	106.6	0.61	2.28	81.4	0.66	2.77	64.3	0.69	3.34	52.0	0.74	3.88	42.4	0.79	4.42	33.9	0.85	5.18	27.5	0.91	5.93
110	239.9	0.53	1.28	157.7	0.57	1.81	117.0	0.61	2.28	89.3	0.66	2.78	70.6	0.69	3.35	57.1	0.74	3.88	46.5	0.79	4.45	37.2	0.84	5.20	30.2	0.91	5.95
120	260.8	0.53	1.29	171.5	0.57	1.82	127.3	0.61	2.29	97.2	0.66	2.79	76.9	0.69	3.35	62.2	0.74	3.89	50.7	0.79	4.44	40.5	0.84	5.22	32.9	0.91	5.96
130	281.5	0.53	1.29	185.3	0.57	1.82	137.6	0.61	2.29	105.1	0.66	2.79	83.1	0.69	3.36	67.2	0.74	3.90	54.8	0.79	4.46	43.8	0.84	5.23	35.6	0.91	5.96
140	302.1	0.53	1.30	199.0	0.57	1.83	147.8	0.61	2.30	112.9	0.66	2.80	89.3	0.69	3.36	72.3	0.74	3.90	58.9	0.79	4.46	47.1	0.84	5.24	38.3	0.91	5.97
150	322.6	0.53	1.30	212.6	0.57	1.83	157.9	0.61	2.30	120.7	0.66	2.80	95.5	0.69	3.37	77.3	0.74	3.91	63.0	0.79	4.47	50.4	0.84	5.24	40.9	0.91	6.01
160	342.9	0.53	1.30	226.1	0.57	1.84	168.0	0.61	2.31	128.5	0.66	2.80	101.7	0.69	3.37	82.3	0.74	3.91	67.1	0.79	4.47	53.7	0.85	5.24	43.6	0.91	6.00
170	363.1	0.53	1.31	239.6	0.57	1.84	178.0	0.61	2.32	136.2	0.66	2.81	107.8	0.69	3.37	87.3	0.74	3.91	71.2	0.79	4.47	56.9	0.84	5.27	46.3	0.91	5.99
180	383.1	0.53	1.31	253.0	0.57	1.84	188.0	0.61	2.32	143.9	0.66	2.81	113.9	0.69	3.38	92.2	0.74	3.93	75.2	0.79	4.49	60.2	0.85	5.26	48.9	0.91	6.02
190	403.0	0.53	1.32	266.3	0.57	1.85	197.9	0.61	2.33	151.5	0.66	2.82	120.0	0.70	3.38	97.1	0.74	3.94	79.3	0.79	4.49	63.4	0.84	5.28	51.6	0.91	6.01
200	422.7	0.53	1.32	279.5	0.57	1.85	207.8	0.61	2.33	159.1	0.66	2.82	126.1	0.70	3.39	102.1	0.74	3.93	83.3	0.79	4.50	66.7	0.85	5.27	54.2	0.91	6.03
220	463.4	0.53	1.32	306.6	0.57	1.86	228.0	0.61	2.33	174.6	0.66	2.83	138.4	0.70	3.39	112.0	0.74	3.95	91.5	0.79	4.50	73.2	0.85	5.29	59.6	0.91	6.02
240	503.8	0.53	1.33	333.6	0.57	1.86	248.1	0.61	2.34	190.1	0.66	2.83	150.7	0.70	3.40	122.0	0.74	3.95	99.7	0.80	4.50	79.8	0.85	5.28	64.9	0.91	6.04
260	543.8	0.53	1.33	360.4	0.57	1.86	268.1	0.61	2.34	205.4	0.66	2.84	162.9	0.70	3.40	131.9	0.74	3.96	107.8	0.80	4.51	86.3	0.85	5.29	70.2	0.91	6.05
280	583.6	0.53	1.34	387.0	0.57	1.87	287.9	0.61	2.35	220.7	0.66	2.85	175.1	0.70	3.41	141.8	0.74	3.96	115.9	0.80	4.51	92.8	0.85	5.30	75.5	0.91	6.05
300	623.2	0.53	1.34	413.5	0.58	1.87	307.7	0.61	2.36	235.9	0.66	2.85	187.2	0.70	3.41	151.6	0.74	3.97	123.9	0.80	4.53	99.3	0.85	5.30	80.8	0.91	6.06

TABLE B.4 *(Continued)* Parabolic Waterway Design

V_1 for *RETARDANCE "D"*. Top Width (T), Depth (D) and V_2 for *RETARDANCE "C"*.

Grade 8.0 Percent

Q cfs	V_1 = 2.0			V_1 = 2.5			V_1 = 3.0			V_1 = 3.5			V_1 = 4.0			V_1 = 4.5			V_1 = 5.0			V_1 = 5.5			V_1 = 6.0		
	T	D	V_2	T	D	V_2	T	D	V_2	T	D	V_2	T	D	V_2	T	D	V_2	T	D	V_2	T	D	V_2	T	D	V_2
15	37.0	0.47	1.26	26.6	0.51	1.65	18.7	0.54	2.19	15.3	0.57	2.54	11.7	0.61	3.08	9.4	0.65	3.61	7.9	0.70	3.99	6.4	0.74	4.62	5.3	0.80	5.18
20	49.2	0.47	1.26	35.3	0.50	1.66	24.9	0.54	2.19	20.4	0.57	2.54	15.6	0.61	3.08	12.5	0.65	3.65	10.4	0.68	4.17	8.5	0.74	4.70	7.1	0.80	5.16
25	61.2	0.47	1.27	44.0	0.50	1.67	31.0	0.54	2.21	25.4	0.57	2.56	19.4	0.61	3.12	15.6	0.65	3.67	13.0	0.68	4.17	10.6	0.73	4.75	8.8	0.79	5.32
30	73.2	0.47	1.28	52.6	0.50	1.67	37.2	0.54	2.19	30.4	0.57	2.57	23.3	0.61	3.10	18.7	0.65	3.67	15.6	0.68	4.17	12.7	0.73	4.77	10.5	0.78	5.41
35	85.1	0.47	1.28	61.2	0.50	1.68	43.2	0.54	2.21	35.3	0.57	2.59	27.1	0.61	3.12	21.7	0.64	3.72	18.1	0.68	4.23	14.8	0.73	4.78	12.2	0.77	5.48
40	96.9	0.47	1.28	69.7	0.50	1.68	49.3	0.54	2.21	40.3	0.57	2.59	30.9	0.61	3.13	24.8	0.64	3.71	20.7	0.68	4.20	16.9	0.73	4.78	13.9	0.77	5.53
45	108.6	0.47	1.29	78.2	0.51	1.68	55.3	0.54	2.22	45.2	0.57	2.60	34.6	0.61	3.16	27.8	0.64	3.73	23.2	0.68	4.24	19.0	0.73	4.78	15.7	0.78	5.45
50	120.2	0.48	1.29	86.5	0.50	1.69	61.2	0.54	2.23	50.0	0.57	2.62	38.4	0.61	3.15	30.9	0.65	3.71	25.8	0.68	4.21	21.0	0.73	4.84	17.4	0.78	5.48
55	131.8	0.48	1.30	94.9	0.51	1.70	67.1	0.54	2.24	54.9	0.57	2.62	42.1	0.61	3.17	33.9	0.65	3.73	28.3	0.68	4.23	23.1	0.73	4.83	19.1	0.78	5.50
60	143.2	0.48	1.30	103.1	0.51	1.70	73.0	0.54	2.24	59.7	0.57	2.62	45.9	0.61	3.16	36.9	0.65	3.73	30.8	0.68	4.24	25.1	0.73	4.87	20.8	0.78	5.51
65	154.6	0.48	1.30	111.4	0.51	1.71	78.9	0.54	2.24	64.5	0.57	2.63	49.6	0.61	3.16	39.9	0.65	3.74	33.3	0.68	4.25	27.2	0.73	4.85	22.5	0.78	5.52
70	165.9	0.48	1.31	119.5	0.51	1.71	84.7	0.54	2.25	69.3	0.57	2.63	53.2	0.61	3.19	42.8	0.64	3.76	35.8	0.68	4.26	29.2	0.73	4.88	24.1	0.77	5.59
75	177.1	0.48	1.31	127.6	0.51	1.72	90.5	0.54	2.25	74.1	0.57	2.63	56.9	0.61	3.19	45.8	0.65	3.76	38.2	0.68	4.29	31.2	0.73	4.90	25.8	0.77	5.59
80	188.2	0.48	1.32	135.6	0.51	1.72	96.2	0.54	2.26	78.8	0.57	2.64	60.5	0.61	3.20	48.7	0.64	3.78	40.7	0.68	4.29	33.3	0.73	4.87	27.5	0.77	5.58
90	210.9	0.48	1.32	152.1	0.51	1.73	108.0	0.55	2.26	88.4	0.57	2.65	67.9	0.61	3.21	54.7	0.65	3.78	45.7	0.68	4.29	37.4	0.73	4.88	30.9	0.77	5.59
100	233.5	0.48	1.32	168.4	0.51	1.73	119.6	0.55	2.27	98.0	0.57	2.65	75.3	0.61	3.21	60.7	0.65	3.78	50.6	0.68	4.32	41.4	0.73	4.92	34.2	0.77	5.64
110	255.9	0.48	1.33	184.6	0.51	1.74	131.2	0.55	2.28	107.5	0.57	2.66	82.6	0.61	3.22	66.6	0.65	3.79	55.6	0.68	4.32	45.5	0.73	4.91	37.6	0.77	5.63
120	278.1	0.48	1.33	200.7	0.51	1.74	142.7	0.55	2.28	116.9	0.57	2.67	89.9	0.61	3.22	72.5	0.65	3.79	60.5	0.68	4.33	49.5	0.73	4.93	40.9	0.77	5.66
130	300.2	0.48	1.34	216.6	0.51	1.75	154.1	0.55	2.29	126.4	0.57	2.67	97.1	0.61	3.24	78.3	0.65	3.81	65.4	0.68	4.34	53.6	0.73	4.92	44.3	0.77	5.64
140	322.1	0.48	1.34	232.5	0.51	1.76	165.5	0.55	2.29	135.7	0.57	2.67	104.4	0.62	3.23	84.2	0.65	3.81	70.3	0.68	4.34	57.6	0.73	4.93	47.6	0.77	5.66
150	343.9	0.48	1.34	248.3	0.51	1.76	176.8	0.55	2.30	145.0	0.57	2.68	111.5	0.61	3.25	90.0	0.65	3.82	75.2	0.68	4.35	61.6	0.73	4.94	50.9	0.77	5.67
160	365.5	0.48	1.35	264.0	0.51	1.76	188.1	0.55	2.30	154.3	0.57	2.68	118.7	0.62	3.25	95.8	0.65	3.82	80.0	0.68	4.36	65.6	0.73	4.94	54.2	0.77	5.68
170	386.9	0.48	1.35	279.6	0.51	1.77	199.3	0.55	2.31	163.5	0.57	2.69	125.8	0.62	3.26	101.6	0.65	3.82	84.8	0.68	4.37	69.5	0.73	4.97	57.5	0.77	5.69
180	408.2	0.48	1.36	295.1	0.51	1.77	210.4	0.55	2.31	172.7	0.57	2.69	132.8	0.62	3.27	107.3	0.65	3.84	89.7	0.68	4.36	73.5	0.73	4.96	60.8	0.77	5.69
190	429.3	0.48	1.36	310.5	0.51	1.78	221.4	0.55	2.32	181.8	0.57	2.70	139.9	0.62	3.27	113.0	0.65	3.84	94.4	0.68	4.38	77.4	0.73	4.98	64.0	0.77	5.71
200	450.2	0.48	1.36	325.7	0.51	1.78	232.4	0.55	2.32	190.8	0.57	2.71	146.9	0.62	3.28	118.7	0.65	3.85	99.2	0.68	4.39	81.4	0.73	4.97	67.3	0.77	5.71
220	493.4	0.48	1.37	357.1	0.51	1.79	254.9	0.55	2.33	209.4	0.58	2.71	161.1	0.62	3.29	130.3	0.65	3.85	108.9	0.68	4.39	89.3	0.73	4.99	73.9	0.77	5.72
240	536.2	0.48	1.37	388.3	0.51	1.79	277.3	0.55	2.34	227.8	0.58	2.71	175.4	0.62	3.29	141.8	0.65	3.86	118.6	0.68	4.40	97.3	0.74	4.99	80.5	0.77	5.72
260	578.8	0.48	1.38	419.3	0.51	1.80	299.5	0.55	2.34	246.2	0.58	2.72	189.5	0.62	3.30	153.3	0.65	3.87	128.2	0.68	4.40	105.2	0.74	5.00	87.0	0.77	5.74
280	621.1	0.48	1.38	450.1	0.51	1.80	321.6	0.55	2.35	264.4	0.58	2.72	203.6	0.62	3.31	164.8	0.65	3.87	137.8	0.68	4.41	113.1	0.74	5.00	93.5	0.77	5.76
300	663.1	0.48	1.39	480.7	0.51	1.81	343.6	0.55	2.35	282.6	0.58	2.73	217.6	0.62	3.31	176.1	0.65	3.88	147.4	0.69	4.41	120.9	0.74	5.02	100.1	0.78	5.75

TABLE B.4 *(Continued)* Parabolic Waterway Design

V_1 for *RETARDANCE "D"*. Top Width (T), Depth (D) and V_2 for *RETARDANCE "C"*.

Grade 10.0 Percent

Q cfs	V_1 = 2.0			V_1 = 2.5			V_1 = 3.0			V_1 = 3.5			V_1 = 4.0			V_1 = 4.5			V_1 = 5.0			V_1 = 5.5			V_1 = 6.0		
	T	D	V_2	T	D	V_2	T	D	V_2	T	D	V_2	T	D	V_2	T	D	V_2	T	D	V_2	T	D	V_2	T	D	V_2
15	45.2	0.43	1.14	32.5	0.45	1.50	22.9	0.49	1.98	16.6	0.52	2.56	13.4	0.55	2.99	10.7	0.58	3.55	9.0	0.61	3.99	7.4	0.65	4.57	6.2	0.70	5.08
20	60.1	0.43	1.14	43.2	0.45	1.50	30.4	0.49	2.00	22.1	0.52	2.56	17.8	0.55	3.02	14.3	0.59	3.52	12.0	0.61	4.00	9.9	0.65	4.55	8.3	0.70	5.06
25	74.8	0.43	1.15	53.8	0.45	1.51	37.9	0.49	2.00	27.5	0.52	2.59	22.3	0.55	2.99	17.8	0.58	3.56	14.9	0.61	4.08	12.3	0.65	4.63	10.3	0.69	5.18
30	89.3	0.43	1.15	64.2	0.45	1.52	45.3	0.49	2.01	33.0	0.52	2.58	26.6	0.55	3.03	21.3	0.58	3.58	17.9	0.61	4.05	14.7	0.64	4.68	12.4	0.70	5.13
35	103.7	0.43	1.16	74.6	0.45	1.53	52.6	0.48	2.03	38.3	0.52	2.60	31.0	0.55	3.02	24.8	0.58	3.59	20.8	0.61	4.09	17.2	0.65	4.63	14.4	0.69	5.19
40	118.0	0.43	1.16	85.0	0.46	1.53	59.9	0.48	2.04	43.7	0.52	2.60	35.3	0.55	3.04	28.3	0.58	3.59	23.8	0.61	4.06	19.6	0.65	4.66	16.4	0.69	5.24
45	132.2	0.43	1.17	95.2	0.45	1.53	67.2	0.49	2.04	49.0	0.52	2.61	39.6	0.55	3.05	31.7	0.58	3.62	26.7	0.61	4.08	22.0	0.65	4.68	18.4	0.69	5.27
50	146.3	0.43	1.17	105.3	0.45	1.54	74.4	0.49	2.04	54.3	0.52	2.61	43.9	0.55	3.05	35.2	0.58	3.61	29.6	0.61	4.09	24.4	0.65	4.68	20.4	0.69	5.28
55	160.2	0.43	1.17	115.4	0.46	1.55	81.5	0.49	2.06	59.5	0.52	2.63	48.2	0.55	3.05	38.6	0.58	3.62	32.4	0.61	4.13	26.7	0.64	4.74	22.4	0.69	5.29
60	174.0	0.43	1.18	125.3	0.46	1.55	88.6	0.49	2.06	64.7	0.52	2.63	52.4	0.55	3.07	42.0	0.53	3.63	35.3	0.61	4.12	29.1	0.65	4.73	24.4	0.69	5.30
65	187.6	0.43	1.18	135.2	0.46	1.56	95.6	0.49	2.07	69.9	0.52	2.64	56.6	0.55	3.08	45.4	0.58	3.64	38.2	0.61	4.12	31.5	0.65	4.72	26.4	0.69	5.30
70	201.2	0.43	1.19	145.0	0.46	1.57	102.6	0.49	2.08	75.1	0.52	2.64	60.8	0.55	3.08	48.7	0.58	3.66	41.0	0.61	4.14	33.8	0.65	4.75	28.4	0.69	5.29
75	214.6	0.43	1.19	154.7	0.46	1.57	109.6	0.49	2.08	80.2	0.52	2.65	65.0	0.55	3.08	52.1	0.58	3.66	43.8	0.61	4.15	36.2	0.65	4.74	30.3	0.69	5.34
80	227.9	0.43	1.20	164.3	0.46	1.58	116.4	0.49	2.09	85.3	0.52	2.65	69.1	0.55	3.09	55.4	0.58	3.67	46.6	0.61	4.16	38.5	0.65	4.76	32.3	0.69	5.33
90	255.2	0.43	1.20	184.1	0.46	1.58	130.5	0.49	2.10	95.6	0.52	2.67	77.5	0.55	3.11	62.1	0.58	3.69	52.3	0.61	4.17	43.2	0.65	4.78	36.3	0.69	5.33
100	282.4	0.43	1.20	203.7	0.46	1.59	144.5	0.49	2.10	105.9	0.52	2.67	85.9	0.56	3.11	68.8	0.58	3.70	58.0	0.61	4.18	47.9	0.65	4.78	40.2	0.69	5.36
110	309.2	0.43	1.21	223.2	0.46	1.60	158.4	0.49	2.11	116.2	0.52	2.68	94.3	0.56	3.11	75.5	0.58	3.71	63.7	0.61	4.18	52.6	0.65	4.79	44.2	0.69	5.34
120	335.9	0.43	1.21	242.4	0.46	1.60	172.2	0.49	2.12	126.4	0.52	2.68	102.5	0.56	3.12	82.2	0.58	3.71	69.3	0.61	4.19	57.3	0.65	4.79	48.1	0.69	5.36
130	362.3	0.43	1.22	261.6	0.46	1.61	185.9	0.49	2.12	136.5	0.52	2.69	110.8	0.56	3.13	88.8	0.58	3.72	74.9	0.61	4.20	61.9	0.65	4.81	52.0	0.69	5.37
140	388.4	0.44	1.22	280.5	0.46	1.61	199.5	0.49	2.13	146.5	0.52	2.70	119.0	0.56	3.13	95.4	0.58	3.73	80.5	0.61	4.20	66.5	0.65	4.82	55.9	0.69	5.38
150	414.4	0.44	1.23	299.3	0.46	1.62	213.0	0.49	2.13	156.5	0.52	2.71	127.1	0.56	3.14	101.9	0.58	3.74	86.0	0.61	4.22	71.1	0.65	4.83	59.7	0.69	5.41
160	440.1	0.44	1.23	318.1	0.46	1.62	226.5	0.49	2.14	166.4	0.53	2.71	135.2	0.56	3.15	108.4	0.58	3.75	91.6	0.62	4.21	75.7	0.65	4.83	63.6	0.69	5.41
170	465.5	0.44	1.24	336.7	0.46	1.63	239.8	0.49	2.14	176.3	0.53	2.72	143.3	0.56	3.15	114.9	0.58	3.76	97.1	0.62	4.22	80.3	0.65	4.83	67.4	0.69	5.42
180	490.8	0.44	1.24	355.1	0.46	1.63	253.1	0.49	2.15	186.1	0.53	2.73	151.3	0.56	3.16	121.3	0.58	3.77	102.5	0.62	4.24	84.9	0.65	4.83	71.3	0.69	5.42
190	515.7	0.44	1.25	373.5	0.46	1.64	266.3	0.49	2.15	195.9	0.53	2.73	159.3	0.56	3.17	127.7	0.58	3.78	108.0	0.62	4.24	89.4	0.65	4.84	75.1	0.69	5.43
200	540.5	0.44	1.25	391.6	0.46	1.64	279.4	0.49	2.16	205.6	0.53	2.74	167.2	0.56	3.17	134.1	0.58	3.79	113.4	0.62	4.25	93.9	0.65	4.85	78.9	0.69	5.43
220	592.0	0.44	1.26	429.2	0.46	1.65	306.3	0.49	2.17	225.4	0.53	2.75	183.5	0.56	3.18	147.1	0.58	3.80	124.4	0.62	4.26	103.1	0.65	4.85	86.6	0.69	5.45
240	643.0	0.44	1.26	466.4	0.46	1.65	333.1	0.49	2.17	245.2	0.53	2.75	199.6	0.56	3.19	160.0	0.58	3.81	135.4	0.62	4.27	112.2	0.65	4.87	94.3	0.69	5.45
260	693.6	0.44	1.27	503.4	0.46	1.66	359.6	0.49	2.18	264.8	0.53	2.76	215.6	0.56	3.19	172.9	0.58	3.82	146.4	0.62	4.27	121.3	0.65	4.87	101.9	0.69	5.47
280	743.7	0.44	1.27	540.0	0.46	1.66	386.0	0.49	2.18	284.4	0.53	2.77	231.6	0.56	3.20	185.8	0.59	3.82	157.3	0.62	4.28	130.4	0.65	4.88	109.6	0.69	5.47
300	793.4	0.44	1.28	576.4	0.46	1.67	412.3	0.49	2.19	303.8	0.53	2.77	247.5	0.56	3.21	198.6	0.59	3.83	168.1	0.62	4.29	139.4	0.65	4.89	117.2	0.69	5.48

TABLE B.5 Trapezoidal Channel Design
"C" Retardance Grade 0.25 Percent Side Slope = 2:1
Bottom Width, b, in Feet

Q cfs	b = 2 D	b = 2 V	b = 4 D	b = 4 V	b = 6 D	b = 6 V	b = 8 D	b = 8 V	b = 10 D	b = 10 V	b = 12 D	b = 12 V	b = 14 D	b = 14 V	b = 16 D	b = 16 V
15	2.1	1.1	1.8	1.1	1.6	1.0	1.5	1.0	1.3	0.9	1.3	0.8	1.2	0.8	1.2	0.7
20	2.3	1.3	1.9	1.3	1.7	1.2	1.5	1.2	1.4	1.1	1.4	1.0	1.3	0.9	1.2	0.9
25	2.4	1.5	2.1	1.5	1.8	1.4	1.7	1.3	1.5	1.2	1.4	1.2	1.4	1.1	1.3	1.0
30	2.5	1.7	2.2	1.6	1.9	1.6	1.7	1.5	1.6	1.4	1.5	1.3	1.4	1.2	1.4	1.2
35	2.7	1.8	2.3	1.8	2.0	1.7	1.8	1.6	1.7	1.5	1.6	1.5	1.5	1.4	1.4	1.3
40	2.8	1.9	2.4	1.9	2.1	1.8	1.9	1.8	1.8	1.7	1.7	1.6	1.6	1.5	1.5	1.4
45	2.9	2.0	2.5	2.0	2.2	1.9	2.0	1.9	1.8	1.8	1.7	1.7	1.6	1.6	1.6	1.5
50	2.9	2.1	2.6	2.1	2.3	2.1	2.1	2.0	1.9	1.9	1.8	1.8	1.7	1.7	1.6	1.6
55	3.0	2.2	2.7	2.2	2.4	2.2	2.2	2.1	2.0	2.0	1.9	1.9	1.8	1.8	1.7	1.7
60	3.1	2.3	2.8	2.3	2.5	2.2	2.2	2.2	2.0	2.1	1.9	2.0	1.8	1.9	1.7	1.8
65	3.2	2.4	2.8	2.4	2.5	2.3	2.3	2.2	2.1	2.2	2.0	2.1	1.9	2.0	1.8	1.9
70	3.3	2.5	2.9	2.4	2.6	2.4	2.4	2.3	2.2	2.2	2.0	2.2	1.9	2.1	1.8	2.0
75	3.4	2.5	3.0	2.5	2.7	2.5	2.4	2.4	2.2	2.3	2.1	2.2	1.9	2.1	1.8	2.1
80	3.4	2.6	3.1	2.6	2.7	2.5	2.5	2.5	2.3	2.4	2.1	2.3	2.0	2.2	1.9	2.1
90	3.6	2.7	3.2	2.7	2.9	2.7	2.6	2.6	2.4	2.5	2.2	2.4	2.1	2.3	2.0	2.3
100	3.7	2.8	3.3	2.8	3.0	2.8	2.7	2.7	2.5	2.7	2.3	2.6	2.2	2.5	2.1	2.4
110	3.9	2.9	3.5	2.9	3.1	2.9	2.8	2.8	2.6	2.8	2.4	2.7	2.3	2.6	2.2	2.5
120	4.0	3.1	3.5	3.0	3.2	3.0	2.9	3.0	2.7	2.9	2.5	2.8	2.4	2.7	2.2	2.6
130	4.1	3.1	3.7	3.1	3.3	3.1	3.0	3.0	2.8	3.0	2.6	2.9	2.5	2.8	2.3	2.7
140	4.2	3.2	3.8	3.2	3.4	3.2	3.1	3.1	2.9	3.1	2.7	3.0	2.5	2.9	2.4	2.8
150	4.3	3.3	3.9	3.3	3.5	3.3	3.2	3.2	3.0	3.1	2.8	3.1	2.6	3.0	2.5	2.9
160	4.4	3.4	4.0	3.4	3.6	3.4	3.3	3.3	3.1	3.2	2.9	3.2	2.7	3.1	2.5	3.0
170	4.5	3.5	4.1	3.5	3.7	3.4	3.4	3.4	3.1	3.3	2.9	3.2	2.8	3.2	2.6	3.1
180	4.6	3.6	4.1	3.6	3.8	3.5	3.5	3.5	3.2	3.4	3.0	3.3	2.8	3.2	2.7	3.2
190	4.6	3.6	4.2	3.6	3.9	3.6	3.6	3.5	3.3	3.5	3.1	3.4	2.9	3.3	2.7	3.2
200	4.7	3.7	4.3	3.7	3.9	3.7	3.6	3.6	3.4	3.5	3.2	3.5	3.0	3.4	2.8	3.3
220	4.9	3.8	4.5	3.8	4.1	3.8	3.8	3.7	3.5	3.7	3.3	3.6	3.1	3.5	2.9	3.4
240	5.0	3.9	4.6	3.9	4.2	3.9	3.9	3.8	3.6	3.8	3.4	3.7	3.2	3.6	3.0	3.6
260	5.2	4.0	4.8	4.0	4.4	4.0	4.1	4.0	3.8	3.9	3.5	3.8	3.3	3.8	3.2	3.7
280	5.3	4.1	4.9	4.1	4.5	4.1	4.2	4.1	3.9	4.0	3.7	4.0	3.6	3.9	3.3	3.8
300	5.5	4.2	5.0	4.2	4.7	4.2	4.3	4.2	4.0	4.1	3.8	4.1	3.6	4.0	3.4	3.9

Q = Flow, Cubic Feet per Second, V = Velocity, Feet per Second, b = Bottom Width, Feet, D = Depth, Feet.

Source: U.S. Department of Agriculture, Soil Conservation Service, Standards and Specifications for Soil Erosion and Sediment Control in Developing Areas, U.S.D.A., SCS, College Park, Maryland, 1975.

TABLE B.5 *(Continued)* Trapezoidal Channel Design
"C" Retardance Grade 0.5 Percent Side Slope = 2:1
Bottom Width, b, in Feet

Q cfs	b = 2 D	b = 2 V	b = 4 D	b = 4 V	b = 6 D	b = 6 V	b = 8 D	b = 8 V	b = 10 D	b = 10 V	b = 12 D	b = 12 V	b = 14 D	b = 14 V	b = 16 D	b = 16 V
15	1.7	1.6	1.5	1.5	1.3	1.4	1.1	1.3	1.1	1.1	1.0	1.0	1.0	1.0	0.9	0.9
20	1.9	1.8	1.6	1.8	1.4	1.7	1.2	1.5	1.1	1.4	1.1	1.3	1.0	1.2	1.0	1.1
25	2.0	2.1	1.7	2.0	1.5	1.9	1.3	1.8	1.2	1.6	1.2	1.5	1.1	1.4	1.0	1.3
30	2.1	2.3	1.8	2.2	1.6	2.1	1.4	1.9	1.3	1.8	1.2	1.7	1.2	1.6	1.1	1.5
35	2.2	2.5	1.9	2.4	1.6	2.3	1.5	2.1	1.4	2.0	1.3	1.9	1.2	1.8	1.2	1.7
40	2.3	2.6	2.0	2.5	1.7	2.5	1.6	2.3	1.4	2.2	1.3	2.0	1.3	1.9	1.2	1.8
45	2.4	2.8	2.0	2.7	1.8	2.6	1.6	2.5	1.5	2.3	1.4	2.2	1.3	2.1	1.2	2.0
50	2.5	2.9	2.1	2.8	1.9	2.7	1.7	2.6	1.5	2.5	1.4	2.3	1.4	2.2	1.3	2.1
55	2.6	3.0	2.2	2.9	1.9	2.9	1.8	2.7	1.6	2.6	1.5	2.5	1.4	2.3	1.3	2.2
60	2.6	3.1	2.3	3.1	2.0	2.9	1.8	2.9	1.7	2.7	1.5	2.6	1.5	2.4	1.4	2.3
65	2.7	3.2	2.4	3.1	2.1	3.1	1.9	2.9	1.7	2.8	1.6	2.7	1.5	2.6	1.4	2.4
70	2.8	3.3	2.4	3.3	2.1	3.2	2.0	3.1	1.8	2.9	1.6	2.8	1.5	2.7	1.5	2.6
75	2.8	3.4	2.5	3.4	2.2	3.3	2.0	3.1	1.8	3.0	1.7	2.9	1.6	2.8	1.5	2.7
80	2.9	3.5	2.5	3.4	2.3	3.4	2.0	3.3	1.9	3.1	1.7	3.0	1.7	2.9	1.5	2.7
90	3.0	3.6	2.7	3.6	2.4	3.5	2.1	3.4	2.0	3.3	1.8	3.2	1.7	3.0	1.6	2.9
100	3.1	3.8	2.8	3.8	2.5	3.7	2.2	3.6	2.1	3.5	1.9	3.3	1.8	3.2	1.7	3.1
110	3.3	4.0	2.9	3.9	2.6	3.8	2.3	3.7	2.1	3.6	2.0	3.5	1.9	3.3	1.8	3.2
120	3.4	4.1	3.0	4.0	2.7	4.0	2.4	3.9	2.2	3.7	2.1	3.6	1.9	3.5	1.8	3.4
130	3.5	4.2	3.1	4.1	2.8	4.1	2.5	4.0	2.3	3.9	2.1	3.7	2.0	3.6	1.9	3.5
140	3.6	4.3	3.2	4.3	2.8	4.2	2.6	4.1	2.4	4.0	2.2	3.9	2.1	3.8	1.9	3.6
150	3.7	4.4	3.3	4.4	2.9	4.3	2.7	4.2	2.4	4.1	2.3	4.0	2.1	3.9	2.0	3.7
160	3.7	4.5	3.3	4.5	3.0	4.4	2.7	4.4	2.5	4.2	2.3	4.1	2.2	4.0	2.1	3.9
170	3.8	4.6	3.4	4.6	3.1	4.5	2.8	4.4	2.6	4.3	2.4	4.2	2.2	4.1	2.1	4.0
180	3.9	4.7	3.5	4.7	3.1	4.6	2.9	4.5	2.7	4.4	2.5	4.3	2.3	4.2	2.2	4.1
190	4.0	4.8	3.6	4.8	3.2	4.7	2.9	4.7	2.7	4.5	2.5	4.4	2.4	4.3	2.2	4.2
200	4.1	4.9	3.6	4.9	3.3	4.8	3.0	4.7	2.8	4.6	2.6	4.5	2.4	4.4	2.3	4.2
220	4.2	5.1	3.8	5.0	3.4	5.0	3.1	4.9	2.9	4.8	2.7	4.7	2.5	4.6	2.4	4.4
240	4.3	5.2	3.9	5.2	3.6	5.1	3.3	5.1	3.0	4.9	2.8	4.9	2.6	4.7	2.5	4.6
260	4.4	5.4	4.0	5.3	3.7	5.3	3.4	5.2	3.1	5.1	2.9	5.0	2.7	4.9	2.6	4.8
280	4.6	5.5	4.1	5.5	3.8	5.4	3.5	5.4	3.2	5.2	3.0	5.1	2.8	5.0	2.7	4.9
300	4.7	5.6	4.3	5.6	3.9	5.6	3.6	5.5	3.3	5.4	3.1	5.3	2.9	5.2	2.8	5.1

Q = Flow, Cubic Feet per Second, V = Velocity, Feet per Second, b = Bottom Width, Feet, D = Depth, Feet.

TABLE B.5 *(Continued)* Trapezoidal Channel Design
"C" Retardance Grade 1.0 Percent Side Slope 2:1
Bottom Width, b, in Feet

Q cfs	b = 2		b = 4		b = 6		b = 8		b = 10		b = 12		b = 14		b = 16	
	D	V	D	V	D	V	D	V	D	V	D	V	D	V	D	V
15	1.4	2.2	1.2	2.0	1.0	1.8	0.9	1.7	0.9	1.5	0.8	1.4	0.8	1.2	0.7	1.1
20	1.5	2.5	1.3	2.4	1.1	2.2	1.0	2.0	0.9	1.8	0.9	1.7	0.8	1.5	0.8	1.4
25	1.7	2.8	1.4	2.7	1.2	2.5	1.1	2.3	1.0	2.1	0.9	2.0	0.9	1.8	0.8	1.7
30	1.8	3.1	1.5	3.0	1.3	2.8	1.1	2.6	1.0	2.4	1.0	2.2	0.9	2.0	0.9	1.9
35	1.9	3.3	1.6	3.2	1.3	3.0	1.2	2.8	1.1	2.6	1.0	2.4	1.0	2.3	0.9	2.1
40	1.9	3.5	1.6	3.4	1.4	3.2	1.3	3.0	1.1	2.8	1.1	2.6	1.0	2.5	1.0	2.3
45	2.0	3.7	1.7	3.6	1.5	3.4	1.3	3.2	1.2	3.0	1.1	2.8	1.1	2.7	1.0	2.5
50	2.1	3.9	1.8	3.7	1.5	3.6	1.4	3.4	1.3	3.2	1.2	3.0	1.1	2.8	1.0	2.7
55	2.2	4.0	1.8	3.9	1.6	3.7	1.4	3.5	1.3	3.3	1.2	3.2	1.1	3.0	1.1	2.8
60	2.2	4.2	1.9	4.1	1.7	3.9	1.5	3.7	1.3	3.5	1.3	3.3	1.2	3.1	1.1	3.0
65	2.3	4.3	1.9	4.2	1.7	4.0	1.5	3.9	1.4	3.6	1.3	3.5	1.2	3.3	1.1	3.1
70	2.4	4.4	2.0	4.3	1.8	4.2	1.6	4.0	1.4	3.8	1.3	3.6	1.2	3.4	1.2	3.3
75	2.4	4.5	2.1	4.5	1.8	4.3	1.6	4.1	1.5	3.9	1.4	3.7	1.3	3.5	1.2	3.4
80	2.5	4.6	2.1	4.5	1.9	4.4	1.7	4.2	1.5	4.0	1.4	3.9	1.3	3.7	1.2	3.5
90	2.6	4.9	2.2	4.8	2.0	4.6	1.7	4.5	1.6	4.3	1.5	4.1	1.4	3.9	1.3	3.7
100	2.7	5.1	2.3	5.0	2.0	4.9	1.8	4.7	1.7	4.5	1.5	4.3	1.4	4.1	1.4	3.9
110	2.8	5.2	2.4	5.2	2.1	5.0	1.9	4.9	1.7	4.7	1.6	4.5	1.5	4.3	1.4	4.1
120	2.9	5.4	2.5	5.4	2.2	5.2	2.0	5.0	1.8	4.9	1.7	4.7	1.6	4.5	1.5	4.3
130	3.0	5.6	2.6	5.5	2.3	5.4	2.1	5.2	1.9	5.0	1.7	4.9	1.6	4.7	1.5	4.5
140	3.0	5.7	2.7	5.6	2.4	5.5	2.1	5.4	1.9	5.2	1.8	5.0	1.7	4.8	1.6	4.7
150	3.1	5.9	2.7	5.8	2.4	5.6	2.2	5.5	2.0	5.4	1.8	5.2	1.7	5.0	1.6	4.8
160	3.2	6.0	2.8	6.0	2.5	5.8	2.2	5.7	2.1	5.5	1.9	5.3	1.8	5.1	1.7	4.9
170					2.6	6.0	2.3	5.8	2.1	5.6	2.0	5.5	1.8	5.2	1.7	5.1
180							2.4	5.9	2.2	5.8	2.0	5.6	1.9	5.4	1.8	5.2
190									2.2	5.9	2.1	5.7	1.9	5.5	1.8	5.4
200									2.3	6.0	2.1	5.8	2.0	5.6	1.9	5.4
220													2.1	5.9	2.0	5.7
240															2.0	5.9

Q = Flow, Cubic Feet per Second, V = Velocity, Feet per Second, b = Bottom Width, Feet, D = Depth, Feet.

TABLE B.5 *(Continued)* Trapezoidal Channel Design
"C" Retardance Grade 2.0 Percent Side Slope = 2:1
Bottom Width, b, in Feet

Q cfs	b = 2		b = 4		b = 6		b = 8		b = 10		b = 12		b = 14		b = 16	
	D	V	D	V	D	V	D	V	D	V	D	V	D	V	D	V
15	1.2	3.0	0.9	2.7	0.8	2.4	0.7	2.1	0.7	1.9	0.7	1.7	0.6	1.6	0.6	1.4
20	1.3	3.4	1.0	3.2	0.9	2.9	0.8	2.6	0.7	2.3	0.7	2.1	0.7	1.0	0.6	1.8
25	1.4	3.8	1.1	3.6	1.0	3.3	0.9	3.0	0.8	2.7	0.7	2.5	0.7	2.3	0.7	2.1
30	1.5	4.2	1.2	3.9	1.0	3.6	0.9	3.3	0.8	3.1	0.8	2.8	0.7	2.6	0.7	2.4
35	1.6	4.4	1.3	4.2	1.1	3.9	1.0	3.6	0.9	3.3	0.8	3.1	0.8	2.9	0.7	2.7
40	1.6	4.7	1.3	4.5	1.1	4.2	1.0	3.9	0.9	3.6	0.9	3.4	0.8	3.1	0.8	2.9
45	1.7	4.9	1.4	4.7	1.2	4.5	1.1	4.2	1.0	3.9	0.9	3.6	0.8	3.4	0.8	3.2
50	1.8	5.2	1.5	5.0	1.2	4.7	1.1	4.4	1.0	4.1	0.9	3.9	0.9	3.6	0.8	3.4
55	1.8	5.4	1.5	5.1	1.3	4.9	1.2	4.6	1.0	4.3	1.0	4.0	0.9	3.8	0.9	3.6
60	1.9	5.5	1.6	5.3	1.4	5.1	1.2	4.8	1.1	4.5	1.0	4.2	0.9	4.0	0.9	3.8
65	1.9	5.7	1.6	5.5	1.4	5.3	1.2	5.0	1.1	4.7	1.0	4.4	1.0	4.2	0.9	4.0
70	2.0	5.9	1.7	5.7	1.4	5.5	1.3	5.2	1.2	4.9	1.1	4.6	1.0	4.3	1.0	4.1
75			1.7	5.9	1.5	5.6	1.3	5.3	1.2	5.0	1.1	4.7	1.0	4.5	1.0	4.3
80					1.5	5.8	1.4	5.5	1.2	5.2	1.1	4.9	1.1	4.7	1.0	4.4
90							1.4	5.8	1.3	5.5	1.2	5.2	1.1	5.0	1.1	4.7
100									1.4	5.8	1.3	5.5	1.2	5.2	1.1	5.0
110									1.4	6.0	1.3	5.8	1.2	5.5	1.1	5.2
120											1.4	6.0	1.3	5.7	1.2	5.4
130													1.3	5.9	1.2	5.7
140															1.3	5.9

Q = Flow, Cubic Feet per Second, V = Velocity, Feet per Second, b = Bottom Width, Feet, D = Depth, Feet.

TABLE B.5 *(Continued)* Trapezoidal Channel Design
"C" Retardance   Grade 3 Percent   Side Slope = 2:1
Bottom Width, b, in Feet

Q cfs	b = 2		b = 4		b = 6		b = 8		b = 10		b = 12		b = 14		b = 16	
	D	V	D	V	D	V	D	V	D	V	D	V	D	V	D	V
15	1.0	3.5	0.8	3.2	0.7	2.8	0.6	2.5	0.6	2.2	0.6	2.0	0.5	1.8	0.5	1.7
20	1.1	4.1	0.9	3.7	0.8	3.4	0.7	3.0	0.7	2.7	0.6	2.5	0.6	2.3	0.6	2.1
25	1.2	4.5	1.0	4.2	0.8	3.8	0.8	3.5	0.7	3.2	0.7	2.9	0.6	2.6	0.6	2.5
30	1.3	4.9	1.1	4.6	0.9	4.2	0.8	3.8	0.7	3.5	0.7	3.2	0.7	3.0	0.6	2.8
35	1.4	5.3	1.1	4.9	1.0	4.6	0.9	4.2	0.8	3.9	0.7	3.6	0.7	3.3	0.7	3.1
40	1.5	5.6	1.2	5.3	1.0	4.9	0.9	4.5	0.8	4.2	0.8	3.9	0.7	3.6	0.7	3.3
45	1.5	5.9	1.2	5.6	1.1	5.2	0.9	4.8	0.9	4.5	0.8	4.2	0.8	3.9	0.7	3.6
50			1.3	5.9	1.1	5.4	1.0	5.1	0.9	4.7	0.8	4.4	0.8	4.1	0.7	3.9
55					1.2	5.7	1.0	5.4	0.9	5.0	0.9	4.7	0.8	4.4	0.8	4.1
60					1.2	5.9	1.2	5.6	1.0	5.2	0.9	4.9	0.8	4.6	0.8	4.3
65							1.1	5.8	1.0	5.4	0.9	5.1	0.9	4.8	0.8	4.5
70							1.1	6.0	1.0	5.6	1.0	5.3	0.9	5.0	0.8	4.7
75									1.1	5.8	1.0	5.5	0.9	5.1	0.9	4.9
80											1.0	5.6	0.9	5.3	0.9	5.0
90													1.0	5.6	1.0	5.4
100													1.0	6.0	1.0	5.7
110															1.0	6.0

Q = Flow, Cubic Feet per Second, V = Velocity, Feet per Second, b = Bottom Width, Feet, D = Depth, Feet.

TABLE B.5 *(Continued)* Trapezoidal Channel Design
"C" Retardance   Grade 4 Percent   Side Slope = 2:1
Bottom Width, b, in Feet

Q cfs	b = 2		b = 4		b = 6		b = 8		b = 10		b = 12		b = 14		b = 16	
	D	V	D	V	D	V	D	V	D	V	D	V	D	V	D	V
15	1.0	4.0	0.8	3.6	0.7	3.2	0.6	2.8	0.5	2.4	0.5	2.2	0.5	2.0	0.5	1.8
20	1.1	4.6	0.8	4.2	0.7	3.8	0.6	3.3	0.6	3.0	0.6	2.7	0.5	2.5	0.5	2.3
25	1.1	5.1	0.9	4.8	0.8	4.3	0.7	3.8	0.6	3.5	0.6	3.1	0.6	2.9	0.5	2.7
30	1.2	5.6	1.0	5.2	0.8	4.7	0.7	4.3	0.7	3.9	0.6	3.6	0.6	3.3	0.6	3.0
35	1.3	5.9	1.0	5.6	0.9	5.1	0.8	4.7	0.7	4.3	0.7	3.9	0.6	3.6	0.6	3.4
40			1.1	5.9	0.9	5.4	0.8	5.0	0.8	4.6	0.7	4.2	0.6	3.9	0.6	3.7
45					1.0	5.8	0.9	5.4	0.8	4.9	0.7	4.6	0.7	4.3	0.6	4.0
50							0.9	5.6	0.8	5.2	0.8	4.9	0.7	4.6	0.7	4.2
55							0.9	5.9	0.8	5.5	0.8	5.1	0.7	4.8	0.7	4.5
60									0.9	5.8	0.8	5.4	0.8	5.0	0.7	4.8
65									0.9	6.0	0.8	5.6	0.8	5.3	0.7	5.0
70											0.9	5.9	0.9	5.5	0.8	5.2
75													0.9	5.7	0.8	5.4
80													0.9	5.9	0.8	5.6
90															0.9	5.9

Q = Flow, Cubic Feet per Second, V = Velocity, Feet per Second, b = Bottom Width, Feet, D = Depth, Feet.

TABLE B.5 *(Continued)* Trapezoidal Channel Design
"C" Retardance Grade 5 Percent Side Slope 2:1
Bottom Width, b, in Feet

Q cfs	b = 2		b = 4		b = 6		b = 8		b = 10		b = 12·		b = 14		b = 16	
	D	V	D	V	D	V	D	V	D	V	D	V	D	V	D	V
15	0.9	4.4	0.7	3.9	0.6	3.4	0.6	3.0	0.5	2.6	0.5	2.3	0.5	2.1	0.5	2.0
20	1.0	5.1	0.8	4.6	0.7	4.1	0.6	3.6	0.6	3.2	0.5	2.9	0.5	2.7	0.5	2.4
25	1.1	5.6	0.9	5.2	0.7	4.6	0.6	4.2	0.6	3.7	0.6	3.4	0.5	3.1	0.5	2.9
30			0.9	5.6	0.8	5.1	0.7	4.6	0.6	4.2	0.6	3.9	0.6	3.5	0.5	3.3
35					0.8	5.5	0.7	5.1	0.7	4.6	0.6	4.2	0.6	3.9	0.6	3.6
40					0.9	5.9	0.8	5.4	0.7	5.0	0.7	4.6	0.6	4.3	0.6	4.0
45							0.8	5.7	0.7	5.3	0.7	5.0	0.6	4.6	0.6	4.3
50									0.8	5.6	0.7	5.3	0.7	4.9	0.6	4.6
55									0.9	6.0	0.7	5.5	0.7	5.2	0.7	4.9
60											0.8	5.8	0.7	5.4	0.7	5.1
65													0.7	5.7	0.7	5.3
70													0.8	5.9	0.7	5.6
75															0.7	5.8
80															0.8	6.0

Grade 6 Percent

Q cfs	b = 2		b = 4		b = 6		b = 8		b = 10		b = 12		b = 14		b = 16	
	D	V	D	V	D	V	D	V	D	V	D	V	D	V	D	V
15	0.9	4.8	0.7	4.2	0.6	3.6	0.5	3.2	0.5	2.8	0.5	2.5	0.4	2.3	0.4	2.1
20	0.9	5.4	0.7	4.9	0.6	4.4	0.6	3.8	0.5	3.4	0.5	3.1	0.5	2.8	0.5	2.6
25			0.8	5.6	0.7	4.9	0.6	4.4	0.6	4.0	0.5	3.6	0.5	3.3	0.5	3.1
30					0.7	5.5	0.7	4.9	0.6	4.5	0.6	4.1	0.5	3.7	0.5	3.5
35					0.8	5.9	0.7	5.4	0.6	4.9	0.6	4.5	0.6	4.2	0.5	3.8
40							0.7	5.8	0.7	5.3	0.6	4.9	0.7	4.5	0.6	4.2
45									0.7	5.6	0.6	5.2	0.6	4.9	0.6	4.6
50											0.7	5.6	0.6	5.2	0.6	4.9
55											0.7	5.8	0.7	5.5	0.6	5.1
60													0.7	5.8	0.6	5.4
65															0.7	5.7
70															0.7	5.9

Q = Flow, Cubic Feet per Second, V = Velocity, Feet per Second, b = Bottom Width, Feet, D = Depth, Feet.

TABLE B.5 *(Continued)* Trapezoidal Channel Design
"C" Retardance Grade 8 Percent Side Slope = 2:1
Bottom Width, b, in Feet

Q cfs	b = 2		b = 4		b = 6		b = 8		b = 10		b = 12		b = 14		b = 16	
	D	V	D	V	D	V	D	V	D	V	D	V	D	V	D	V
15	0.8	5.3	0.6	4.7	0.5	4.1	0.5	3.5	0.4	3.1	0.4	2.8	0.4	2.5	0.4	2.3
20			0.7	5.5	0.6	4.8	0.5	4.2	0.5	3.8	0.5	3.4	0.4	3.1	0.4	2.8
25					0.6	5.5	0.6	4.9	0.5	4.4	0.5	4.0	0.5	3.6	0.4	3.3
30							0.6	5.5	0.5	4.9	0.5	4.5	0.5	4.2	0.5	3.8
35							0.6	6.0	0.6	5.5	0.5	5.0	0.5	4.6	0.5	4.2
40									0.6	5.9	0.6	5.4	0.5	5.0	0.5	4.6
45											0.6	5.7	0.6	5.4	0.5	5.0
50													0.6	5.7	0.5	5.3
55															0.6	5.7
60															0.6	6.0

Grade 10 Percent

Q cfs	b = 2		b = 4		b = 6		b = 8		b = 10		b = 12		b = 14		b = 16	
	D	V	D	V	D	V	D	V	D	V	D	V	D	V	D	V
15	0.7	5.9	0.6	5.1	0.5	4.4	0.4	3.8	0.4	3.4	0.4	3.0	0.4	2.6	0.4	2.4
20			0.6	6.0	0.5	5.2	0.5	4.6	0.4	4.1	0.4	3.7	0.4	3.4	0.4	3.1
25					0.6	6.0	0.5	5.3	0.5	4.7	0.4	4.3	0.4	3.9	0.4	3.6
30							0.6	5.9	0.5	5.3	0.5	4.9	0.5	4.4	0.4	4.1
35									0.5	5.8	0.5	5.4	0.5	4.9	0.5	4.6
40											0.5	5.8	0.5	5.3	0.5	5.0
45													0.5	5.7	0.5	5.4
50															0.5	5.7

Q = Flow, Cubic Feet per Second, V = Velocity, Feet per Second, b = Bottom Width, Feet, D = Depth, Feet.

TABLE B.6 Diversion Design Table—"D" Retardance (V and Trapezoidal Section)

(Based on Handbook of Channel Design, SCS-TP-61)

● 3:1 Side Slopes
"D" Retardance

Grade	Triangular 0.2		Triangular 0.3		Triangular 0.4		Triangular 0.5		6' bottom width 0.2		6' bottom width 0.3		6' bottom width 0.4		6' bottom width 0.5	
Q-cfs	d	A	d	A	d	A	d	A	d	A	d	A	d	A	d	A
10	1.9	11	1.8	10	1.7	9	1.6	8	1.3	12	1.1	10	1.0	9	0.9	8
20	2.2	15	2.1	13	1.9	11	1.8	10	1.5	16	1.4	14	1.2	12	1.1	10
30	2.5	19	2.3	16	2.2	15	2.0	12	1.8	21	1.6	17	1.5	16	1.3	13
40	2.6	20	2.5	19	2.3	16	2.2	15	2.0	24	1.8	21	1.7	19	1.5	16
60	3.0	27	2.9	24	2.7	22	2.5	19	2.3	30	2.1	28	1.9	22	1.7	19
80			3.1	29	2.9	25	2.7	22	2.5	34	2.3	30	2.1	26	1.9	22
100					3.1	29	2.9	25	2.8	40	2.5	34	2.3	30	2.1	26
120							3.0	27	3.0	45	2.8	40	2.5	34	2.3	30
140											2.9	43	2.6	36	2.4	32
160											3.0	45	2.8	40	2.6	36
180																
200																
220																

Grade	8' bottom width 0.2		8' bottom width 0.3		8' bottom width 0.4		8' bottom width 0.5		10' bottom width 0.2		10' bottom width 0.3		10' bottom width 0.4		10' bottom width 0.5		12' bottom width 0.2		12' bottom width 0.3		12' bottom width 0.4		12' bottom width 0.5	
Q-cfs	d	A	d	A	d	A	d	A	d	A	d	A	d	A	d	A	d	A	d	A	d	A	d	A
10	1.2	13	1.1	11	1.0	10	0.9	9	1.1	14	1.0	12	0.9	11	0.8	10	1.0	14	0.9	12	0.8	11	0.7	10
20	1.4	17	1.3	16	1.2	14	1.1	12	1.3	18	1.2	16	1.0	13	0.9	11	1.2	19	1.1	17	1.0	15	0.9	13
30	1.7	22	1.5	19	1.4	17	1.2	14	1.5	22	1.4	20	1.2	16	1.1	15	1.3	21	1.2	19	1.2	19	1.1	17
40	1.8	24	1.7	22	1.5	19	1.4	17	1.7	26	1.5	22	1.4	20	1.2	16	1.6	27	1.5	25	1.3	21	1.2	19
60	2.1	30	1.9	26	1.8	24	1.6	21	2.0	32	1.8	28	1.7	26	1.5	22	1.8	31	1.6	27	1.5	25	1.3	21
80	2.4	37	2.2	32	2.0	28	1.8	24	2.3	39	2.1	34	1.9	30	1.7	26	2.1	38	1.9	34	1.8	31	1.6	27
100	2.6	41	2.4	37	2.2	32	2.0	28	2.4	41	2.2	37	2.1	34	1.9	30	2.3	44	2.1	38	1.9	34	1.7	29
120	2.8	46	2.5	39	2.3	34	2.1	30	2.6	46	2.4	41	2.2	37	2.0	32	2.5	50	2.3	44	2.1	38	1.9	34
140	2.9	48	2.7	44	2.5	39	2.3	34	2.7	49	2.5	44	2.3	39	2.1	34	2.6	52	2.4	46	2.2	41	2.0	36
160	3.1	51	2.9	48	2.7	44	2.5	39	2.9	54	2.7	49	2.5	44	2.3	39	2.7	54	2.5	50	2.3	44	2.1	38
180							2.6	41	3.0	57	2.8	51	2.6	46	2.4	41	2.8	57	2.6	52	2.4	46	2.2	41
200															2.5	44	2.9	60	2.7	54	2.5	50	2.3	44
220																	3.0	63	2.8	57	2.7	54	2.5	50

NOTE: For diversions built on slopes under 2% the available cross-sectional area above normal ground will allow a reduction in design depth as follows:

For land slopes of 1% or less reduce depth of flow (taken from Design Table) 20%.
For land slopes of 1% to 2% reduce depth of flow (taken from Design Table) 10%.
For land slopes greater than 2% use depth of flow taken from Design Table.

For Example: A diversion 6 feet wide with a 2.5 foot depth of flow is required to remove 120 c.f.s. on a 0.4% grade. If this is built on a 1% slope the depth may be reduced 20% thus obtaining a flow depth of 2.0 feet. The required cross-sectional area of the channel plus that above normal ground line will be 34 square feet corresponding to the 2.5 foot depth. The overall height of diversion will be 2.0 feet plus 0.5 foot freeboard or 2.5 feet, instead of the original 3.0 feet.

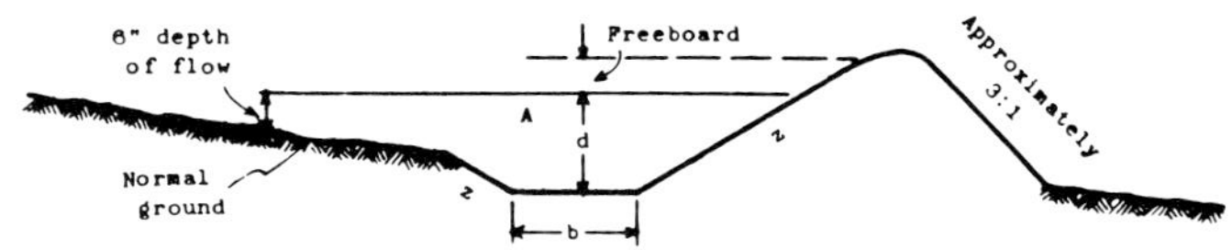

d — depth of flow, feet
b — bottom width of channel, feet
A — channel capacity, sq. ft., including area below 0.5' freeboard and excluding any area less than 0.5' depth of flow
z — side slope of channel (horizontal to vertical)

IMPORTANT: To all designed depths of flow add freeboard required by State Standards and Specifications to obtain overall height of terrace above bottom of channel. For final check on cross-sectional area subtract required freeboard from settled height of diversion and provide for cross-sectional area shown in table.

TABLE B.6 *(Continued)* Diversion Design Table—"D" Retardance (V and Trapezoidal Section)

(Based on Handbook of Channel Design, SCS-TP-61)

● 4:1 Side Slopes
"D" Retardance

	Triangular								6' bottom width							
Grade	0.2		0.3		0.4		0.5		0.2		0.3		0.4		0.5	
Q-cfs	d	A	d	A	d	A	d	A	d	A	d	A	d	A	d	A
10	1.8	13	1.7	12	1.6	11	1.5	10	1.2	13	1.1	11	1.0	10	0.9	9
20	2.1	18	2.0	16	1.8	13	1.7	12	1.5	18	1.4	16	1.2	13	1.1	11
30	2.4	23	2.2	19	2.1	18	1.9	14	1.8	24	1.6	20	1.5	13	1.3	15
40	2.5	25	2.4	23	2.2	19	2.1	18	1.9	26	1.8	24	1.6	20	1.5	18
60	2.8	31	2.6	27	2.5	25	2.3	21	2.2	33	2.0	28	1.9	26	1.7	22
80	3.1	38	2.9	34	2.7	29	2.5	25	2.4	37	2.2	33	2.1	30	1.9	26
100			3.1	38	2.9	34	2.7	29	2.7	45	2.5	40	2.3	35	2.1	30
120							2.9	31	2.9	51	2.7	45	2.4	37	2.2	33
140									3.0	54	2.9	48	2.6	43	2.4	37
160									3.1	57	2.9	51	2.8	48	2.6	43
180																
200																
220																

	8' bottom width								10' bottom width								12' bottom width							
Grade	0.2		0.3		0.4		0.5		0.2		0.3		0.4		0.5		0.2		0.3		0.4		0.5	
Q-cfs	d	A	d	A	d	A	d	A	d	A	d	A	d	A	d	A	d	A	d	A	d	A	d	A
10	1.1	14	1.0	13	0.9	11	0.8	10	1.1	14	1.0	13	0.9	12	0.8	11	1.0	15	0.9	14	0.8	13	0.7	12
20	1.4	19	1.3	17	1.2	15	1.1	14	1.3	20	1.2	18	1.0	14	0.9	12	1.2	20	1.1	18	1.0	16	0.9	14
30	1.7	25	1.5	21	1.4	19	1.2	15	1.5	24	1.4	22	1.2	18	1.1	16	1.3	22	1.2	20	1.2	20	1.1	18
40	1.8	27	1.7	25	1.5	21	1.4	19	1.6	26	1.5	24	1.3	20	1.2	18	1.6	29	1.5	27	1.3	22	1.2	20
60	2.0	32	1.9	30	1.7	25	1.6	23	1.9	33	1.8	31	1.6	26	1.5	24	1.7	32	1.6	29	1.4	25	1.3	22
80	2.3	40	2.1	34	2.0	32	1.9	27	2.2	41	2.0	36	1.8	31	1.6	26	2.0	40	1.9	37	1.7	32	1.6	29
100	2.5	45	2.3	40	2.1	34	1.9	30	2.3	44	2.1	39	2.0	36	1.9	31	2.2	46	2.0	40	1.9	37	1.7	32
120	2.7	51	2.5	45	2.3	40	2.1	34	2.5	50	2.3	44	2.2	41	2.0	36	2.4	52	2.2	46	2.0	40	1.8	35
140	2.8	54	2.6	48	2.5	45	2.3	40	2.6	53	2.4	47	2.3	44	2.1	39	2.5	55	2.3	49	2.2	46	2.0	40
160	3.0	60	2.8	54	2.6	48	2.4	42	2.8	59	2.6	53	2.5	50	2.3	44	2.7	62	2.5	55	2.3	49	2.1	43
180							2.5	45	2.9	63	2.7	56	2.6	53	2.4	47	2.8	65	2.6	58	2.4	52	2.2	46
200															2.5	50	2.9	68	2.7	62	2.5	55	2.3	49
220																	3.0	72	2.8	65	2.6	58	2.4	52

● 6:1 Side Slopes
"D" Retardance

	Triangular								6' bottom width							
Grade	0.2		0.3		0.4		0.5		0.2		0.3		0.4		0.5	
Q-cfs	d	A	d	A	d	A	d	A	d	A	d	A	d	A	d	A
10	1.6	15	1.5	14	1.4	13	1.3	11	1.2	16	1.1	14	1.0	12	0.9	10
20	1.9	22	1.8	19	1.6	15	1.5	14	1.5	23	1.4	20	1.2	16	1.1	14
30	2.1	27	2.0	24	1.8	19	1.7	17	1.7	28	1.5	23	1.4	20	1.2	16
40	2.3	32	2.2	29	2.0	24	1.8	19	1.8	30	1.7	28	1.5	23	1.4	20
60	2.5	38	2.3	32	2.2	26	2.0	24	2.0	36	1.9	33	1.7	28	1.6	25
80	2.7	44	2.5	38	2.4	35	2.2	29	2.2	42	2.1	39	1.9	33	1.8	30
100	2.9	51	2.7	44	2.6	41	2.4	35	2.4	49	2.2	42	2.1	39	1.9	33
120	3.0	54	2.8	47	2.7	44	2.5	38	2.6	56	2.4	49	2.3	46	2.1	39
140									2.7	61	2.6	56	2.4	49	2.3	46
160									2.9	68	2.8	64	2.6	56	2.5	53
180																
200																
220																
240																

	8' bottom width								10' bottom width								12' bottom width							
Grade	0.2		0.3		0.4		0.5		0.2		0.3		0.4		0.5		0.2		0.3		0.4		0.5	
Q-cfs	d	A	d	A	d	A	d	A	d	A	d	A	d	A	d	A	d	A	d	A	d	A	d	A
10	1.1	16	1.0	14	0.9	13	0.8	11	1.1	17	1.0	15	0.9	13	0.8	12	1.0	17	0.9	15	0.8	14	0.7	12
20	1.4	23	1.3	21	1.1	16	1.0	14	1.3	21	1.2	20	1.1	13	1.0	16	1.3	25	1.2	23	1.0	17	0.9	16
30	1.6	28	1.5	26	1.3	21	1.2	18	1.4	26	1.3	23	1.2	20	1.1	18	1.4	29	1.3	27	1.1	20	1.0	18
40	1.7	31	1.6	28	1.4	23	1.3	21	1.5	29	1.4	26	1.3	23	1.2	20	1.5	32	1.4	29	1.3	27	1.2	22
60	1.9	37	1.8	34	1.6	28	1.5	26	1.8	38	1.7	34	1.5	29	1.4	26	1.6	34	1.5	32	1.4	29	1.3	27
80	2.1	43	2.0	40	1.8	34	1.7	31	2.0	44	1.9	41	1.7	34	1.6	31	1.8	41	1.7	37	1.6	34	1.5	32
100	2.3	50	2.1	43	2.0	40	1.8	34	2.2	51	2.0	44	1.9	41	1.7	34	2.1	51	1.9	45	1.8	41	1.6	34
120	2.5	58	2.3	50	2.2	47	2.0	40	2.3	55	2.2	51	2.0	44	1.9	41	2.2	55	2.0	48	1.9	45	1.7	37
140	2.6	61	2.5	58	2.3	50	2.2	47	2.5	63	2.3	55	2.2	51	2.0	44	2.4	64	2.2	55	2.1	51	1.9	45
160	2.8	67	2.6	61	2.5	58	2.3	50	2.7	71	2.5	63	2.4	59	2.2	51	2.6	70	2.4	64	2.3	59	2.1	51
180	2.9	71	2.7	64	2.6	61	2.4	54	2.8	75	2.6	67	2.5	63	2.3	55	2.7	76	2.5	68	2.4	63	2.2	55
200	3.0	72	2.8	67	2.6	61	2.4	54	2.9	79	2.7	77	2.6	67	2.4	59	2.8	81	2.6	72	2.5	68	2.3	59
220							2.5	58							2.4	59	2.9	85	2.7	76	2.5	68	2.3	59
240															2.5	63							2.4	64

TABLE B.7 Diversion Design Table—"C" Retardance (V and Trapezoidal Section)

(Based on Handbook of Channel Design, SCS-TP-61)

● 3:1 Side Slopes
"C" Retardance

Triangular

Grade	0.2		0.3		0.4		0.5	
Q	d	A	d	A	d	A	d	A
20	2.5	19	2.3	16	2.1	13	1.9	11
30			2.5	19	2.3	16	2.2	15
40					2.5	19	2.4	17
50							2.5	19
60								
80								
100								
120								
140								
160								
180								
200								
220								

6' bottom

Grade	0.2		0.3		0.4		0.5	
Q	d	A	d	A	d	A	d	A
20	1.9	22	1.7	19	1.5	16	1.4	14
30	2.0	24	1.8	21	1.7	19	1.5	16
40	2.2	28	2.0	24	1.9	22	1.7	19
50	2.3	30	2.1	26	2.0	24	1.8	21
60	2.5	34	2.3	30	2.1	26	1.9	22
80			2.5	34	2.3	30	2.1	26
100					2.5	34	2.3	30
120							2.5	34
140								
160								
180								
200								
220								

8' bottom

Grade	0.2		0.3		0.4		0.5	
Q	d	A	d	A	d	A	d	A
20	1.7	22	1.5	19	1.4	17	1.3	15
30	1.9	26	1.7	22	1.5	19	1.4	17
40	2.0	28	1.8	24	1.6	21	1.5	19
50	2.2	32	2.0	28	1.8	24	1.6	21
60	2.3	34	2.1	30	1.9	26	1.8	24
80	2.5	39	2.3	34	2.1	30	1.9	26
100			2.5	39	2.3	34	2.1	30
120					2.5	39	2.3	34
140							2.4	36
160								
180								
200								
220								

10' bottom

Grade	0.2		0.3		0.4		0.5	
Q	d	A	d	A	d	A	d	A
20	1.6	24	1.4	20	1.3	18	1.2	16
30	1.7	26	1.5	22	1.4	20	1.3	18
40	1.9	30	1.7	26	1.5	22	1.4	20
50	2.0	32	1.8	28	1.6	24	1.5	22
60	2.2	37	2.0	32	1.8	28	1.6	24
80	2.4	41	2.2	37	2.0	32	1.8	28
100			2.4	41	2.2	37	2.0	32
120					2.3	39	2.1	34
140					2.5	44	2.3	39
160							2.4	41
180							2.5	44
200								
220								

12' bottom

Grade	0.2		0.3		0.4		0.5	
Q	d	A	d	A	d	A	d	A
20								
30	1.6	27	1.4	23	1.3	21	1.2	19
40	1.8	31	1.6	27	1.5	25	1.4	23
50	1.9	34	1.7	29	1.6	27	1.5	25
60	2.0	36	1.8	31	1.7	29	1.6	27
80	2.2	41	2.0	36	1.9	34	1.7	29
100	2.5	49	2.2	41	2.0	36	1.8	31
120			2.4	46	2.2	41	2.0	36
140					2.3	43	2.1	38
160					2.4	46	2.2	41
180							2.4	46
200							2.5	49
220							2.6	51

● 4:1 Side Slopes
"C" Retardance

Triangular

Grade	0.2		0.3		0.4		0.5	
Q	d	A	d	A	d	A	d	A
30	2.5	25	2.4	23	2.3	21	2.1	18
40			2.5	25	2.4	23	2.2	19
50					2.5	25	2.4	23
60							2.5	25
80								
100								
120								
140								
160								
180								
200								
220								

6' bottom width

Grade	0.2		0.3		0.4		0.5	
Q	d	A	d	A	d	A	d	A
30	2.0	28	1.8	24	1.6	20	1.5	18
40	2.1	30	1.9	26	1.7	22	1.6	20
50	2.3	35	2.1	30	1.9	26	1.7	22
60	2.4	37	2.2	33	2.0	28	1.9	26
80			2.4	37	2.2	33	2.0	28
100			2.5	40	2.3	35	2.2	33
120					2.5	40	2.4	37
140							2.5	40
160								
180								
200								
220								

8' bottom width

Grade	0.2		0.3		0.4		0.5	
Q	d	A	d	A	d	A	d	A
30	1.8	27	1.7	25	1.5	21	1.4	19
40	2.0	32	1.8	27	1.6	23	1.5	21
50	2.1	34	1.9	30	1.7	25	1.6	23
60	2.3	40	2.1	34	1.9	30	1.7	25
80	2.5	45	2.3	40	2.1	34	1.9	30
100			2.5	45	2.3	40	2.0	32
120					2.4	42	2.2	37
140					2.5	45	2.3	40
160							2.5	45
180								
200								
220								

10' bottom width

Grade	0.2		0.3		0.4		0.5	
Q	d	A	d	A	d	A	d	A
30	1.8	31	1.6	26	1.4	22	1.3	20
40	1.9	33	1.7	29	1.5	24	1.4	22
50	2.0	36	1.8	31	1.6	26	1.5	24
60	2.1	39	1.9	34	1.7	29	1.6	26
80	2.3	44	2.1	39	1.9	34	1.8	31
100	2.5	50	2.3	44	2.1	39	1.9	34
120			2.5	50	2.3	44	2.1	39
140					2.4	47	2.2	41
160					2.5	50	2.3	44
180							2.4	47
200							2.5	50
220								

12' bottom width

Grade	0.2		0.3		0.4		0.5	
Q	d	A	d	A	d	A	d	A
30	1.7	32	1.6	29	1.4	25	1.2	20
40	1.8	35	1.7	32	1.5	27	1.3	22
50	1.9	37	1.8	35	1.6	29	1.4	25
60	2.0	40	1.9	37	1.7	32	1.5	27
80	2.2	46	2.0	40	1.8	35	1.7	32
100	2.3	49	2.1	43	2.0	40	1.8	35
120	2.5	55	2.3	49	2.2	46	2.0	40
140			2.5	55	2.3	49	2.1	43
160					2.4	52	2.2	46
180					2.5	55	2.3	49
200							2.4	52
220							2.5	55

10-26-55 4-L-10122-1

TABLE B.7 *(Continued)* Diversion Design Table—"C" Retardance (V and Trapezoidal Section)

(Based on Handbook of Channel Design, SCS-TP-61)

● 6:1 Side Slopes
"C" Retardance

Grade	Triangular 0.2		0.3		0.4		0.5	
Q	d	A	d	A	d	A	d	A
20	2.2	29	2.1	26	1.9	22	1.8	19
30	2.4	35	2.2	29	2.1	26	1.9	22
40	2.5	38	2.3	32	2.2	29	2.0	24
50			2.5	38	2.3	32	2.1	26
60					2.4	35	2.2	29
80					2.5	38	2.3	32
100							2.5	38

Grade	6' bottom width 0.2		0.3		0.4		0.5	
Q	d	A	d	A	d	A	d	A
20	1.8	30	1.7	28	1.6	25	1.4	20
30	2.0	36	1.9	33	1.7	28	1.5	23
40	2.1	39	2.0	36	1.8	30	1.6	25
50	2.2	42	2.1	39	1.9	33	1.7	28
60	2.3	46	2.2	42	2.0	36	1.8	30
80	2.5	53	2.3	46	2.1	39	1.9	33
100			2.5	53	2.3	46	2.1	39
120					2.4	49	2.2	42
140					2.5	53	2.3	46
160							2.4	49
180							2.5	53

Grade	8' bottom width 0.2		0.3		0.4		0.5	
Q	d	A	d	A	d	A	d	A
20	1.7	31	1.6	28	1.5	26	1.3	21
30	1.8	34	1.7	31	1.6	28	1.4	23
40	2.0	40	1.9	37	1.7	31	1.5	26
50	2.1	43	2.0	40	1.8	34	1.6	28
60	2.2	47	2.1	43	1.9	37	1.7	31
80	2.4	54	2.2	47	2.0	40	1.8	34
100	2.5	58	2.4	54	2.2	47	2.0	40
120			2.5	58	2.3	50	2.1	43
140					2.4	54	2.2	47
160					2.5	58	2.3	50
180							2.4	54
200							2.5	58

Grade	10' bottom width 0.2		0.3		0.4		0.5	
Q	c	A	d	A	d	A	d	A
20	1.6	31	1.5	29	1.4	26	1.2	20
30	1.8	38	1.7	34	1.5	29	1.3	23
40	1.9	41	1.8	38	1.6	31	1.4	26
50	2.0	44	1.9	41	1.7	34	1.5	29
60	2.1	47	2.0	44	1.8	38	1.6	31
80	2.3	55	2.1	47	1.9	41	1.7	34
100	2.4	59	2.3	55	2.1	47	1.9	41
120	2.5	63	2.4	59	2.2	51	2.0	44
140			2.5	63	2.3	55	2.1	47
160					2.4	59	2.2	51
180					2.5	63	2.3	55
200							2.4	59
220							2.5	63

Grade	12' bottom width 0.2		0.3		0.4		0.5	
Q	d	A	d	A	d	A	d	A
20	1.5	32	1.4	29	1.3	27	1.2	22
30	1.7	37	1.6	34	1.4	29	1.3	27
40	1.8	41	1.7	37	1.5	32	1.3	27
50	1.9	45	1.8	41	1.6	34	1.4	29
60	2.0	48	1.9	45	1.8	41	1.6	34
80	2.2	55	2.1	52	1.9	45	1.7	37
100	2.3	59	2.2	55	2.0	48	1.8	41
120	2.4	64	2.3	59	2.1	52	1.9	45
140	2.5	68	2.4	64	2.2	55	2.0	48
160			2.5	68	2.3	59	2.1	52
180					2.4	64	2.2	55
200					2.5	68	2.2	55
220							2.3	59

4-L-10122-2

appendix C

Solutions to Calculation Problems

Chapter 4, Review Question 16

$$C = 0.35$$
$$T_c = 12 \text{ min}$$

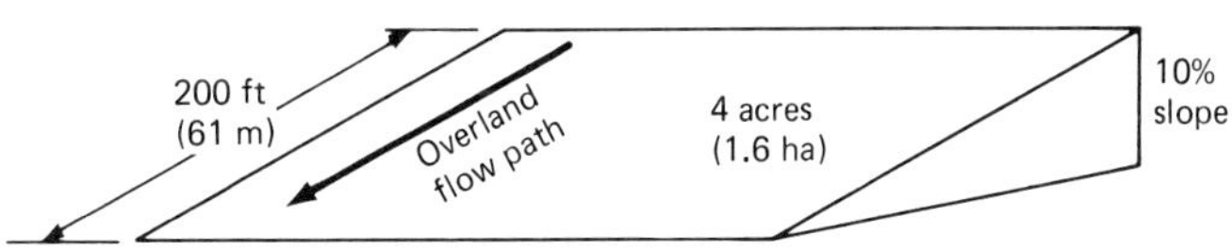

10-year, 30-min rainfall = 1.6 in (40.6 mm)
10-year, 15-min rainfall = (1.6 in)(0.9) = 1.44 in (36.6 mm)
10-year, 7.5-min rainfall = (1.44 in)(0.9) = 1.3 in (33.0 mm)
Interpolating to 12 min = 1.38 in/12 mm (35.1 mm/12 min)

$$i_{\text{peak}} = 6.9 \text{ in/hr} = 175 \text{ mm/hr}$$
$$Q_{\text{peak}} = C \times i \times A = 0.35\,(6.9)(4) = 9.7 \text{ ft}^3\text{/sec } (0.27 \text{ m}^3\text{/hr})$$

Review Question 17

10-year, 6-hr rainfall = 3.5 in (88.9 mm)

$$i_{\text{avg}} = 3.5 \text{ in/6 hr } (88.9 \text{ mm/6 hr}) = 0.58 \text{ in/hr } (14.7 \text{ mm/hr})$$
$$Q_{\text{avg}} = 0.35\,(0.58)(4) = 0.81 \text{ ft}^3\text{/sec } (0.023 \text{ m}^3\text{/sec})$$

Chapter 5, Review Question 17

1. R factor, from Fig. 5.2 = 175
2. K factor, from Fig. 5.6 = 0.46

   Since the very fine sand content is less than 15 percent, no adjustment is necessary (step 1, Table 5.3). You enter the nomograph, Fig. 5.6, with total sand and total silt 11.3 and 65.9 percent, respectively, to find a K value of 0.46 (by interpolation). *Note:* You need only two of three components to enter Fig. 5.6. The clay content is not an essential piece of information here.
3. LS factor, from Table 5.5
   2 percent, 120 ft (36.6 m) = 0.22
   3:1, 20 ft (6.1 m) = 4.22
4. C factor, from Table 5.6
   10 acres (4 ha), housepads, no cover = 1.0
   5 acres (2 ha), seeded slopes = 0.5
   Note: It is unlikely that a 90 percent cover will be obtained, so don't use the C values for "temporary seedings" in Table 5.6. Instead, to be conservative, assume 75 percent cover and use 0.5.
5. P factor, from Table 5.7
   Trackwalked slopes = 0.9
   Bare, smooth housepads = 1.3
6. Soil loss must be calculated separately for each area of a site in which soil surface condition, slope, and vegetation are substantially different. In this problem the housepads are treated as one area and the seeded slopes are treated as a second area. Soil loss estimates for the areas are then summed to provide a total soil loss estimate.

$$\text{Soil loss} = R \times K \times \text{LS} \times C \times P \times \text{area}$$

Soil loss [10-acre (4.0-ha) portion]

$$= 175\,(0.46)(0.22)(1.0)(1.3)(10) = 230 \text{ tons/yr } (209 \text{ t/yr})$$

Soil loss [5-acre (2.0-ha) portion]

$$= 175\,(0.46)(4.22)(0.5)(0.9)(5) = 764 \text{ tons/yr } (693 \text{ t/yr})$$

$$\text{Total soil loss} = 230 + 764 = 994 \text{ tons/year } (902 \text{ t/yr})$$

Chapter 7, Review Question 16

1. The permissible velocity V_1 is determined from Table 6.7 for the given slope and grass lining:

$$V_1 = 5 \text{ ft/sec } (1.5 \text{ m/sec})$$

   Fully grown Kentucky bluegrass has a retardance of B (Table B-1).
2. By Table B-3, retardances B and D for a 3 percent slope, V_1 = 5 ft/sec (1.5 m/sec), and a flow of 50 ft^3/sec (1.4 m^3/sec), the following channel dimensions are found:

$$\text{Top width } T = 16.3 \text{ ft } (5.0 \text{ m})$$
$$\text{Depth of channel } D = 1.45 \text{ ft } (0.44 \text{ m})$$

A minimum of 0.3 ft (0.1 m) of freeboard must be added to the channel depth.

3. The velocity V_2 through a tall stand of grass will be 3.14 ft/sec (0.96 m/sec).

Review Question 17

1. One-half the pipe diameter = 1 ft (0.3 m).
The maximum tailwater depth will not exceed 1.45 ft (0.44 m) [design depth of channel for 50 ft^3/sec (1.4 m^3/sec) flow].
Under minimum tailwater conditions, depth of flow in the waterway would be less than 1 ft (0.3 m), since, by definition, TW_{min} is less than one-half the pipe diameter. Size apron for both conditions.
2. Solving for maximum tailwater, TW > 1.0 ft (0.3 m), from Fig. 7.46:

$$\text{Riprap size } d_{50} = 0.1 \text{ ft } (0.03 \text{ m})$$
$$\text{Apron length } L_a = 8 \text{ ft } (2.4 \text{ m})$$
$$\text{Upstream apron width} = 3 \times \text{diam} = 6 \text{ ft } (1.8 \text{ m})$$
$$\text{Downstream apron width } W = \text{diam} + 0.4L_a = 2 + 3.2 = 5.2 \text{ ft } (1.6 \text{ m})$$

3. Solving for minimum tailwater, TW < 1.0 ft (0.3 m), from Fig. 7.45:

$$d_{50} = 0.4 \text{ ft } (0.1 \text{ m})$$
$$L_a = 13 \text{ ft } (4.0 \text{ m})$$
$$W = \text{diam} + L_a = 15 \text{ ft } (4.6 \text{ m})$$

4. Note that the minimum tailwater apron is longer and wider than the apron designed for maximum tailwater conditions. Because both minimum and maximum tailwater conditions may occur at different times in the same channel, we design the apron for the larger dimensions: 13 ft (4.0 m) long and 15 ft (4.6 m) wide. Similarly, we choose the larger diameter riprap size of 0.4 ft (0.1 m).
5. The design apron width of 15 ft (4.6 m) is almost the full width of the channel. For simplicity in installation, the entire perimeter of the channel should be lined with rock. The apron should flare out from the pipe until it reaches the top of the channel bank. It should then cover the entire channel until it reaches the calculated length. Extra stability can be obtained by increasing the rock size and the thickness of the rock layer.
Note that the riprap cannot simply be placed in the channel, because the capacity of the channel would be reduced. Also, the discharge from the pipe cannot be allowed to drop off the edge of the apron and erode the channel at that point. To install riprap properly, the apron area must be excavated and the rock installed to form a surface level with or below the bottom of the existing channel. By making the riprap surface lower than the channel and grading back to channel level gradually, an extra measure of protection can be obtained. The effect is like that of a preformed scour hole that increases energy dissipation at the outlet.

Chapter 8, Review Question 19

1. Calculate the average runoff by using the rational method and the 10-year, 6-hr rainfall intensity:

$$Q_{avg} = C \times i \times A$$

C factor: From Table 4.1, choose 0.45 for housepads (bare, packed soil, smooth) and 0.5 for the cut and fill slopes. (Because of steepness, use factor for barren slopes, >30 percent, rough surface.) Calculate C based on a weighted average of the two types of surfaces:

$$C = \frac{(0.45 \times 10 \text{ acres}) + (0.5 \times 5 \text{ acres})}{15 \text{ acres}} = 0.47$$

i factor: 10-year, 6-hr rainfall = 3.5 in (89 mm) (from Fig. 4.4)
A: A = 15 acres (6.1 ha)
Average runoff:

$$Q_{avg} = C \times i \times A = 0.47\ (0.58)(15) = 4.1 \text{ ft}^3\text{/sec } (0.12 \text{ m}^3\text{/sec})$$

2. Particle settling velocity V_s for the 0.02-mm particle = 0.00096 ft/sec (0.00029 m/sec) (from Fig. 8.12).
3. The minimum required surface area is

$$\begin{aligned} A_s &= \frac{1.2Q_{avg}}{V_s} \\ &= \frac{1.2(4.1 \text{ ft}^3\text{/sec})}{0.00096 \text{ ft/sec}} \quad \left[\frac{1.2(0.12 \text{ m}^3\text{/sec})}{0.00029 \text{ m/sec}}\right] \\ &= 5125 \text{ ft}^2 \quad (476 \text{ m}^2) \end{aligned}$$

4. The minimum settling depth is 2 ft (0.6 m).
5. The storage depth is in addition to the settling depth, and it is estimated by using the USLE and an expected basin efficiency.
Basin efficiency: The ideal efficiency is equal to the percent by weight of soil particles larger than the design particle size. For the soil in Review Question 17, Chap. 5,
Sand (2 to 0.05 mm) = 11.3%
Silt (0.05 to 0.002 mm) = 65.9%
Fine silt (0.02 to 0.002 mm) = 41.7% (note overlap with silt content)

$$11.3 + 65.9 - 41.7 = 35.5\% \text{ basin efficiency}$$

The clay fraction was ignored because it would not be captured in a sediment basin.
Predicted soil loss: From Review Question 17, Chap. 5,

$$\text{Soil loss} = 994 \text{ tons/year } (902 \text{ t/year})$$

Volume of soil captured: Assume 1 ton of sediment occupies 1 yd^3.

$$\text{Predicted soil loss} \times \text{basin efficiency} = \text{volume captured}$$
$$(994 \text{ tons})(0.355) = 353 \text{ yd}^3 \ (270 \text{ m}^3)$$

Storage depth: Calculate required depth by using the minimum surface area determined in step 3:

$$\frac{(353 \text{ yd}^3)(27 \text{ ft}^3/\text{yd}^3)}{5125 \text{ ft}^2} = 1.86 \text{ ft } (0.57 \text{ m})$$

6. Total minimum basin depth = 3.9 ft (1.2 m) plus freeboard.

Comments: The volume of the basin is calculated under the assumption that the ryegrass is successfully established and reduces erosion by at least one-half (C factor in USLE = 0.5). The use of a C factor of 0.5 instead of 0.1 is a conservative estimate to account for potential events such as early rains, uneven seed coverage or germination, and slow growth due to low fertility. The C factor was not reduced to 0.1 by the wood fiber mulch because wood fiber mulch is not very effective at reducing raindrop impact or preventing soil loss on the steeper slopes.

The low ideal basin efficiency of 35.5 percent has some significant implications. Not only is soil in this problem severely erodible; it also contains such a high proportion of fine particles that the sediment is difficult to remove from suspension. A sediment basin alone cannot prevent off-site movement of these soils if the soils are allowed to erode. On this site it is far better to concentrate on preventing erosion with mulch, grass, shortened slopes, and diversions while providing a sediment basin or several traps as additional backup protection.

A reduction in soil loss can be obtained by trackwalking, seeding, and mulching the house pads. This would also reduce runoff. (The runoff coefficient used in the rational method would become smaller and so, therefore, would the required surface area of the basin.) Briefly, average runoff Q_{avg}:

New C for housepads = 0.3

We recalculate a weighted average C factor:

$$C = \frac{(0.3 \times 10 \text{ acres}) + (0.5 \times 5 \text{ acres})}{15 \text{ acres}} = 0.37$$

$$Q_{avg} = C \times i \times A = 0.37(0.58)(15) = 3.2 \text{ ft}^3/\text{sec } (0.09 \text{ m}^3/\text{sec})$$

Basin area:

$$A_s = \frac{1.2Q_{avg}}{V_s}$$

$$= \frac{1.2(3.2 \text{ ft}^3/\text{sec})}{0.00096 \text{ ft/sec}} = 4000 \text{ ft}^2 \ (372 \text{ m}^2)(22\% \text{ smaller})$$

Soil loss:

New C = 0.5

New P = 0.9

New soil loss, 10 acres (4 ha),

$$R \times K \times \text{LS} \times C \times P = 175(0.46)(0.22)(0.5)(0.9)(10)$$
$$= 79.7 \text{ tons/year } (72.5 \text{ t/yr})$$
$$\text{New total soil loss} = 79.7 + 764 = 844 \text{ tons/year } (767 \text{ t/yr}) \quad (15\% \text{ less})$$
$$\text{Basin capture} = 0.355(844) = 300 \text{ tons/year } (273 \text{ t/yr})$$

Storage depth:

$$\frac{(300 \text{ yd}^3)(27 \text{ ft}^3/\text{yd}^3)}{4000 \text{ ft}^2} = 2 \text{ ft } (0.6 \text{ m})$$

The reduced runoff resulted in a 22 percent smaller required basin surface area; soil loss will be 15 percent less. Because of the reduced surface area, the storage depth is increased by 0.14 ft (4 cm) to 2 ft (0.61 m).

Index

ABOUT THE AUTHORS

Steven J. Goldman has over 10 years' experience in erosion and sediment control and water quality management. He coordinated the development of surface runoff management plans in eight San Francisco Bay Area counties for the Association of Bay Area Governments. Working with his two co-authors, he developed the standards and specifications for erosion and sediment control measures that are now used throughout Northern California. He also developed a technical training seminar and handbook on erosion and sediment control, and teaches the seminar to professional audiences on a regular basis. He has assisted over 100 cities and counties in implementing erosion control ordinances and programs. Mr. Goldman now coordinates the erosion control program of the California Tahoe Conservancy and also consults in erosion and sediment control.

Katharine Jackson has worked as an environmental engineer and soil scientist for a number of private firms and public agencies. She has prepared course materials on runoff and soil-loss prediction for erosion control, and on design of erosion and sediment control measures. She also adapted the universal soil loss equation for use on construction sites. Ms. Jackson has a master's degree in soil science and environmental toxicology.

Taras Bursztynsky has been manager of the water quality program of the Association of Bay Area Governments since 1978. He worked with Mr. Goldman and Ms. Jackson on the development of the standards and specifications for erosion and sediment control measures and on technical training materials. He also developed improved procedures for designing sediment basins and traps based on particle settling properties. Mr. Bursztynsky is a registered civil engineer in California and Florida.